T0314040

Inverse Heat Conduction

Inverse Heat Conduction

Ill-Posed Problems

Second Edition

Keith A. Woodbury
University of Alabama
Tuscaloosa, AL, USA

Hamidreza Najafi
Florida Institute of Technology
Melbourne, FL, USA

Filippo de Monte
University of L'Aquila
L'Aquila, Italy

James V. Beck
Michigan State University
East Lansing, MI, USA

Edition History
First edition published in 1985

Published by John Wiley & Sons, Inc., Hoboken, New Jersey.
Published simultaneously in Canada.

For general information on our other products and services or for technical support, please contact our Customer Care Department within the United States at (800) 762-2974, outside the United States at (317) 572-3993 or fax (317) 572-4002.

Wiley also publishes its books in a variety of electronic formats. Some content that appears in print may not be available in electronic formats. For more information about Wiley products, visit our web site at www.wiley.com.

Library of Congress Cataloging-in-Publication Data:
Names: Woodbury, Keith A., author. | Najafi, Hamidreza, author. | De Monte,
    Filippo, author. | Beck, J. V. (James Vere), 1930– author.
Title: Inverse heat conduction : ill-posed problems / Keith Woodbury,
    University of Alabama, Hamidreza Najafi, Florida Institute of
    Technology, Filippo de Monte, University L'Aquila, James V. Beck,
    Michigan State University.
Description: Second edition. | Hoboken, NJ : John Wiley & Sons, Inc., 2023.
    | Preceded by: Inverse heat conduction / James V. Beck, Ben Blackwell,
    Charles R. St. Clair, Jr. c1985. | Includes bibliographical references.
Identifiers: LCCN 2022039047 (print) | LCCN 2022039048 (ebook) | ISBN
    9781119840190 (cloth) | ISBN 9781119840206 (adobe pdf) | ISBN
    9781119840213 (epub)
Subjects: LCSH: Heat–Conduction. | Numerical analysis–Improperly posed
    problems.
Classification: LCC QC320 .B4 2023 (print) | LCC QC320 (ebook) | DDC
    536/.23–dc23/eng20230106
LC record available at https://lccn.loc.gov/2022039047
LC ebook record available at https://lccn.loc.gov/2022039048

Cover Design: Wiley
Cover Images: © AnitaVDB/Getty Images; bischy/Getty Images; Bernt Ove Moss/Getty Images; SasinT Gallery/Getty Images; Courtesy of Hamidreza Najafi

Set in 9.5/12.5pt STIXTwoText by Straive, Pondicherry, India

To engineers and researchers who thirst for knowledge and pass this way, with undying gratitude for those who came before and illuminated the path, and with special appreciation to my colleague, mentor, and co-author: James V. Beck.

K. A. W.

To my wife, Mehrasa; and to my mother and father, Kobra and Mohammadsadegh.

H. N.

To my wife Emanuela and my children Federica Flora, Matilde, and Fabrizio Nicola; to my sisters Saveria and Alessia.

F. d. M.

To my wife, Barbara; children, Sharon and Douglas; and father and mother, Peter and Louise Beck

J. V. B.

In Memorium
James V. Beck (1930–2022)

Prof. Beck was the inspirational and motivational force in the development of the two editions of Inverse Heat Conduction: Ill-posed problems. Sadly, he passed away during the last stages of copyediting of this second edition. His spirited enthusiasm and boundless energy for scientific research will not soon be forgotten.

Contents

List of Figures

Nomenclature

1	vector of all ones
a	a constant
A	counting integer for number of decimal places of precision in building block solution $(-)$
A,B,C	coefficient matrices from numerical discretization of heat conduction equation
b	a constant
Bi	Biot number, hL/k $(-)$
c	specific heat capacity (J kg^{-1}K^{-1})
C	volumetric heat capacity, ρc (J m^{-3}K^{-1})
ceil(.)	MATLAB function that rounds the argument to the nearest integer greater than or equal to it $(-)$
d	deviation distance (m)
$\tilde{d}$	dimensionless deviation distance, d/L $(-)$
D_M^2	square of deterministic error for a single heat flux (Eq. 6.13)
E	relative error based on maximum temperature variation $(-)$
$E(.)$	expected value operator
E	depth, $0 < E < L$ (m)
$E_{q,\text{bias}}^2$	square of the expected value of the bias error
$E_{q,\text{rand}}^2$	square of the expected value of the random (variance) error
$E_{q,\text{RMS}}$	root mean squared (RMS) error
f	filter coefficient
$f(.)$	arbitrary function
F	eigenfunction $(-)$
F	filter matrix
Fo	Fourier number, at/L^2 $(-)$
g	filter coefficients (Chapter 5)
$g(.)$	arbitrary function
G	one-dimensional Green's function (m^{-1})
G	filter matrix (X12 case in Chapter 5)
h	heat transfer coefficient (W m^{-2}K^{-1})
$H(.)$	Heaviside unit step function $(-)$
H	temperature correction kernel (Chapter 10) $(-)$
H	vector of temperature correction kernel values $(-)$
H$_i$	discrete Tikhonov regularization matrix $(i = 0,1,2)$
i	counting integer for time steps $(-)$
I	identity matrix
j	counting integer for space steps $(-)$
k	thermal conductivity (W m^{-1}K^{-1})
J	number of sensors
K	gain coefficients in Future Times regularization
K	Kalman gain matrix

L	length (m)
$\mathcal{L}(.)$	Laplace transform operator
m	counting integer in the x–direction (−)
m_f	number of required data points from future time steps for filter form solutions
$m_{\max}$	maximum number of terms in summations along x (−)
m_p	number of required data points from previous time steps for filter form solutions
M	M-th time step and maximum number of time steps (−); constant in Eq. (4.7)
n	counting integer in the y–direction; also number of data points (−)
n_p	number of pulses
$n_{\max}$	maximum number of terms in summations along y (−)
N	N-th space step; also maximum number of steps or times (−)
p	power-law exponent (−)
P	constant in Eq. (4.8)
$\mathbf{P}$	covariance matrix in Kalman filter solution
q	heat flux (W m^{-2})
$\tilde{q}$	dimensionless heat flux, q/q_0 (−)
$\mathbf{q}, \mathbf{q}_0$	vector of surface heat-flux components/values (W m^{-2})
Q^2	process noise variance
r	number of future times
R^2	measurement noise variance
R_q^2	mean squared error for vector of heat flux estimates (Eq. 6.9)
R_q	root mean squared error of heat flux estimates, $R_q = \sqrt{R_q^2}$
R_T^2	mean squared error for vector of measured temperature estimates (Eq. 6.36)
R_M^2	mean squared error of a single heat flux value
s	an integer
s_q	standard error
S	total number of terms in the computational analytical solution
S	sum of squared errors, e.g. $(\mathbf{Y}\text{–}\mathbf{T})^T(\mathbf{Y}\text{–}\mathbf{T})$
$\mathbf{S}$	measurement covariance in Kalman filter solution
t	time (s)
$\tilde{t}$	dimensionless time, at/L^2 (−)
T	temperature (K)
$\tilde{T}$	dimensionless temperature, $(T - T_{in})/(T_0 - T_{in})$ or $(T - T_{in})/(q_0 L/k)$ in Chapter 2 (−)
$\mathbf{T}$	vector of temperatures (K)
$tr[.]$	trace of a matrix
u	cotime, $t - \tau$ (s)
u_i	random error with zero mean and unity standard deviation
$\mathbf{v}$	measurement noise vector
$V(.)$	variance operator
V_α	the GCV function (Eq. 6.49)
$\mathbf{w}^k$	descent direction for iteration k in conjugate gradient method
$\mathbf{w}$	process noise vector
W	width (m)
$\tilde{W}$	aspect ratio, W/L (−)
W_0	width of "active" (heated or cooled) boundary (m)
$\tilde{W}_0$	dimensionless width of "active" boundary, W_0/L (−)
x	rectangular space coordinate (m)
$\tilde{x}$	dimensionless rectangular space coordinate, x/L (−)
x'	dummy variable for space (m)
X	sensitivity coefficient

$X_{M,i}$ heat flux-based sensitivity coefficient, $\partial T_M(x)/\partial q_{0,i}$ (°C m^2 W^{-1})

$X_{M,i,j}$ heat flux-based sensitivity coefficient, $\partial T_M(x, y)/\partial q_{0,i,j}$ (°C m^2 W^{-1})

X   matrix of sensitivity coefficients $X_{M,i}$ (1D case) and $X_{M,i,j}$ (2D case) (°C m^2 W^{-1})

X$_0$ vector of sensitivity coefficients $X_{M,0} = \partial T_M/\partial q_{0,0}$ for $q_{0,0}$ (Chapter 3) (°C m^2 W^{-1})

X$_0$ matrix of sensitivity coefficients evaluated at surface $x = 0$ (Chapter 10) (°C m^2 W^{-1})

y rectangular space coordinate (m)

$\tilde{y}$ dimensionless rectangular space coordinate, y/L (−)

Y measured temperature (°C)

Y   vector of measured temperatures (°C)

y   vector of measured temperatures by a second sensor on the remote boundary (°C)

Z_h step sensitivity for heat transfer coefficient (Chapter 9) (m^2 °C^2 W^{-1})

Z   vector of step sensitivity coefficients for h (m^2 °C^2 W^{-1})

Z   sensitivity matrix for measured temperature boundary condition at remote boundary (Chapter 5)

Greek Symbols

α thermal diffusivity, k/C (m^{2}s^{-1})

α_i Tikhonov regularization coefficient ($i = 0,1,2$)

α_T Tikhonov regularization parameter for the time term (Chapter 8)

α_s Tikhonov regularization parameter for the space term (Chapter 8)

β eigenvalue in the x−direction; also a parameter (−)

β ratio of thermal properties (Eq. 10.20)

β   vector of parameters

$\delta(.)$ Dirac delta function

δ level of error in measured temperatures for Morozov discrepancy principle

γ temperature rise per unit linear-in-time increase of surface heat-flux at $x = 0$ (with $x = L$ kept insulated) over the time $t_{ref} = \Delta t$, that is, $\tilde{T}_{X22B20T0} \times L/k$ with $\tilde{t}_{ref} = \Delta \tilde{t}$ (°C m^2 W^{-1})

Γ input gain in Kalman filter

Δ variation of a function (e.g. ΔT is the variation of the temperature function)

$\Delta \gamma_i$ forward difference of the γ temperature at time $i\Delta t$, $\gamma_{i+1} - \gamma_i$ (°C m^2 W^{-1})

Δt time step (s)

$\Delta \tilde{t}$ dimensionless time step, $\alpha \Delta t/L^2$ (−)

Δy space step along y (m)

$\Delta \tilde{y}$ dimensionless space step along y, $\Delta y/L$ (−)

$\Delta \phi_i$ forward difference of the ϕ temperature at time $i\Delta t$, $\phi_{i+1} - \phi_i$ (°C m^2 W^{-1})

$\Delta \varphi_i$ forward difference of the φ temperature at time $i\Delta t$, $\varphi_{i+1} - \varphi_i$ (−)

$\Delta(\Delta \gamma)_i$ central difference of the γ temperature at time $i\Delta t$, $\Delta \gamma_i - \Delta \gamma_{i-1}$ (°C m^2 W^{-1})

$\Delta_i \lambda$ forward difference at time $i\Delta t$ of the λ temperature, $\lambda_{i+1} - \lambda_i$ (°C m^2 W^{-1})

$\Delta_j \lambda$ backward difference at the j−th grid point of the λ temperature, $\lambda_j - \lambda_{j-1}$ (°C m^2 W^{-1})

$\Delta_i \Delta_j(\lambda)$ second cross difference in space (Δ_j) and time (Δ_i) at the j−th grid point and at time $i\Delta t$ of the λ temperature, $\Delta_j(\lambda_{i+1}) - \Delta_j(\lambda_i) = \Delta_j \Delta_i(\lambda)$ (°C m^2 W^{-1})

$\Delta \mu_i$ forward difference of the μ temperature at time $i\Delta t$, $\mu_{i+1} - \mu_i$ (−)

$\Delta(\Delta \mu)_i$ central difference of the μ temperature at time $i\Delta t$, $\Delta \mu_i - \Delta \mu_{i-1}$ (−)

ε_a absolute error (°C)

ε_r relative error (−)

ε_i a single random error

ε   vector of random error

η eigenvalue in the y−direction (−)

∇ gradient operator

$\nabla_{\mathbf{q}}$	gradient with respect to the components of vector $\mathbf{q}$
θ_0	surface temperature rise, $T_0(t) - T_{in}$ (K)
$\boldsymbol{\theta}_0$	vector of surface temperature-rise components/values (°C)
λ	temperature rise per unit of surface heat flux of the X22B(y1pt1)0Y22B00T0 heat conduction case, that is, $\tilde{T}_{\text{X22B(y1pt1)0Y22B00T0}} \times L/k$ (°C m^2 W^{-1})
Λ_j	heat flux-based sensitivity coefficient, $\partial T(x,y,t)/\partial q_{0,j}$ (°C m^2 W^{-1})
$\boldsymbol{\Lambda}$	vector of sensitivity coefficients Λ_j (°C m^2 W^{-1})
μ	temperature rise per unit linear-in-time increase of surface temperature at $x = 0$ being $x = L$ insulated over the time $t_{ref} = \Delta t$, that is, $\tilde{T}_{\text{X12B20T0}}$ with $\tilde{t}_{ref} = \Delta \tilde{t}$ (−)
ρ	density (kg m^{-3})
ρ^k	descent direction on iteration k for conjugate gradient method
σ	standard deviation
τ	dummy variable for time (s)
ϕ	temperature rise per unit step change of surface heat flux at $x = 0$ with $x = L$ kept insulated, that is, $\tilde{T}_{\text{X22B10T0}} \times L/k$ (°C m^2 W^{-1})
$\phi(.)$	generic influence function – response in domain due to step change in surface condition
$\boldsymbol{\Phi}$	state transition matrix
φ	temperature rise per unit step change of surface temperature at $x = 0$ being $x = L$ insulated, that is, $\tilde{T}_{\text{X12B10T0}}$ (−)
ψ	Lagrange multiplier
$\Psi_{M,i}$	temperature-based sensitivity coefficient, $\partial T_M(x)/\partial \theta_{0,i}$
$\boldsymbol{\Psi}$	matrix of sensitivity coefficients $\Psi_{M,i}$ for $\boldsymbol{\theta}_0$ (−)
$\boldsymbol{\Psi}_0$	vector of sensitivity coefficients $\Psi_{M,0} = \partial T_M(x)/\partial \theta_{0,0}$ for $\theta_{0,0}$ (−)

Subscripts

0	boundary surface at $x = 0$; also, initial time $t = 0$
∞	undisturbed fluid
α	pertaining to regularization (Tikhonov or other)
b	body
c	computational
c	constant
ct	complementary transient
d	deviation
δ	second deviation
e	exact
f	final time
i	i-th time step; also, i-th component in time of surface temperature or surface heat flux
j	j-th space step ($y_N = W$); also, j-th component in space of surface heat flux
in	initial (time $t = 0$)
L-curve	pertaining to the L-curve (Section 6.6)
m	counting integer in the x−direction
M	M-th time step
n	counting integer in the y−direction
n_s	number of singular values removed in TSVD
N	N-th space step
opt	optimal
p, pen	penetration time (Eq. 4.1)
p	"look-ahead" time, $p = r_{opt}\Delta t$ (Chapter 7) (s)
p	sensor measurement location (Chapter 10)
qs	quasi steady
ref	reference

ss	steady state
us	unsteady
w	wire
x	along the x–direction
X12	heat conduction along x with BCs of the 1st and 2nd kind
X22	heat conduction along x with BCs of the 2nd kind on both sides
XI0BKT0	heat conduction along x for a semi-infinite body with BC of the 1st kind (I = 1) or of the 2nd kind (I = 2); $K = 1$ or 2 for a constant or linear-in-time BC
y	along the y–direction
Y	pertaining to measurement

Superscripts

~, +	dimensionless quantity
‾	mean or average
‾	Laplace transform of function (Chapter 10)
^	estimated value
T	transpose of a matrix
(A)	counting integer for number of decimal places of precision in solution
(L)	large-time form
(S)	short-time form

Acronyms

1D	one-dimensional
2D	two-dimensional
3D	three-dimensional
BC	boundary condition
BEM	boundary element method
CD	central difference
CGM	conjugate gradient method
DHCP	direct heat conduction problem
FD	forward difference
FR	Fletcher-Reeves conjugate gradient method
FS	function Specification
GF	Green's function
GFSE	Green's function solution equation
IHCP	inverse heat conduction problem
KF	Kalman filter
LHS	left-hand side
LT	Laplace transform
RHS	right-hand side
SD	Steepest Descent iteration method
SOV	separation of variables
SVD	singular value decomposition
TSVD	truncated singular value decomposition
TR	Tikhonov regularization
USEM	unsteady surface element method

Preface to First Edition

This book presents a study of the inverse heat conduction problem (IHCP), which is the estimation of the surface heat flux history of a heat conducting body. Transient temperature measurements *inside* the body are utilized in the calculational procedure. The presence of errors in the measurements as well as the ill-posed nature of the problem lead to "estimates" rather than the "true" surface heat flux and/or temperature.

This book was written because of the importance and practical nature of the IHCP; furthermore, at the time of writing, there is no available book on the subject written in English. The specific problem treated is only one of many ill-posed problems but the techniques discussed herein can be applied to many others. The basic objective is to estimate a function given measurements that are "remote" in some sense. Other applications include remote sensing, oil exploration, nondestructive evaluation of materials, and determination of the Earth's interior structure.

The authors became interested in the IHCP over two decades ago while employed in the aerospace industry. One of the applications was the determination of the surface heat flux histories of reentering heat shields.

This book is written as a textbook in engineering with numerical examples and exercises for students. These examples will be useful to practicing engineers who use the book to become acquainted with the problem and methods of solution. A companion book, *Parameter Estimation in Engineering and Science* by J. V. Beck and K. J. Arnold (Wiley 1977), discusses the estimation of certain constants or parameters rather than functions as in the IHCP. Though many of the ideas relating to least squares and sensitivity coefficients are present in both books, the present book does not require a mastery of parameter estimation.

The book is written at the advanced BS or the MS level. A course in heat conduction at the MS level or courses in partial differential equations and numerical methods are recommended as prerequisite materials.

Our philosophy in writing this book was to emphasize general techniques rather than specialized procedures unique to the IHCP. For example, basic techniques developed in Chapter 4 can be applied either to integral equation representations of the heat diffusion phenomena or to finite difference (or element) approximations of the heat conduction equation. The basic procedures in Chapter 4 can treat nonlinear cases, multiple sensors, nonhomogeneous media, multidimensional bodies, and many equations, in addition to the transient heat conduction equation.

The two general procedures that are used are called (a) function specification and (b) regularization. A method of combining these (the trial function method) is also suggested. One of the important contributions of this book is the demonstration that all of these methods can be implemented in a sequential manner. The sequential method in some case gives nearly the same result as whole domain estimation and yet is much more computationally efficient.

One of our goals was to provide the reader with an insight into the basic procedures that provide analytical tools to compare various procedures. We do this by using the concepts of sensitivity coefficients, basic test cases, and the mean squared error. The reader is also shown that optimal estimation involves the compromise between minimum sensitivity to random measurement errors and the minimum bias.

Preliminary notes have been used for an ASME short course and for a graduate course at Michigan State University.

There are many people who have helped in the preparation of this text and to whom we express our appreciation. These include D. Murio, M. Raynaud, other colleagues, and students who have read and commented on the notes. Thanks are also due to Judy Duncan, Phyllis Murph, Terese Stuckman, Alice Montoya, and Jeana Pineau, who have aided in typing the manuscript.

James V. Beck wishes to express appreciation for the contributions to his education made by Kenneth Astill of Tufts University, Warren Rohsenow of Massachusetts Institute of Technology, and A. M. Dhanak of Michigan State University.

Ben Blackwell would like to acknowledge the contributions that several people made to his heat transfer education: H. Wolf of the University of Arkansas, M. W. Wildin of the University of New Mexico, and W. M. Kays of Stanford University.

A special and deep appreciation is extended to George A. Hawkins for the education and philosophy that he imparted to Charles R. St. Clair, Jr. as his graduate student.

James V. Beck
Ben Blackwell
Charles R. St. Clair, Jr.
East Lansing, Michigan
Albuquerque, New Mexico
East Lansing, Michigan
August 1985

Preface to Second Edition

The first edition of this textbook was published in 1985. At that time, the IBM PC had been available for about four years, and release of the IBM PS/2 was still two years in the future. These historical milestones frame the thinking of the times about computing in the mid-1980s. It is often said, and is demonstrably true, that any common pocket cell phone of the 2020s has significantly more computation and storage capability than early scientific computing machines. While the first edition of the book continues to be used by researchers and professionals as a major reference on inverse heat conduction problems (IHCPs), with the major advancements in solution techniques for IHCPs and the vast progress in use of computer programs for solving engineering problems over the past few decades, the need for writing the second edition of the book was long due.

Over the past 30 years, the authors of this second edition have collaborated on an array of research projects on exact solutions to heat conduction problems and solution of inverse heat conduction problems (IHCPs), and much of that research forms the basis for this book.

This text aims to present the methodology of many IHCP solution techniques (Chapter 4) and outline approaches for optimizing the degree of regularization in an IHCP solution (Chapter 6). These discussions will be presented primarily in the framework of the filter matrix concept presented in Chapter 5.

Almost 40 years have passed since publication of the first edition of this book, and much research on IHCPs and their solutions has been performed since that time. Chapter 1 presents an extensive literature review of IHCP work over the past decades in the areas of manufacturing, aerospace, biomedical, electronics cooling, instrumentation, nondestructive testing, and other areas of engineering applications. Throughout the text, additional references are offered to guide the interested reader to sources for additional information.

Because computing power today is great, many (most?) direct heat conduction problems are solved using commercial software based on finite element or finite difference methodology. However, analytical solutions provide highly accurate results that can be computed efficiently. This second edition continues the emphasis from the first edition on utilization of exact solutions for linear heat transfer problems – those with thermophysical properties that are independent of temperature. Chapter 2 presents exact solutions to several fundamental problems, and superposition of these solutions provides a gateway to solution of a wide array of problems. Also discussed in Chapter 2 is the concept of the computational analytical solution which provides efficient and accurate computation of the solution.

Chapter 3 outlines approximate solution methods for linear problems which form the basis for the IHCP solutions in the remainder of the book. These concepts are presented using ideas of superposition of exact solutions; however, a more rigorous treatment using Green's functions is given in Appendix.

Chapter 4 retains detailed description of the Stolz method, the Function Specification (Beck's) Method, and Tikhonov regularization. New to this second edition are descriptions of the Conjugate Gradient Method and Singular Value Decomposition. Additionally, two methods that are amenable to nonlinear problems are explained: the Adjoint Method, which is used in conjunction with the Conjugate Gradient Method, and the Kalman Filter approach to solving the IHCP.

Chapter 6 is new to the second edition and focuses on selection of the optimal degree of regularization in solution of IHCPs. Because IHCPs are inherently ill-posed, some form of regularization is necessary. Regularization decreases sensitivity to noise in measurements at the expense of introducing bias into the estimates. Striking the appropriate balance between these competing factors is an important, and often overlooked, part of analyzing an IHCP.

Because of the relative costliness of computer resources at the time, a significant focus of the first edition was computational efficiency of IHCP algorithms. In this second edition, more attention is devoted to comparing accuracy of IHCP solution methods than to their computational efficiency. Chapter 7 contains head-to-head comparison of many IHCP solution methods using a common suite of test problems.

The filter coefficient concept is used throughout the text. Chapter 5 explores the nature of the filter matrix for many IHCP methods presented in Chapter 4. The character of many of these filter matrices is such that a truncated filter vector can be extracted from the matrix to use in sequential online (near real-time) estimation. Additionally, Chapter 5 illustrates how linear IHCP concepts can be applied to problems with temperature-dependent thermophysical properties. Chapter 8 outlines application of the filter concepts to two-dimensional transient IHCP problems with multiple unknown heat fluxes.

Chapter 9 addresses the problem of estimating the heat transfer coefficient, h. Much of the chapter focuses on the case of lumped capacitance bodies (with negligible internal temperature gradients) and outlines several approaches for estimating h under those conditions. Application to bodies with temperature gradients is also discussed.

Chapter 10 is new to this edition and addresses bias in temperature measurements caused by sensor installation. Two situations are explored: correction of measured temperature to remove the bias and utilization of the biased measurements directly in the IHCP solution. Both situations typically require a computational model which incorporates the sensor in the domain and can be used to quantify the bias.

The second edition of the book also comes with additional materials that are available on a companion website. These materials include MATLAB codes for the examples that are solved in the book. Furthermore, course slides, additional problems, and suggested course syllabus will be available for instructors.

Keith A. Woodbury
Hamidreza Najafi
Filippo de Monte
James V. Beck
Tuscaloosa, AL, USA
Melbourne, FL, USA
L'Aquila, Italy
East Lansing, MI, USA
June 2022

1

Inverse Heat Conduction Problems: An Overview

1.1 Introduction

Heat conduction problems can be categorized as direct (or forward) problems and inverse problems. If the heat flux or temperature histories at the surface of a solid are known as functions of time, then the temperature distribution can be found from knowing the thermophysical properties of the material by solving the heat diffusion equation. This is termed a direct problem (Beck et al. 1985). The direct problem can be solved using analytical techniques (Hahn and Ozisik 2012) such as Green's functions, separation of variables, and Laplace transform or using numerical techniques such as finite difference method and finite element method.

In many dynamic heat transfer situations, the surface heat flux and surface temperature histories of a solid must be determined from the transient temperature measurements at one or more interior locations; this is the inverse heat conduction problem (IHCP). There are other inverse problems in heat transfer (e.g. parameters estimation (Beck and Arnold 1977)), but the IHCP is the main subject of this book.

The IHCP is much more difficult to solve analytically than the direct problem. But in the direct problem many experimental impediments may arise in measuring or producing given boundary conditions. The physical situation at the surface may be unsuitable for attaching a sensor, or the accuracy of a surface measurement may be seriously impaired by the presence of the sensor. Although it is often difficult to measure the temperature history of the heated surface of a solid, it is easier to measure accurately the temperature history at an interior location or at an insulated surface of the body. Thus, there is a choice between relatively inaccurate measurements or a difficult analytical problem. An accurate and tractable inverse problem solution would thus minimize both disadvantages at once.

The problems of determining the surface temperature and the surface heat flux histories are equivalent in the sense that if one is known the other can be found in a straightforward fashion. They cannot be independently found since in direct heat conduction problems only one boundary condition can be imposed at a given time and boundary. Even though this is true, the following seemingly contradictory statement can be made: the heat flux is more difficult to calculate accurately than the surface temperature. For this reason, the emphasis in this book is on the calculation of the surface heat flux history. (The surface temperature is a by-product of the heat flux calculations.)

For the purposes of this book, the IHCP is defined as the estimation of the surface heat flux history given one or more measured temperature histories inside a heat-conducting body. The word "estimation" is used because in measuring the internal temperatures, errors are always present to some extent and they affect the accuracy of the heat flux calculation. Furthermore, even if discrete data accurate to a large but finite number of significant figures are used, the heat flux cannot be exactly determined.

Figure 1.1a, b depict a direct heat conduction problem and an IHCP in a one-dimensional (1D) domain respectively. The domain is heated at $x = 0$ and insulated at $x = L$. For the direct problem (Figure 1.1a), the surface heat flux is known and the temperature distribution within the domain with respect to time and space is to be determined. For the IHCP (Figure 1.1b), the temperature measurement at an interior location ($0 < x_1 < L$) is available and the active boundary condition (the cause for the temperature response in the domain, which is surface heat flux in this case) is to be determined. The location, x_1, of the sensor is assumed to be measured and to have negligible error. The thickness of the plate, L, is also known and considered errorless.

Figure 1.2 shows representations of the measured temperature values (Fig 1.2a) at an interior location (x_1) as well as the calculated surface heat flux (Fig 1.2b), $\hat{q}_i$. The IHCP is one of many mathematically "ill-posed" problems that are difficult to

Inverse Heat Conduction: Ill-Posed Problems, Second Edition. Keith A. Woodbury, Hamidreza Najafi, Filippo de Monte, and James V. Beck.
© 2023 John Wiley & Sons, Inc. Published 2023 by John Wiley & Sons, Inc.

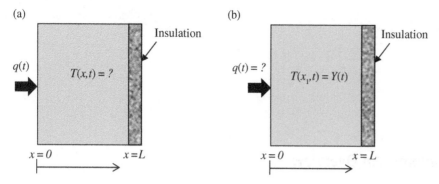

Figure 1.1 Schematic of 1D slab subject to heat flux at $x = 0$. (a) Direct heat conduction problem. (b) Inverse heat conduction problem (IHCP).

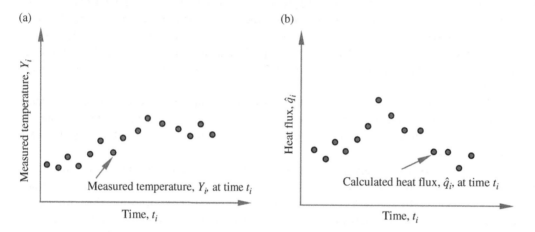

Figure 1.2 (a) Measured temperatures at discrete times. (b) Representations of calculated surface heat fluxes.

solve. A problem can be considered ill-posed if it doesn't meet the three criteria for well-posed problems. According to Hadamard (1923), a well-posed problem must have:

- A solution
- A unique solution
- A solution that depends continuously on the parameters or input data.

In case of IHCP, the problem is ill-posed since the solution is extremely sensitive to measurement errors and therefore does not satisfy the above mentioned third criteria. Interior measurements contain much less information than given for classical direct problems where the surface conditions are continuous, errorless relations. The difficulties are particularly pronounced as one tries to obtain the maximum amount of information from the data. For the 1D-IHCP when discrete values of the $q(t)$ curve are estimated, maximizing the amount of information implies small time-steps between q_i values (see Figure 1.2b). However, the use of small time-steps generally introduces instabilities in the solution of the IHCP which can be countermanded using regularization techniques discussed in Chapter 4. Notice the condition of small time-steps has the opposite effect in the IHCP compared to that in the numerical solution of the heat conduction equation. In the latter, stability problems often can be corrected by reducing the size of the time steps.

Another complexity of IHCPs is associated with what's known as "damping" effect and "lagging" effect. The transient temperature response of an internal point in an opaque, heat-conducting body is quite different from that of a point at the surface. The internal temperature excursions are much diminished internally compared to the surface temperature changes. This is a damping effect. A large time delay in the internal response can also be noted that is known as lagging effect. These damping and lagging effects for the direct problem are important to study because they provide engineering insight into the difficulties encountered in the inverse problems as will be discussed in next chapters.

1.2 Basic Mathematical Description

Considering Figure 1.1, a brief mathematical description of the direct problem and IHCP follows. The 1D transient heat diffusion equation can be given as:

$$\frac{1}{\alpha}\frac{\partial T}{\partial t} = \frac{\partial^2 T}{\partial x^2} \tag{1.1}$$

Here, α is thermal diffusivity of the material (m/s^2) which is defined as $k/\rho c$, where, k, ρ, and c are thermal conductivity (W/m-K), density (kg/m^3) and specific heat (J/kg-K) of the material, respectively. Note that that k, ρ, and c are assumed known and constant. If any one of these thermal properties varies with temperature, the IHCP becomes nonlinear.

The initial condition and boundary condition at $x = L$ are:

$$T(x, 0) = T_0, \tag{1.2}$$

$$-k\frac{\partial T}{\partial x}\bigg|_{x=L} = 0. \tag{1.3}$$

For the direct problem, the boundary condition at $x = 0$ is given as below while the temperature distribution is to be determined:

$$-k\frac{\partial T}{\partial x}\bigg|_{x=0} = q(t), \tag{1.4}$$

$$T(x, t) = ? \tag{1.5}$$

For the IHCP, the temperature measurement values are available at interior location(s) and the active boundary condition (i.e. surface heat flux) is to be determined. Consider the one-dimensional IHCP given in Figure 1.1b. The heat flux at the heated surface is needed, $q(0,t)$ or simply $q(t)$. The net surface heat flux as a function of time is estimated from measurements obtained from an interior temperature sensor at position x_1 as shown in Figure 1.1b:

$$T(x_1, t) = Y(t), \tag{1.6}$$

$$q(0, t) = -k\frac{\partial T}{\partial x}\bigg|_{x=0} = ? \tag{1.7}$$

The measurements are made at discrete times, t_1, $t_{2,...}$ or in general at time t_i at which the temperature measurement is denoted Y_i. (The word "discrete" means at several particular times, such as 1 second, 2 seconds, 3 seconds, etc., but not continuously.) Figure 1.2 is an illustration of postulated values. An estimated surface heat flux, denoted $\hat{q}_i$, is associated with the time t_i at which the corresponding temperature measurement, Y_i, is made. The *true* value of the surface heat flux can be denoted q_i. In general, the heat flux can rise and fall abruptly and can be both positive and negative where negative values indicate heat losses from the surface.

The source of heating is immaterial to the IHCP procedures. Convective sources include high-temperature fluids that flow in reactor heat exchangers, over reentry vehicle surfaces, or across turbine blades. The heating can also be by radiation from any source or by conduction from an adjacent solid that is in thermal contact with the boundary in question. In every case, however, the heat flux estimate $\hat{q}_i$ is the *net* energy absorbed at the surface of the heat conducting body.

The known boundary condition of perfect insulation given by Eq. (1.3) is only one of many that can be prescribed at $x = L$. There can be a convective and/or a radiation condition at $x = L$. For an IHCP with a single unknown heat flux, it is only necessary that the boundary condition at $x = L$ be known. For a temperature-dependent heat transfer coefficient h and also for a radiation condition, the inverse heat conduction problem again becomes nonlinear.

For the case of a single interior temperature history, the problem can be envisioned as two separate problems, one of which is a direct problem, as shown in Figure 1.3. The portion of the body from $x = x_1$ to $x = L$, body 2, can be analyzed as a direct problem because there are known boundary conditions at both boundaries [$T(t) = Y(t)$ at $x = x_1$, $\partial T/\partial x = 0$ at $x = L$]. From this direct problem the heat flux at x_1 can be found from the solution for the temperature distribution in $x_1 \leq x \leq L$ by using

$$\hat{q}_{x_1}(t) = -k\frac{\partial T}{\partial x}\bigg|_{x=x_1}. \tag{1.8}$$

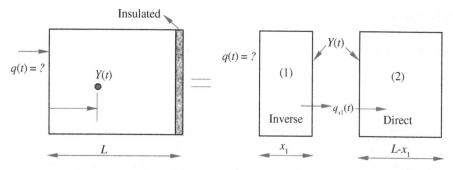

Figure 1.3 Subdivision of a single interior sensor IHCP into inverse and direct problems.

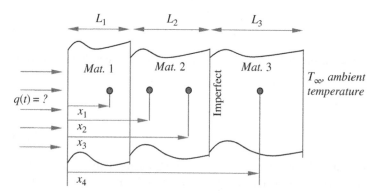

Figure 1.4 Composite plate with multiple temperature sensors.

This same heat flux must *leave* body 1 ($0 \leq x < x_1$). Consequently, *two* conditions are specified at $x - x_1$ in body 1 and none at $x = 0$. Such a set of boundary conditions for the transient heat conduction equation, Eq. (1.1), is related to the mathematical problem being ill-posed.

The IHCP for a single unknown surface heat flux can be complicated in many ways, some of which are illustrated in Figure 1.4. There are four temperature sensors shown which preclude the simple subdivision shown in Figure 1.3 because the heat flux calculated to leave one subdivision would not in general equal the calculated heat flux entering the next subdivision. Another complication is the composite body of three different materials which may be joined together with either perfect or imperfect contacts. In addition, the plates might not be flat but rather be parts of a cylindrical wall. A satisfactory solution of the inverse heat conduction problem should permit treatment of each of these complicating factors.

1.3 Classification of Methods

A number of procedures have been advanced for the solution of ill-posed problems in general including gradient-based methods and non-gradient-based methods (e.g. evolutionary algorithms). In using all of these methods, the main approach is to minimize a function that represents the error (often sum of the square of the errors) between the measured ($\mathbf{Y}$) and calculated ($\mathbf{T}$) temperature:

$$S = (\mathbf{Y} - \mathbf{T})^T (\mathbf{Y} - \mathbf{T})$$

(1.9)

The exact solution can be obtained when S becomes equal to zero. However, given the inevitable noise in the measured data, the minimum of S is not truly zero.

Some of the most notable IHCP solution techniques such as function specification method (FSM) (Beck 1970; Beck et al. 1985), Tikhonov regularization method (Tikhonov 1963a, 1963b; Tikhonov and Arsenin 1977), singular value decomposition (SVD) method (Lawson and Hanson 1974; Mandel 1982), conjugate-gradient method and adjoint method will be discussed in this book (Chapter 4). Brief discussions of the Kalman filter approach is also included (Chapter 4). Additionally, the filter coefficient form of IHCP solution will be discussed in detail (Chapter 5).

The methods for solving the inverse heat conduction problem can be classified in several ways, some of which are discussed in this section.

One classification relates to the ability of a method to treat nonlinear as well as linear IHCPs. This book outlines basic algorithms that can be employed for both linear and nonlinear problems, but emphasis is placed on linear problems that can advantage the principle of superposition. Nonlinear aspects can be addressed using the filter coefficient concepts (see Chapter 5).

The method of solution of the heat conduction equation is another way to classify the IHCP. Methods of solution include the use of superposition of exact solutions, finite differences, finite elements and finite control volumes. The use of superposition techniques restricts the IHCP algorithm to the linear case, whereas the other procedures can treat the nonlinear problem. Superposition principles are used frequently in this book because the basic IHCP algorithms are easier to use and program for simple calculations. Moreover, for linear problems, the answers for the surface heat flux are nearly identical for all methods mentioned, provided the heat conduction equation is solved accurately. Consequently, experience acquired using superposition concepts incorporated in a basic IHCP algorithm is also relevant to the other methods when used for linear IHCPs.

The time domain utilized in the IHCP can also be used to classify the method of solution. Three time domains have been proposed: (i) only to the present time, (ii) to the present time plus a few time stops, and (iii) the complete time domain. The use of measurements only to the present time with a single temperature sensor allows the calculated temperatures to match the corresponding measured temperatures in an exact manner; that is, the calculated temperatures equal the measured values. This is called the Stolz method (Stolz 1960a). Such exact matching is intuitively appealing but the algorithms based on it frequently are extremely sensitive to measurement errors. In the second method, a few future temperatures (associated with future times) are used; the associated algorithms are called "sequential." Greatly improved algorithms are obtained compared with exact matching. The improvements are noted in the considerably reduced sensitivity to measurement errors and in the much smaller time steps that are possible. Small time-steps permit more detailed information regarding the time variation of the surface heat flux to be found. The whole domain estimation procedure is also very powerful because very small time steps can be taken. Many of the whole-domain techniques introduced in this text can be considered as sequential through the concept of filter coefficients (see Chapter 5).

The last classification to be mentioned is relative to the dimensionality of the IHCP. If a single heat flux history is to be determined, the IHCP can be considered as one dimensional. In the use of superposition concepts, the physical dimensions of the problem are not of concern; that is, the same procedure is used for physically one-, two-, or three-dimensional bodies provided a single heat flux history is to be estimated. If two or more heat flux histories are estimated and superposition is used, the problem is multidimensional. When the finite difference or the other methods for nonlinear problems are employed, the dimensionality of the problem depends on the number of space coordinates needed to describe the heat-conducting body; one coordinate would give a one-dimensional problem, two-coordinates a two-dimensional problem, and so on.

1.4 Function Estimation Versus Parameter Estimation

The words "function estimation" were used in the previous section in connection with the IHCP. In the IHCP, the heat flux is found as an arbitrary, single-valued function of time. The heat flux can be positive or negative, constant or abruptly changing, periodic or nonperiodic, and so on. It may be influenced by human decisions. For example, the pilot of a shuttle can change the reentry trajectory. In the IHCP problem the surface heat flux is a function of time and may require hundreds of individually estimated heat-flux components, $\hat{q}_i$, to define it adequately.

Related estimation problems are those called "parameter estimation" problems, which are also inverse problems but with the emphasis on the estimation of certain "parameters" or constants or physical properties. In the context of heat conduction one might be interested in determining the thermal conductivity of a solid given some internal temperature histories *and* the surface heat flux and other boundary conditions (Beck and Arnold 1977). The thermal conductivity of ARMCO iron near room temperature, for example, could be a parameter; it is not a function and does not require hundreds of values of k_i to describe it. The parameter estimation and function estimation problems start to merge if estimates are made of the thermal conductivity, k, as a function of temperature, T. However, the $k(T)$ function is not arbitrary and is not adjustable by humans.

A background in parameter estimation is not required to understand this book. The subject of parameter estimation has been built on a statistical base, but that is not as true for function estimation problems. This book stresses the numerical and mathematical aspects of function estimation rather than the statistical aspects.

1.5 Other Inverse Function Estimation Problems

In addition to the problem of estimating the surface heat flux using internal temperature measurements which is the focus of this book, there are several other problems related to the inverse heat conduction problem. The IHCP involves estimation of the surface heat-flux time-function utilizing measured interior temperature histories. It is a linear problem if the thermal properties are independent of temperature and the boundary condition at the "known" boundary is linear. A closely related problem involves the convective boundary condition:

$$-k\frac{\partial T}{\partial x}\Big|_{x=0} = h[T_\infty(t) - T(0,\ t)]. \tag{1.10}$$

If the heat transfer coefficient, h, is known either as a constant or as a function of time, the estimation of the ambient temperature $T_\infty(t)$ from given internal temperature measurements is a linear, inverse function estimation problem. If h is a known function of T, then the inverse problem becomes nonlinear (Beck and Arnold 1977).

Another important function estimation problem is the determination of h as a function of time. This is a *nonlinear* problem even if the differential equation is linear. The determination of the transient heat transfer coefficient is an important technique. The estimation of $h(t)$ is discussed in Chapter 9.

An interface contact conductance, $h_c(t)$, is often used to model imperfect contact. For the interface in Figure 1.4 the heat flux is related to h_c by

$$-k\frac{\partial T}{\partial x}\Big|_{x=(L_1+L_2)^-} = h_c\left[T\Big|_{x=(L_1+L_2)^-} - T\Big|_{x=(L_1+L_2)^+}\right] = -k\frac{\partial T}{\partial x}\Big|_{x=(L_1+L_2)^+}, \tag{1.11}$$

where the sign + refers to the right hand side (material 3) of the interface and the sign − refers to the left hand side (material 2) of the interface. The problem of estimating $h_c(t)$ is very similar to that for the convective heat transfer coefficient.

Endothermic or exothermic chemical reactions can occur inside materials. These can be of unknown magnitudes. Also, there can be an energy source due to electric heating, a nuclear source, or frictional heating. In these cases an appropriate describing equation for one-dimensional plane geometries is

$$\frac{\partial}{\partial x}\left(k\frac{\partial T}{\partial x}\right) + g(x,t) = pc\frac{\partial T}{\partial t}, \tag{1.12}$$

where $g(x,\ t)$ is a volume energy source term. If g is a function of time only, then estimation of $g(t)$ from transient interior temperature measurements is quite similar to the one-dimensional IHCP. If Eq. (1.12) is linear and the boundary conditions are linear, estimation of $g(t)$ is a linear problem; however, if $k = k(T)$, the problem of estimating $g(t)$ becomes nonlinear. When g is a function of both x and t, the estimation of $g(x,\ t)$ is similar to that of a two-dimensional IHCP.

Two inverse function estimation problems that have received a great deal of attention from mathematicians are called the Cauchy problem for the two-dimensional Laplace's equation (Cannon and Douglas 1967; Payne 1975; Tikhonov and Arsenin 1977) and the initial-boundary value problem for the backward heat equation (John 1955; Payne 1975).

One form of the Cauchy problem for the equation,

$$\frac{\partial^2 T}{\partial x^2} + \frac{\partial^2 T}{\partial y^2} = 0, \tag{1.13}$$

is for incomplete specification of the boundary conditions but some interior *measurements* of temperature are given. The objective is to obtain an estimate of $T(x, y)$ for the complete domain including the boundaries.

One example of a backward heat equation problem is the determination of the initial temperature distribution, $T_0(x)$, in a finite body given the boundary conditions and some internal transient measurements of temperature.

The inverse problems mentioned become more complex as more functions are determined simultaneously. For example, one might attempt to simultaneously estimate for Figure 1.4 the heat flux $q(t)$ on the left boundary of the body and $T_\infty(t)$ on the right. This would involve simultaneous estimation of two time functions.

If the surface heat flux is a function of position across the surface as shown in Figure 1.5, a number of heat flux components would be simultaneously estimated; this is the two-dimensional IHCP and is discussed further in Chapter 8.

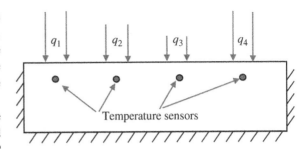

Figure 1.5 Surface heat flux as a function of position for a flat plate.

1.6 Early Works on IHCPs

One of the earliest papers on the IHCP was published by Stolz in 1960 which addressed calculation of heat transfer rates during quenching of bodies of simple finite shapes. Stolz (1960a) claimed that his method was in use as early as June 1957. For semi-infinite geometries, Mirsepassi (1959a) maintained that he had used the same technique both numerically (Mirsepassi 1959a) and graphically (Mirsepassi 1959b) for several years prior to 1960. A Russian paper by Shumakov (1957) on the IHCP was translated in 1957. The space program, starting about 1956, gave considerable impetus to the study of the inverse heat conduction problem. The applications therein were related to nose cones of missiles and probes, to rocket nozzles, and other devices. Beck also initiated his work on the IHCP about that time and developed the basic concepts (Beck 1962, 1968, 1970, 1979; Beck and Wolf 1965; Beck et al. 1982) that permitted much smaller time steps than the Stolz (1960b) method. The IHCP solution approach developed by Beck (1970) was used in developing several computer programs that was used by various industries (Muzzy et al. 1975; Bass 1979, 1980; Snider 1981).

Others whose work had application to the space program included Blackwell (1968, 1981), Imber (1973, 1974a, 1974b, 1975a, 1975b, 1979), Imber and Khan (1972), Mulholland and San Martin (1973), Mulholland et al. (1975), Mulholland and San Martin (1971), Mulholland and Cobble (1972), and Williams and Curry (1977). Another research area that extensively required solutions of the IHCP was the testing of nuclear reactor components (France et al. 1978).

Other early applications reported for the IHCP included: (i) periodic heating in combustion chambers of internal combustion engines (Alkidas 1980), (ii) solidification of glass (Howse et al. 1971), (iii) indirect calorimetry for laboratory use (Sparrow et al. 1964), and (iv) transient boiling curve studies (Lin and Westwater 1982).

1.7 Applications of IHCPs: A Modern Look

In general, where direct measurement of surface condition (i.e. heat flux or temperature) is either difficult or not practical, solving the associated IHCP can facilitate the estimation of surface condition using temperature measurement from interior locations. Having the knowledge of surface condition is critical in wide ranges of applications and therefore examples of IHCPs are present in numerous areas including manufacturing processes, aerospace industry, biomedical applications, electronics cooling, and more. Some of these examples are briefly discussed in this section.

1.7.1 Manufacturing Processes

Measurement of heat flux is of immense importance, but difficult, in several manufacturing processes including cutting, quenching, machining, casting, welding, spray cooling, and more. Over the past several decades, developing solutions to IHCPs facilitated the improvement in these manufacturing processes. The IHCPs associated with manufacturing applications have been discussed in numerous references.

1.7.1.1 Machining Processes
Machining processes involve major heat generation in a rather small area that causes increase of temperature in the cutting tool and workpiece. Therefore, heat transfer in metal cutting is known as a critical aspect which affects machining efficiency, cutting tool life, and quality of the products (Ståhl et al. 2012). It is not possible to directly measure the surface heat

flux on the cutting tool and therefore the solution must be developed for an IHCP. Kusiak et al. (2005) used an inverse method to estimate the heat flux in the tool during machining using temperature measurements at one or several locations in the tool. Using their approach, they conducted a comparison between the effect of various coatings on the heat transfer in the tool during machining. Kryzhanivskyy et al. (2017) developed a method for solving the IHCP associated with metal cutting tool via an iterative approach and through transforming the problem into a constrained optimization problem. A high-speed steel (HSS) cutting tool with eight embedded thermocouples was used for experimental validation and the numerical simulation was performed in COMSOL Multiphysics and MATLAB. In another study by Kryzhanivskyy et al. (2018), they established a proof for the transient behavior of heat flux into cutting tool and developed a two-stage solution for an IHCP associated with this application based on the sequential function specification method (Figure 1.6a). They showed that the transient nature of the heat flux can be best fitted by a reducing power function. Gostimirovic et al. (2011) used the temperature measurement from the workpiece to calculate the heat flux and the temperature field in

(a)

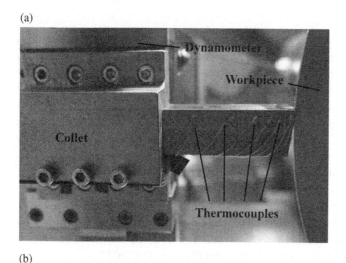

(b)

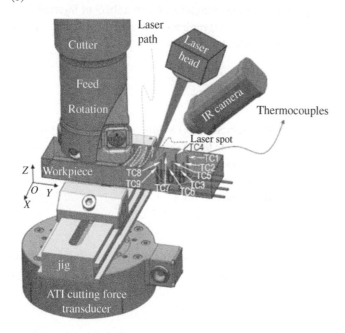

Figure 1.6 Examples of studies involving solution of IHCP in manufacturing. (a) Metal cutting (b) Laser assisted machining. *Source:* (a) Kryzhanivskyy et al. (2018)/with permission of Elsevier. (b) Shang et al. (2019)/with permission of Elsevier.

the cutting zone through solution of IHCP. This information is then used to optimize the machining process in order to maintain the best use of the tool and highest quality of the workpiece. Modeling and optimization of the electrical discharge machining (EDM) process were performed by Gostimirovic et al. (2018) through solution of IHCP based on the desirable temperature distribution in the workpiece. Their approach facilitates calculation of temperature and heat flux in electrodes which allows estimating the dimension of crater, the material removal rate, and surface roughness. Application of IHCPs is demonstrated in innovative use of laser assisted machining (LAM) by Shang et al. (2019) to address the restrictions associated with the common practice where the laser is focused on a small area which is at a fixed location with respect to the cutting tool all the time. They used the solution of IHCP to optimize the laser power and laser path leading to the desired temperature distribution in the product (Figure 1.6b). They showed that by applying this approach the peak and mean principal cutting forces can be substantially reduced compared with the conventional technique. An improvement in surface roughness was also achieved and a microstructure analysis revealed that overheating was successfully avoided below the intended depth of cut when the new method was used.

1.7.1.2 Milling and Hot Forming

IHCPs are also prevalent in manufacturing processes such as milling and hot forming. Application of IHCP in calculating the optimal heat flux for achieving desired temperature field in highspeed milling is studied by Wei et al. (2016). They developed a temperature measuring tool holder system to continuously measure the temperature at the tip of the milling tool. Particle swarm optimization algorithm is used to use the temperature measurements in evaluating the rake face heat flux. Evaluation of thermal boundary conditions during the solidification of metallic alloys through the investment casting process is studied by O'Mahoney and Browne (2000) for different types of alloys. Their solution approach was proved to be effective through a test case that involved phase change and sensor noise. The solution of IHCP associated with transient solidification of a Sn–Sb peritectic solder alloy on distinct substrates is studied by Curtulo et al. (2019). The temperature profiles that are measured experimentally through the solidification of a Sn5.5 wt% Sb alloy casting are compared against the values generated using the analytical model and the metal-substrate (mold) heat transfer coefficient and the cooling rate at the solid-liquid interface are calculated accordingly. Pandey et al. (2020) studied calculation of interfacial heat transfer coefficient (thermal resistance between the die and alloy) during hot forming process. For this purpose, they established an experimental apparatus with three thermocouples on the die and three thermocouples on the alloy and solved the associated IHCP to calculate the heat flux and temperature at the interface between them. A similar study was performed by Liu et al. (2020) and good agreement were observed between the experimental measurements and the calculated temperatures and heat flux using the IHCP solution through sequential function specification method (Figure 1.7a).

1.7.1.3 Quenching and Spray Cooling

Solving IHCPs can also help with improving quenching and spray cooling processes. Quenching is the sudden cooling of a workpiece to achieve desired material properties. Soejima et al. (2015) investigated the proper spray quenching conditions including spray flow rate and nozzles arrangement and compared the use of dip and spray quenches for a A6061 hollow cylinder. They measured the temperature on the cylinder and calculated the interior temperature distribution through the solution of IHCP. Calculating the quenching rate, as a critical factor in product quality in heat treatment of metals, through solution of IHCP is assessed by Pati et al. (2019). They conducted a parametric study to understand the effect of various parameters such as plate thickness, number of temperature sensors, distance of thermocouple hole from quenched surface, temperature dependent material properties, etc. on the accuracy of their solution. Evaluation of surface temperature and heat flux during cryogenic quenching process for stainless steel rodlets with different coating layers in liquid nitrogen is explored by Xu and Zhang (2020). They used the transient central temperatures of the rodlets and solved the associated IHCP to calculate the surface conditions (Figure 1.7b).

Spray cooling is commonly used in manufacturing processes such as castings as well as other applications including cryogenic freezing and electronics cooling. Somasundaram and Tay (2013) studied intermittent spray cooling (ISC) process, involving pulsed spray, on a copper block and compared the ISC performance in low and high heat flux rates. They measured the transient temperature below the surface of the copper block and calculated the surface heat flux and heat transfer coefficient through the solution of the associated IHCP using the sequential function specification method. An investigation of using air-atomized spray cooling with aqueous polymer additive for hot steel plate is performed by Ravikumar et al. (2014a). The transient temperature is measured at three sub-surface locations and this data is used to evaluate the surface

(a)

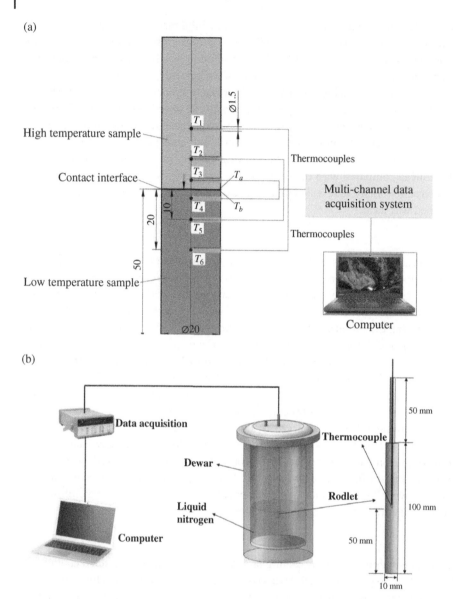

(b)

Figure 1.7 Examples of studies involving solution of IHCP in manufacturing. (a) Hot forming. (b) Hryogenic quenching. *Source:* (a) Liu et al. (2020) / with permission of Elsevier (b) Xu and Zhang (2020) / with permission of Elsevier.

heat flux through the solution of IHCP by INTEMP. INTEMP is an IHCP solver developed by Trujillo and Busby (1997). Evaluating the heat transfer between a hot metallic surface and spray droplets using transient temperature data at interior locations of a test plate is performed by Zhang et al. (2014) through solving an IHCP using INTEMP. They proposed a number of correlations to calculate heat flux in different boiling regimes as a function of for water flow rate.

Ravikumar et al. (2014b) used the transient temperature measurement by three subsurface thermocouples for a test plate of AISI 304 during cooling and evaluated the surface heat flux and temperatures through solving an IHCP using INTEMP. The results allowed them to assess the effect of mixed-surfactants on cooling performance. An investigation of spray cooling by means of nanofluid (Al_2O_3 particles and water) is conducted by Ravikumar et al. (2015) for high heat flux applications. A pre-heated steel plate is used as the cooling target with initial temperature of 900 °C and various cooling fluids are used for spray cooling purposes. Internal temperatures are measured using sub surface thermocouples and an IHCP is then solved to calculate the surface temperature and heat flux accordingly. The results showed that the cooling performance substantially enhanced when nanofluids are used as cooling fluid for the air-atomized spray process. Application of IHCP solution through function specification method in evaluating instantaneous surface heat flux using internal temperature measurement in a copper block that is cooled by liquid nitrogen spray is studied by Somasundaram and Tay (2017).

1.7.1.4 Jet Impingement

Water jet impinging is known as a commonly used process in manufacturing of iron and steel and other applications including nuclear power plants. The cooling rate at the surface is of great significance and cannot be directly measured and therefore the IHCP concept can become very useful. In a study by Dou et al. (2014), a stainless-steel plate is heated to 1000 °C and high-pressure water (up to 2 MPa) is used for water jet cooling. Temperature data from an interior location is used to evaluate surface temperature and heat flux through a solution of an IHCP using an iterative algorithm (perturbation method). The results from the IHCP solution were compared against experimentally measured data and showed a very good agreement with less than 5% error. Volle et al. (2009) assessed application of IHCP solutions (sequential function specification method) for calculating the wall heat flux of a cylindrical geometry using transient temperature measurements by subsurface thermocouples.

Flame jet impingement is frequently used in various heating applications. The use of IHCP in evaluating the heat flux distribution on a flat plate exposed to methane-air premixed flame jet impingement is studied by Hindasageri et al. (2014, 2015). The temperature measurements were conducted by an infrared camera and the associated IHCP is solved accordingly.

1.7.1.5 Other Manufacturing Applications

IHCP has been also investigated for other manufacturing applications. Evaluating the heat flux conducted into a wall using temperature data from interior location in laser cladding application is performed by Lin and Steen (1998) through sequential function specification method. The simultaneous calculation of heat fluxes on all sides of a steel slab in a reheating furnace is studied by Li et al. (2015). Yang et al. (2017) studied estimating laser-induced heat generation in a porous medium subjected to laser heating using temperature measurements from interior points. They solved the IHCP using the conjugate gradient method and the discrepancy principle. Ji et al. (2018) investigated evaluating the surface heat flux distribution on a sample of Ti-6wt%Al-4wt%V (Ti64) alloy subjected to industrial plasma torch. The temperature data were collected from 15 thermocouples installed on the sub-surface of the sample and the IHCP is solved using the future time-step approach to calculate the surface heat flux distribution. Magalhães et al. (2018) proposed an approach to predict the temperature and the weld bead in laser welding. A solution algorithm for a three-dimensional IHCP is developed based on the time traveling regularization with the Golden Section method (Vanderplaats 2005) to estimate the heat flux under various heat distributions.

1.7.2 Aerospace Applications

Aerospace applications are among classic examples of IHCPs. One good example is estimating the surface heating history experienced by a shuttle or missile during atmospheric entry.

Figure 1.8a depicts a reentering body and Fig. 1.8b is an enlarged section of its skin. Though the heat flux, denoted q, may be in general a function of both position y and time t, it is assumed at present that lateral conduction can be neglected compared to the heat flow normal to the surface. Thus, the net surface heat flux as a function of time is estimated from measurements obtained from an interior temperature sensor at position x_1 as shown in Figure 1.8b. The measurements are made at discrete times, t_1, $t_2, ...$ or in general at time t_i at which the temperature measurement is denoted Y_i. (The word

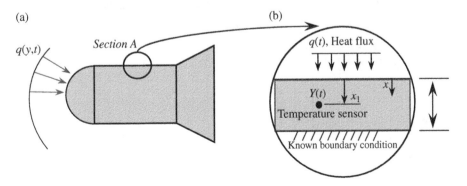

Figure 1.8 Example of reentering vehicle for which the surface heat flux is needed. (a) Reentering vehicle schematic; (b) Section A.

"discrete" means at several particular times, such as 1 second, 2 seconds, 3 seconds, etc., but not continuously.) It should be noted that as previously mentioned, the source of heating is immaterial to the IHCP procedures.

Alifanov et al. (2009) used IHCP solution to study multi-layer insulation (MLI) for spacecrafts. They estimated the emissivity and thermal conductance of a MLI blanket based on the solution of IHCP using temperature and heat flux measurement data. Nakamura et al. (2014) used sequential function specification method and truncated singular value decomposition to solve the transient IHCP associated with the surface heat flux estimation for atmospheric reentry of a capsule using sub-surface temperature measurements. Cheng et al. (2016) used the outer wall temperatures of a supersonic combustor to calculate the heat flux and temperature on the inner wall through solution of the IHCP based on the conjugate gradient method. They verified their approach through both numerical test cases and experiments with Mach 3 supersonic combustor.

Accurate measurement of surface condition in a scramjet combustor is necessary for optimal design of the cooling system. Cui et al. (2018) solved a three-dimensional IHCP to determine the surface heat flux in a scramjet combustor with regenerative cooling system using temperature measurements from interior locations. The direct problem was solved using ABAQUS and a gradient-based method was used to solve the IHCP.

Albano et al. (2019) used conjugate gradient method to solve an IHCP and evaluate the heat flux on the surface of a carbon/carbon high thickness shell for a space vehicle using internal temperature measurements. Uyanna et al. (2021) developed a filter-form solution based on Tikhonov regularization method for solving IHCP in a one-dimensional multi-layer medium representing the thermal protection system of a space vehicle (Figure 1.9). Their approach allows near-real time estimation of the surface heat flux using temperature measurement from two interior points. They also explored the impact of sensor placement and the temperature dependent material properties.

1.7.3 Biomedical Applications

Inverse heat transfer problems are abundant in biomedical applications. Most references that assess thermal transport in biomedical applications use the Pennes bioheat transfer equation (Pennes 1948) as the governing equation for the direct problem to describe the heat conduction in blood perfused tissues. The transient one-dimensional Pennes bioheat transfer equation which presents temperature of the tissue (T) as a function of time (t) and space (x) can be given as (Shih et al. 2007):

$$\rho_t c_t \frac{\partial T}{\partial t} + W_b c_b (T - T_a) = k_t \frac{\partial^2 T}{\partial x^2} \tag{1.14}$$

where ρ_t, c_t, and k_t are the density, specific heat, and thermal conductivity of tissue, respectively and W_b and c_b are the perfusion rate and the specific heat of blood. The IHCP for this category of applications usually involves estimation of surface heat flux, temperature distribution or the source term identification (i.e. blood perfusion). Some examples are discussed as follows.

Real-time monitoring of human body's interior temperatures is critical in medical applications. Jin et al. (2016) studied reconstruction of disease-associated heat source distribution through solution of an IHCP. Particularly, they considered a hypothetical tumor in a sphere and a digital human head. They used finite element method to solve the direct problem which was described by the steady-state form of the Pennes' equation and used Tikhonov regularization method to solve the inverse problem to estimate the heat generation term and the three-dimensional temperature distribution. Lee et al. (2013) used a conjugate gradient method and the discrepancy principle to solve the IHCP for estimating the transient surface

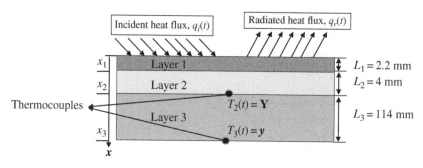

Figure 1.9 Schematic of a three-layer thermal protection system. *Source:* Uyanna et al. (2021).

heat flux in a skin tissue from the internal temperature measurements. They further studied the surface heat flux estimation problem using conjugate-gradient method for a multi-layer tissue (Lee et al. 2015).

Hyperthermia is a well-known technique for destroying cancerous cells by exposing the body tissue to elevated temperatures. Optimization of the heating process for this application is of extreme importance to ensure the cancerous cells will reach to the desired temperature and the thermal damage will remain isolated to the pre-determined cancerous area. Loulou et al. (2018) studied the optimal transient heat (thermal dose) in treatment of cancerous tumors. They used finite difference approach for solving the one-dimensional Pennes' bio heat equation to calculate the temperature distribution in the tissue and applied conjugate-gradient method for solving the IHCP leading to estimation of the time-dependent heat source. Gayzik et al. (2006) also investigated optimization of a hyperthermia treatment protocol using the conjugate-gradient method with the adjoint problem. Jalali et al. (2014) used solution of IHCP for estimating the external heat source and heat transfer coefficient at the skin surface using during hyperthermia. A finite volume approach was implemented to calculate the temperature distribution by solving the bio heat equation and the inverse problem is solved via conjugate-gradient method to evaluate the transient external heat source and heat transfer coefficient at the skin surface.

Inverse problems in other forms (e.g. parameter estimation, shape estimation, etc.) have also been used for numerous biomedical applications, including estimation of blood perfusion rate in tissues (Bezerra et al. 2013; Iljaž and Škerget 2014), overall heat transfer coefficient for temperature distribution calculation in hyperthermia (Aghayan et al. 2013), and estimation of tumor characteristics using skin temperature (Das and Mishra 2013; Hossain and Mohammadi 2016). A comprehensive review on the inverse heat transfer in biomedical application is presented by Scott (2018).

1.7.4 Electronics Cooling

IHCP methods can be instrumental in estimating the location and intensity of heat sources in electronics (Krane et al. 2019). Application of IHCP techniques for temperature monitoring of integrated circuits is explored by Janicki et al. (1998). They considered one-dimensional heat conduction in a semiconductor slab. They used temperature measurement data to evaluate the power dissipated in heat sources through the solution of an IHCP and then calculated the temperature distribution through the entire medium accordingly. In another study by Janicki and Napieralski (2004) assessed the heat transfer process in microelectronic circuits and the use of IHCP in such applications. Krane et al. (2019) investigated application of the IHCP for locating hotspots and estimating their temperatures for an electronic package. The direct problem is solved using FloTHERM and the temperature values are used as inputs for the IHCP solution through Tikhonov regularization method. The solution of IHCP resulted in the magnitude and distribution of the heat generation rate in the system.

Hsu and Chu (2004) developed solution to an inverse non-Fourier heat conduction problem to evaluate the surface heat flux of an electronic device using temperature measurement from interior points. It should be noted that while the Fourier's law is usually accurate to model the heat transfer in most engineering applications, in the case of exceedingly low temperatures, extremely high temperature gradients, or very short times it no longer provides an accurate representation of the heat transfer process and a hyperbolic equation must be used. Therefore, the very fast transient change in temperature in electronics chip studied in this paper requires the use of non-Fourier heat conduction equation.

1.7.5 Instrumentation, Measurement, and Non-Destructive Testing

The inverse problems have been used for developing various innovative techniques and devices for measurement purposes, and particularly a basis for non-intrusive and non-destructive testing.

Lu et al. (2010) studied estimation of the fluid temperature near the inner wall of a pipe elbow through solution of an IHCP in a two-dimensional domain. They used temperature measurements on the outer wall and implemented the conjugate-gradient method for estimating the fluid temperature near the inner wall of the pipe. In another study by Lu et al. (2011), a three-dimensional IHCP was solved using conjugate-gradient method to estimate the inner wall temperature of a pipe elbow using temperature data from the outer surface of the pipe. Jaremkiewicz and Taler (2016) used solution of IHCP to calculate the transient temperature of the fluid using a cylindrical thermometer with a temperature sensor on its axis. Abreu et al. (2016) investigated the evaluation of adhesion failures between layers of composite materials by estimating the temperature difference, heat flux, and the thermal contact conductance at their interface through solving an IHCP.

Cebula (2016) developed a device for measuring the heat flux on the surface of a cylindrical element. The measurement device is made of stainless-steel pad which covers a 94° cylindrical sector. Fourteen thermocouples are placed on sub-surface locations and the resulting temperature measurements are used to calculate the surface heat flux through the solution of an

IHCP. Zhou et al. (2016) explored various solution algorithms for IHCPs for measurement of the surface heat flux in response to ultra-fast surface cooling. Najafi et al. (2015) developed a digital filter form of the Tikhonov regularization method to solve an IHCP associated with a three layer domain with temperature dependent material properties for a directional flame thermometer (DFT) that can be used for near real-time measurement of surface heat flux in harsh environments (inside furnaces, wild fires, etc.). The DFT consist of two layers of Inconel with a ceramic fiber layer in between. Two thermocouples are sandwiched between the layers that actively measure the temperature on internal locations. It was shown that the filter-based approach can accurately evaluate the surface heat flux with a relatively small delay.

The use of the temperature measurement from one sub-surface thermocouple for estimating wall heat flux through solution of an IHCP is performed by Singh et al. (2017). Asif et al. (2019) used conjugate gradient method to calculate the contact thermal conductance between metallic bodies. Taler et al. (2018) studied thermal stress monitoring in thick-walled components of a thermal power unit based on the solutions of the inverse heat conduction problems.

Taler et al. (2019) assessed two different approaches based on solution of IHCPs for thermal stress monitoring in thick walled pressure components of steam boilers. In the first approach, they used inverse space marching method for calculating the temperature distribution using measurement of the outer surface temperature to solve a three-dimensional IHCP. In the second approach, the temperature of the fluid and the component near the internal surface are used to evaluate the temperature distribution within the component. They presented a comparison between the results generated using the two proposed methods versus the artificial and measurement data and good agreements were observed. Borazjani et al. (2015) used the solution to an IHCP for detecting defects in composite and multi-layer materials through thermographic nondestructive tests.

1.7.6 Other Applications

Besides the previously discussed references, there are numerous other examples of using IHCP solutions to address challenging industry-related problems from nuclear reactors to combustion chambers.

Cebula and Taler (2014) investigated application of IHCP solution to calculate the heat flux on the outer surface of a control rod in a nuclear reactor. The control rod is merged in a mixing region with both hot and cold streams that create thermal fluctuations. They developed numerical tests to assess the performance of their develop solution and provided a detailed design for a device to implement the approach in practice. Estimating temperature fluctuations on the inner wall of a horizontal T-junction pipe under turbulent penetration in a nuclear reactor from measurements of outer wall temperatures is performed through solution of IHCP by Guo et al. (2017). They considered both a one-dimensional and two-dimensional IHCP and compared the results for four different cases based on the maximum disturbance of temperature inside the pipe and concluded that the simpler approach with one-dimensional IHCP solution in most cases lead to sufficiently accurate results with less than 5% error.

Huang and Lee (2015) solved a three-dimensional IHCP through conjugate gradient method to estimate six internal heat fluxes in a square combustion chamber with irregular external fins using external temperature measurements by thermal camera.

Luo and Yang (2017) explored estimation of the total heat exchange factor that is necessary for obtaining reference trajectories in reheat furnace control. They proposed an approach based on developing a solution for the one-dimensional IHCP with temperature dependent material properties using conjugate gradient method. They used temperature measurement data on the slab and the furnace at the top and bottom surface of the slab and evaluated the total heat exchange factor accordingly.

Bozzoli et al. (2019) developed a solution for an IHCP in order to calculate the distribution of convective heat flux at the internal wall surface in ducts using temperature measurement data at the exterior boundary. Their approach was based on the singular value decomposition method which was smoothed by the Gaussian filter (GFSVD). The results from the GFSVD approach was compared against results generated by other IHCP solution techniques including Tikhonov Regularization and Damped SVD (DSVD).

The measurement of the heat conduction from concrete to liquified natural gas (LNG) is critical to assess the risk from LNG spill. Li et al. (2017) calculated the conducted heat flux to liquid nitrogen (as a similar material to LNG) through solution of an IHCP using conjugate gradient method. They conducted an experiment through which liquid nitrogen was spilled on the surface of a concrete sample with embedded thermocouples at different depths. The thermocouples measured the transient temperature during vaporization of the liquid nitrogen. The temperature data were then used to evaluate the surface heat flux and the evaporation rate accordingly.

Shao et al. (2021) used the solution of IHCP for assessing the transient liquid nitrogen jet impingement boiling on concrete surface. They established an experimental apparatus and used a thermocouple to measure the surface temperature of the concrete that is subjected to the liquid nitrogen (LN_2) impingement. The resulting problem is a one-dimensional IHCP to use the temperature data from the thermocouple and calculate the heat flux on the surface accordingly. They used the transfer function method to solve the IHCP and achieved good results.

1.8 Measurements

1.8.1 Description of Measurement Errors

In the inverse heat conduction problem, there are a number of measured quantities in addition to temperature; such as time, sensor location, and specimen thickness. Each is assumed to be accurately known except the temperature. If this is not true, then it may be necessary, for example, to simultaneously estimate sensor location and the surface heat flux. The latter problem would involve both the inverse heat conduction and parameter estimation problems and is beyond the scope of this book. If the thermal properties are not accurately known, they should be determined as accurately as possible using parameter estimation techniques.

The temperature measurements are assumed to contain the major sources of error or uncertainty. Any known systematic effects due to calibration errors, presence of the sensor, conduction and convection losses or whatever are assumed to be removed to the extent that the remaining errors may be considered to be *random*. These random errors can then be statistically described.

The information provided by the sensors inside the heat-conducting body is incomplete in several respects. First, these measurements are at discrete locations. There is only a finite number of sensors, sometimes only one. Hence the spatial variation of temperature is quite incompletely known. Moreover, the measurements obtained from any sensor are available only at discrete times, rather than continuously. Due to the nature of the measurement errors, a *continuous* temperature record might contribute little more information than the discrete values, however.

1.8.2 Statistical Description of Errors

A set of eight standard statistical assumptions regarding the temperature measurements is given in this section. These are *standard* assumptions and may not be valid for a particular case. These eight assumptions do provide a yardstick with which to compare the actual conditions. The random errors in the temperature measurements cause random errors in the surface heat flux values. The standard assumptions permit simplifications in the analysis of random errors. The eight standard assumptions discussed in Beck and Arnold (1977) are:

1) The first standard assumption is that the errors are additive or

$$Y_i = T_i + \varepsilon_i \text{ (additive errors)},\tag{1.15}$$

where Y_i is the temperature measurement at time t_i, T_i is the "true" temperature at time t_i, and ε_i is the random error at time t_i.

2) The second standard assumption is that the temperature errors, ε_i, have a zero mean (a theoretical quantity),

$$E(\varepsilon_i) = 0 \text{ (zero mean errors)},\tag{1.16}$$

where $E(\cdot)$ is the "expected value operator" (Beck and Arnold 1977). A random error is one that varies as the measurement is repeated but the theoretical mean does not have to be equal to zero. There can be a *bias;* that is, the error might tend to be positive. It is frequently possible to calculate and remove the bias.

A *sample* mean is the average that is based on actual measurements. The true average of ε_i cannot be determined because ε_i is unknown; a typical equation for finding the sample mean of a random variable such as Y_i is

$$\overline{Y} = \frac{1}{J} \sum_{j=1}^{J} Y_{ji},\tag{1.17}$$

where Y_{ji} is the jth measurement at time t_i and there are J measurements at time t_i. If Eq. (1.16) is true, the expected value of $\overline{Y}_i$ is T_i. The expression given by Eq. (1.17) is called the *sample* mean and can sometimes be used to check the assumption given by Eq. (1.16).

3) The third standard assumption is that of a constant variance,

$$V(Y_i) = \sigma^2 \text{ (constant variance error)}, \tag{1.18}$$

where $V(\cdot)$ is the "variance operator" and is related to the expected value operator by.

$$V(Y_i) = E\left\{[Y_i - E(Y_i)]^2\right\} \tag{1.19}$$

The symbol σ^2 does not contain an i subscript, thus Eq. (1.18) means that the variance of Yi is independent of time t_i and is a constant. If the constant variance assumption embodied in Eq. (1.18) is valid and there is only a single sensor, an estimate of the "variance of Y," denoted s^2, is

$$s^2 = \frac{1}{n-p}\sum_{i=1}^{n}\left(Y_i - \hat{Y}_i\right)^2 = \frac{1}{n-p}\sum_{i=1}^{n}e_i^2 \tag{1.20}$$

for n measurement times; p is the number of parameters being used to estimate T_i, the estimate of which is denoted $\hat{Y}_i$ and e_i is the residual defined by

$$e_i = Y_i - \hat{Y}_i \tag{1.21}$$

Expressions of the type given by Eq. (1.20) can be employed to investigate the validity of the constant variance assumption given by Eq. (1.18)

4) The fourth standard assumption relates to the correlations among measurements. For two measurement errors ε_i and ε_j where $i \neq j$, the two errors are uncorrelated if the covariance of ε_i and ε_j is zero or

$$\text{cov}(\varepsilon_i - \varepsilon_j) \equiv E\left\{\left[\varepsilon_i - E(\varepsilon_i)\right]\left[\varepsilon_j - E(\varepsilon_j)\right]\right\} = 0 \text{ for } i \neq j \text{ (uncorrelated errors)}. \tag{1.22}$$

The different errors ε_i and ε_j are uncorrelated if each has no effect on or relationship to the other. An example of correlated errors is $\varepsilon_i = p\varepsilon_{i-1} + u_i$ where u_i is uncorrelated to the ε_is and p is a constant. As the sampling rate of an automatic data acquisition system increases, the errors tend to become more correlated. High correlation between succeeding temperature measurements indicates that each new measurement is contributing much less information than if the correlation were zero. Very high sampling rates (which approach continuous measurements) may contribute little more information than considerably lower rates; that is, larger time steps, Δt, between the measurements.

A measure of the correlation between the two succeeding data points Y_i and Y_{i+1} is the sample correlation coefficient, $\hat{p}_i$, defined as:

$$\hat{p} = \frac{\sum_{i=1}^{n-1} e_i e_i + 1}{\sum_{i=1}^{n-1} e_i^2} \tag{1.23}$$

(This is an appropriate estimator for p if $\varepsilon_i = p\varepsilon_{i-1} + u_i$ which was mentioned previously.) A low correlation is near zero and a high correlation is near ± 1.

5) The fifth standard assumption is that the temperature measurement errors (ε_i) have a normal (that is, Gaussian) distribution.

If the second, third, and fourth standard assumptions are valid, the probability density of ε_i is given by

$$f(\varepsilon_i) = \frac{1}{\sigma\sqrt{2\pi}}\exp\left(\frac{-\varepsilon_i^2}{2\sigma^2}\right) \tag{1.24}$$

The assumption of normality is frequently valid even if standard assumptions 2, 3, and 4 are not; in that case a joint probability density for the errors is needed.

6) The sixth standard assumption is that the statistical parameters such as σ^2 and p are known.

7) The seventh standard assumption is that the times $t_1, t_2, ..., t_n$, positions $x_1, x_2, ..., x_j$, specimen dimensions, and thermal properties are accurately known.

In other words, the only source of error is in the measured temperatures. In statistical terms, the variances of time, and so on, are zero.

8) The last standard assumption is that there is no prior information regarding the shape of the surface heat flux.

"Prior information" means information known before any temperature measurements are made for a particular case. If prior information exists, then it can be utilized to obtain better estimates. If, for example, from experience with previous similar tests the heat flux is constant over some time period or is periodic, this information can be used to improve upon the estimators given in this book. It is assumed herein that little is known about the surface heat flux except that it can vary abruptly with time. High-frequency fluctuations of $\hat{q}_i$, that is, those that vary significantly between successive values, are not permitted, however.

1.9 Criteria for Evaluation of IHCP Methods

In order to evaluate the several IHCP procedures, various criteria are needed (Beck 1979):

1) The predicted temperatures and heat fluxes should be accurate if the measured data are of high accuracy.
2) The method should be insensitive to measurement errors.
3) The method should be stable for small time-steps or intervals. This permits the extraction of more information regarding the time variation of surface conditions than is permitted by large time-steps.
4) Temperature measurements from one or more sensors should be permitted.
5) The method should not require continuous first-time derivatives of the surface heat flux. Furthermore, step changes or even more abrupt changes in the surface heat fluxes should be permitted.
6) Knowledge of the precise starting time of the application of the surface heat flux should not be required. The start of heating is frequently not synchronized with the discrete times that temperatures are measured. Reasons for this might be that the starting time is not accurately known or is difficult to measure. Precise times at which abrupt changes in the heat flux occur may also be unknown.
7) The method should not be restricted to any fixed number of observations.
8) Composite solids should be permitted.
9) Temperature-variable properties should be permitted.
10) Contact conductances should not be excluded.
11) The method should be easy to program.
12) The computer cost should be moderate.
13) The user should not have to be highly skilled in mathematics in order to use the method or to adapt it to other geometries.
14) The method should be capable of treating various one-dimensional coordinate systems.
15) The method should permit extension to more than one heating surface.
16) The method should have a statistical basis and permit various statistical assumptions for the measurement errors.

The function specification and regularization methods are capable of satisfying all these criteria provided the nonlinear heat conduction equation is approximated using methods such as the finite difference, finite element, or finite control volume methods.

If the heat conduction equation is solved by Duhamel's theorem and the function specification or regularization method is used, all the criteria can be satisfied except that of treating the nonlinear problem (criterion number 9).

1.10 Scope of Book

The scope of the book is limited to the inverse heat conduction problem. However, many of the techniques given herein apply to a wide variety-of-ill-posed problems. A number of solution methods are presented and the emphasis is on general methods that can meet the criteria in Section 1.9:

Chapter 1 has given an introduction to the subject including a literature review of the use of IHCP solutions in various applications.

Chapter 2 discusses and presents the exact analytical solution to several direct heat conduction problems, which will be the primary modeling method for solution of IHCPs in the remainder of the text.

Chapter 3 uses superposition principles to fashion the analytical solutions from Chapter 2 into building blocks for solution of IHCPs.

Chapter 4 discusses the challenge of the IHCP solution presented by damping and lagging of information from subsurface temperatures. A classic, but impractical, exact IHCP solution is presented and used to demonstrate the ill-posedness of the IHCP. Several methods for approximate solution of the IHCP are the main content of this chapter.

Chapter 5 introduces and presents the idea of a filter coefficient solution to the IHCP. This concept is used extensively in Chapters 6, 7, and 8.

Chapter 6 presents several techniques for optimizing the regularization in IHCPs.

Chapter 7 presents several test cases for benchmarking IHCPs and exercises each of the methods presented in Chapter 4. The chapter concludes with a comparison of the performance of all the methods in the benchmark solution.

Chapter 8 addresses two-dimensional IHCPs and uses the filter coefficient concept to address the solution.

Chapter 9 discusses methods to estimate the heat transfer coefficient in an IHCP.

Chapter 10 describes the problem of deterministic measurement errors and some approaches for correcting these errors or accounting for their effects in the IHCP solution.

1.11 Chapter Summary

The IHCP is the task of estimating an unknown external heating action (the cause) from interior or subsurface temperature measurements (the effects). This problem contrasts the direct or forward heat conduction problem of determining the interior temperatures in response to known external heating action.

This text focuses on linear IHCPs which can leverage principles of superposition for solutions of the forward problem. Both sequential and whole domain estimation techniques will be presented.

IHCPs have been developed and utilized for over 70 years and have been applied to a wide range of problems. This chapter provides a rich review of literature on IHCP solutions in manufacturing, aerospace, biomedical, and electronics cooling applications.

Temperature measurements inherently bear random errors. The standard statistical assumptions for measurement errors are reviewed in this chapter.

References

Abreu, L. A. S. *et al.* (2016) Thermography detection of contact failures in double layered materials using the reciprocity functional approach, *Applied Thermal Engineering*, 100, pp. 1173–1178. https://doi.org/10.1016/j.applthermaleng.2016.02.078.

Aghayan, S. A. *et al.* (2013) An inverse problem of temperature optimization in hyperthermia by controlling the overall heat transfer coefficient, *Journal of Applied Mathematics*, 2013, 734020. https://doi.org/10.1155/2013/734020.

Albano, M. *et al.* (2019) Carbon/carbon high thickness shell for advanced space vehicles, *International Journal of Heat and Mass Transfer*, 128, pp. 613–622. doi: https://doi.org/10.1016/j.ijheatmasstransfer.2018.05.106.

Alifanov, O. M., Nenarokomov, A. V. and Gonzalez, V. M. (2009) Study of multilayer thermal insulation by inverse problems method, *Acta Astronautica*, 65(9–10), pp. 1284–1291. https://doi.org/10.1016/j.actaastro.2009.03.053.

Alkidas, A. L. (1980) Heat transfer characteristics of a spark-ignition engine, *Journal of Heat Transfer*, 102, pp. 189–193. https://doi.org/10.1115/1.3244258.

Asif, M., Tariq, A. and Singh, K. M. (2019) Estimation of thermal contact conductance using transient approach with inverse heat conduction problem, *Heat and Mass Transfer/Waerme- und Stoffuebertragung*, 55(11), pp. 3243–3264. doi: https://doi.org/10.1007/s00231-019-02617-x.

Bass, B. R. (1979) Incap: A Finite Element Program for One-Dimensional Nonlinear Inverse Heat Conduction Analysis, Analysis, Technical Report NRC/NUREG/CSD/TM-8. Oak Ridge, TN, USA.

Bass, B. R. (1980) Applications of the finite element to the inverse heat conduction problem using Beck's second method, *Journal of Engineering for Industry*, 102, pp. 168–176. https://doi.org/10.1115/1.3183849.

Beck, J. V. (1962) Calculation of surface heat flux from an internal temperature history. ASME Paper No. 62-HT-46. ASME.

Beck, J. V. (1968) Surface heat flux determination using an integral method, *Nuclear Engineering and Design*, 7(2), pp. 170–178. doi: https://doi.org/10.1016/0029-5493(68)90058-7.

Beck, J. V. (1970) Nonlinear estimation applied to the nonlinear inverse heat conduction problem, *International Journal of Heat and Mass Transfer*, 13(4), pp. 703–716. doi: https://doi.org/10.1016/0017-9310(70)90044-X.

Beck, J. V. (1979) Criteria for comparison of methods of solution of the inverse heat conduction problem, *Nuclear Engineering and Design*, 53(1), pp. 11–22. doi: https://doi.org/10.1016/0029-5493(79)90035-9.

Beck, J. V. and Arnold, K. J. (1977) *Parameter Estimation in Engineering and Science*. New York: John Wiley and Sons.

Beck, J. V., Litkouhi, B. and St. Clair, C. R. (1982) Efficient sequential solution of the nonlinear inverse heat conduction problem, *Numerical Heat Transfer*, 5, pp. 275–286. https://doi.org/10.1080/10407788208913448.

Beck, J. V, Blackwell, B. and St Clair Jr, C. R. (1985) *Inverse Heat Conduction: Ill-Posed Problems*. New York: A Wiley-Interscience.

Beck, J. V and Wolf, H. (August 8–11, 1965) The nonlinear inverse heat conduction problem, ASME Paper, 65-HT-40.

Bezerra, L. A. *et al.* (2013) Estimation of breast tumor thermal properties using infrared images, *Signal Processing*, 93(10), pp. 2851–2863. doi: https://doi.org/10.1016/J.SIGPRO.2012.06.002.

Blackwell, B. F. (1968) *A New Iterative Technique for Solving the Implicit Finite-Difference Equations for the Inverse Problem of Heat Conduction, unpublished technical report*. Albuquerque, NM.

Blackwell, B. F. (1981) An efficient technique for the numerical solution of the one-dimensional inverse problem of heat donduction, *Numerical Heat Transfer A*, 4, pp. 229–239. https://doi.org/10.1080/01495728108961789.

Borazjani, E., Spinello, D. and Necsulescu, D. S. (2015) Determination of the modulation frequency for thermographic non-destructive testing, *NDT and E International*, 70(1), pp. 1–8. doi: https://doi.org/10.1016/j.ndteint.2014.12.001.

Bozzoli, F. *et al.* (2019) A novel method for estimating the distribution of convective heat flux in ducts: Gaussian filtered singular value decomposition, *Inverse Problems in Science and Engineering*, 27(11), pp. 1595–1607. doi: https://doi.org/10.1080/17415977.2018.1540615.

Cannon, J. R. and Douglas, J. (1967) The cauchy problem for the heat equation, *SIAM Journal on Numerical Analysis*, 4(3), pp. 317–336. http://www.jstor.org/stable/2949400.

Cebula, A. (2016) A device for measuring the heat flux on the cylinder outer surface in a cross-flow, *Procedia Engineering*, 157, pp. 264–270. doi: https://doi.org/10.1016/j.proeng.2016.08.365.

Cebula, A. and Taler, J. (2014) Determination of transient temperature and heat flux on the surface of a reactor control rod based on temperature measurements at the interior points, *Applied Thermal Engineering*, 63(1), pp. 158–169. doi: https://doi.org/10.1016/j.applthermaleng.2013.10.066.

Cheng, L. *et al.* (2016) Application of conjugate gradient method for estimation of the wall heat flux of a supersonic combustor, *International Journal of Heat and Mass Transfer*, 96, pp. 249–255. doi: https://doi.org/10.1016/j.ijheatmasstransfer.2016.01.036.

Cui, M. *et al.* (2018) Inverse identification of boundary conditions in a scramjet combustor with a regenerative cooling system, *Applied Thermal Engineering*, 134(November 2017), pp. 555–563. doi: https://doi.org/10.1016/j.applthermaleng.2018.02.038.

Curtulo, J. P. *et al.* (2019) 'The application of an analytical model to solve an inverse heat conduction problem: Transient solidification of a Sn-Sb peritectic solder alloy on distinct substrates', *Journal of Manufacturing Processes*, 48(October), pp. 164–173. doi: https://doi.org/10.1016/j.jmapro.2019.10.029.

Das, K. and Mishra, S. C. (2013) Estimation of tumor characteristics in a breast tissue with known skin surface temperature', *Journal of Thermal Biology*, 38(6), pp. 311–317. doi: https://doi.org/10.1016/J.JTHERBIO.2013.04.001.

Dou, R. *et al.* (2014) Experimental study on heat-transfer characteristics of circular water jet impinging on high-temperature stainless steel plate, *Applied Thermal Engineering*, 62(2), pp. 738–746. doi: https://doi.org/10.1016/j.applthermaleng.2013.10.037.

France, D. M. *et al.* (1978) *Measurements and Correlation of Critical Heat Flux in a Sodium Heated Steam Generator Tube, Technical Memorandum, ANL-CT-78-15*. Argonne, IL.

Vanderplaats, G. N. (2005) *Numerical Optimization Techniques for Engineering Design*. 4th edn. Vanderplaats Research and Development Inc.

Gayzik, F. S., Scott, E. P. and Loulou, T. (2006) Experimental validation of an inverse heat transfer algorithm for optimizing hyperthermia treatments, *Journal of Biomechanical Engineering*, 128(4), pp. 505–515. doi: https://doi.org/10.1115/1.2205375.

Gostimirovic, M. *et al.* (2018) M-Machining-2018-Inverse electro-thermal analysis of the material removal mechanism in electrical discharge machining.pdf, *The International Journal of Advanced Manufacturing Technology*, 97, pp. 1861–1871. doi: https://doi.org/10.1007/s00170-018-2074-y.

Gostimirovic, M., Kovac, P. and Sekulic, M. (2011) An inverse heat transfer problem for optimization of the thermal process in machining, *Sadhana – Academy Proceedings in Engineering Sciences*, 36(4), pp. 489–504. doi: https://doi.org/10.1007/s12046-011-0034-4.

Guo, Z., Lu, T. and Liu, B. (2017) Inverse heat conduction estimation of inner wall temperature fluctuations under turbulent penetration, *Journal of Thermal Science*, 26(2), pp. 160–165. doi: https://doi.org/10.1007/s11630-017-0925-8.

Hadamard, J. (1923) *Lectures on Cauchy's Problem in Linear Partial Differential Equations*. New Haven, CT: Yale University Press (Yale University. Mrs. Hepsa Ely Silliman memorial lectures). http://libdata.lib.ua.edu/login?url=https://search.ebscohost.com/login.aspx?direct=true&db=cat00456a&AN=ua.3873704&site=eds-live&scope=site.

Hahn, D. W. and Ozisik, N. M. (2012) *Heat Conduction*. 3rd edn. John Wiley and Sons, Inc.

Hindasageri, V., Vedula, R. P. and Prabhu, S. V. (2014) Heat transfer distribution for impinging methane-air premixed flame jets, *Applied Thermal Engineering*, 73(1), pp. 461–473. doi: https://doi.org/10.1016/j.applthermaleng.2014.08.002.

Hindasageri, V., Vedula, R. P. and Prabhu, S. V. (2015) Heat transfer distribution for three interacting methane-air premixed impinging flame jets, *International Journal of Heat and Mass Transfer*, 88, pp. 914–925. doi: https://doi.org/10.1016/j.ijheatmasstransfer.2015.04.098.

Hossain, S. and Mohammadi, F. A. (2016) Tumor parameter estimation considering the body geometry by thermography, *Computers in Biology and Medicine*, 76, pp. 80–93. doi: https://doi.org/10.1016/J.COMPBIOMED.2016.06.023.

Howse, T. K. J., Kent, R. and Rawson, H. (1971) The determination of glass-mould heat fluxes from mould temperature measurements, *Glass Technology*, 12, pp. 91–93.

Hsu, P. T. and Chu, Y. H. (2004) An inverse non-Fourier heat conduction problem approach for estimating the boundary condition in electronic device, *Applied Mathematical Modelling*, 28(7), pp. 639–652. doi: https://doi.org/10.1016/j.apm.2003.10.010.

Huang, C. H. and Lee, C. T. (2015) An inverse problem to estimate simultaneously six internal heat fluxes for a square combustion chamber, *International Journal of Thermal Sciences*, 88, pp. 59–76. doi: https://doi.org/10.1016/j.ijthermalsci.2014.08.021.

Iljaž, J. and Škerget, L. (2014) Blood perfusion estimation in heterogeneous tissue using BEM based algorithm, *Engineering Analysis with Boundary Elements*, 39(1), pp. 75–87. doi: https://doi.org/10.1016/J.ENGANABOUND.2013.11.002.

Imber, M. (1973) A temperature extrapolation method for hollow cylinders, *AIAA Journal*, 11, pp. 117–118. https://doi.org/10.2514/3.6684.

Imber, M. (1974a) A temperature extrapolation mechanism for two-dimensional heat flow, *AIAA Journal*, 12, pp. 1087–1093. https://doi.org/10.2514/3.49417.

Imber, M. (1974b) The two dimensional inverse problem in heat conduction. *Fifth International Heat Transfer Conference*. Tokyo, Japan (September 3–7, 1974).

Imber, M. (1975a) Comment on "On transient cylindrical surface heat flux predicted from interior temperature responses", *AIAA Journal*, 14, pp. 542–543. https://doi.org/10.2514/3.61393.

Imber, M. (1975b) Two-dimensional inverse conduction problem—further observations, *AIAA Journal*, 13, pp. 114–115. https://doi.org/10.2514/3.49645.

Imber, M. (1979) Nonlinear heat transfer in planar solids: Direct and inverse applications, *AIAA Journal*, 17, pp. 204–212. https://doi.org/10.2514/3.61096.

Imber, M. and Khan, J. (1972) Prediction of transient temperature distributions with embedded thermocouples, *AIAA Journal*, 10, pp. 784–789. https://doi.org/10.2514/3.50211.

Jalali, A., Ayani, M. B. and Baghban, M. (2014) Simultaneous estimation of controllable parameters in a living tissue during thermal therapy, *Journal of Thermal Biology*, 45, pp. 37–42. doi: https://doi.org/10.1016/J.JTHERBIO.2014.07.008.

Janicki, M. and Napieralski, A. (2004) Inverse heat conduction problems in electronic circuits, *International Semiconductor Conference*, pp. 455–458. https://doi.org/10.1109/SMICND.2004.1403047.

Janicki, M., Zubert, M. and Napieralski, A. (1998) Application of inverse heat conduction methods in temperature monitoring of integrated circuits, *Sensors and Actuators, A: Physical*, 71(1–2), pp. 51–57. doi: https://doi.org/10.1016/S0924-4247(98)00171-X.

Jaremkiewicz, M. and Taler, J. (2016) Measurement technique of transient fluid temperature in a pipeline, *Procedia Engineering*, 157, pp. 58–65. doi: https://doi.org/10.1016/j.proeng.2016.08.338.

Ji, S. *et al.* (2018) Quantification of the heat transfer during the plasma arc re-melting of titanium alloys, *International Journal of Heat and Mass Transfer*, 119, pp. 271–281. doi: https://doi.org/10.1016/j.ijheatmasstransfer.2017.11.064.

Jin, C., He, Z.-Z. and Liu, J. (2016) Finite element method based three-dimensional thermal tomography for disease diagnosis of human body, *Journal of Heat Transfer*, 138(10). doi: https://doi.org/10.1115/1.4033612.

John, F. (1955) Numerical solution of the heat equation for preceeding times, *Annali di Matematica Pura ed Applicata*, 40, pp. 129–142.

Krane, P. *et al.* (2019) Identifying hot spots in electronics packages with a sensitivity-coefficient based inverse heat conduction method, *InterSociety Conference on Thermal and Thermomechanical Phenomena in Electronic Systems, ITHERM*, 2019-May, pp. 504–510. doi: https://doi.org/10.1109/ITHERM.2019.8757292.

Kryzhanivskyy, V. *et al.* (2017) Computational and experimental inverse problem approach for determination of time dependency of heat flux in metal cutting, *Procedia CIRP*, 58, pp. 122–127. doi: https://doi.org/10.1016/j.procir.2017.03.204.

Kryzhanivskyy, V. *et al.* (2018) Heat flux in metal cutting: Experiment, model, and comparative analysis, *International Journal of Machine Tools and Manufacture*, 134(July), pp. 81–97. doi: https://doi.org/10.1016/j.ijmachtools.2018.07.002.

Kusiak, A., Battaglia, J. L. and Rech, J. (2005) Tool coatings influence on the heat transfer in the tool during machining, *Surface and Coatings Technology*, 195(1), pp. 29–40. doi: https://doi.org/10.1016/j.surfcoat.2005.01.007.

Lawson, C. L. and Hanson, R. J. (1974) *Solving Least Squares Problems.* Englewood Cliffs, NJ: Prentice-Hall.

Lee, H. L. *et al.* (2013) An inverse hyperbolic heat conduction problem in estimating surface heat flux of a living skin tissue, *Applied Mathematical Modelling*, 37(5), pp. 2630–2643. doi: https://doi.org/10.1016/j.apm.2012.06.025.

Lee, H. L. *et al.* (2015) Estimation of surface heat flux and temperature distributions in a multilayer tissue based on the hyperbolic model of heat conduction, *Computer Methods in Biomechanics and Biomedical Engineering.* Taylor & Francis, pp. 1525–1534. doi: https://doi.org/10.1080/10255842.2014.925108.

Li, R. *et al.* (2017) Study of the conductive heat flux from concrete to liquid nitrogen by solving an inverse heat conduction problem, *Journal of Loss Prevention in the Process Industries*, 48, pp. 48–54. doi: https://doi.org/10.1016/j.jlp.2017.04.001.

Li, Y., Wang, G. and Chen, H. (2015) Simultaneously estimation for surface heat fluxes of steel slab in a reheating furnace based on DMC predictive control, *Applied Thermal Engineering*, 80, pp. 396–403. doi: https://doi.org/10.1016/j.applthermaleng.2015.01.069.

Lin, D. Y. T. and Westwater, J. W. (1982) Effect of metal thermal properties on boiling curves obtained by the quenching method, In: *Heat Transfer 1982—Munchen Conference Proceedings.* New York, NY, USA: Hemisphere Publ. Corp, pp. 155–160. http://dx.doi.org/10.1615/IHTC7.3930.

Lin, J. and Steen, W. M. (1998) An in-process method for the inverse estimation of the powder catchment efficiency during laser cladding, *Optics and Laser Technology*, 30(2), pp. 77–84. doi: https://doi.org/10.1016/S0030-3992(98)00007-3.

Liu, Z., Wang, G. and Yi, J. (2020) Study on heat transfer behaviors between Al-Mg-Si alloy and die material at different contact conditions based on inverse heat conduction algorithm, *Journal of Materials Research and Technology*, 9(2), pp. 1918–1928. doi: https://doi.org/10.1016/j.jmrt.2019.12.024.

Loulou, T. and Scott, E. (2006). An inverse heat conduction problem with heat flux measurements, *International Journal for Numerical Methods in Engineering*, 67 (11), pp. 1587–1616. https://doi.org/10.1002/nme.1674.

Lu, T. *et al.* (2010) A two-dimensional inverse heat conduction problem in estimating the fluid temperature in a pipeline, *Applied Thermal Engineering*, 30(13), pp. 1574–1579. doi: https://doi.org/10.1016/j.applthermaleng.2010.03.011.

Lu, T., Liu, B. and Jiang, P. X. (2011) Inverse estimation of the inner wall temperature fluctuations in a pipe elbow, *Applied Thermal Engineering*, 31(11–12), pp. 1976–1982. doi: https://doi.org/10.1016/j.applthermaleng.2011.03.002.

Luo, X. and Yang, Z. (2017) A new approach for estimation of total heat exchange factor in reheating furnace by solving an inverse heat conduction problem, *International Journal of Heat and Mass Transfer*, 112, pp. 1062–1071. doi: https://doi.org/10.1016/j.ijheatmasstransfer.2017.05.009.

dos Magalhães, E. S. *et al.* (2018) A thermal analysis in laser welding using inverse problems, *International Communications in Heat and Mass Transfer*, 92(March), pp. 112–119. doi: https://doi.org/10.1016/j.icheatmasstransfer.2018.02.014.

Mandel, J. (1982) Use of the singular value decomposition in regression analysis, *The American Statistician*, 36(1), pp. 15–24. https://doi.org/10.1080/00031305.1982.10482771.

Mirsepassi, T. (1959a) Heat-transfer charts for time-variable boundary conditions, *British Chemical Engineering*, 4, pp. 130–136.

Mirsepassi, T. J. (1959b) Graphical evaluation of a convolution integral, *Mathematical Tables and Other Aides to Computation*, 13, pp. 130–136.

Mulholland, G. P. and Cobble, M. H. (1972) Diffusion through composite media, *International Journal of Heat and Mass Transfer*, 15, pp. 147–152. https://doi.org/10.1016/0017-9310(72)90172-X.

Mulholland, G. P., Gupta, B. P. and San Martin, R. L. (1975) Inverse problem of heat conduction in composite media, ASME Paper, 75-WA/HT-83.

Mulholland, G. P. and San Martin, R. L. (1971) Inverse problem of heat conduction in composite media. In: Glockner, P. G. (ed.) *Third Canadian Congress of Applied Mechanics.* Calgary: Alberta, Canada.

Mulholland, G. P. and San Martin, R. L. (1973) Indirect thermal sensing in composite media, *International Journal of Heat and Mass Transfer*, 16, pp. 1056–1060. https://doi.org/10.1016/0017-9310(73)90046-X.

Muzzy, R. J., Avila, J. H. and Root, R. E. (1975) *Topical Report: Determination of Transient Heat Transfer Coefficients and the Resultant Surface Heat Flux from Internal Temperature Measurements.* San Jose, CA, USA: Boiling Water Reactor Systems Dept., General Electric Co.

Najafi, H. *et al.* (2015) Real-time heat flux measurement using directional flame thermometer, *Applied Thermal Engineering*, 86, pp. 229–237. doi: https://doi.org/10.1016/j.applthermaleng.2015.04.053.

Nakamura, T. *et al.* (2014) Inverse analysis for transient thermal load identification and application to aerodynamic heating on atmospheric reentry capsule, *Aerospace Science and Technology*, 38, pp. 48–55. doi: https://doi.org/10.1016/j.ast.2014.07.015.

O'Mahoney, D. and Browne, D. J. (2000) Use of experiment and an inverse method to study interface heat transfer during solidification in the investment casting process, *Experimental Thermal and Fluid Science*, 22(3–4), pp. 111–122. doi: https://doi.org/10.1016/S0894-1777(00)00014-5.

Pandey, V. *et al.* (2020) AZ31-alloy, H13-die combination heat transfer characteristics by using inverse heat conduction algorithm, *Materials Today: Proceedings*, 44, pp. 4762–4766. doi: https://doi.org/10.1016/j.matpr.2020.11.258.

Pati, A. R. *et al.* (2019) The discrepancy in the prediction of surface temperatures by inverse heat conduction models for different quenching processes from very high initial surface temperature, *Inverse Problems in Science and Engineering*, 27(6), pp. 808–835. doi: https://doi.org/10.1080/17415977.2018.1501369.

Payne, L. E. (1975) Improperly posed problems in partial differential equations, *Improperly Posed Problems in Partial Differential Equations*. doi: https://doi.org/10.1137/1.9781611970463.

Pennes, H. H. (1948) Analysis on tissue arterial blood temperature in the resting human forearm, *Journal of Applied Physiology*, 1, pp. 93–122. https://doi.org/10.1152/jappl.1948.1.2.93.

Ravikumar, S. V., Jha, J. M., Tiara, A. M., *et al.* (2014a) Experimental investigation of air-atomized spray with aqueous polymer additive for high heat flux applications, *International Journal of Heat and Mass Transfer*, 72, pp. 362–377. doi: https://doi.org/10.1016/j.ijheatmasstransfer.2014.01.024.

Ravikumar, S. V., Jha, J. M., Sarkar, I., *et al.* (2014b) Mixed-surfactant additives for enhancement of air-atomized spray cooling of a hot steel plate, *Experimental Thermal and Fluid Science*, 55, pp. 210–220. doi: https://doi.org/10.1016/j.expthermflusci.2014.03.007.

Ravikumar, S. V. *et al.* (2015) Heat transfer enhancement using air-atomized spray cooling with water-Al2O3 nanofluid, *International Journal of Thermal Sciences*, 96, pp. 85–93. doi: https://doi.org/10.1016/j.ijthermalsci.2015.04.012.

Scott, E. P. (2018) Inverse heat transfer for biomedical applications, In: Shrivastava, D. (ed.) *Theory and Applications of Heat Transfer in Humans.* (Wiley Online Books), pp. 133–152. https://doi.org/10.1002/9781119127420.ch8.

Shang, Z. *et al.* (2019) 'On modelling of laser assisted machining: Forward and inverse problems for heat placement control', *International Journal of Machine Tools and Manufacture*, 138(December 2018), pp. 36–50. doi: https://doi.org/10.1016/j.ijmachtools.2018.12.001.

Shao, X. *et al.* (2021) Experimental study of transient liquid nitrogen jet impingement boiling on concrete surface using inverse conduction problem algorithm, *Process Safety and Environmental Protection*, 147, pp. 45–54. doi: https://doi.org/10.1016/j.psep.2020.09.032.

Shih, T. C. *et al.* (2007) Analytical analysis of the Pennes bioheat transfer equation with sinusoidal heat flux condition on skin surface, *Medical Engineering & Physics*, 29(9), pp. 946–953. doi: https://doi.org/10.1016/J.MEDENGPHY.2006.10.008.

Shumakov, N. V. (1957) A method for the experimental study of the process of heating a solid body, *Soviet Physics-Technical Physics (Translated by American Institute of Physics)*, 2, p. 771.

Singh, S. K. *et al.* (2017) Estimation of time-dependent wall heat flux from single thermocouple data, *International Journal of Thermal Sciences*, 115, pp. 1–15. doi: https://doi.org/10.1016/j.ijthermalsci.2017.01.010.

Snider, D. M. (1981) *INVERT 1.0—A Program for Solving the Nonlinear Inverse Heat Conduction Problem for One-Dimensional Solids.* Idaho Falls, Idaho, USA: U.S. Department of Energy, Idaho Operations Office.

Soejima, H. *et al.* (2015) Application of the spray quenching to T6 heat treatment of thick A6061 hollow cylinders, *Procedia Engineering*, 105(Icte 2014), pp. 776–786. doi: https://doi.org/10.1016/j.proeng.2015.05.070.

Somasundaram, S. and Tay, A. A. O. (2013) Comparative study of intermittent Spray cooling in single and two phase regimes, *International Journal of Thermal Sciences*, 74(C), pp. 174–182. doi: https://doi.org/10.1016/j.ijthermalsci.2013.06.008.

Somasundaram, S. and Tay, A. A. O. (2017) Methodology to construct full boiling curves for refrigerant spray cooling, *Applied Thermal Engineering*, 111, pp. 369–376. doi: https://doi.org/10.1016/j.applthermaleng.2016.09.112.

Sparrow, E. M., Haji-Sheikh, A. and Lundgren, T. S. (1964) The inverse problem in transient heat conduction, *Journal of Applied Mechanics, Transactions. ASME, Series E*, 86, pp. 369–375. https://doi.org/10.1115/1.3629649.

Ståhl, J.-E. (2012) *Metal Cutting — Theories and Models.* Lund, Sweden: Lund University Press, Division of Production and Materials Engineering, Lund University

Stolz, G. (1960a) Numerical solutions to an inverse problem of heat conduction for simple shapes, *Journal of Heat Transfer*, 82(1), pp. 20–25. https://doi.org/10.1115/1.3679871.

Taler, J. *et al.* (2018) Thermal stress monitoring in thick-walled pressure components based on the solutions of the inverse heat conduction problems, *Journal of Thermal Stresses*, 41(10–12), pp. 1501–1524. doi: https://doi.org/10.1080/01495739.2018.1520621.

Taler, J. *et al.* (2019) Thermal stress monitoring in thick walled pressure components of steam boilers, *Energy*, 175, pp. 645–666. doi: https://doi.org/10.1016/j.energy.2019.03.087.

Tikhonov, A. N. (1963a) On the regularization of ill-posed problems, *Doklady Akademii Nauk SSSR*, 153, pp. 49–52.

Tikhonov, A. N. (1963b) On the solution of ill-posed problems and the method of regularization, *Doklady Akademii Nauk SSSR*, 151, pp. 501–504.

Tikhonov, A. N. and Arsenin, V. Y. (1977) *Solutions of Ill-Posed Problems*. Washington DC: V.H. Winston & Sons.

Trujillo, D. M. and Busby, H. R. (1997) *Practical Inverse Analysis in Engineering*. CRC Press. doi: https://doi.org/10.1201/9780203710951.

Uyanna, O., Najafi, H. and Rajendra, B. (2021) An inverse method for real-time estimation of aerothermal heating for thermal protection systems of space vehicles, *International Journal of Heat and Mass Transfer*, 177, p. 121482. doi: https://doi.org/10.1016/j.ijheatmasstransfer.2021.121482.

Volle, F. *et al.* (2009) Practical application of inverse heat conduction for wall condition estimation on a rotating cylinder, *International Journal of Heat and Mass Transfer*, 52(1–2), pp. 210–221. doi: https://doi.org/10.1016/j.ijheatmasstransfer.2008.05.025.

Wei, B. *et al.* (2016) Research on inverse problems of heat flux and simulation of transient temperature field in high-speed milling, *International Journal of Advanced Manufacturing Technology*, 84, pp. 2067–2078. https://doi.org/10.1007/s00170-015-7850-3.

Williams, S. D. and Curry, D. M. (1977) An analytical experimental study for surface heat flux determination, *Journal of Spacecraft*, 14, pp. 632–637. https://doi.org/10.2514/3.27987.

Xu, W. and Zhang, P. (2020) Cryogenic quenching of a stainless steel rodlet with various coatings, *International Journal of Heat and Mass Transfer*, 154. doi: https://doi.org/10.1016/j.ijheatmasstransfer.2020.119642.

Yang, Y. C. *et al.* (2017) Function estimation of laser-induced heat generation in a gas-saturated powder layer heated by a short-pulsed laser, *International Communications in Heat and Mass Transfer*, 81, pp. 56–63. doi: https://doi.org/10.1016/j.icheatmasstransfer.2016.12.013.

Zhang, X. *et al.* (2014) Experimental study of the air-atomized spray cooling of high-temperature metal, *Applied Thermal Engineering*, 71(1), pp. 43–55. doi: https://doi.org/10.1016/j.applthermaleng.2014.06.026.

Zhou, Z. F., Xu, T. Y. and Chen, B. (2016) Algorithms for the estimation of transient surface heat flux during ultra-fast surface cooling, *International Journal of Heat and Mass Transfer*, 100, pp. 1–10. doi: https://doi.org/10.1016/j.ijheatmasstransfer.2016.04.058.

2

Analytical Solutions of Direct Heat Conduction Problems

2.1 Introduction

In this chapter some exact analytical solutions to direct transient heat conduction problems are given. These solutions are the basic "building blocks" for solving direct heat conduction problems (DHCPs) when dealing with nonhomogeneous boundary conditions that vary arbitrarily with space and time, which will be shown in Chapter 3. These direct approximate techniques are the first stage of procedures for solving inverse heat conduction problems (IHCPs). The second stage, which involves specific algorithms for the IHCP, is developed in Chapter 4.

The development of exact analytical solutions is relevant not only as they are basic building blocks for problems involving arbitrary boundary functions, but also because (i) they provide more accurate values of temperature than a numerical solution, (ii) their correctness can be checked by means of symbolic intrinsic verification methods (Cole et al. 2014; D'Alessandro and de Monte 2018), (iii) they give considerable insight into the physical significance of various parameters affecting a given problem than a purely numerical solution, and (iv) they provide standards for verification purposes of large numerically based codes (Beck et al. 2004; ASME 2021; McMasters et al. 2021). Also, the large improvements in accuracy can be obtained without excessive computational costs. This contrasts significantly with the additional effort required to reduce errors using the numerical finite element method (FEM) though this method allows solving problems with irregular geometry and nonlinear material properties.

Exact analytical solutions are of special interest since the superposition-based numerical approximation proposed in Chapter 3 for solving problems with an arbitrary variation of the boundary conditions is based on differences (very closely subtractions) of the building block solutions in time (1D problems) and in both space and time (2D problems). These differences, in fact, require very accurate values of the building block solutions when the space and time intervals chosen for approximating the surface heat flux or temperature are very small, a feature which is not possible with numerical solutions. Also, IHCPs are extremely sensitive to errors in the mathematical models that are being fitted to experimental measurements. Numerical excursions and fluctuations in the unknown functions are in fact inherent not only in ill-posed problems but also in inaccuracy of the direct problem solution. The same is true when dealing with parameter estimation problems (Mishra et al. 2017). Lastly, the computational time is greatly reduced by using analytical solutions (Fernandes et al. 2010).

Throughout the chapter, the thermal properties (thermal conductivity k and volumetric heat capacity $C = \rho c$ or thermal diffusivity $\alpha = k/C$) are assumed independent of temperature. Section 2.2 and Appendix A present a number system for heat conduction for which the number itself contains a great deal of information. Such a system both simplifies the construction of a computer data base such as the *Exact Analytical Conduction Toolbox* (ExACT) (Cole et al. 2014) and makes locating existing solutions less tedious. Section 2.3 considers one-dimensional (1D) transient problems in which the surface temperature and heat flux are constant and linearly-in-time variable. In addition, the concept of computational analytical solution associate to the exact one is given as well as the related relative errors when using the approximate or computational solution. The short- and large-time forms of the series solution are also given along with the definition of deviation and quasi-steady (or steady state) times. Section 2.4 and Appendix B present the computational development of an exact analytical solution for a two-dimensional (2D) transient rectangular problem with constant partial heating by defining deviation and quasi-steady times as well as deviation distances along the two directions. The short- and large-time forms of the series solution are both considered along with two different expressions of the quasi-steady part of the same solution. An example problem (Example 2.1) is presented to show how the rectangular body temperature may be computed at early times. Lastly, the number of terms needed to obtain a certain truncation error in the series solution is defined for both the 1D and 2D transient problems.

Inverse Heat Conduction: Ill-Posed Problems, Second Edition. Keith A. Woodbury, Hamidreza Najafi, Filippo de Monte, and James V. Beck.
© 2023 John Wiley & Sons, Inc. Published 2023 by John Wiley & Sons, Inc.

2.2 Numbering System

Since the fields of heat conduction and diffusion are relatively mature, many exact analytical solutions exist in literature. For example, Carslaw and Jaeger (1959) and, then, Özişik (1993) provide a very large number of solutions, which are organized by analytical method. Thambynayagam (2011) produced a monumental handbook containing over one thousand transient diffusion solutions. To some extent, Thambynayagam solved the index problem by using sketches of the geometries with the boundary conditions indicated by a D, N, or R for conditions of Dirichlet, Neumann, or Robin types, respectively. This helps considerably but sketches cannot be readily indexed. Polyanin (2002) provides an extensive handbook of solutions of linear partial differential equations, including the diffusion equation, and this handbook is organized by the type of partial differential equation. In addition, a significant number of solutions has been published in the open specialized literature in the past 50 years.

To make the exact analytical solutions more accessible and, hence, to organize and index the same solutions, a consistent numbering system is needed. The numbering system used in this book, introduced by Beck and Litkouhi (1988), covers basic geometries such as plates, cylinders, and spheres. Irregular geometries and heat conduction problems with temperature-dependent properties are not covered in the numbering system. Recently, however, an extension to nonlinearities caused by temperature-variable properties has been proposed by Toptan et al. (2020).

The number system has several components to identify the elements of a conventional heat conduction problem including dimensionality, coordinate system, boundary conditions, initial temperature distribution, volumetric heat source/sink term, fin-type, and moving body terms. For the sake of brevity, only the rectangular coordinate system is considered here. For cylindrical and spherical frames of reference as well as an exhaustive treatment on this subject, the reader can refer to chapter 2 of the book by Cole et al. (2011).

A comprehensive explanation of the solution numbering system is given in Appendix A of this book. Appendix A provides details on numbering for solutions involving first, second, and third kind boundary conditions, as well as the important case of "zero" kind. Also considered in the appendix are descriptors for partial heating/cooling in time, multi-dimensional problems, and partial heating/cooling on a boundary for multi-dimensional cases.

Four specific 1D cases are considered in this text which can be described using XI2BK0T0, with I = 1 or 2 and K = 1 or 2. "X" indicates the coordinate system is rectangular. Here the "I" indicates the type of boundary condition at the surface (1 = first kind, 2 = second kind). "B" signifies boundary information. The "K" indicator is K = 1 for a constant boundary condition, and K = 2 for a linearly-in-time variable boundary condition. The trailing "T0" in the descriptor indicates a transient temperature solution with zero initial condition.

2.3 One-Dimensional Temperature Solutions

2.3.1 Generalized One-Dimensional Heat Transfer Problem

Consider a flat plate of thickness L with temperature-independent properties, initially at a uniform temperature T_{in}, and subject to a nonhomogeneous time-variable but uniform boundary condition of the third kind at $x = 0$ ("active" boundary), as depicted in Figure 2.1.

The problem may be denoted by X32B-0T1. Its mathematical formulation is

$$\frac{\partial^2 T}{\partial x^2} = \frac{1}{\alpha} \frac{\partial T}{\partial t} \qquad (0 < x < L; t > 0) \tag{2.1a}$$

$$-k\left(\frac{\partial T}{\partial x}\right)_{x=0} + h_0 T(x = 0, t) = f_0(t) \quad (t > 0) \tag{2.1b}$$

$$\left(\frac{\partial T}{\partial x}\right)_{x=L} = 0 \quad (t > 0) \tag{2.1c}$$

$$T(x, t = 0) = T_{in} \quad (0 < x < L) \tag{2.1d}$$

where the heat transfer coefficient h_0 is considered constant and $f_0(t)$ is the boundary function (see Appendix A).

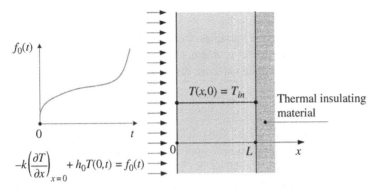

Figure 2.1 Schematic of the 1D transient X32B-0T1 problem.

This problem is very general and may have different forms according to the definition of h_0 and $f_0(t)$:

- $f_0(t) = 0$. This happens when the convective environment in contact with the slab at $x = 0$ has a constant temperature whose value is zero, that is, $T_{\infty,0}(t) = 0$. In this case, the third kind boundary condition is homogeneous and the problem notation is X32B00T1.
- $f_0(t) = h_0 T_{\infty,0}(t)$. This occurs when the body can exchange heat by convection (characterized by h_0) with the surrounding fluid at temperature $T_{\infty,0}(t) \neq 0$. In this case, the third kind boundary condition is nonhomogeneous and the problem notation is X32B-0T1.
- $f_0(t) = h_0 T_{\infty,0}(t)$ with $h_0 \to \infty$. This convection at $x = 0$, with a fluid at $T_{\infty,0}(t) \neq 0$ and a very high convective heat transfer coefficient, approximates a nonhomogeneous boundary condition of the first kind, that is, $T(0, t) = T_0(t)$, with $T_0(t) = T_{\infty,0}(t)$, and the problem notation becomes X12B-0T1.
- $f_0(t) = q_0(t)$ with $h_0 = 0$. This happens when the slab is in direct and perfect thermal contact ($h_0 = 0$) with a heat source/sink (heater/cooler) supplying or removing heat ($q_0(t) > 0$ or $q_0(t) < 0$), respectively, to the flat plate. In this case, the boundary condition Eq. (2.1b) reduces to a nonhomogeneous boundary condition of the second kind. Also, the problem notation becomes X22B-0T1.
- $f_0(t) = q_0(t) + h_0 T_{\infty,0}(t)$. This is the most general case, and the related nonhomogeneous boundary condition of the third kind is termed as "*generalized*" (Cole et al. 2016; McMasters et al. 2019). This can happen when the body is exposed to a net radiant heat flux and exchanges heat by convection with the surrounding fluid at $x = 0$. Also, it can occur when the slab is in direct but imperfect thermal contact ($h_0 > 0$) with a heat source/sink. The problem notation is X32B-0T1 but, sometimes, to avoid a possible misunderstanding with the "*pure*" boundary condition of the third kind ($f_0(t) = h_0 T_{\infty,0}(t)$), it is modified as X32B-0q-0T1, where "q" signifies heat flux information. If the surrounding fluid has zero temperature, the third kind boundary condition becomes homogeneous and the problem notation becomes X32B00q-0T1.

2.3.2 Cases of Interest

Solutions in this text will be focus on application of boundary conditions of the first and second kind at $x = 0$ with an insulated boundary at $x = L$. Consequently, the cases here of interest are X12B-0T1 and X22B-0T1 listed before. Also, both cases will be analyzed afterwards by assuming constant (B1) and linear with time (B2) variations. A general polynomial form for these boundary conditions can be asserted. For the X12B-0T1 case, Eq. (2.1b) may be taken as

$$T(x = 0, t) = T_{in} + (T_0 - T_{in})\left(\frac{t}{t_{ref}}\right)^p \quad (t > 0; p = 0 \text{ or } 1) \tag{2.2}$$

and for the X22B-0T1

$$-k\left(\frac{\partial T}{\partial x}\right)_{x = 0} = q_0\left(\frac{t}{t_{ref}}\right)^p \quad (t > 0; p = 0 \text{ or } 1) \tag{2.3}$$

where t_{ref} is an arbitrary reference time that affects the surface temperature and heat flux variation rates.

2.3.3 Dimensionless Variables

Most of the solutions in this text will be described in terms of dimensionless variables. Dimensionless variables are used to reduce the number of dimensional variables involved in the heat conduction problem solution.

In particular, the dimensionless space and time coordinates which will be used are defined as

$$\tilde{x} = \frac{x}{L}, \quad \tilde{t} = \frac{\alpha t}{L^2} \tag{2.4}$$

where $\tilde{x} \in [0, 1]$ and $\tilde{t} \geq 0$ (sometimes called the Fourier number) is based on the slab thickness. In other sections of this text when special non-dimensionalization is utilized, a superscript "+" is used and the applicable definition given.

As far as the temperature variation $T - T_{in}$ is concerned (positive when heating and negative when cooling), it can be made dimensionless by using the variables involved in the boundary condition applied at $x = 0$, that is, Eq. (2.2) or (2.3). Therefore, two different situations are possible:

- X12B10T1 and X12B20T1 cases. The dimensionless temperature may be defined as

$$\tilde{T} = \frac{T - T_{in}}{T_0 - T_{in}} \tag{2.5}$$

Equation (2.5) makes the initial condition defined by Eq. (2.1d) homogeneous, that is, $\tilde{T}(\tilde{x}, \tilde{t} = 0) = 0$ with $\tilde{x} \in (0, 1)$, and, hence, the notations for these two problems become: X12B10T0 and X12B20T0. Notice that the X12B10T0 case described by six variables, namely $x, t, L, \alpha, (T_0 - T_{in})$, and $(T - T_{in})$, can be expressed simply by using three dimensionless groups, namely $\tilde{T}, \tilde{x}$, and $\tilde{t}$. For the companion X12B20T0 case, there are seven variables due to t_{ref} and, hence, there are four dimensionless groups adding $\tilde{t}_{ref} = \alpha t_{ref}/L^2$ to the groups listed before. Also, the dimensionless temperature is positive for both heating ($T_0 > T_{in}$) and cooling ($T_0 < T_{in}$) of the plate. In the former, in fact, $T > T_{in}$, while in the latter $T < T_{in}$.

- X22B10T1 and X22B20T1 cases. The dimensionless temperature may be taken as

$$\tilde{T} = \frac{T - T_{in}}{(q_0 L/k)} \tag{2.6}$$

Equation (2.6) still makes the initial condition homogeneous, and, hence, the notations become X22B10T0 and X22B20T0. The X22B10T0 problem, which is described by seven variables, namely x, t, L, α, k, q_0, and $(T - T_{in})$, can also be described simply with three dimensionless quantities, that is, $\tilde{T}, \tilde{x}$, and $\tilde{t}$. For the companion X22B20T0 case, the variables are eight due to t_{ref} and, hence, there are still four dimensionless groups adding $\tilde{t}_{ref} = \alpha t_{ref}/L^2$. Also, the dimensionless temperature is positive for both heating ($q_0 > 0$) and cooling ($q_0 < 0$) of the plate. In the former, in fact, $T > T_{in}$, while in the latter $T < T_{in}$.

Notice that these two forms of non-dimensionalization can be compared directly when $(T_0 - T_{in})$ is equal to $(q_0 L/k)$.

Before proceeding to analyze the four problems stated before, possible forms of their "*exact*" analytical solutions will be examined and the so-called "*computational*" analytical solution associate with the "*exact*" one will be defined.

2.3.4 Exact Analytical Solution

Exact analytical temperature solutions to the four problems of interest are well-established in the heat conduction literature. Importantly, these solutions can take different forms depending on the time and/or location of interest for evaluation. The exact solution is denoted by $\tilde{T}_e$ (where the subscript "e" indicates "*exact*") and is unique according to the inherently well-posed nature of direct problems, but two forms of it are available and can be used:

- For early times, the *short-time* form, which comes from the application of Laplace transform (LT) approach (Özişik 1993; chapter 7) as solution method of the governing equations, can be used. This form can also come from the application of Green's function method when using "*short-cotime*" Green's functions (Cole et al. 2011). The short-time form is valid at any time, but it is computationally advantageous at short times. The form of the short-time solution is

$$\tilde{T}_e(\tilde{x}, \tilde{t}) = \tilde{T}_{\text{XI0BKT0}}(\tilde{x}, \tilde{t}) + \sum_{m=1}^{\infty} \tilde{T}_m^{(S)}(\tilde{x}, \tilde{t}) \tag{2.7a}$$

The first part, denoted by $\tilde{T}_{\text{XI0BKT0}}(\tilde{x}, \tilde{t})$ (where the first part of the subscript "XI0" indicates 1D "*semi-infinite*" body with I = 1 or 2 according to the surface boundary condition), is the solution of the associate 1D semi-infinite problem (where K = 1 or 2 for a constant or linear-in-time boundary condition. See Section 2.2). This part is unaffected by the boundary condition at $x = L$. Also, when K = 2, this part depends on the dimensionless reference time defined in the previous section.

The second part accounts for the thermal deviation effects on temperature due to the homogeneous boundary at $x = L$. For this reason, it may be termed as *"inactive boundary"* disturbance part. This part is in an infinite series consisting of terms denoted by $\tilde{T}_m^{(S)}(\tilde{x}, \tilde{t})$ (where the superscript *"S"* indicates *"short-time"*). The mth term in Eq. (2.7a) consists of two parts as

$$\tilde{T}_m^{(S)}(\tilde{x}, \tilde{t}) = \tilde{T}^{(S^-)}(2m - \tilde{x}, \tilde{t}) + \tilde{T}^{(S^+)}(2m + \tilde{x}, \tilde{t}) \tag{2.7b}$$

where $\tilde{T}^{(S^-)}$ and $\tilde{T}^{(S^+)}$ involve complementary error functions integrals i^n erfc(.), with $n = 0, 1, 2, 3, ...,$ (Cole et al. 2011, appendix E) and depend on both boundary condition type and boundary function prescribed. When the boundary function changes linearly with time, they also depend on the dimensionless reference time.

- For longer times, the *large-time* form of the solution must be used. This form can be obtained from the application of separation-of-variables (SOV) method (Özişik 1993, chapter 2). Also, from the application of Green's function method when using *"large-cotime"* Green's functions (Cole et al. 2011). The large-time solution is valid at any time, but it is computationally efficient at large times. The form of this solution is

$$\tilde{T}_e(\tilde{x}, \tilde{t}) = \tilde{T}_{qs}(\tilde{x}, \tilde{t}) + \underbrace{\sum_{m=1}^{\infty} \tilde{T}_m^{(L)}(\tilde{x}, \tilde{t})}_{\tilde{T}_{ct}(\tilde{x}, \tilde{t})} \tag{2.8a}$$

The first part, denoted by $\tilde{T}_{qs}(\tilde{x}, \tilde{t})$ (where the subscript *"qs"* indicates *"quasi-steady"*), occurs when the time goes to infinity. This part is rightly called quasi-steady solution when its temperature gradient is steady, that is, the temperature gradient $(\partial \tilde{T}_{qs}/\partial \tilde{x})$ and, hence, the heat flux does not change with time at any given point. This happens when dealing with the X12B20T0 and X22B10T0 cases. Also, it depends on two variables, $\tilde{x}$ and $\tilde{t}$. However, when the boundary function changes linearly with time, it also depends on the dimensionless reference time defined in the previous section. In addition, it reduces to a steady-state solution, denoted by $\tilde{T}_{ss}(\tilde{x})$ (where the subscript *"ss"* indicates *"steady-state"*), when the active surface is subject to a time-independent boundary condition with the only exception of the X22B10T0 case (see Section 2.3.8). While it complicates to an unsteady solution (the subscript *"us"* will be used to denote *"unsteady"*) when the temperature gradient (i.e., the heat flux) varies with time at any location. See, for example, the X22B20T0 case of Section 2.3.9. The second part, $\tilde{T}_{ct}(\tilde{x}, \tilde{t})$, is the so-called *"complementary"* transient part that becomes negligible for large values of the time. It is in a Fourier-series consisting of an infinite number of terms denoted by $\tilde{T}_m^{(L)}(\tilde{x}, \tilde{t})$ (where the superscript *"L"* indicates *"large-time"*). This mth term has the form:

$$\tilde{T}_m^{(L)}(\tilde{x}, \tilde{t}) = c_m F_{x,m}(\tilde{x}) e^{-\beta_m^2 \tilde{t}} \tag{2.8b}$$

where β_m is the mth dimensionless eigenvalue (mth root of the eigencondition associate to the problem) and $F_{x,m}(\tilde{x})$ is the corresponding eigenfunction that can oscillate between –1 and 1. Also, c_m is an integration constant that depends on the dimensionless reference time when the boundary condition varies linearly with time.

When dealing with a time-dependent boundary condition as defined by Eqs. (2.2) and (2.3) for $p = 1$, both the above short- and large-time forms of the exact analytical temperature solution require the use of Duhamel's integral (Özişik 1993, chapter 5) or Green's functions (Cole et al. 2011) for their derivation. If the heating/cooling process at the $x = 0$ boundary of the flat plate occurs at time $t = t_0 > 0$, the temperature solution can still be derived by using the short- and large-time forms of the solution listed before, where $\tilde{t}$ has however to be replaced with $\tilde{t} - \tilde{t}_0$. Therefore, $\tilde{T}_e(\tilde{x}, \tilde{t}) \rightarrow \tilde{T}_e(\tilde{x}, \tilde{t} - \tilde{t}_0)$, with $\tilde{t}_0 = \alpha t_0 / L^2$. This is of great concern when using the principle of superposition related to the piecewise-constant and piecewise-linear approximations of the time-dependent boundary function as proposed in Sections 3.3 and 3.4 of Chapter 3, respectively.

2.3.5 The Concept of Computational Analytical Solution

Both the short- and large-time forms of the *"exact"* analytical solution $\tilde{T}_e$ expressed by Eqs. (2.7a)-(2.7b) and (2.8a)-(2.8b), respectively, are in a series-form that requires an infinite number of terms. As no calculator can account for an infinite number of terms, the exact analytical solution needs to be replaced with an approximate one consisting of a finite number

m_{max} of terms. This solution is termed as "*computational*" analytical solution and is denoted by $\tilde{T}_c$ (where the subscript "*c*" indicates precisely "*computational*").

Therefore, Eqs. (2.7a) and (2.8a) become, respectively,

$$\tilde{T}_c(\tilde{x}, \tilde{t}, m_{max}) = \tilde{T}_{X10BKT0}(\tilde{x}, \tilde{t}) + \sum_{m=1}^{m_{max}} \tilde{T}_m^{(S)}(\tilde{x}, \tilde{t}) \tag{2.9a}$$

$$\tilde{T}_c(\tilde{x}, \tilde{t}, m_{max}) = \tilde{T}_{qs}(\tilde{x}, \tilde{t}) + \sum_{m=1}^{m_{max}} \tilde{T}_m^{(L)}(\tilde{x}, \tilde{t}) \tag{2.9b}$$

where $\tilde{T}_m^{(S)}(\tilde{x}, \tilde{t})$ and $\tilde{T}_m^{(L)}(\tilde{x}, \tilde{t})$ are given by Eqs. (2.7b) and (2.8b), respectively.

When using a computational solution in place of an exact one, the numerical value of temperature at location and time of interest is not the "exact" value because of the truncation error in the series solution with the only exception of those boundaries where the temperature value is prescribed (boundary condition of the first kind). Therefore, it is important to define absolute and relative errors, as shown in the next section.

2.3.5.1 Absolute and Relative Errors

The absolute error ε_a (units of °C) between exact and computational solutions may be calculated as

$$\varepsilon_a = |[T_e(x, t) - T_{in}] - [T_c(x, t, m_{max}) - T_{in}]| = |T_e(x, t) - T_c(x, t, m_{max})| \tag{2.10a}$$

In dimensionless form,

$$\tilde{\varepsilon}_a = |\tilde{T}_e(\tilde{x}, \tilde{t}) - \tilde{T}_c(\tilde{x}, \tilde{t}, m_{max})| \tag{2.10b}$$

where $\tilde{\varepsilon}_a = \tilde{\varepsilon}_a(\tilde{x}, \tilde{t}, m_{max})$ is made dimensionless according to Eqs. (2.5) and (2.6), and depends on the location, the time, and the number m_{max} of terms used in the computational series-solution as well as the form of the solution (short-time or large-time). Also, it depends on the reference time when the boundary condition changes linearly with time. Therefore, the exact numerical value of temperature (that remains unknown) falls between $\tilde{T}_c(\tilde{x}, \tilde{t}, m_{max}) - \tilde{\varepsilon}_a$ and $\tilde{T}_c(\tilde{x}, \tilde{t}, m_{max}) + \tilde{\varepsilon}_a$, that is,

$$\tilde{T}_e(\tilde{x}, \tilde{t}) = \tilde{T}_c(\tilde{x}, \tilde{t}, m_{max}) \pm \tilde{\varepsilon}_a \tag{2.10c}$$

A relative error ε_r (dimensionless) can also be defined as

$$\varepsilon_r = \frac{\varepsilon_a}{T_c(x, t, m_{max}) - T_{in}} = \frac{\tilde{\varepsilon}_a}{\tilde{T}_c(\tilde{x}, \tilde{t}, m_{max})} = \frac{|\tilde{T}_e(\tilde{x}, \tilde{t}) - \tilde{T}_c(\tilde{x}, \tilde{t}, m_{max})|}{\tilde{T}_c(\tilde{x}, \tilde{t}, m_{max})} \tag{2.11a}$$

and gives an estimate of the accuracy of the computational solution at the location and time of interest with respect to the computational temperature value at the same space and time coordinates. Therefore, it results in

$$\tilde{T}_e(\tilde{x}, \tilde{t}) = \tilde{T}_c(\tilde{x}, \tilde{t}, m_{max}) \pm [\varepsilon_r \tilde{T}_c(\tilde{x}, \tilde{t}, m_{max})] \tag{2.11b}$$

However, for short times the temperature value can be very small providing misleadingly a very large value of the relative error as it also appears at the denominator of Eq. (2.11a). For this reason, it is here preferred to normalize the absolute error ε_a by using the maximum temperature variation (up to and including the time of interest) that, in the four 1D current cases, always occurs at the "active" boundary surface $x = 0$ (D'Alessandro and de Monte 2018).

Thus, the relative error E (dimensionless) is defined as

$$E = \frac{\varepsilon_a}{\max\{[T_c(0, t, m_{max}) - T_{in}]\}} = \frac{|\tilde{T}_e(\tilde{x}, \tilde{t}) - \tilde{T}_c(\tilde{x}, \tilde{t}, m_{max})|}{\max\{\tilde{T}_c(0, \tilde{t}, m_{max})\}} \tag{2.12a}$$

Therefore, it is found that

$$\tilde{T}_e(\tilde{x}, \tilde{t}) = \tilde{T}_c(\tilde{x}, \tilde{t}, m_{max}) \pm [E \cdot \max\{\tilde{T}_c(0, \tilde{t}, m_{max})\}] \tag{2.12b}$$

Note that, for the four 1D transient cases here of interest, it results in: $\max\{\tilde{T}_c(0, \tilde{t}, m_{max})\} = \tilde{T}_c(0, \tilde{t}, m_{max})$ for $\tilde{t} \in [0, \tilde{t}]$ as the dimensionless temperature at the active surface of the plate always increases monotonically with time.

Substituting Eqs. (2.7a) and (2.9a) in Eq. (2.12a) yields

$$E = \frac{\left|\sum_{m=m_{max}+1}^{\infty} \tilde{T}_m^{(S)}(\tilde{x}, \tilde{t})\right|}{\tilde{T}_{X10BKT0}(0, \tilde{t}) + \sum_{m=1}^{m_{max}} \tilde{T}_m^{(S)}(0, \tilde{t})} \tag{2.13a}$$

Similarly, substituting Eqs. (2.8a) and (2.9b) in Eq. (2.12a) gives

$$
E = \frac{\left| \sum_{m=m_{\max}+1}^{\infty} \tilde{T}_m^{(L)}(\tilde{x}, \tilde{t}) \right|}{\tilde{T}_{qs}(0, \tilde{t}) + \sum_{m=1}^{m_{\max}} \tilde{T}_m^{(L)}(0, \tilde{t})}
\tag{2.13b}
$$

The numerator of Eqs. (2.13a) and (2.13b) represents the "tail" of the series that may be evaluated in a conservative way by using the Euler–Maclaurin formula (Oldham et al. 2009, see Eq. (4.14.1), p. 43). This formula can also be applied to the finite series appearing at the denominator of the same equations.

The relative error E depends on the maximum number $m_{\max}$ of terms used in the series-solution, the location, the time of interest and, hence, the form of the solution (short-time or large-time). However, contrary to the absolute error, it is independent of the reference time as it cancels out. For small values of the time, it is computationally convenient to use the short-time solution defined by Eq. (2.9a) as it requires only a few terms. On the contrary, for large values of the time, it is computationally more efficient employing the large-time solution defined by Eq. (2.9b) as it requires right a few terms.

By setting $E = 10^{-A}$ in Eqs. (2.13a) and (2.13b), where $A = 2, 3, ..., 15$ is the counting integer for the accuracy desired, and solving the resulting equations, the maximum number of terms $m_{\max}^{(A)}$ to get an accuracy of one part in 10^A (with respect to the maximum temperature rise at the heated surface up to and including the time of interest) may be calculated as a function of time, location and accuracy. In particular, $A = 2$ (accuracy of 1%) is for visual comparison and is acceptable in many engineering applications, while $A = 10$ or even 15 is mainly for verification purposes of large numerical codes (Beck et al. 2004, 2006; ASME 2021). Note that the floating-point relative accuracy in double precision within MATLAB software is of 2.2204×10^{-16}. When dealing with $A = 15$, the computational analytical solution may be considered in every practical respect as the "*exact*" one.

2.3.5.2 Deviation Time

For very short times less than the so-called "*deviation*" time, the solution of Eq. (2.13a) for $E = 10^{-A}$ with $\tilde{T}_m^{(S)}(\tilde{x}, \tilde{t})$ defined by Eq. (2.7b) gives $m_{\max}^{(A)} = 0$ for any location. Therefore, the exact solution Eq. (2.7a) reduces, with a relative error less than or equal to 10^{-A}, with $A = 2, 3, ..., 15$, to Eq. (2.9a) with only the first part,

$$
\tilde{T}_c(\tilde{x}, \tilde{t}) = \tilde{T}_{\text{XI0BKT0}}(\tilde{x}, \tilde{t}) \quad (0 \le \tilde{x} \le 1; 0 \le \tilde{t} \le \tilde{t}_d)
\tag{2.14}
$$

where the dimensionless deviation time $\tilde{t}_d$ may be taken as (de Monte et al. 2008, 2012; Cole et al. 2014):

$$
\tilde{t}_d = \frac{1}{10A}(2 - \tilde{x})^2 \quad (0 \le \tilde{x} \le 1; A = 2, 3, ..., 15)
\tag{2.15}
$$

Deviation time is defined as the time that it takes for the temperature variation at a point of a finite solid heated or cooled at a boundary to be just affected (one part in 10^A with respect to the maximum temperature variation at the active surface up to and including that time) by the presence of the homogeneous boundary condition ("inactive" boundary). Equation (2.15) is a conservative equation leading to the smallest deviation time; it is valid for any kind of boundary condition and any boundary function $f_0(t)$. As an example, for $\tilde{x} = 1/2$ (middle plane) and $A = 3$, it is found that $\tilde{t}_d = 0.075$. This indicates that, for dimensionless times less than or equal to 0.075, the body behaves at $\tilde{x} = 1/2$ as a semi-infinite one with relative errors E less than or equal to $10^{-3} = 0.001$. (The related dimensional deviation time at that location is: $t_d = 0.075 L^2/\alpha$.) The deviation time will be shorter if a higher accuracy is desired.

The computational solution Eq. (2.14) exhibits only one short-time term. It also depends on the dimensionless reference time when the driving force at the active boundary changes linearly with time.

2.3.5.3 Second Deviation Time

For short times less than the so-called "*second deviation*" time but greater than the (first) deviation time, the solution of Eq. (2.13a) for $E = 10^{-A}$ gives $m_{\max}^{(A)} = 1$ for any location when considering only the first part $\tilde{T}^{(S^-)}$ of $\tilde{T}_m^{(S)}(\tilde{x}, \tilde{t})$ defined by Eq. (2.7b). In such a case, the exact solution Eq. (2.7a) reduces, with a relative error less than or equal to 10^{-A}, with $A = 2, 3, ..., 15$, to Eq. (2.9a) with $m_{\max}^{(A)} = 1$,

$$
\tilde{T}_c(\tilde{x}, \tilde{t}) = \tilde{T}_{\text{XI0BKT0}}(\tilde{x}, \tilde{t}) + \tilde{T}^{(S^-)}(2 - \tilde{x}, \tilde{t}) \quad (0 \le \tilde{x} \le 1; \tilde{t}_d < \tilde{t} \le \tilde{t}_\delta)
\tag{2.16}
$$

where the dimensionless second deviation time $\tilde{t}_\delta$ may be taken as (de Monte et al. 2023)

$$
\tilde{t}_\delta = \frac{1}{10A}(2 + \tilde{x})^2 \quad (0 \le \tilde{x} \le 1; A = 2, 3, ..., 15)
\tag{2.17}
$$

Equation (2.17) is a conservative equation leading to the smallest second deviation time; it is valid for any kind of boundary condition and any boundary function $f_0(t)$. For $\tilde{x} = 0$, $\tilde{t}_\delta = \tilde{t}_d$ but, for $\tilde{x} = 1$, $\tilde{t}_\delta$ is nine times larger than $\tilde{t}_d$ for any accuracy.

The computational solution Eq. (2.16) has two short-time terms. It also depends on the reference time when the boundary condition varies linearly with time according to Eqs. (2.2) and (2.3).

2.3.5.4 Quasi-Steady, Steady-State and Unsteady Times

For very large times greater than the so-called "*quasi-steady*" time (or "*steady-state*" time when $\tilde{T}_{qs}(\tilde{x}, \tilde{t}) \rightarrow \tilde{T}_{ss}(\tilde{x})$), the solution of Eq. (2.13b) for $E = 10^{-A}$ with $\tilde{T}_m^{(L)}(\tilde{x}, \tilde{t})$ defined by Eq. (2.8b) gives $m_{max}^{(A)} = 0$. In this case, the exact solution Eq. (2.8a) reduces, with a relative error $E \leq 10^{-A}$, with $A = 2, 3, ..., 15$, to Eq. (2.9b) with only the first part,

$$\tilde{T}_c(\tilde{x}, \tilde{t}) = \tilde{T}_{qs}(\tilde{x}, \tilde{t}) \quad \left(0 \leq \tilde{x} \leq 1; \tilde{t} \geq \tilde{t}_{qs}\right) \tag{2.18}$$

where the dimensionless quasi-steady time $\tilde{t}_{qs}$ (or steady-state time $\tilde{t}_{ss}$) may be taken as

$$\tilde{t}_{qs} = \frac{A \ln(10) + \ln\left(2/\beta_1^b\right)}{\beta_1^2} \quad (0 \leq \tilde{x} \leq 1; A = 2, 3, ..., 15) \tag{2.19}$$

being $b = 2p + 1$ for the X12 slab and $b = 2(p + 1)$ for the X22 one. Also, $p = 0$ or 1 according to Eqs. (2.2) and (2.3). Moreover, Eq. (2.19) is a modified expression of the original equation given by de Monte and Beck (2009; see Eq. (26), p. 5571).

In the above equation, β_1 is the first eigenvalue of the eigencondition associate to the heat conduction problem (Özişik 1993; see chapter 2). It depends on the boundary condition type but always falls into the range $[\pi/2, \pi]$, where $\beta_1 = \pi/2$ for the X12 case, while $\beta_1 = \pi$ for the X22 one. It is hence independent of the boundary function, for example constant or linear in time.

The quasi-steady (or steady-state) time may be defined as the time that it takes for the temperature at a point of a finite body heated or cooled at a boundary to reach the quasi-steady (or steady-state) condition with a relative error of one part in 10^A with respect to the maximum temperature variation at the active surface up to and including that time. Also, Eq. (2.19) is a conservative equation leading to the largest quasi-steady (or steady state) time; it is valid for any location within the flat plate. As an example, for $0 \leq \tilde{x} \leq 1$ and $A = 3$, when the plate is subject to a time-independent boundary condition ($p = 0$), it is found that $\tilde{t}_{qs} \rightarrow \tilde{t}_{ss} \approx 2.9$ for X12B10T0 and $\tilde{t}_{qs} \approx 0.54$ for X22B10T0 (more than five times lower). This indicates that, after a dimensionless time of 2.9, the X12B10T0 slab has practically reached the steady-state solution. In other words, by neglecting the complementary transient part of the solution given by Eq. (2.9b), the relative error E is always less than or equal to $10^{-3} = 0.001$. (The related dimensional steady-state time is: $t_{ss} \approx 2.9L^2/\alpha$.) If a higher accuracy is desired, say $A = 6$, it results in $\tilde{t}_{ss} \approx 5.7$ for X12B10T0 and $\tilde{t}_{qs} \approx 1.24$ for X22B10T0.

The computational solution Eq. (2.18) has only the quasi-steady (or steady-state) part. This part also depends on the dimensionless reference time when the driving force at the active boundary changes linearly with time. In addition, it complicates to an unsteady part when the temperature gradient changes with time at any location (as occurs for the X22B20T0 case of Section 2.3.9). In such a case, the corresponding characteristic time still defined by the generalized equation (2.19) is called "*unsteady*" time t_{us}.

2.3.5.5 Solution for Large Times

For large times greater than the second deviation time given by Eq. (2.17) but less than the "quasi-steady" (or steady-state) time defined by Eq. (2.19), that is, $\tilde{t}_\delta < \tilde{t} < \tilde{t}_{qs}$, the solution of Eq. (2.13b) for $E = 10^{-A}$ with $\tilde{T}_m^{(L)}(\tilde{x}, \tilde{t}_L)$ defined by Eq. (2.8b) gives the maximum number of terms $m_{max}^{(A)}$ as a function of time $\tilde{t}$ and accuracy A desired for any location within the plate ($0 \leq \tilde{x} \leq 1$). It is found that

$$m_{max}^{(A)} = \text{ceil}\left(a + \sqrt{\frac{A \ln(10) + \ln\left(2/\beta_1^b\right)}{\pi^2 \tilde{t}}}\right) \quad (A = 2, 3, ..., 15) \tag{2.20a}$$

where $a = 1/2$ and $b = 2p + 1$ for the X12 slab; while $a = 0$ and $b = 2(p + 1)$ for the X22 one. Also, $p = 0$ or 1 according to Eqs. (2.2) and (2.3). Moreover, Eq. (2.20a) is an extended expression of the original equation given by Woodbury et al. (2017). Lastly the function "ceil(z)" is a MATLAB function that rounds the number z to the nearest integer greater than or equal to z.

The exact analytical solution Eq. (2.8a) with Eq. (2.8b) reduces, with a relative error $E \leq 10^{-A}$, for $0 \leq \tilde{x} \leq 1$ and $\tilde{t}_\delta < \tilde{t} < \tilde{t}_{qs}$, to

$$\tilde{T}_c\left(\tilde{x}, \tilde{t}, m_{\max}^{(A)}\right) = \tilde{T}_{qs}(\tilde{x}, \tilde{t}) + \sum_{m=1}^{m_{\max}^{(A)}} c_m F_{x,m}(\tilde{x})e^{-\beta_m^2 \tilde{t}} \tag{2.20b}$$

where $\tilde{T}_{qs}(\tilde{x}, \tilde{t}) \to \tilde{T}_{ss}(\tilde{x})$ when the flat plate can reach the steady state. In such a case, $\tilde{t}_{qs} \to \tilde{t}_{ss}$. Similarly, when the slab is characterized by an unsteady behavior for large times (greater than the unsteady time).

The computational solution consists of $m_{\max}^{(A)}$ large-time terms (that represent the complementary transient part of the solution) plus the quasi-steady or steady-state or unsteady part. It also depends on the reference time when the boundary condition varies linearly with time according to Eqs. (2.2) and (2.3). In addition, the time $\tilde{t}_\delta$ may be considered as a partitioning time, that is, the time matching the computational short-time form Eq. (2.16) and the computational large-time form Eq. (2.20b) (Yen et al. 2002). However, for locations $\tilde{x}$ close to the back side of the slab and for very low accuracies, $\tilde{t}_\delta$ defined by Eq. (2.17) increases, while $\tilde{t}_{qs}$ (or $\tilde{t}_{ss}$) defined by Eq. (2.19) decreases. Therefore, it might occur that $\tilde{t}_\delta > \tilde{t}_{qs}$ for $\tilde{x} \in (\tilde{x}_c, 1]$, being $\tilde{x}_c$ a critical distance from the active surface $\tilde{x} = 0$ defined afterwards. In such a case, Eq. (2.20b) vanishes and Eq. (2.16) is valid in the time range $\tilde{t}_d < \tilde{t} < \tilde{t}_{qs}$. Then, for $\tilde{t} \geq \tilde{t}_{qs}$, Eq. (2.18) still applies. The above critical distance may be derived setting $\tilde{t}_\delta = \tilde{t}_{qs}$. It results in $\tilde{x}_c = \left(\sqrt{10A\tilde{t}_{qs}} - 2\right)$. This possibility is discussed in Sections 2.3.6–2.3.9.

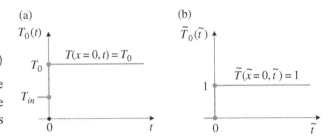

Figure 2.2 Nonhomogeneous constant boundary condition of the first kind at $x = 0$ for a slab initially at T_{in} and insulated on the back side $x = L$: (a) dimensional (X12B10T1); and (b) dimensionless (X12B10T0).

2.3.5.6 Intrinsic Verification

The computational analytical solution defined through Eqs. (2.14), (2.16), (2.18), and (2.20b) was verified by using the concept of symbolic intrinsic verification (Cole et al. 2014; D'Alessandro and de Monte 2018). The related concept of numerical intrinsic verification was utilized to verify the related computer code.

The computational analytical solution combines four different expressions in such a way as to be the most efficient solution computationally. These expressions match each other thorough three different characteristic times. Of course, at very early times, the large-time form Eq. (2.20b) can also be used but the number of terms might be very high, in particular when dealing with a high accuracy.

2.3.6 X12B10T0 Case

This case is characterized by a step change in temperature, $T_{in} \to T_0$, at $x = 0$ and $t = 0$, with T_0 time-independent ($T_0 > T_{in}$ for heating and $T_0 < T_{in}$ for cooling), as given by Eq. (2.2) for $p = 0$ and depicted in Figure 2.2a for $T_0 > T_{in}$. In dimensionless form, the boundary condition simplifies to $\tilde{T}(\tilde{x} = 0, \tilde{t}) = 1$, with $\tilde{t} > 0$, where the nondimensional temperature is defined by Eq. (2.5) (see Figure 2.2b), and the initial condition becomes homogeneous. The slab reaches a steady-state condition when the time goes to infinity.

The *short-time* form of the solution to the current problem is given by Carslaw and Jaeger (1959; Eq. (3), p. 309) for the related X21B01T0 case. To obtain the solution of the X12B10T0 case here of interest, the space variable x of the X21B01T0 problem has simply to be replaced with $L - x$. In dimensionless form, the X12B10T0 exact analytical temperature solution $\tilde{T}_e$ is

$$\tilde{T}_e(\tilde{x}, \tilde{t}) = \underbrace{\text{erfc}\left(\frac{\tilde{x}}{2\sqrt{\tilde{t}}}\right)}_{\tilde{T}_{X10B1T0}(\tilde{x}, \tilde{t})} + \sum_{m=1}^{\infty} (-1)^{m-1}\underbrace{\left[\text{erfc}\left(\frac{2m-\tilde{x}}{2\sqrt{\tilde{t}}}\right) - \text{erfc}\left(\frac{2m+\tilde{x}}{2\sqrt{\tilde{t}}}\right)\right]}_{\tilde{T}_m^{(S)}(\tilde{x}, \tilde{t})} \tag{2.21}$$

where the semi-infinite temperature solution, $\tilde{T}_{X10B1T0}(\tilde{x}, \tilde{t})$, depends only on one variable, that is, $\tilde{x}/\sqrt{\tilde{t}} = x/\sqrt{\alpha t} = 1/\sqrt{\tilde{t}_x}$, being $\tilde{t}_x = \alpha t/x^2$ the dimensionless time based on x.

The *large-time* form of the solution is still given by Carslaw and Jaeger (1959; Eq. (2), p. 100) for the X11B11T0 case in the space domain $-L \leq x \leq L$, which is thermally equivalent to the X21B01T0 problem in the spatial domain $0 \leq x \leq L$. To derive the solution to the current X12B10T0 problem, the space variable x of the X21B01T0 problem has to be replaced with $L - x$. Therefore, in dimensionless form, the X12B10T0 solution $\tilde{T}_e$ is

$$\tilde{T}_e(\tilde{x}, \tilde{t}) = \underbrace{1}_{\tilde{T}_{ss}(\tilde{x})} - \sum_{m=1}^{\infty} 2\underbrace{\frac{e^{-[(m-1/2)\pi]^2\tilde{t}}}{(m-1/2)\pi}\sin\left[\left(m-\frac{1}{2}\right)\pi\tilde{x}\right]}_{-\tilde{T}_m^{(L)}(\tilde{x}, \tilde{t})} \tag{2.22}$$

where $(m - 1/2)\pi$ is the mth eigenvalue, β_m, for the X12 case, while $\sin[(m - 1/2)\pi\tilde{x}]$ is the corresponding mth eigenfunction $F_{x,m}(\tilde{x})$. Also, the steady-state part is characterized by a unit uniform distribution of temperature.

2.3.6.1 Computational Analytical Solution

This solution is given in the following for a relative error $E \leq 10^{-A}$, with $A = 2, 3, ..., 15$, for $0 \leq \tilde{x} \leq 1$ and four different time intervals. In detail,

- $0 \leq \tilde{t} \leq \tilde{t}_d$. Short-time form with $m_{\max}^{(A)} = 0$:

$$\tilde{T}_c(\tilde{x}, \tilde{t}) = \text{erfc}\left(\frac{\tilde{x}}{2\sqrt{\tilde{t}}}\right) \tag{2.23}$$

where $\tilde{t}_d$ is the dimensionless deviation time defined through Eq. (2.15).

- $\tilde{t}_d < \tilde{t} \leq \tilde{t}_\delta$. Short-time form with $m_{\max}^{(A)} = 1$, when considering only the first part of $\tilde{T}_m^{(S)}(\tilde{x}, \tilde{t})$ defined by Eq. (2.21):

$$\tilde{T}_c(\tilde{x}, \tilde{t}) = \text{erfc}\left(\frac{\tilde{x}}{2\sqrt{\tilde{t}}}\right) + \text{erfc}\left(\frac{2 - \tilde{x}}{2\sqrt{\tilde{t}}}\right) \tag{2.24}$$

where $\tilde{t}_\delta$ is the dimensionless second deviation time as defined through Eq. (2.17).

- $\tilde{t}_\delta < \tilde{t} < \tilde{t}_{ss}$. Large-time form with $m_{\max}^{(A)} \geq 1$:

$$\tilde{T}_c(\tilde{x}, \tilde{t}) = 1 - 2\sum_{m=1}^{m_{\max}^{(A)}} \frac{e^{-[(m-1/2)\pi]^2\tilde{t}}}{(m - 1/2)\pi} \sin\left[\left(m - \frac{1}{2}\right)\pi\tilde{x}\right] \tag{2.25a}$$

where, using Eqs. (2.19) and (2.20a),

$$\tilde{t}_{ss} = \frac{A\ln(10) + \ln(4/\pi)}{\pi^2/4}, \qquad m_{\max}^{(A)} = \text{ceil}\left(\frac{1}{2} + \sqrt{\frac{A\ln(10) + \ln(4/\pi)}{\pi^2\tilde{t}}}\right) \tag{2.25b}$$

The maximum number of terms is required for $\tilde{t} \to \tilde{t}_\delta$ and for interior points close to the heated (or cooled) surface as the second deviation time is lower for those points. However, it is always less than or equal to 9 terms for an accuracy of $A = 10$ (to be used for verification purposes of large numerical codes) and less than or equal to 3 terms for $A = 2$ (to be used for visual comparison) depending on the time of interest.

- $\tilde{t} \geq \tilde{t}_{ss}$. Large-time form with $m_{\max}^{(A)} = 0$, that is, steady-state solution:

$$\tilde{T}_c(\tilde{x}, \tilde{t}) = 1 \tag{2.26}$$

In the current case, for $A = 2$, the critical location $\tilde{x}_c$ defined in Section 2.3.5.5 is of about 4.6 and, hence, is not consistent with the physical domain of the plate. Therefore, it results in $\tilde{t}_\delta < \tilde{t}_{ss}$ for any location within the body and accuracy A.

2.3.6.2 Computer Code and Plots

A MATLAB function called `fdX12B10T0.m` allows the dimensionless temperature to be computed for the current case. It is available for ease of use on the web site related to this book.

By using the `fdX12B10T0.m` function, temperature plots can be obtained for any accuracy though $A = 3$ is enough for visual comparison. They are shown in Figure 2.3a versus space and in Figure 2.3b as a function of time.

Both figures indicate that, for an interior point close to the back side of the flat plate, the response is slow, being both damped and lagged. For $\tilde{t} = 0.05$, as an example, $\tilde{T}(0.25, \tilde{t}) = 0.4292$, whereas $\tilde{T}(1, \tilde{t}) = 0.0031$, a factor of almost 100 smaller. (These numerical values can be obtained by using the `fdX12B10T0.m` function.) Such a factor increases as $\tilde{t}$ becomes smaller. On the other hand, for sufficiently large times greater than the steady-state time (note that $\tilde{t}_{ss} \approx 2.9$ for $A = 3$), the factor approaches unity.

2.3.7 X12B20T0 Case

This problem is characterized by a linear-in-time variation of the surface temperature at $x = 0$, according to Eq. (2.2) for $p = 1$ and depicted in Figure 2.4a for $T_0 > T_{in}$. The $(T_0 - T_{in})/t_{ref}$ gives the heating rate at the active boundary surface when positive ($T_0 > T_{in}$); while it gives the cooling rate at the same boundary when negative ($T_0 < T_{in}$). In dimensionless form, the boundary condition simplifies to $\tilde{T}(\tilde{x} = 0, \tilde{t}) = \tilde{t}/\tilde{t}_{ref}$, with $\tilde{t} > 0$, where the nondimensional temperature is defined

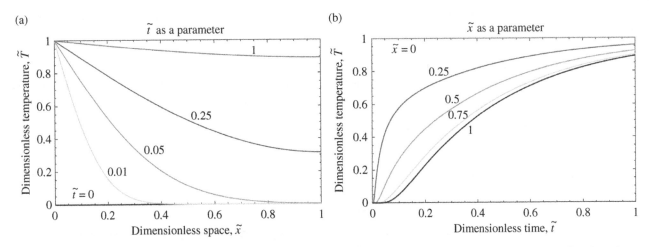

Figure 2.3 Dimensionless temperature of the X12B10T0 problem: (a) versus space with time as a parameter; (b) as a function of time for different locations.

Figure 2.4 Nonhomogeneous linearly-in-time variable boundary condition of the first kind at $x = 0$ for a slab initially at T_{in} and insulated on the back side $x = L$: (a) dimensional (X12B20T1); and (b) dimensionless (X12B20T0).

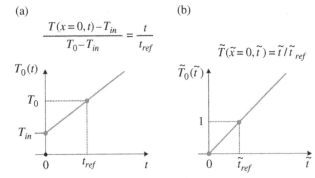

by Eq. (2.5) (see Figure 2.4b), and the initial condition becomes homogeneous. When the time goes to infinity, the slab reaches a quasi-steady condition (in fact, its temperature gradient is constant) whose temperature is the same as the linear-in-time forcing function at $x = 0$.

The *short-time* form of the solution to the current problem is given by Luikov (1968; Eq. (7.1.16), p. 304) for the related X21B02T0 case. To obtain the solution of the X12B20T0 case here of interest, the space variable x of the X21B02T0 problem has to be replaced with $L - x$. In dimensionless form, the X12B20T0 exact analytical temperature solution $\tilde{T}_e$ is

$$\tilde{T}_e\left(\tilde{x}, \tilde{t}, \tilde{t}_{ref}\right) = \underbrace{4\left(\frac{\tilde{t}}{\tilde{t}_{ref}}\right) i^2 \text{erfc}\left(\frac{\tilde{x}}{2\sqrt{\tilde{t}}}\right)}_{\tilde{T}_{X10B2T0}(\tilde{x},\,\tilde{t})}$$

$$+ \underbrace{\sum_{m=1}^{\infty} (-1)^{m-1} 4\left(\frac{\tilde{t}}{\tilde{t}_{ref}}\right)\left[i^2\text{erfc}\left(\frac{2m-\tilde{x}}{2\sqrt{\tilde{t}}}\right) - i^2\text{erfc}\left(\frac{2m+\tilde{x}}{2\sqrt{\tilde{t}}}\right)\right]}_{\tilde{T}_m^{(S)}(\tilde{x},\,\tilde{t})} \tag{2.27a}$$

where (Cole *et al.* 2011; Eq. (E.14b), p. 501)

$$i^2\text{erfc}\left(\frac{\tilde{x}}{2\sqrt{\tilde{t}}}\right) = \frac{1}{4}\left[1 + 2\left(\frac{\tilde{x}^2}{4\tilde{t}}\right)\right]\text{erfc}\left(\frac{\tilde{x}}{2\sqrt{\tilde{t}}}\right) - \frac{1}{2\sqrt{\pi}}\left(\frac{\tilde{x}}{2\sqrt{\tilde{t}}}\right)e^{-\frac{\tilde{x}^2}{4\tilde{t}}} \tag{2.27b}$$

In particular, $i^2\,\text{erfc}(0) = 1/4$.

The *large-time* form of the solution is given by Beck et al. (2008; Eq. (24)) in dimensionless form as

$$\tilde{T}_e\left(\tilde{x}, \tilde{t}, \tilde{t}_{ref}\right) = \underbrace{\frac{1}{\tilde{t}_{ref}}\left[\tilde{t} + \left(\frac{1}{2}\tilde{x}^2 - \tilde{x}\right)\right]}_{\tilde{T}_{qs}(\tilde{x},\,\tilde{t})} + \underbrace{\sum_{m=1}^{\infty} \frac{2}{\tilde{t}_{ref}}\frac{e^{-[(m-1/2)\pi]^2\tilde{t}}}{[(m-1/2)\pi]^3}\sin\left[\left(m-\frac{1}{2}\right)\pi\tilde{x}\right]}_{\tilde{T}_m^{(L)}(\tilde{x},\,\tilde{t})} \tag{2.28}$$

where the quasi-steady part is a linear function of time at any point of the plate, while its distribution over the thickness is parabolic at any time.

2.3.7.1 Computational Analytical Solution

This solution is given in the following for a relative error $E \leq 10^{-A}$, with $A = 2, 3, ..., 15$, for $0 \leq \tilde{x} \leq 1$ and four different time intervals. In detail,

- $0 \leq \tilde{t} \leq \tilde{t}_d$. Short-time form with $m_{\max}^{(A)} = 0$:

$$\tilde{T}_c(\tilde{x}, \tilde{t}, \tilde{t}_{ref}) = 4\left(\frac{\tilde{t}}{\tilde{t}_{ref}}\right) i^2 \mathrm{erfc}\left(\frac{\tilde{x}}{2\sqrt{\tilde{t}}}\right) \tag{2.29}$$

where $\tilde{t}_d$ is the dimensionless deviation time defined through Eq. (2.15).

- $\tilde{t}_d < \tilde{t} \leq \tilde{t}_\delta$. Short-time form with $m_{\max}^{(A)} = 1$, when considering only the first part of $\tilde{T}_m^{(S)}(\tilde{x}, \tilde{t})$ defined by Eq. (2.27a):

$$\tilde{T}_c(\tilde{x}, \tilde{t}, \tilde{t}_{ref}) = 4\left(\frac{\tilde{t}}{\tilde{t}_{ref}}\right) i^2 \mathrm{erfc}\left(\frac{\tilde{x}}{2\sqrt{\tilde{t}}}\right) + 4\left(\frac{\tilde{t}}{\tilde{t}_{ref}}\right) i^2 \mathrm{erfc}\left(\frac{2-\tilde{x}}{2\sqrt{\tilde{t}}}\right) \tag{2.30}$$

where $\tilde{t}_\delta$ is the dimensionless second deviation time as defined through Eq. (2.17).

- $\tilde{t}_\delta < \tilde{t} < \tilde{t}_{qs}$. Large-time form with $m_{\max}^{(A)} \geq 1$:

$$\tilde{T}_c(\tilde{x}, \tilde{t}, \tilde{t}_{ref}) = \frac{1}{\tilde{t}_{ref}}\left[\tilde{t} + \left(\frac{1}{2}\tilde{x}^2 - \tilde{x}\right)\right] + \frac{2}{\tilde{t}_{ref}} \sum_{m=1}^{m_{\max}^{(A)}} \frac{e^{-[(m-1/2)\pi]^2 \tilde{t}}}{[(m-1/2)\pi]^3} \sin\left[\left(m - \frac{1}{2}\right)\pi\tilde{x}\right] \tag{2.31a}$$

where, using Eqs. (2.19) and (2.20a),

$$\tilde{t}_{qs} = \frac{A\ln(10) + \ln(16/\pi^3)}{\pi^2/4}, \qquad m_{\max}^{(A)} = \mathrm{ceil}\left(\frac{1}{2} + \sqrt{\frac{A\ln(10) + \ln(16/\pi^3)}{\pi^2 \tilde{t}}}\right) \tag{2.31b}$$

The maximum number of terms is required for $\tilde{t} \rightarrow \tilde{t}_\delta$ and $\tilde{x}$ close to the heated (or cooled) surface as occurs for the X12B10T0 case discussed in Section 2.3.6. It is always less than or equal to 8 terms for an accuracy of $A = 10$ and less than or equal to 2 terms for $A = 2$ depending on the time of evaluation.

- $\tilde{t} \geq \tilde{t}_{qs}$. Large-time form with $m_{\max}^{(A)} = 0$, that is, quasi-steady solution:

$$\tilde{T}_c(\tilde{x}, \tilde{t}, \tilde{t}_{ref}) = \frac{1}{\tilde{t}_{ref}}\left[\tilde{t} + \left(\frac{1}{2}\tilde{x}^2 - \tilde{x}\right)\right] \tag{2.32}$$

In the current case, for $A = 2$, the critical location $\tilde{x}_c$ defined in Section 2.3.5.5 is of about 3.7 and, hence, is not consistent with the physical domain of the plate. Therefore, it results in $\tilde{t}_\delta < \tilde{t}_{qs}$ for any location within the body and accuracy A.

2.3.7.2 Computer Code and Plots

A MATLAB function called fdX12B20T0.m allows the dimensionless temperature to be computed for the current case. It is available for ease of use on the web site related to this book.

By using the fdX12B20T0.m function, temperature plots can be obtained for any accuracy though $A = 3$ is enough for visual comparison. They are shown for $\tilde{t}_{ref} = 1$ in Figure 2.5a versus space and in Figure 2.5b as a function of time.

Both figures indicate that, for an interior point close to the back side of the flat plate, the response is slow, being both damped and lagged. For $\tilde{t} = 0.05$, as an example, $\tilde{T}(0.25, \tilde{t}) = 0.01180$, whereas $\tilde{T}(1, \tilde{t}) = 0.00002$, a factor of almost 500 smaller. (These numerical values can be derived by using the fdX12B20T0.m function.) This factor increases as $\tilde{t}$ becomes smaller. On the other hand, for sufficiently large times greater than the quasi-steady time (note that $\tilde{t}_{qs} \approx 2.5$ for $A = 3$), the factor decreases approaching unity. This can be demonstrated by using Eq. (2.32) for large times with $\tilde{t}_{ref} = 1$ so that

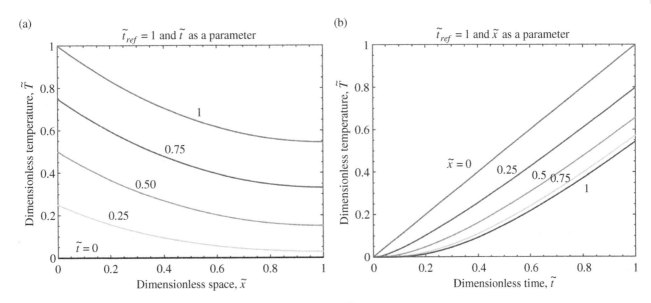

Figure 2.5 Dimensionless temperature of the X12B20T0 problem for $\tilde{t}_{ref} = 1$: (a) versus space with time as a parameter; (b) as a function of time for different locations.

$$\tilde{T}_c(0.25, \tilde{t}) = \tilde{t} - \frac{7}{32}, \quad \tilde{T}_c(1, \tilde{t}) = \tilde{t} - \frac{1}{2} \tag{2.33a}$$

Hence, the temperature ratio is

$$\frac{\tilde{T}_c(0.25, \tilde{t})}{\tilde{T}_c(1, \tilde{t})} = \frac{\tilde{t} - 7/32}{\tilde{t} - 1/2} = 1 + \frac{9/32}{\tilde{t} - 1/2} \tag{2.33b}$$

When $\tilde{t} = \tilde{t}_{qs} \approx 2.5$, the ratio is of about 1.15.

A comparison with the companion X12B10T0 problem, when $\tilde{t}_{ref} = 1$, indicates that $\tilde{T}(0.25, \tilde{t}) = 0.01180$ vs. 0.4292 for $\tilde{t} = 0.05$ and $\tilde{x} = 0.25$ (about 36 times smaller) but the ratio with respect to the maximum temperature variation which occurs at the heated (or cooled) boundary is only two times smaller. In fact, it is found that $\tilde{T}(0.25, \tilde{t})/\tilde{T}(0, \tilde{t}) = \tilde{T}(0.25, \tilde{t})/\tilde{t} = 0.23595$ against $\tilde{T}(0.25, \tilde{t})/\tilde{T}(0, \tilde{t}) = 0.4292$ for $\tilde{t} = 0.05$ of the X12B10T0 case as there $\tilde{T}(0, \tilde{t}) = 1$. Similarly, when considering the insulated back side, $\tilde{T}(1, \tilde{t}) = 0.00002$ vs. 0.0031 of the X12B10T0 case for $\tilde{t} = 0.05$ (155 times smaller). However, $\tilde{T}(1, \tilde{t})/\tilde{T}(0, \tilde{t}) = 0.0004$ vs. $\tilde{T}(1, \tilde{t})/\tilde{T}(0, \tilde{t}) = 0.0031$ for $\tilde{t} = 0.05$ of the X12B10T0 case that indicates a factor of almost 8 vs. 155.

2.3.8 X22B10T0 Case

This case is characterized by a step change in heat flux, $0 \rightarrow q_0$, at $x = 0$ and $t = 0$, with q_0 time-independent ($q_0 > 0$ for heating and $q_0 < 0$ for cooling), as given by Eq. (2.3) for $p = 0$ and depicted in Figure 2.6a for $q_0 > 0$. In dimensionless form, the boundary condition simplifies to $-\left(\partial \tilde{T}/\partial \tilde{x}\right)_{\tilde{x}=0} = 1$, with $\tilde{t} > 0$, where the nondimensional temperature is defined by Eq. (2.6) (see Figure 2.6b), and the initial condition becomes homogeneous. Also, the dimensionless heat flux is defined as $\tilde{q} = q/q_0$. The plate reaches a quasi-steady condition when the time goes to infinity, as the temperature gradient and, hence, the heat flux is time-independent at any location (see afterwards).

The *short-time* form of the solution to the current problem is given by Carslaw and Jaeger (1959; Eq. (4), p. 112) for the related X22B01T0 case. To obtain the solution of the X22B10T0 case here of interest, the space variable x of the X22B01T0 problem has to be replaced with $L - x$. In dimensionless form, the X22B10T0 exact analytical temperature solution $\tilde{T}_e$ is

$$\tilde{T}_e(\tilde{x}, \tilde{t}) = \underbrace{2\sqrt{\tilde{t}}\,\mathrm{ierfc}\left(\frac{\tilde{x}}{2\sqrt{\tilde{t}}}\right)}_{\tilde{T}_{\mathrm{X20B1T0}}(\tilde{x},\,\tilde{t})} + \underbrace{\sum_{m=1}^{\infty} 2\sqrt{\tilde{t}}\left[\mathrm{ierfc}\left(\frac{2m-\tilde{x}}{2\sqrt{\tilde{t}}}\right) + \mathrm{ierfc}\left(\frac{2m+\tilde{x}}{2\sqrt{\tilde{t}}}\right)\right]}_{\tilde{T}_m^{(S)}(\tilde{x},\,\tilde{t})} \tag{2.34a}$$

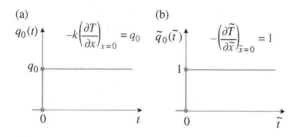

Figure 2.6 Nonhomogeneous constant boundary condition of the second kind at $x = 0$ for a slab initially at T_{in} and insulated on the back side $x = L$: (a) dimensional (X22B10T1); and (b) dimensionless (X22B10T0).

where (Cole *et al.* 2011; Eq. (E.14a), p. 501)

$$\text{ierfc}\left(\frac{\tilde{x}}{2\sqrt{\tilde{t}}}\right) = \frac{1}{\sqrt{\pi}}e^{-\frac{\tilde{x}^2}{4\tilde{t}}} - \left(\frac{\tilde{x}}{2\sqrt{\tilde{t}}}\right)\text{erfc}\left(\frac{\tilde{x}}{2\sqrt{\tilde{t}}}\right) \qquad (2.34b)$$

In particular, $\text{ierfc}(0) = 1/\sqrt{\pi}$.

The *large-time* form of the solution (Beck et al. 2008; Cole et al. 2011; Woodbury and Beck 2013) in dimensionless form is

$$\tilde{T}_e(\tilde{x}, \tilde{t}) = \underbrace{\tilde{t} + \left(\frac{\tilde{x}^2}{2} - \tilde{x} + \frac{1}{3}\right)}_{\tilde{T}_{qs}(\tilde{x},\,\tilde{t})} - \underbrace{\sum_{m=1}^{\infty} 2\frac{e^{-(m\pi)^2\tilde{t}}}{(m\pi)^2}\cos\left(m\pi\tilde{x}\right)}_{-\tilde{T}_m^{(L)}(\tilde{x},\,\tilde{t})}$$

$$(2.35)$$

where $m\pi$ is the mth eigenvalue, β_m, for the X22 case, while $\cos\left(m\pi\tilde{x}\right)$ is the corresponding mth eigenfunction $F_x(\beta_m\tilde{x})$. Also, the quasi-steady part is a linear function of time at any point of the plate, while the temperature distribution over the domain is quadratic.

2.3.8.1 Computational Analytical Solution

This solution is given in the following for a relative error $E \le 10^{-A}$, with $A = 2, 3, ..., 15$, for $0 \le \tilde{x} \le 1$ and four different time intervals. In detail,

- $0 \le \tilde{t} \le \tilde{t}_d$. Short-time form with $m_{max}^{(A)} = 0$:

$$\tilde{T}_c(\tilde{x}, \tilde{t}) = 2\sqrt{\tilde{t}}\,\text{ierfc}\left(\frac{\tilde{x}}{2\sqrt{\tilde{t}}}\right) \qquad (2.36)$$

where $\tilde{t}_d$ is the dimensionless deviation time defined through Eq. (2.15).

- $\tilde{t}_d < \tilde{t} \le \tilde{t}_\delta$. Short-time form with $m_{max}^{(A)} = 1$, when considering only the first part of $\tilde{T}_m^{(S)}(\tilde{x}, \tilde{t})$ defined by Eq. (2.34a):

$$\tilde{T}_c(\tilde{x}, \tilde{t}) = 2\sqrt{\tilde{t}}\,\text{ierfc}\left(\frac{\tilde{x}}{2\sqrt{\tilde{t}}}\right) + 2\sqrt{\tilde{t}}\,\text{ierfc}\left(\frac{2-\tilde{x}}{2\sqrt{\tilde{t}}}\right) \qquad (2.37)$$

where $\tilde{t}_\delta$ is the dimensionless second deviation time as defined through Eq. (2.17).

- $\tilde{t}_\delta < \tilde{t} < \tilde{t}_{qs}$. Large-time form with $m_{max}^{(A)} \ge 1$:

$$\tilde{T}_c(\tilde{x}, \tilde{t}) = \tilde{t} + \left(\frac{\tilde{x}^2}{2} - \tilde{x} + \frac{1}{3}\right) - 2\sum_{m=1}^{m_{max}^{(A)}} \frac{e^{-(m\pi)^2\tilde{t}}}{(m\pi)^2}\cos\left(m\pi\tilde{x}\right) \qquad (2.38a)$$

where, using Eqs. (2.19) and (2.20a),

$$\tilde{t}_{qs} = \frac{A\ln(10) - \ln(\pi^2/2)}{\pi^2}, \quad m_{max}^{(A)} = \text{ceil}\left(\sqrt{\frac{A\ln(10) - \ln(\pi^2/2)}{\pi^2\tilde{t}}}\right) \qquad (2.38b)$$

The maximum number of terms is required for $\tilde{t} \to \tilde{t}_\delta$ and, hence, for locations close to the heated (or cooled) surface as $\tilde{t}_\delta$ decreases when $\tilde{x} \to 0$. However, it is always less than or equal to 8 terms for an accuracy of $A = 10$ and less than or equal to 2 terms for $A = 2$ depending on the time of interest.

- $\tilde{t} \ge \tilde{t}_{qs}$. Large-time form with $m_{max}^{(A)} = 0$, that is, quasi-steady solution:

$$\tilde{T}_c(\tilde{x}, \tilde{t}) = \tilde{t} + \left(\frac{\tilde{x}^2}{2} - \tilde{x} + \frac{1}{3}\right) \qquad (2.39)$$

In the current case, but only for $A = 2$, the critical location $\tilde{x}_c$ defined in Section 2.3.5.5 is less than 1 and of about 0.45. This indicates that, for $\tilde{x} \in [0, \tilde{x}_c]$, $\tilde{t}_\delta \leq \tilde{t}_{qs}$. But for $\tilde{x} \in (\tilde{x}_c, 1]$ it occurs that $\tilde{t}_\delta > \tilde{t}_{qs}$. Therefore, for those locations and $A = 2$, Eq. (2.38a) vanishes and Eq. (2.39) applies also for $\tilde{t} > \tilde{t}_\delta$.

2.3.8.2 Computer Code and Plots

A MATLAB function called fdX22B10T0.m allows the dimensionless temperature to be computed for the current case. It is available for ease of use on the web site related to this book.

By using the fdX22B10T0.m function, temperature graphics can be obtained for any accuracy. They are shown in Figure 2.7a versus space and in Figure 2.7b as a function of time.

Both figures indicate that, for an interior point close to the back side of the flat plate, the response is slow, being both damped and lagged. For $\tilde{t} = 0.05$, as an example, $\tilde{T}(0, \tilde{t}) = 0.2523$, whereas $\tilde{T}(1, \tilde{t}) = 0.000269$, a factor of almost 1000 smaller. (These numerical values can be derived by using the fdX22B10T0.m function.) This factor increases as $\tilde{t}$ becomes smaller. On the other hand, for sufficiently large times (much greater than the quasi-steady time of $\tilde{t}_{qs} \approx 0.54$ for $A = 3$), the factor approaches unity. This can be proven by using the same procedure shown in Section 2.3.7.2. In this case, the temperature ratio is

$$\frac{\tilde{T}_c(0, \tilde{t})}{\tilde{T}_c(1, \tilde{t})} = \frac{\tilde{t} + 1/3}{\tilde{t} - 1/6} = 1 + \frac{1}{2\tilde{t} - 1/3} \tag{2.40}$$

which simplifies to $1 + 1/(2\tilde{t})$ for $\tilde{t} \gg 1$. When $\tilde{t} = \tilde{t}_{qs} \approx 0.54$, the ratio defined by Eq. (2.40) is of about 2.34. But when $\tilde{t} = 5$, the ratio is of about 1.1.

A comparison with the X12B10T0 case of Section 2.3.6 indicates that $\tilde{T}(0.25, \tilde{t}) = 0.0773$ vs. 0.4292 for $\tilde{t} = 0.05$ and $\tilde{x} = 0.25$ (about six times smaller) but the ratio with respect to the maximum temperature variation which occurs at the heated or cooled boundary is only 1 and ½ times smaller. In fact, it is found that $\tilde{T}(0.25, \tilde{t})/\tilde{T}(0, \tilde{t}) = 0.3064$ against $\tilde{T}(0.25, \tilde{t})/\tilde{T}(0, \tilde{t}) = 0.4292$ for $\tilde{t} = 0.05$ of the X12B10T0 case as there $\tilde{T}(0, \tilde{t}) = 1$. Similarly, when considering the insulated back side, $\tilde{T}(1, \tilde{t}) = 0.000269$ vs. 0.0031 of the X12B10T0 case for $\tilde{t} = 0.05$ (12 times smaller). However, $\tilde{T}(1, \tilde{t})/\tilde{T}(0, \tilde{t}) = 0.00107$ vs. $\tilde{T}(1, \tilde{t})/\tilde{T}(0, \tilde{t}) = 0.0031$ for $\tilde{t} = 0.05$ of the X12B10T0 case that indicates a factor of almost 3 vs. 12.

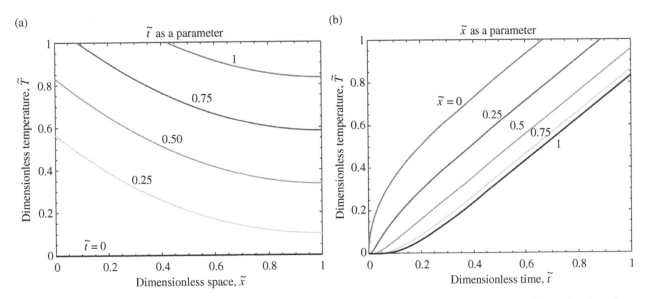

Figure 2.7 Dimensionless temperature of the X22B10T0 problem: (a) versus space with time as a parameter; (b) as a function of time for different locations.

(a)

(b)

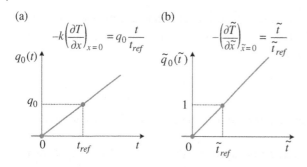

Figure 2.8 Nonhomogeneous linearly-in-time variable boundary condition of the second kind at $x = 0$ for a slab initially at T_{in} and insulated on the back side: (a) dimensional (X22B20T1); and (b) dimensionless (X22B20T0).

2.3.9 X22B20T0 Case

This problem is characterized by a linear-in-time variation of the surface heat flux at $x = 0$, according to Eq. (2.3) for $p = 1$ and shown in Figure 2.8a for $q_0 > 0$ (heating). When $q_0 < 0$, it results in a cooling of the flat plate. In dimensionless form, the boundary condition simplifies to $-\left(\partial \tilde{T}/\partial \tilde{x}\right)_{\tilde{x}=0} = \tilde{t}/\tilde{t}_{ref}$, with $\tilde{t} > 0$, where the nondimensional temperature is defined by Eq. (2.6) (see Figure 2.8b), and the initial condition becomes homogeneous. Also, the dimensionless heat flux is defined as $\tilde{q} = q/q_0$. When the time goes to infinity, the slab reaches neither a steady-state condition nor a quasi-steady condition. In fact, the heat flux is the same as the linear-in-time forcing function at $x = 0$, so determing an unsteady thermal behavior of the plate.

The *short-time* form of the solution to the current problem is given by Carslaw and Jaeger (1959; Eq. (7), p. 113) for the related X22B03T0 case, where "3" in B03 indicates a power time variation other than 1 according to Table A.3 of Appendix A. To obtain the solution of the current X22B20T0 problem, the space variable x of the X22B03T0 problem has simply to be replaced by $L - x$ and the time variation has to be considered linear. In dimensionless form, the X22B20T0 exact analytical temperature solution $\tilde{T}_e$ is

$$\tilde{T}_e\left(\tilde{x}, \tilde{t}, \tilde{t}_{ref}\right) = \underbrace{\frac{(4\tilde{t})^{3/2}}{\tilde{t}_{ref}} \mathrm{i}^3\mathrm{erfc}\left(\frac{\tilde{x}}{2\sqrt{\tilde{t}}}\right)}_{\tilde{T}_{X20B2T0}(\tilde{x}, \tilde{t})} + \sum_{m=1}^{\infty} \underbrace{\frac{(4\tilde{t})^{3/2}}{\tilde{t}_{ref}}\left[\mathrm{i}^3\mathrm{erfc}\left(\frac{2m-\tilde{x}}{2\sqrt{\tilde{t}}}\right) + \mathrm{i}^3\mathrm{erfc}\left(\frac{2m+\tilde{x}}{2\sqrt{\tilde{t}}}\right)\right]}_{\tilde{T}_m^{(S)}(\tilde{x}, \tilde{t})} \tag{2.41a}$$

where (Cole *et al.* 2011; Eq. (E.14c), p. 501)

$$\mathrm{i}^3\mathrm{erfc}\left(\frac{\tilde{x}}{2\sqrt{\tilde{t}}}\right) = \frac{1}{6\sqrt{\pi}}\left[1 + \left(\frac{\tilde{x}^2}{4\tilde{t}}\right)\right]e^{-\frac{\tilde{x}^2}{4\tilde{t}}} - \frac{1}{12}\left[3\left(\frac{\tilde{x}}{2\sqrt{\tilde{t}}}\right) + 2\left(\frac{\tilde{x}}{2\sqrt{\tilde{t}}}\right)^3\right]\mathrm{erfc}\left(\frac{\tilde{x}}{2\sqrt{\tilde{t}}}\right) \tag{2.41b}$$

In particular, $\mathrm{i}^3\mathrm{erfc}(0) = 1/(6\sqrt{\pi})$.

The *large-time* form of the solution (Özişik 1993; Beck et al. 2008; Woodbury and Beck 2013) in dimensionless form is

$$\tilde{T}_e\left(\tilde{x}, \tilde{t}, \tilde{t}_{ref}\right) = \underbrace{\frac{1}{\tilde{t}_{ref}}\left[\frac{\tilde{t}^2}{2} + \left(\frac{\tilde{x}^2}{2} - \tilde{x} + \frac{1}{3}\right)\tilde{t} + \left(\frac{\tilde{x}^4}{24} - \frac{\tilde{x}^3}{6} + \frac{\tilde{x}^2}{6} - \frac{1}{45}\right)\right]}_{\tilde{T}_{us}(\tilde{x}, \tilde{t})}$$
$$+ \sum_{m=1}^{\infty} \underbrace{\frac{2}{\tilde{t}_{ref}}\frac{e^{-(m\pi)^2\tilde{t}}}{(m\pi)^4}\cos(m\pi\tilde{x})}_{\tilde{T}_m^{(L)}(\tilde{x},\tilde{t})} \tag{2.42}$$

where the unsteady part is a quadratic function of time at any point of the plate, while its distribution over the domain is polynomial of the fourth order at any time.

2.3.9.1 Computational Analytical Solution

This solution is given in the following for a relative error $E \leq 10^{-A}$, with $A = 2, 3, ..., 15$, for $0 \leq \tilde{x} \leq 1$ and four different time intervals. In detail,

- $0 \leq \tilde{t} \leq \tilde{t}_d$. Short-time form with $m_{\max}^{(A)} = 0$:

$$\tilde{T}_c\left(\tilde{x}, \tilde{t}, \tilde{t}_{ref}\right) = \frac{(4\tilde{t})^{3/2}}{\tilde{t}_{ref}}\mathrm{i}^3\mathrm{erfc}\left(\frac{\tilde{x}}{2\sqrt{\tilde{t}}}\right) \tag{2.43}$$

where $\tilde{t}_d$ is the dimensionless deviation time defined through Eq. (2.15).

- $\tilde{t}_d < \tilde{t} \leq \tilde{t}_\delta$. Short-time form with $m_{max}^{(A)} = 1$, when considering only the first part of $\tilde{T}_m^{(S)}(\tilde{x}, \tilde{t})$ defined by Eq. (2.41a):

$$\tilde{T}_c(\tilde{x}, \tilde{t}, \tilde{t}_{ref}) = \frac{(4\tilde{t})^{3/2}}{\tilde{t}_{ref}} \, i^3 \text{erfc}\left(\frac{\tilde{x}}{2\sqrt{\tilde{t}}}\right) + \frac{(4\tilde{t})^{3/2}}{\tilde{t}_{ref}} \, i^3 \text{erfc}\left(\frac{2-\tilde{x}}{2\sqrt{\tilde{t}}}\right) \qquad (2.44)$$

where $\tilde{t}_\delta$ is the dimensionless second deviation time as defined through Eq. (2.17).

- $\tilde{t}_\delta < \tilde{t} < \tilde{t}_{us}$. Large-time form with $m_{max}^{(A)} \geq 1$:

$$\tilde{T}_c(\tilde{x}, \tilde{t}, \tilde{t}_{ref}) = \frac{1}{\tilde{t}_{ref}} \left[\frac{\tilde{t}^2}{2} + \left(\frac{\tilde{x}^2}{2} - \tilde{x} + \frac{1}{3}\right)\tilde{t} + \left(\frac{\tilde{x}^4}{24} - \frac{\tilde{x}^3}{6} + \frac{\tilde{x}^2}{6} - \frac{1}{45}\right) \right]$$
$$+ \frac{2}{\tilde{t}_{ref}} \sum_{m=1}^{m_{max}^{(A)}} \frac{e^{-(m\pi)^2 \tilde{t}}}{(m\pi)^4} \cos(m\pi\tilde{x}) \qquad (2.45a)$$

where, using Eqs. (2.19) and (2.20a),

$$\tilde{t}_{us} = \frac{A \ln(10) - \ln(\pi^4/2)}{\pi^2}, \quad m_{max}^{(A)} = \text{ceil}\left(\sqrt{\frac{A \ln(10) - \ln(\pi^4/2)}{\pi^2 \tilde{t}}}\right) \qquad (2.45b)$$

The maximum number of terms is required for $\tilde{t} \to \tilde{t}_\delta$ and, hence, for locations close to the heated (or cooled) surface as $\tilde{t}_\delta$ decreases when $\tilde{x} \to 0$. However, it is always less than or equal to 7 terms for an accuracy of $A = 10$ and less than or equal to 2 terms for $A = 3$ depending on the time of evaluation. For an accuracy of $A = 2$, Eq. (2.45a) vanishes, as shown afterwards.

- $\tilde{t} \geq \tilde{t}_{us}$. Large-time form with $\dot{m}_{max}^{(A)} = 0$, that is, unsteady solution:

$$\tilde{T}_c(\tilde{x}, \tilde{t}, \tilde{t}_{ref}) = \frac{1}{\tilde{t}_{ref}} \left[\frac{\tilde{t}^2}{2} + \left(\frac{\tilde{x}^2}{2} - \tilde{x} + \frac{1}{3}\right)\tilde{t} + \left(\frac{\tilde{x}^4}{24} - \frac{\tilde{x}^3}{6} + \frac{\tilde{x}^2}{6} - \frac{1}{45}\right) \right] \qquad (2.46)$$

In the current case, but only for $A = 2$, the critical location $\tilde{x}_c$ defined in Section 2.3.5.5 is negative. This indicates that, for any $\tilde{x} \in [0, 1]$, $\tilde{t}_\delta > \tilde{t}_{us}$. Therefore, when $A = 2$, Eq. (2.45a) vanishes for any location of the flat plate and Eq. (2.46) applies also for $\tilde{t} > \tilde{t}_\delta$.

2.3.9.2 Computer Code and Plots

A MATLAB function called fdX22B20T0.m allows the dimensionless temperature to be computed for the current case. It is available for ease of use on the web site related to this book.

By using the fdX22B20T0.m function, temperature graphics can be obtained for any accuracy. They are illustrated in Figure 2.9a versus space and in Figure 2.9b as a function of time, both for $\tilde{t}_{ref} = 1$.

Both figures indicate that, for an interior point close to the back side of the flat plate, the response is slow, being both damped and lagged. For $\tilde{t} = 0.05$, as an example, $\tilde{T}(0, \tilde{t}) = 0.00841$, whereas $\tilde{T}(1, \tilde{t}) = 0.0000017$, a factor of nearly 5000 smaller. (These numerical values can be derived by using the fdX22B20T0.m function.) This factor increases as $\tilde{t}$ becomes smaller but approaches unity for sufficiently large times. This can be demonstrated by using the same procedure shown in Sections 2.3.7.2 and 2.3.8.2.

A comparison with the companion X22B10T0 case, when $\tilde{t}_{ref} = 1$, indicates that $\tilde{T}(0, \tilde{t}) = 0.00841$ vs. 0.2523 for $\tilde{t} = 0.05$ and $\tilde{x} = 0$, that is, the dimensionless surface temperature is 30 times smaller. As regards the insulated back side, $\tilde{T}(1, \tilde{t}) = 0.0000017$ vs. 0.000269 of the X22B10T0 case for $\tilde{t} = 0.05$ (about 160 times smaller). However, the ratio with respect to the maximum temperature variation which occurs at the heated (or cooled) boundary reduces to 5. In fact, it is found that $\tilde{T}(1, \tilde{t})/\tilde{T}(0, \tilde{t}) = 0.0002$ vs. $\tilde{T}(1, \tilde{t})/\tilde{T}(0, \tilde{t}) = 0.001$ for $\tilde{t} = 0.05$ of the X22B10T0 problem.

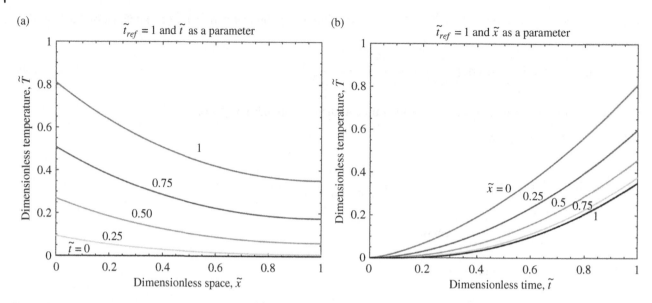

Figure 2.9 Dimensionless temperature of the X22B20T0 problem for $\tilde{t}_{ref} = 1$: (a) versus space with time as a parameter; (b) as a function of time for different locations.

2.4 Two-Dimensional Temperature Solutions

Consider a flat plate of length L and width W with temperature-independent properties, initially at a uniform temperature T_{in}. The rectangular body is subject to a nonhomogeneous boundary condition of the second kind at $x = 0$ ("active" boundary) that is constant in time and uniform in space but with a step change in heat flux at $y = W_0$, as depicted in Figure 2.10. Adiabatic conditions exist over the remainder of the boundaries ("inactive" boundaries).

Notwithstanding the initial temperature is uniform, and the two boundaries along y are both thermally insulated, the above problem is a 2D diffusion problem along x and y. In fact, the step change in heat flux along y at the boundary surface $x = 0$ causes 2D effects within the plate when the time increases (de Monte et al. 2008, 2012). Using the notation of Section 2.2 and Appendix A, this problem is described by X22B(y1pt1)0Y22B00T1, where "X22" indicates that the boundary conditions along x are both of the second kind; "B(y1pt1)0" denotes a piecewise-uniform along y ("y1p") and constant-in-time ("t1") heat flux at the $x = 0$ boundary, and a zero heat flux ("0") on the plate back side. Similarly, "Y22" indicates that the boundary conditions at $y = 0$ and $y = W$ are of the second kind; and "B00" denotes that both are homogeneous. Lastly, "T1" indicates a uniform initial temperature distribution.

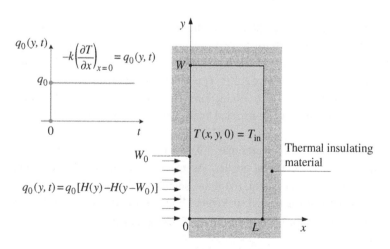

Figure 2.10 Schematic of the 2D transient X22B(y1pt1)0Y22B00T1 problem.

Mathematically, the problem is formulated as

$$\frac{\partial^2 T}{\partial x^2} + \frac{\partial^2 T}{\partial y^2} = \frac{1}{\alpha}\frac{\partial T}{\partial t} \quad (0 < x < L; 0 < y < W; t > 0) \tag{2.47a}$$

$$-k\left(\frac{\partial T}{\partial x}\right)_{x=0} = q_0(y, t) \quad (0 < y < W; t > 0) \tag{2.47b}$$

$$\left(\frac{\partial T}{\partial x}\right)_{x=L} = 0 \quad (0 < y < W; t > 0) \tag{2.47c}$$

$$\left(\frac{\partial T}{\partial y}\right)_{y=0} = 0 \quad (0 < x < L; t > 0) \tag{2.47d}$$

$$\left(\frac{\partial T}{\partial y}\right)_{y=W} = 0 \quad (0 < x < L; t > 0) \tag{2.47e}$$

$$T(x, y, t = 0) = T_{in} \quad (0 < x < L; 0 < y < W) \tag{2.47f}$$

where the surface heat flux $q_0(y, t)$ is time-independent and uniform in space but with a step change ($q_0 \rightarrow 0$) at $y = W_0$ as

$$q_0(y, t) = q_0[H(y) - H(y - W_0)] = \begin{cases} q_0, & 0 < y < W_0 \\ 0, & W_0 < y < W \end{cases} \quad (0 < y < W; t > 0) \tag{2.48}$$

In the above equation, q_0 is constant and $H(.)$ is the Heaviside unit step function (Cole *et al.* 2011; see p. 12). Also, $q_0 > 0$ indicates a heating of the body, while $q_0 < 0$ denotes a cooling.

Contrary to Section 2.3 where four 1D transient problems were considered, in the current section only the 2D transient problem described by Eqs. (2.47a)–(2.47f) and (2.48) is analyzed. Also, when $W_0 \rightarrow W$, the surface heat flux becomes uniform, and the problem reduces to the 1D case denoted by X22B10T1 and already analyzed in Section 2.3.8.

The exact analytical temperature solution (denoted by $\tilde{T}_e$) to the defining Eqs. (2.47a)–(2.47f) and (2.48) is unique according to the inherently well-posed nature of direct problems, though two different forms of it can exist, that is, the short- and the large-time forms, as occurs for 1D transient cases. However, before proceeding to analyze the two forms of the temperature solution, dimensionless variables are defined.

2.4.1 Dimensionless Variables

Dimensionless variables are used to reduce the number of dimensional variables involved in the current 2D transient, rectangular, heat conduction case. They are $x, y, t, L, W, W_0, \alpha, k, q_0$, and $(T - T_{in})$ (10 variables). As four basic dimensions, mass $[M]$, length $[L]$, time $[t]$, and temperature $[T]$, are involved in these variables, according to the Buckingham pi-theorem (Lienhard and Lienhard 2012; p. 151) a reduction of up to four may be hoped for in the number of the dimensional variables. Therefore, there is a total of six dimensionless quantities that may be defined as

$$\tilde{x} = \frac{x}{L}, \quad \tilde{y} = \frac{y}{L}, \quad \tilde{t} = \frac{\alpha t}{L^2}, \quad \tilde{W} = \frac{W}{L}, \quad \tilde{W}_0 = \frac{W_0}{L}, \quad \tilde{T} = \frac{T - T_{in}}{q_0 L / k} \tag{2.49}$$

where $\tilde{x} \in [0, 1]$, $\tilde{y} \in [0, \tilde{W}]$, and $\tilde{t} \geq 0$ is based on the plate length.

The definition of dimensionless temperature makes the initial condition Eq. (2.47f) homogeneous, that is, $\tilde{T}(\tilde{x}, \tilde{y}, \tilde{t} = 0) = 0$ with $\tilde{x} \in (0, 1)$ and $\tilde{y} \in (0, \tilde{W})$. Hence, the notation for the problem becomes X22B(y1pt1)0Y22B00T0.

2.4.2 Exact Analytical Solution

Just as in the case of the 1D solutions, the solution to the problem of Eqs. (2.47a)–(2.47f) and (2.48) has representations suitable for short and long times. Additionally, the appropriate form of solution that is most efficient for evaluation depends on the time variable and the coordinates of the point of evaluation relative to the heating source/sink near the corner of the domain.

A comprehensive presentation of the various forms of the solution is presented in Appendix B of this book. Appropriate results from that appendix are cited in the following section describing the Computational Analytical Solution for this 2D conduction problem.

2.4.3 Computational Analytical Solution

The exact analytical solution discussed in Sections B.1 and B.2 of Appendix B exhibits several single and double summations that require an infinite number of terms. As no calculator can account for an infinite number of terms, these summations need to be replaced with approximate ones consisting of a finite number of terms, say m_{max} along x and n_{max} along y, depending on the summation considered. In such a way, the exact solution reduces to the so-called "*computational*" (or approximate) analytical solution denoted by $\tilde{T}_c$, as defined in Section 2.3.5.

Therefore, the short-time form of the exact solution (Appendix B, Section B.1) becomes

- short-time form with semi-infinite solution along y, defined by Eqs. (B.1a) and (B.2):

$$\tilde{T}_c(\tilde{x}, \tilde{y}, \tilde{t}, S) = \tilde{T}_{X22Y20}(\tilde{x}, \tilde{y}, \tilde{t}, S_x) + \tilde{T}_{x,X22Y22}(\tilde{x}, \tilde{y}, \tilde{t}, S_{xy}) \tag{2.50}$$

where $S = S_x + S_{xy}$, being S_x and S_{xy} the total numbers of terms in either part of the computational solution. In detail, bearing also in mind Eq. (B.4),

$$\tilde{T}_{X22Y20}(\tilde{x}, \tilde{y}, \tilde{t}, S_x) = \underbrace{\frac{1 + \text{sign}(\tilde{W}_0 - \tilde{y})}{2} \tilde{T}_{X20B1T0}(\tilde{x}, \tilde{t}) - \tilde{T}_{2D}(\tilde{x}, \tilde{y}, \tilde{t})}_{\tilde{T}_{X20Y20}(\tilde{x}, \tilde{y}, \tilde{t})}$$
$$+ \sum_{m=1}^{m_{max}^{(1)}} \tilde{T}_{x,m}^{(S)}(\tilde{x}, \tilde{y}, \tilde{t}) \tag{2.51a}$$

$$\tilde{T}_{x,X22Y22}(\tilde{x}, \tilde{y}, \tilde{t}, S_{xy}) = \sum_{m=1}^{m_{max}^{(2)}} \sum_{n=1}^{n_{max}^{(1)}(m)} \tilde{T}_{x,mn}^{(S)}(\tilde{x}, \tilde{y}, \tilde{t}) \tag{2.51b}$$

where $S_x = m_{max}^{(1)}$ and $S_{xy} = \left[m_{max}^{(2)} \cdot n_{max}^{(1)}(m)\right]_{max}$, being S_{xy} the maximum number of terms used in the double summation;

- short-time form with semi-infinite solution along x, defined by Eqs. (B.1b) and (B.3):

$$\tilde{T}_c(\tilde{x}, \tilde{y}, \tilde{t}, S) = \tilde{T}_{X20Y22}(\tilde{x}, \tilde{y}, \tilde{t}, S_y) + \tilde{T}_{y,X22Y22}(\tilde{x}, \tilde{y}, \tilde{t}, S_{yx}) \tag{2.52}$$

where $S = S_y + S_{yx}$, being S_y and S_{yx} the total numbers of terms in either part of the computational solution. In detail, bearing also in mind Eq. (B.4),

$$\tilde{T}_{X20Y22}(\tilde{x}, \tilde{y}, \tilde{t}, S_y) = \underbrace{\frac{1 + \text{sign}(\tilde{W}_0 - \tilde{y})}{2} \tilde{T}_{X20B1T0}(\tilde{x}, \tilde{t}) - \tilde{T}_{2D}(\tilde{x}, \tilde{y}, \tilde{t})}_{\tilde{T}_{X20Y20}(\tilde{x}, \tilde{y}, \tilde{t})}$$
$$+ \sum_{n=1}^{n_{max}^{(1)}} \tilde{T}_{y,n}^{(S)}(\tilde{x}, \tilde{y}, \tilde{t}) \tag{2.53a}$$

$$\tilde{T}_{y,X22Y22}(\tilde{x}, \tilde{y}, \tilde{t}, S_{yx}) = \sum_{m=1}^{m_{max}^{(1)}} \sum_{n=1}^{n_{max}^{(2)}(m)} \tilde{T}_{y,mn}^{(S)}(\tilde{x}, \tilde{y}, \tilde{t}) \tag{2.53b}$$

where $S_y = n_{max}^{(1)}$ and $S_{yx} = \left[m_{max}^{(1)} \cdot n_{max}^{(2)}(m)\right]_{max}$, being S_{yx} the maximum number of terms used in the double summation.

As stated in Section B.1 and discussed herein, the use of the short-time forms of the solution is limited to only $\tilde{T}_{X20Y20}(\tilde{x}, \tilde{y}, \tilde{t})$ at $\tilde{x} = 0$ for which an exact analytical solution is available as given by Eq. (B.5a).

Similarly, the large-time form of the exact solution defined by Eqs. (B.6), (B.7), (B.11), (B.15), and (B.17) becomes

$$\tilde{T}_c(\tilde{x}, \tilde{y}, \tilde{t}, S) = \tilde{T}_{qs}(\tilde{x}, \tilde{y}, \tilde{t}, S_{qs}) + \tilde{T}_{ct}(\tilde{x}, \tilde{y}, \tilde{t}, S_{ct}) \tag{2.54}$$

where $S = S_{qs} + S_{ct}$, being S_{qs} and S_{ct} the total numbers of terms in the quasi-steady and complementary transient parts, respectively, of the computational solution.

The complementary transient part defined by Eq. (B.17) may concisely be taken as

$$\tilde{T}_{ct}(\tilde{x}, \tilde{y}, \tilde{t}, S_{ct}) = -\sum_{m=1}^{m_{\max}^{(1)}} \tilde{T}_{xt,m}^{(L)}(\tilde{x}, \tilde{t}) - \sum_{n=1}^{n_{\max}^{(1)}} \tilde{T}_{yt,n}^{(L)}(\tilde{y}, \tilde{t}) - \sum_{m=1}^{m_{\max}^{(2)}} \sum_{n=1}^{n_{\max}^{(2)}(m)} \tilde{T}_{xyt,mn}^{(L)}(\tilde{x}, \tilde{y}, \tilde{t}) \tag{2.55}$$

where $n_{\max}^{(2)}(m)$ depends on the counting integer m, and decreases as m increases; while $S_{ct} = m_{\max}^{(1)} + n_{\max}^{(1)} + \left[m_{\max}^{(2)} n_{\max}^{(2)}(m)\right]_{\max}$, being $\left[m_{\max}^{(2)} n_{\max}^{(2)}(m)\right]_{\max}$ the maximum number of terms used in the double summation (see Section 2.4.3.8).

As regards the quasi-steady part, there are two different forms that can affect the number of terms S_{qs}. They are

- the large-time form with eigenvalues in the homogeneous direction (y), defined by Eq. (B.7) along with Eq. (B.11), that concisely is given by

$$\tilde{T}_{qs,y}(\tilde{x}, \tilde{y}, \tilde{t}, S_{qs,y}) = \frac{\tilde{W}_0}{\tilde{W}}\tilde{t} + \frac{\tilde{W}_0}{\tilde{W}}\tilde{T}_x^{(L)}(\tilde{x}) + \sum_{n=1}^{n_{\max}^{(3)}} \tilde{T}_{xy,n}^{(L)}(\tilde{x}, \tilde{y}) + \sum_{n=1}^{n_{\max}^{(4)}} \tilde{T}_{xy,n}^{(L)}(2-\tilde{x}, \tilde{y}) \tag{2.56a}$$

where $S_{qs,y} = n_{\max}^{(3)} + n_{\max}^{(4)}$;

- the large-time form with eigenvalues in the nonhomogeneous direction (x), defined by Eq. (B.7) along with Eq. (B.15), that concisely may be taken as

$$\begin{aligned}
\tilde{T}_{qs,x}(\tilde{x}, \tilde{y}, \tilde{t}, S_{qs,x}) = {} & \frac{\tilde{W}_0}{\tilde{W}}\tilde{t} + \frac{1-\text{sign}(\tilde{y}-\tilde{W}_0)}{2}\tilde{T}_x^{(L)}(\tilde{x}) + \tilde{T}_y^{(L)}(\tilde{y}) \\
& + \text{sign}(\tilde{y}-\tilde{W}_0)\sum_{m=1}^{m_{\max}^{(3)}} \tilde{T}_{xy,m}^{(L)}\left(\tilde{x}, |\tilde{y}-\tilde{W}_0|\right) \\
& - \text{sign}(\tilde{y}-\tilde{W}_0)\sum_{m=1}^{m_{\max}^{(4)}} \tilde{T}_{xy,m}^{(L)}\left(\tilde{x}, 2\tilde{W}-|\tilde{y}-\tilde{W}_0|\right) \\
& - \sum_{m=1}^{m_{\max}^{(5)}} \tilde{T}_{xy,m}^{(L)}\left(\tilde{x}, \tilde{y}+\tilde{W}_0\right) + \sum_{m=1}^{m_{\max}^{(6)}} \tilde{T}_{xy,m}^{(L)}\left(\tilde{x}, 2\tilde{W}-\tilde{y}-\tilde{W}_0\right)
\end{aligned} \tag{2.56b}$$

where $S_{qs,x} = \sum_{j=3}^{6} m_{\max}^{(j)}$. Also, for $\tilde{y} = 0$, $m_{\max}^{(3)} = m_{\max}^{(5)}$, and $m_{\max}^{(4)} = m_{\max}^{(6)}$.

When using a computational solution in place of an exact one, the numerical value of temperature at location and time of interest is not the "exact" value because of the truncation error in the various summations. Therefore, as in Section 2.3.5.1 for 1D cases, it is important to define absolute and relative errors.

2.4.3.1 Absolute and Relative Errors

When dealing with a 2D computational solution, a relative error E can be defined normalizing the absolute error ε_a with respect to the maximum temperature variation (up to and including the time of interest) that, in the current case, always occurs at the corner point $(x, y) = (0, 0)$ of the heated or cooled surface. Therefore,

$$\begin{aligned}
E &= \frac{\varepsilon_a}{\max\{[T_c(0, 0, t, S) - T_{in}]\}} = \frac{|[T_e(x, y, t) - T_{in}] - [T_c(x, y, t, S) - T_{in}]|}{\max\{[T_c(0, 0, t, S) - T_{in}]\}} \\
&= \frac{|\tilde{T}_e(\tilde{x}, \tilde{y}, \tilde{t}) - \tilde{T}_c(\tilde{x}, \tilde{y}, \tilde{t}, S)|}{\max\{\tilde{T}_c(0, 0, \tilde{t}, S)\}}
\end{aligned} \tag{2.57}$$

where $E = E(\tilde{x}, \tilde{y}, \tilde{t}, S)$.

Thus, the exact numerical value of temperature (that remains unknown) falls into the range

$$\tilde{T}_e(\tilde{x}, \tilde{y}, \tilde{t}) = \tilde{T}_c(\tilde{x}, \tilde{y}, \tilde{t}, S) \pm \left[E \cdot \max\{\tilde{T}_c(0, 0, \tilde{t}, S)\}\right] \tag{2.58}$$

In the current case, as the rectangle of Figure 2.10 is subject to a time-independent boundary condition of the second kind, the dimensionless temperature at the corner point increases monotonically with time and, hence, it results in: $\max\left\{\tilde{T}_c(0, 0, \tilde{t}, S)\right\} = \tilde{T}_c(0, 0, \tilde{t}, S)$ for $\tilde{t} \in [0, \tilde{t}]$.

The relative error E depends on the total number S of terms used in the computational solution that in turn depends on the maximum numbers of terms, namely $m_{\max}$ along x and $n_{\max}$ along y, used in either summation of the computational solution, as shown by Eqs. (2.50)–(2.56b). Also, it depends on the location, the time of interest and the form of the solution (short-form or large-time form).

By setting $E = 10^{-A}$ in Eq. (2.57), where $A = 2, 3, ..., 15$ is the counting integer for the accuracy desired, and solving the resulting equation, the maximum number of terms $S^{(A)}$ to get an accuracy of one part in 10^A (with respect to the maximum temperature variation) when using the computational solution may in general be calculated as a function of time, location and accuracy. When dealing with $A = 15$, the computational analytical solution may be considered in every practical respect as the "*exact*" one.

2.4.3.2 One- and Two-Dimensional Deviation Times

The 1D semi-infinite X20B1T0 solution appearing in Eqs. (2.50) and (2.52) by means of Eqs. (2.51a) and (2.53a), respectively, can be the only significant part of the temperature solution at early times. In fact, for small values of the time less than the so-called "*1D deviation*" time denoted by $\tilde{t}_d$, the thermal disturbances due to the presence of the homogeneous boundary conditions at $\tilde{x} = 1$ and at $\tilde{y} = \tilde{W}$ are negligible (de Monte et al. 2008, 2012) and, hence, the body can be considered semi-infinite along x and y. In addition, at early times, less than the so-called "*2D deviation*" time denoted by $\tilde{t}_{d,xy}$, the 2D effects due to the sudden variation of heat flux ($q_0 \rightarrow 0$) along y at $y = W_0$ are also negligible (de Monte et al. 2008, 2012) and, hence, the thermal field can be considered not only semi-infinite but also 1D in the x-direction. Therefore, for short times, both the computational solutions Eqs. (2.50) and (2.52) reduce to

$$\tilde{T}_c(\tilde{x}, \tilde{y}, \tilde{t}) = \frac{1 + \text{sign}\left(\tilde{W}_0 - \tilde{y}\right)}{2}\, \tilde{T}_{\text{X20B1T0}}(\tilde{x}, \tilde{t}) = \begin{cases} 2\sqrt{\tilde{t}}\,\text{ierfc}\left(\dfrac{\tilde{x}}{2\sqrt{\tilde{t}}}\right), & \tilde{y} < \tilde{W}_0 \\[2mm] 0, & \tilde{y} > \tilde{W}_0 \\[2mm] \sqrt{\tilde{t}}\,\text{ierfc}\left(\dfrac{\tilde{x}}{2\sqrt{\tilde{t}}}\right), & \tilde{y} = \tilde{W}_0 \end{cases} \tag{2.59}$$

In such a case, the relative error E defined by Eq. (2.57) becomes

$$E = \frac{\left| \tilde{T}_e(\tilde{x}, \tilde{y}, \tilde{t}) - \dfrac{1 + \text{sign}\left(\tilde{W}_0 - \tilde{y}\right)}{2}\, \tilde{T}_{\text{X20B1T0}}(\tilde{x}, \tilde{t}) \right|}{2\sqrt{\tilde{t}/\pi}} \tag{2.60}$$

The above error will be less than or equal to 10^{-A} (with $A = 2, 3, ..., 15$) for $\tilde{t} \leq \min\left\{\tilde{t}_d, \tilde{t}_{d,xy}\right\}$. In detail, $\tilde{t}_d = \min\left\{\tilde{t}_{d,x}, \tilde{t}_{d,y}\right\}$ and $\tilde{t}_{d,x}, \tilde{t}_{d,y}$, and $\tilde{t}_{d,xy}$ are given by de Monte et al. (2008, 2012) and McMasters et al. (2018)

$$\tilde{t}_{d,x} = \frac{1}{10A}(2 - \tilde{x})^2 \quad (0 \leq \tilde{x} \leq 1) \tag{2.61a}$$

$$\tilde{t}_{d,y} = \frac{1}{10A}\left(2\tilde{W} - \tilde{W}_0 - \tilde{y}\right)^2 \quad (0 \leq \tilde{y} \leq \tilde{W}) \tag{2.61b}$$

$$\tilde{t}_{d,xy} = \frac{1}{10A}\left[\tilde{x}^2 + \left(\tilde{y} - \tilde{W}_0\right)^2\right] \quad (0 \leq \tilde{x} \leq 1; 0 \leq \tilde{y} \leq \tilde{W}) \tag{2.61c}$$

In detail, $\tilde{t}_{d,x}$ is the 1D deviation time accounting (at a level of $E = 10^{-A}$) for the thermal disturbance due to the inactive boundary at $\tilde{x} = 1$. Similarly, $\tilde{t}_{d,y}$ is the 1D deviation time accounting (at the same level of relative error) for the thermal disturbance due to the other inactive boundary at $\tilde{y} = \tilde{W}$. Note that Eqs. (2.61a)–(2.61c) are very conservative equations providing the smallest deviation times for any location within the rectangle.

Now, if $\tilde{t}_{d,xy}$ is less than $\tilde{t}_d = \min\left\{\tilde{t}_{d,x}, \tilde{t}_{d,y}\right\}$ at the point of interest, that is, the 2D deviation effects due to the heating edge rise up at that point before the thermal deviation effects caused by the inactive boundaries, the thermal field can be considered 2D semi-infinite for $\tilde{t}_{d,xy} < \tilde{t} \leq \tilde{t}_d$. Therefore, the computational solution becomes

$$\tilde{T}_c(\tilde{x}, \tilde{y}, \tilde{t}) = \tilde{T}_{X20Y20}(\tilde{x}, \tilde{y}, \tilde{t}) = \frac{1 + \text{sign}\left(\tilde{W}_0 - \tilde{y}\right)}{2} \tilde{T}_{X20B1T0}(\tilde{x}, \tilde{t}) - \tilde{T}_{2D}(\tilde{x}, \tilde{y}, \tilde{t}) \tag{2.62}$$

where the 2D part, $\tilde{T}_{2D}(\tilde{x}, \tilde{y}, \tilde{t})$, is given in an integral form by Eq. (B.4).

For the special case of $\tilde{x} = 0$, the algebraic expression defined by Eq. (B.5a) is applicable. Thus, Eq. (2.62) will be utilized only for $\tilde{x} = 0$. In such a case, the relative error E defined by Eq. (2.57) becomes

$$E = \frac{\left|\tilde{T}_e(0, \ \tilde{y}, \ \tilde{t}) - \tilde{T}_{X20Y20}(0, \tilde{y}, \tilde{t})\right|}{\tilde{T}_{X20Y20}(0, 0, \tilde{t})} \tag{2.63}$$

where $\tilde{T}_{X20Y20}(0, 0, \tilde{t})$ is defined by Eq. (B.5b). Also, this error is less than or equal to 10^{-A}, for $\tilde{t}_{d,xy} < \tilde{t} \le \tilde{t}_d$, provided $\tilde{t}_{d,xy}$ and $\tilde{t}_d = \min\{\tilde{t}_{d,x}, \tilde{t}_{d,y}\}$ are computed by using Eqs. (2.61a)–(2.61c). In other words, within this time range, the rectangle behaves at the surface location $(\tilde{x}, \tilde{y}) = (0, \tilde{y})$ as a 2D semi-in finite body with an accuracy of one part in 10^A with respect to the maximum temperature variation that, as stated before, occurs at the corner point (0,0) of the rectangular domain and at the time t of interest.

As an example, for a rectangle having an aspect ratio of $\tilde{W} = 2$ and a heated or cooled region of $\tilde{W}_0 = 1$, the 2D deviation time at the location $(\tilde{x}, \tilde{y}) = (0, 1/2)$ is $\tilde{t}_{d,xy} = 0.025/A$ and, hence, less than $\tilde{t}_d = \min\{\tilde{t}_{d,x} = 0.4/A, \tilde{t}_{d,y} = 0.25/A\} = 0.25/A$. This result indicates that Eq. (2.59) is applicable for $0 \le \tilde{t} \le \tilde{t}_{d,xy}$; while Eq. (2.62) that reduces to Eq. (B.5a) is applicable for $\tilde{t}_{d,xy} < \tilde{t} \le \tilde{t}_d$, both with a relative error less than or equal to one part in 10^A.

On the contrary, if $\tilde{t}_{d,xy}$ is greater than $\tilde{t}_d = \min\{\tilde{t}_{d,x}, \tilde{t}_{d,y}\}$ at the point of interest, that is, the thermal deviation effects caused by the inactive boundaries rise up at that point before the 2D deviation effects due to the heating edge, Eq. (2.59) can be used up to $\tilde{t} = \tilde{t}_d$ in place of the more complicate Eq. (2.62) not only at the surface boundary but at any location $(\tilde{x}, \tilde{y})$. As an example, for a rectangular domain with an aspect ratio of $\tilde{W} = 2$ and a heated or cooled region of $\tilde{W}_0 = 1$, the 2D deviation time at the location $(\tilde{x}, \tilde{y}) = (1, 0)$ is $\tilde{t}_{d,xy} = 0.2/A$ and, hence, greater than $\tilde{t}_d = \min\{\tilde{t}_{d,x} = 0.1/A, \tilde{t}_{d,y} = 0.9/A\} = 0.1/A$. This indicates that Eq. (2.59) is applicable for $0 \le \tilde{t} \le \tilde{t}_d$ at that location.

Example 2.1 A steel plate ($k = 40$ W/m/°C and $\alpha = 10^{-5}$ m^2/s) of thickness $L = 1$ cm and width $W = 5$ cm, initially at a uniform temperature of $T_{in} = 20$ °C, is in perfect contact over a region of $W_0 = 3$ cm with a heating element that can dissipate 4800 W/m. The remaining boundaries of the plate are kept insulated. Suddenly the heating starts. Calculate the temperature of the plate in the middle point of the heated surface region, 1 and 2 s after the start of heating.

Solution

The exact value of temperature at the surface point of interest $(x_P, y_P) = (0, 1.5$ cm$)$ and at time t is $T_e(x_P, y_P, t)$. Therefore,

$$T_e\left(x_P = 0, y_P = 1.5 \text{ cm}, t = 1 \text{ s}\right) = T_{in} + \frac{q_0 L}{k} \tilde{T}_e\left(\tilde{x}_P = 0, \tilde{y}_P = 3/2, \tilde{t} = 0.1\right) \tag{2.64a}$$

$$T_e\left(x_P = 0, y_P = 1.5 \text{ cm}, t = 2 \text{ s}\right) = T_{in} + \frac{q_0 L}{k} \tilde{T}_e\left(\tilde{x}_P = 0, \tilde{y}_P = 3/2, \tilde{t} = 0.2\right) \tag{2.64b}$$

where $q_0 = 160$ kW/m^2.

The dimensionless temperature can be calculated easily by using the computational or approximate Eqs. (2.59) and (2.62) if applicable, that is, if the two dimensionless times of interest, say 0.1 and 0.2, are less than the dimensionless deviation times at the location of interest $(\tilde{x}_P, \tilde{y}_P) = (0, 3/2)$. Bearing in mind that $\tilde{W} = 5$ and $\tilde{W}_0 = 3$, they are

$$\tilde{t}_{d,x}\left(\tilde{x}_P = 0\right) = 0.4/A \tag{2.65a}$$

$$\tilde{t}_{d,y}\left(\tilde{y}_P = 3/2\right) = 3.25/A \tag{2.65b}$$

$$\tilde{t}_{d,xy}\left(\tilde{x}_P = 0, \tilde{y}_P = 3/2\right) = 0.225/A \tag{2.65c}$$

If a relative error of 0.01 (or 1%) is chosen, that is, $A = 2$, it is found that: $\tilde{t}_{d,x}(0) = 0.2$, $\tilde{t}_{d,y}(3/2) = 1.625$, and $\tilde{t}_{d,xy}(0, 3/2) = 0.1125$. Therefore, at $(\tilde{x}_P, \tilde{y}_P) = (0, 3/2)$, $\tilde{t}_{d,xy}$ is less than $\tilde{t}_d = \min\{\tilde{t}_{d,x}, \tilde{t}_{d,y}\} = 0.2$. This result indicates that Eq. (2.59) is applicable for $0 \le \tilde{t} \le 0.1125$; while Eq. (2.62) that reduces to Eq. (B.5a) is applicable for $0.1125 < \tilde{t} \le 0.2$, both with a relative error less than 1%.

As at $\tilde{t} = 0.1$ the body behaves as a 1D semi-infinite body, using Eq. (2.59) yields

$$\tilde{T}_c(0, 3/2, 0.1) = 2\sqrt{\tilde{t}/\pi} = 2\sqrt{0.1/\pi} = 0.357 \tag{2.66a}$$

$$T_c(0, 1.5\,\text{cm}, 1\,\text{s}) = 20°\text{C} + \underbrace{\frac{q_0 L}{k}}_{= 40°\text{C}} \tilde{T}_c(0, 3/2, 0.1) \tag{2.66b}$$

$$= 20 + 40 \times 0.357 = 34.28°\text{C}$$

By applying Eq. (2.58) with $E = 10^{-A} = 10^{-2}$, the exact numerical value of temperature 1 s after the heating start is

$$\tilde{T}_e(0, 3/2, 0.1) = \tilde{T}_c(0, 0.1) \pm 0.01 \cdot \tilde{T}_c(0, 0.1) = 0.357 \pm 0.00357 \tag{2.67a}$$

$$T_e(0, 1.5\,\text{cm}, 1\,\text{s}) = 20 + 40 \times [0.357 \pm 0.00357] = 34.28 \pm 0.1428°\text{C} \tag{2.67b}$$

As at $\tilde{t} = 0.2$ the body behaves as a 2D semi-infinite body, using Eq. (2.62) and, hence, Eq. (B.5a) gives

$$\tilde{T}_c(0, 3/2, 0.2) = \underbrace{2\sqrt{\frac{0.2}{\pi}}}_{= 0.504626}$$

$$- \underbrace{\sqrt{\frac{0.2}{\pi}} \left[\text{erfc}\left(\frac{3}{4\sqrt{0.2}}\right) - \frac{3}{4\sqrt{0.2\pi}} E_1\left(\frac{90}{32}\right) + \text{erfc}\left(\frac{9}{4\sqrt{0.2}}\right) - \frac{9}{4\sqrt{0.2\pi}} E_1\left(\frac{810}{32}\right) \right]}_{= 0.000508} \tag{2.68a}$$

$$= 0.504118$$

$$T_c(0, 1.5\,\text{cm}, 2\,\text{s}) = 20°\text{C} + \underbrace{\frac{q_0 L}{k}}_{= 40°\text{C}} \tilde{T}_c(0, 3/2, 0.2) = 20 + 40 \times 0.504118 \tag{2.68b}$$

$$= 40.16°\text{C}$$

where the exponential integral may be computed using the MATLAB function expint(.). Also, note that the 2D effects due to the heating edge at $(\tilde{x}, \tilde{y}) = (0, \tilde{W}_0 = 3)$ in Eq. (2.68a) are 0.000508 and, hence, 1000 times smaller than 0.504626. Therefore, notwithstanding $\tilde{t} = 0.2$ is greater than $\tilde{t}_{d,xy} = 0.1125$, the body still behaves as a 1D semi-infinite one with a relative error of 0.1% (or 0.001). The reason is that Eqs. (2.61a)–(2.61c) for calculating the deviation times are very conservative equations.

By applying Eq. (2.58) with $E = 10^{-A} = 10^{-2}$, the exact numerical value of temperature 2 s after the heating start is

$$\tilde{T}_e(0, 3/2, 0.2) = \tilde{T}_c(0, 0.2) \pm 0.01 \cdot \tilde{T}_c(0, 0.2) = 0.504118 \pm 0.005041 \tag{2.69a}$$

$$T_e(0, 1.5\,\text{cm}, 1\,\text{s}) = 20 + 40 \times [0.504118 \pm 0.005041] = 40.16 \pm 0.2016°\text{C} \tag{2.69b}$$

2.4.3.3 Quasi-Steady Time

The quasi-steady solution appearing in Eq. (2.54) can be the only significant part of the complete temperature solution at large times. In fact, for large values of the time greater than the so-called "*quasi-steady*" time $\tilde{t}_{qs}$, the complementary transient part in Eq. (2.54) becomes negligible and, hence, the body has practically reached the quasi-steady condition. Therefore, for large times, the computational solution Eq. (2.54) reduces to

$$\tilde{T}_c(\tilde{x}, \tilde{y}, \tilde{t}) = \tilde{T}_{qs}(\tilde{x}, \tilde{y}, \tilde{t}, S_{qs} \to \infty) \tag{2.70}$$

where $\tilde{T}_{qs}(\tilde{x}, \tilde{y}, \tilde{t}, S_{qs} \to \infty)$ is the exact quasi-steady part of the solution, that is, $\tilde{T}_{e,qs}(\tilde{x}, \tilde{y}, \tilde{t})$.

In such a case, the relative error E defined by Eq. (2.57) becomes

$$E = \frac{|\tilde{T}_e(\tilde{x}, \tilde{y}, \tilde{t}) - \tilde{T}_{e,qs}(\tilde{x}, \tilde{y}, \tilde{t})|}{\tilde{T}_{e,qs}(0, 0, \tilde{t})} \tag{2.71}$$

This error will be less than or equal to $10^{-A}/2$, with $A = 2, 3, ..., 15$, for $\tilde{t} \geq \tilde{t}_{qs}$, where the factor of 2 indicates the number of parts appearing in the solution Eq. (2.54). Also, the dimensionless quasi-steady time $\tilde{t}_{qs}$ may be taken as the maximum among three different quasi-steady times, that is, $\tilde{t}_{qs} = \max\{\tilde{t}_{qs,m}, \tilde{t}_{qs,n}, \tilde{t}_{qs,mn}\}$:

1) $\tilde{t}_{qs,m}$ is the time to make negligible the *m*-single summation in Eq. (2.55) at a level of $E = (10^{-A}/2)/3 = 10^{-A}/6$, where the factor of 3 denotes the number of summations appearing in Eq. (2.55). It is given by

$$\tilde{t}_{qs,m} = \frac{A \ln(10) - \ln(\pi^2/12)}{\pi^2} \tag{2.72a}$$

2) $\tilde{t}_{qs,n}$ is the time to make negligible the *n*-single summation in Eq. (2.55) at a level of $E = 10^{-A}/6$. It is

$$\tilde{t}_{qs,n} = \tilde{W}^2 \frac{A \ln(10) - \ln(\pi^2/12) - \ln\left(\pi/\tilde{W}^2\right)}{\pi^2} \tag{2.72b}$$

3) $\tilde{t}_{qs,mn}$ is the time to make negligible the *m* by *n*-double summation appearing in Eq. (2.55) at a level of $E = 10^{-A}/6$. It is

$$\tilde{t}_{qs,mn} = \tilde{W}^2 \frac{A \ln(10) - \ln(\pi^2/12) - \ln\left(\pi/\tilde{W}^2\right) - \ln\left[\left(1 + \tilde{W}^2\right)/2\right]}{\pi^2\left(1 + \tilde{W}^2\right)} \tag{2.72c}$$

As an example, for $\tilde{W} = 1$, it is found that

$$\tilde{t}_{qs,m} = \tilde{t}_{qs,n} + \frac{\ln(\pi)}{\pi^2} = 2\tilde{t}_{qs,mn} + \frac{\ln(\pi)}{\pi^2} \tag{2.73}$$

that is, $\tilde{t}_{qs,m} > \tilde{t}_{qs,n} > \tilde{t}_{qs,mn}$. Hence, the largest time to reach the quasi-steady temperature at a level of one part in 10^A is $\tilde{t}_{qs} = \tilde{t}_{qs,m}$. However, for $\tilde{W} = 2$, it results in

$$\tilde{t}_{qs,m} = \frac{\tilde{t}_{qs,n}}{4} + \frac{\ln(\pi/4)}{\pi^2} = \frac{5}{4}\tilde{t}_{qs,mn} + \frac{\ln(5\pi/8)}{\pi^2} \tag{2.74}$$

In this case, the time to reach the quasi-steady temperature is $\tilde{t}_{qs} = \tilde{t}_{qs,n}$ that is the largest.

The quasi-steady time may be defined as the time that it takes for the temperature at a point of a finite body heated or cooled at a boundary to reach the quasi-steady solution with an error of one part in $2 \cdot 10^A$ with respect to the maximum temperature variation that occurs at the corner point of the heated or cooled surface and at the time of interest. Eqs. (2.72a)–(2.72c) are conservative equations valid for any location within the rectangular domain. Also, these equations are modified expressions of the original equations given by Woodbury et al. (2017).

By using Eq. (2.71) with $E = 10^{-A}/2$, the exact numerical value of temperature for $\tilde{t} \geq \tilde{t}_{qs}$ is

$$\tilde{T}_e(\tilde{x}, \tilde{y}, \tilde{t}) = \tilde{T}_{e,qs}(\tilde{x}, \tilde{y}, \tilde{t}) \pm \frac{10^{-A}}{2}\tilde{T}_{e,qs}(0, 0, \tilde{t}) \tag{2.75}$$

where, however, the exact quasi-steady part of the solution discussed in Sections B.2.1 and B.2.2 exhibits various single summations that require an infinite number of terms.

As no calculator can account for an infinite number of terms, these summations need to be replaced with approximate ones consisting of a finite number of terms. Therefore, for $\tilde{t} \geq \tilde{t}_{qs}$, with $\tilde{t}_{qs} = \max\{\tilde{t}_{qs,m}, \tilde{t}_{qs,n}, \tilde{t}_{qs,mn}\}$, the computational solution Eq. (2.70) reduces to

$$\tilde{T}_c(\tilde{x}, \tilde{y}, \tilde{t}) = \tilde{T}_{qs}(\tilde{x}, \tilde{y}, \tilde{t}, S_{qs}) = \begin{cases} \tilde{T}_{qs,y}(\tilde{x}, \tilde{y}, \tilde{t}, S_{qs,y}) \\ \tilde{T}_{qs,x}(\tilde{x}, \tilde{y}, \tilde{t}, S_{qs,x}) \end{cases} \tag{2.76}$$

where $\tilde{T}_{qs,y}(\tilde{x}, \tilde{y}, \tilde{t}, S_{qs,y})$ and $\tilde{T}_{qs,x}(\tilde{x}, \tilde{y}, \tilde{t}, S_{qs,x})$ are defined by Eqs. (2.56a) and (2.56b), respectively.

In such a case, the relative error E defined by Eq. (2.57) becomes

$$E = \frac{\left|\tilde{T}_{e,qs}(\tilde{x}, \tilde{y}, \tilde{t}) - \tilde{T}_{qs}(\tilde{x}, \tilde{y}, \tilde{t}, S_{qs})\right|}{\tilde{T}_{qs}(0, 0, \tilde{t}, S_{qs})} \tag{2.77}$$

The total number of terms $S_{qs}^{(A)}$ to be used in the single summations to make the above error less than or equal to $10^{-A}/2$ will be evaluated in next sections.

Before proceeding to evaluate it, by using Eq. (2.77) with $E = 10^{-A}/2$, the exact quasi-steady part is

$$\tilde{T}_{e,qs}(\tilde{x}, \tilde{y}, \tilde{t}) = \tilde{T}_{qs}\left(\tilde{x}, \tilde{y}, \tilde{t}, S_{qs}^{(A)}\right) \pm \frac{10^{-A}}{2} \tilde{T}_{qs}\left(0, 0, \tilde{t}, S_{qs}^{(A)}\right) \tag{2.78}$$

Substituting it in Eq. (2.75) yields the exact numerical value of temperature for $\tilde{t} \geq \tilde{t}_{qs}$

$$\tilde{T}_{e}(\tilde{x}, \tilde{y}, \tilde{t}) = \tilde{T}_{qs}\left(\tilde{x}, \tilde{y}, \tilde{t}, S_{qs}^{(A)}\right) \pm 10^{-A}\tilde{T}_{qs}\left(0, 0, \tilde{t}, S_{qs}^{(A)}\right) \tag{2.79}$$

where, during the derivation, the factor of $(1 + 10^{-A}/4)$ was approximated to 1.

2.4.3.4 Number of Terms in the Quasi-Steady Solution with Eigenvalues in the Homogeneous Direction

The total number of terms $S_{qs,y}^{(A)}$ in the quasi-steady part of the solution with eigenvalues along y that appears in Eq. (2.76) through Eq. (2.56a) is given by $S_{qs,y}^{(A)} = n_{\max}^{(3)} + n_{\max}^{(4)}$. In detail,

- $n_{\max}^{(3)}$ is the maximum number of terms to make negligible the "tail" of the former n-summation in Eq. (2.56a) at a level of $E = (10^{-A}/2)/2 = 10^{-A}/4$, where the factor of 2 denotes the number of summations appearing in Eq. (2.56a). It is given by

$$n_{\max}^{(3)} = \begin{cases} \text{ceil}\left(\dfrac{A\ln(10) - \ln\left[\pi^2\left(1 - e^{-2\pi/\tilde{W}}\right)/(8\tilde{W})\right]}{\pi\tilde{x}/\tilde{W}}\right), & 0 < \tilde{x} \leq 1 \\[4mm] \text{ceil}\left(\dfrac{1}{\pi}\sqrt{\dfrac{8\tilde{W}\cdot 10^A}{1 - e^{-2\pi/\tilde{W}}}}\right), & \tilde{x} = 0 \end{cases} \tag{2.80a}$$

Therefore, for $\tilde{x}$ close to the active side of the rectangular domain, the maximum number of terms can be very large for a high accuracy. On the contrary, for $\tilde{x} = 1$, the number of terms $n_{\max}^{(3)}$ will be the lowest.

- $n_{\max}^{(4)}$ is the maximum number of terms to make negligible the "tail" of the latter n-summation in Eq. (2.56a) at a level of $E = 10^{-A}/4$. It is given by

$$n_{\max}^{(4)} = \text{ceil}\left(\dfrac{A\ln(10) - \ln\left[\pi^2\left(1 - e^{-2\pi/\tilde{W}}\right)/(8\tilde{W})\right]}{\pi(2 - \tilde{x})/\tilde{W}}\right), \quad 0 \leq \tilde{x} \leq 1 \tag{2.80b}$$

In this case, for $\tilde{x} = 0$, the maximum number of terms will be the lowest; while, for $\tilde{x} = 1$, the number of terms $n_{\max}^{(4)}$ will be the highest. Also, $n_{\max}^{(4)} < n_{\max}^{(3)}$ for any location $\tilde{x}$ except for $\tilde{x} = 1$ where $n_{\max}^{(4)} = n_{\max}^{(3)}$. Therefore, the latter n-summation in Eq. (2.56a) converges in general faster than the former.

Equations (2.80a) and (2.80b) are valid for any $0 \leq \tilde{y} \leq \tilde{W}$. However, for $\tilde{y} = \tilde{W}/2$, the eigenfunction along y becomes $F_{y,n} = \cos(n\pi/2)$ and, hence, vanishes for odd integers ($n = 1, 3, \ldots$). For this special case, the number of terms is halved. Another special case is for $\tilde{y} = \tilde{W}_0 = \tilde{W}/2$, as shown in Section B.2. In such a case, in fact, $n_{\max}^{(3)} = n_{\max}^{(4)} = 0$ and $S_{qs,y}^{(A)} = 0$.

2.4.3.5 Number of Terms in the Quasi-Steady Solution with Eigenvalues in the Nonhomogeneous Direction

The total number of terms $S_{qs,x}^{(A)}$ in the quasi-steady part of the solution with eigenvalues along x that appears in Eq. (2.76) through Eq. (2.56b) is given by $S_{qs,x}^{(A)} = \sum_{j=3}^{6} m_{\max}^{(j)}$. In detail,

- $m_{\max}^{(3)}$ is the maximum number of terms to make negligible the "tail" of the first m-summation in Eq. (2.56b) at a level of $E = (10^{-A}/2)/4 = 10^{-A}/8$, where the factor of 4 denotes the number of summations appearing in Eq. (2.56b). It is given by

$$m_{\max}^{(3)} = \text{ceil}\left(\frac{A\ln(10) - \ln\left[\pi^2\left(1 - e^{-2\pi\tilde{W}}\right)/8\right]}{\pi\left|\tilde{y} - \tilde{W}_0\right|}\right) \tag{2.81a}$$

where $0 \le \tilde{y} < \tilde{W}_0 \cup \tilde{W}_0 < \tilde{y} \le \tilde{W}$; while for $\tilde{y} = \tilde{W}_0$ the related summation in Eq. (2.56b) vanishes;

- $m_{\max}^{(4)}$ is the maximum number of terms to make negligible the "tail" of the second m-summation in Eq. (2.56b) at a level of $E = 10^{-A}/8$. It is given by

$$m_{\max}^{(4)} = \text{ceil}\left(\frac{A\ln(10) - \ln\left[\pi^2\left(1 - e^{-2\pi\tilde{W}}\right)/8\right]}{\pi\left(2\tilde{W} - \left|\tilde{y} - \tilde{W}_0\right|\right)}\right) \tag{2.81b}$$

where $0 \le \tilde{y} < \tilde{W}_0 \cup \tilde{W}_0 < \tilde{y} \le \tilde{W}$; while for $\tilde{y} = \tilde{W}_0$ the related summation in Eq. (2.56b) vanishes;

- $m_{\max}^{(5)}$ is the maximum number of terms to make negligible the "tail" of the third m-summation in Eq. (2.56b) at a level of $E = 10^{-A}/8$. It is given by

$$m_{\max}^{(5)} = \text{ceil}\left(\frac{A\ln(10) - \ln\left[\pi^2\left(1 - e^{-2\pi\tilde{W}}\right)/8\right]}{\pi\left(\tilde{y} + \tilde{W}_0\right)}\right) \tag{2.81c}$$

where $0 \le \tilde{y} \le \tilde{W}$; however, for $\tilde{y} = \tilde{W}_0$, the above equation can be made less conservative replacing the integer 8 with 4;

- $m_{\max}^{(6)}$ is the maximum number of terms to make negligible the "tail" of the fourth m-summation in Eq. (2.56b) at a level of $E = 10^{-A}/8$. It is given by

$$m_{\max}^{(6)} = \text{ceil}\left(\frac{A\ln(10) - \ln\left[\pi^2\left(1 - e^{-2\pi\tilde{W}}\right)/8\right]}{\pi\left[2\tilde{W} - \left(\tilde{y} + \tilde{W}_0\right)\right]}\right) \tag{2.81d}$$

where $0 \le \tilde{y} \le \tilde{W}$; however, for $\tilde{y} = \tilde{W}_0$, it is convenient to replace 8 with 4 in the above equation.

Equations (2.81a)–(2.81d) are modified expressions of the original equations given by Woodbury et al. (2017) and are conservative equations valid for any $0 \le \tilde{x} \le 1$. However, for $\tilde{x} = 1/2$, the eigenfunction along x becomes $F_{x,m} = \cos(m\pi/2)$ and, hence, vanishes for odd integers ($m = 1, 3, \ldots$). For this special case, the number of terms is halved. Another special case is for $\tilde{y} = \tilde{W}_0 = \tilde{W}/2$, as shown in Section B.2. In such a case, in fact, $m_{\max}^{(j)} = 0$ (for $j = 3, 4, 5, 6$) and $S_{qs,x}^{(A)} = 0$.

A comparison between the total number of terms required from the two different expressions of the quasi-steady solution is shown in Table 2.1.

At the middle point of the active surface, that is, $(\tilde{x}, \tilde{y}) = (0, \tilde{W}_0/2)$, the non-standard quasi-steady solution is very efficient computationally. In fact, for an accuracy of one part in $2 \cdot 10^5$, it requires 21 terms vs. 416 of the standard quasi-steady solution. Also, 38 terms are sufficient to get an error of $E = 10^{-10}/2$ in place of 130176 that is a very large number. On the contrary, at the back side and in correspondence of the heating or cooling edge, say $(\tilde{x}, \tilde{y}) = (1, 99\tilde{W}_0/100)$, the standard quasi-steady solution is more efficient computationally. In fact, for an accuracy of one part in $2 \cdot 10^5$, it requires 16 terms vs. 387 of the non-standard one. In addition, for $E = 10^{-10}/2$, 30 terms are required against 759. Lastly, at the crucial surface point close to the heating/cooling edge, say $(\tilde{x}, \tilde{y}) = (0, 99\tilde{W}_0/100)$, there exists a trade-off. In fact, it is more efficient to use the standard solution for low accuracies as well as it is more convenient to utilize the non-standard one for high accuracies.

2.4.3.6 Deviation Distance Along x

When a quasi-steady condition is reached (for $\tilde{t} \ge \tilde{t}_{qs}$), the 1D solution along x appearing in the complete quasi-steady solution Eq. (2.56a) can be the only significant part of the same solution at large distances from the active boundary. In fact, for large values of the location $\tilde{x}$ greater than the so-called "*deviation*" distance $\tilde{d}_{d,x} = d_{d,x}/L$ along x, the effect of the heating/cooling edge that makes the temperature 2D becomes negligible. From an analytical viewpoint, it means that the former of

Table 2.1 Comparison between the total number of terms required from the two different expressions of the quasi-steady part of the computational solution for different accuracies (to get an error of $E = 10^{-A}/2$) and at three different locations with $\tilde{W} = 2$ and $\tilde{W}_0 = 1$.

$(\tilde{x}, \tilde{y}) = (0, \tilde{W}_0/2)$							
A	$n_{max}^{(3)}$	$n_{max}^{(4)}$	$S_{qs,y}^{(A)}$	$m_{max}^{(3)}$	$m_{max}^{(4)}$	$m_{max}^{(5)}$	$m_{max}^{(6)}$	$S_{qs,x}^{(A)}$
2	14	2	**16**	4	1	2	4	**11**
3	42	3	**45**	5	1	2	5	**13**
5	412	4	**416**	8	2	3	8	**21**
10	130168	8	**130176**	15	3	5	15	**38**

$(\tilde{x}, \tilde{y}) = (1, 99\tilde{W}_0/100)$							
A	$n_{max}^{(3)}$	$n_{max}^{(4)}$	$S_{qs,y}^{(A)}$	$m_{max}^{(3)}$	$m_{max}^{(4)}$	$m_{max}^{(5)}$	$m_{max}^{(6)}$	$S_{qs,x}^{(A)}$
2	4	4	**8**	162	1	1	1	**165**
3	5	5	**10**	236	1	2	2	**241**
5	8	8	**16**	382	1	2	2	**387**
10	15	15	**30**	749	2	4	4	**759**

$(\tilde{x}, \tilde{y}) = (0, 99\tilde{W}_0/100)$							
A	$n_{max}^{(3)}$	$n_{max}^{(4)}$	$S_{qs,y}^{(A)}$	$m_{max}^{(3)}$	$m_{max}^{(4)}$	$m_{max}^{(5)}$	$m_{max}^{(6)}$	$S_{qs,x}^{(A)}$
2	14	2	**16**	162	1	1	1	**165**
3	42	3	**45**	236	1	2	2	**241**
5	412	4	**416**	382	1	2	2	**387**
10	130168	8	**130176**	749	2	4	4	**759**

the two n-summations in Eq. (2.56a) becomes negligible, that is, the maximum number of terms $n_{max}^{(3)} = 0$ (see Section 2.4.3.4). As it is the slower between the two summations, $n_{max}^{(4)} = 0$ too and, hence, $S_{qs,y}^{(A)} = 0$.

In such a case, the body behaves as a 1D one provided the above dimensionless deviation distance be less than or equal to 1 (back side of the plate). This implies that the former of the two expressions of Eq. (2.80a) is applicable only for $0 < \tilde{x} < \tilde{d}_{d,x}$ as, for $\tilde{d}_{d,x} \leq \tilde{x} \leq 1$, it has to be replaced with $n_{max}^{(3)} = 0$. As regards the companion expression defined by Eq. (2.80b), it is no longer valid for any location $0 \leq \tilde{x} \leq 1$ when $\tilde{d}_{d,x} \leq 1$ and, hence, it has to be replaced with $n_{max}^{(4)} = 0$.

Thus, the computational solution Eq. (2.56a) reduces to

$$\tilde{T}_c(\tilde{x}, \tilde{y}, \tilde{t}) = \tilde{T}_{qs,y}\left(\tilde{x}, \tilde{y}, \tilde{t}, S_{qs,y}^{(A)} = 0\right) = \frac{\tilde{W}_0}{\tilde{W}}\tilde{t} + \frac{\tilde{W}_0}{\tilde{W}}\tilde{T}_x^{(L)}(\tilde{x}) \tag{2.82}$$

where $\tilde{T}_x^{(L)}(\tilde{x})$ is defined in Eqs. (B.7) and (B.8).

When dealing with the above equation, the relative error E defined by Eq. (2.77) becomes

$$E = \frac{\left|\tilde{T}_{e,qs}(\tilde{x}, \tilde{y}, \tilde{t}) - \left(\tilde{W}_0/\tilde{W}\right)\left[\tilde{t} + \tilde{T}_x^{(L)}(\tilde{x})\right]\right|}{\left(\tilde{W}_0/\tilde{W}\right)\left(\tilde{t} + 1/3\right)} \tag{2.83}$$

This error will be less than or equal to $10^{-A}/2$ for $\tilde{d}_{d,x} \leq \tilde{x} \leq 1$, where the dimensionless deviation distance $\tilde{d}_{d,x}$ is the minimum distance from the active surface to make negligible the former n-summation in Eq. (2.56a) at a level of $E = (10^{-A}/2)/2 = 10^{-A}/4$, where the factor of 2 denotes the number of summations appearing in Eq. (2.56a). It is given by

$$\tilde{d}_{d,x} = \frac{A\ln(10) + \ln\left[\sin\left(\pi\tilde{W}_0/\tilde{W}\right)\right] - \ln\left[\pi^2\left(1 - e^{-2\pi/\tilde{W}}\right)/(8\tilde{W})\right]}{\pi/\tilde{W}} \tag{2.84}$$

So that $\tilde{d}_{d,x} \leq 1$, the aspect ratio $\tilde{W}$ must in general be less than 1 and $\tilde{W}_0/\tilde{W} \neq 1/2$. In fact, for $\tilde{W}_0/\tilde{W} = 1/2$, the former logarithm is zero and the deviation distance can be maximum and superior to 1. For $0 < \tilde{W}_0/\tilde{W} < 1/2 \cup 1/2 < \tilde{W}_0/\tilde{W} < 1$, the former logarithm is always negative and, hence, the distance is smaller and can also be less than 1.

The deviation distance along x may be defined as the distance measured from the heated or cooled surface of a 2D rectangular body in the x-direction where the quasi-steady temperature becomes 1D with an error of one part in $2 \cdot 10^A$ with respect to the maximum temperature variation (that occurs at the corner point of the active surface and at the time of interest). In other words, for $\tilde{d}_{d,x} \leq \tilde{x} \leq 1$, the location is affected by the heating/cooling edge (that makes the quasi-steady temperature 2D) at a level less than or equal to $10^{-A}/2$ and, hence, negligible.

Also, by using Eq. (2.83), the exact quasi-steady part for $\tilde{d}_{d,x} \leq \tilde{x} \leq 1$ becomes

$$\tilde{T}_{e,qs}(\tilde{x}, \tilde{y}, \tilde{t}) = \frac{\tilde{W}_0}{\tilde{W}} \left[\tilde{t} + \tilde{T}_x^{(L)}(\tilde{x}) \pm \frac{10^{-A}}{2} \left(\tilde{t} + \frac{1}{3} \right) \right] \tag{2.85}$$

Substituting it in Eq. (2.75) yields the exact numerical value of temperature for $\tilde{t} \geq \tilde{t}_{qs}$ and $\tilde{d}_{d,x} \leq \tilde{x} \leq 1$ as

$$\tilde{T}_e(\tilde{x}, \tilde{y}, \tilde{t}) = \frac{\tilde{W}_0}{\tilde{W}} \left[\tilde{t} + \tilde{T}_x^{(L)}(\tilde{x}) \pm 10^{-A}(\tilde{t} + 1/3) \right] \tag{2.86}$$

As an example, for $\tilde{W} = 1/2$, $\tilde{W}_0/\tilde{W} = 1/2$ and $A = 2$, $\tilde{d}_{d,x} \approx 0.48$. Therefore, for $0.48 \leq \tilde{x} \leq 1$ and $\tilde{t} \geq \tilde{t}_{qs}$, Eq. (2.82) is applicable with a relative error less than $10^{-A} = 0.01$ (1%). If a higher accuracy is desired, for example, $A = 4$ (0.01%), the deviation distance is greater than 1 (i.e. $\tilde{d}_{d,x} \approx 1.22$) and Eq. (2.82) cannot be applied. Also, for the former case of $A = 2$, if $\tilde{W}_0/\tilde{W} = 9/10$ (hence, close to a uniform heating or cooling), $\tilde{d}_{d,x} \approx 0.30$.

2.4.3.7 Deviation Distance Along y

When a quasi-steady condition is reached (for $\tilde{t} \geq \tilde{t}_{qs}$), the algebraic terms appearing in the complete quasi-steady solution Eq. (2.56b) can be the only significant part of the same solution at large distances along y from the heating/cooling edge located at $\tilde{y} = \tilde{W}_0$. In fact, for large distances greater than the so-called "*deviation*" distance $\tilde{d}_{d,y} = d_{d,y}/L$ along y, the effect of the above edge on the first of the four m-summations in Eq. (2.56b) becomes negligible, that is, the maximum number of terms $m_{\max}^{(3)} = 0$ (see Section 2.4.3.5). As it is the slowest among the four summations, $m_{\max}^{(j)} = 0$ for $j = 4, 5, 6$ too and, hence, $S_{qs,x}^{(A)} = 0$.

In such a case, Eqs. (2.81a)–(2.81d) become applicable only for $|\tilde{y} - \tilde{W}_0| < \tilde{d}_{d,y}$; while, for $0 \leq \tilde{y} \leq \tilde{W}_0 - \tilde{d}_{d,y}$ and $\tilde{W}_0 + \tilde{d}_{d,y} \leq \tilde{y} \leq \tilde{W}$, they have to be replaced with $m_{\max}^{(j)} = 0$, with $j = 3, 4, 5, 6$, respectively.

Thus, the computational solution Eq. (2.56b) reduces to

$$
\begin{aligned}
\tilde{T}_c(\tilde{x}, \tilde{y}, \tilde{t}) &= \tilde{T}_{qs,x}\left(\tilde{x}, \tilde{y}, \tilde{t}, S_{qs,x}^{(A)} = 0 \right) \\
&= \frac{\tilde{W}_0}{\tilde{W}} \tilde{t} + \frac{1 - \mathrm{sign}(\tilde{y} - \tilde{W}_0)}{2} \tilde{T}_x^{(L)}(\tilde{x}) + \tilde{T}_y^{(L)}(\tilde{y})
\end{aligned}
\tag{2.87}
$$

where $\tilde{T}_x^{(L)}(\tilde{x})$ and $\tilde{T}_y^{(L)}(\tilde{y})$ are defined in Eq. (B.7).

When dealing with the above equation, the relative error E defined by Eq. (2.77) becomes

$$E = \frac{\left| \tilde{T}_{e,qs}(\tilde{x}, \tilde{y}, \tilde{t}) - \left[\frac{\tilde{W}_0}{\tilde{W}} \tilde{t} + \frac{1 - \mathrm{sign}(\tilde{y} - \tilde{W}_0)}{2} \tilde{T}_x^{(L)}(\tilde{x}) + \tilde{T}_y^{(L)}(\tilde{y}) \right] \right|}{\frac{\tilde{W}_0}{\tilde{W}} \tilde{t} + \tilde{T}_x^{(L)}(0) + \tilde{T}_y^{(L)}(0)} \tag{2.88}$$

This error will be less than or equal to $10^{-A}/2$, with $A = 2, 3, ..., 15$, for $0 \leq \tilde{y} \leq \tilde{W}_0 - \tilde{d}_{d,y}$ and $\tilde{W}_0 + \tilde{d}_{d,y} \leq \tilde{y} \leq \tilde{W}$, where the dimensionless deviation distance $\tilde{d}_{d,y}$ is the minimum distance from the heating/cooling edge to make negligible the former m-summation in Eq. (2.56b) at a level of $E = (10^{-A}/2)/4 = 10^{-A}/8$, where the factor of 4 denotes the number of summations appearing in Eq. (2.56b). It is given by

$$\tilde{d}_{d,y} = \frac{A \ln(10) - \ln\left[\pi^2 \left(1 - e^{-2\pi\tilde{W}} \right)/8 \right]}{\pi} \tag{2.89}$$

So that Eq. (2.87) be applicable, $\tilde{d}_{d,y}$ has to be less than or equal to $\tilde{W}_0$ for $\tilde{y} < \tilde{W}_0$; and less than or equal to $\tilde{W} - \tilde{W}_0$ for $\tilde{y} > \tilde{W}_0$.

The deviation distance along y may be defined as the distance measured from the heating/cooling edge of a 2D rectangular body in the y-direction where the quasi-steady temperature can be computed only by using an algebraic expression with an error of one part in $2 \cdot 10^A$ with respect to the maximum temperature variation (that occurs at the corner point of the heated or cooled surface).

Also, by using Eq. (2.88), the exact quasi-steady part for $0 \leq \tilde{y} \leq \tilde{W}_0 - \tilde{d}_{d,y}$ and $\tilde{W}_0 + \tilde{d}_{d,y} \leq \tilde{y} \leq \tilde{W}$ becomes

$$
\tilde{T}_{e,qs}(\tilde{x}, \tilde{y}, \tilde{t}) = \frac{\tilde{W}_0}{\tilde{W}} \tilde{t} + \frac{1 - \mathrm{sign}(\tilde{y} - \tilde{W}_0)}{2} \tilde{T}_x^{(L)}(\tilde{x}) + \tilde{T}_y^{(L)}(\tilde{y})
$$
$$
\pm \frac{10^{-A}}{2} \left[\frac{\tilde{W}_0}{\tilde{W}} \tilde{t} + \tilde{T}_x^{(L)}(0) + \tilde{T}_y^{(L)}(0) \right]
$$
(2.90)

Substituting it in Eq. (2.75) yields the exact numerical value of temperature for $\tilde{t} \geq \tilde{t}_{qs}$ and for $0 \leq \tilde{y} \leq \tilde{W}_0 - \tilde{d}_{d,y}$ and $\tilde{W}_0 + \tilde{d}_{d,y} \leq \tilde{y} \leq \tilde{W}$ as

$$
\tilde{T}_e(\tilde{x}, \tilde{y}, \tilde{t}) = \frac{\tilde{W}_0}{\tilde{W}} \tilde{t} + \frac{1 - \mathrm{sign}(\tilde{y} - \tilde{W}_0)}{2} \tilde{T}_x^{(L)}(\tilde{x}) + \tilde{T}_y^{(L)}(\tilde{y})
$$
$$
\pm 10^{-A} \left[\frac{\tilde{W}_0}{\tilde{W}} \tilde{t} + \tilde{T}_x^{(L)}(0) + \tilde{T}_y^{(L)}(0) \right]
$$
(2.91)

As an example, for $\tilde{W} = 3$ and $A = 2$, $\tilde{d}_{d,y} \approx 1.62$. If $\tilde{W}_0 / \tilde{W} = 1/6$, this indicates that, for $2.12 \leq \tilde{y} \leq 3$ and $\tilde{t} \geq \tilde{t}_{qs}$, where $\tilde{y} > \tilde{W}_0 = 0.5$, Eq. (2.87) is applicable with a relative error less than $10^{-A} = 0.01$ (1%). On the contrary, this equation is not applicable for $\tilde{y} < \tilde{W}_0$ as $\tilde{d}_{d,y} \approx 1.62$ is greater than $\tilde{W}_0 = 0.5$.

2.4.3.8 Number of Terms in the Complementary Transient Solution

When the time of interest is less than the quasi-steady time $\tilde{t}_{qs}$ but greater than the deviation time $\tilde{t}_d$ defined in Section 2.4.3.2, the large-time form of the computational analytical solution defined by Eq. (2.54) can be used. The total number of terms $S_{qs}^{(A)}$ to be used in its quasi-steady part (to get an error of $E = 10^{-A}/2$) has already been discussed in Sections 2.4.3.4 and 2.4.3.5. As regards the total number of terms $S_{ct}^{(A)}$ required in its complementary transient part defined by Eq. (2.55) (to still get an error of $E = 10^{-A}/2$; see Eq. (2.94) below), it is given by $S_{ct} = m_{\max}^{(1)} + n_{\max}^{(1)} + \left[m_{\max}^{(2)} n_{\max}^{(2)}(m) \right]_{\max}$, being $\left[m_{\max}^{(2)} n_{\max}^{(2)}(m) \right]_{\max}$ the maximum number of terms used in the double summation. In detail,

- $m_{\max}^{(1)}$ is the maximum number of terms to make negligible the "tail" of the single m-summation in Eq. (2.55) at a level of $E = (10^{-A}/2)/3 = 10^{-A}/6$, where the factor of 3 denotes the number of summations appearing in Eq. (2.55). It is given by

$$
m_{\max}^{(1)} = \mathrm{ceil}\left(\frac{1}{\pi} \sqrt{\frac{A \ln(10) - \ln(\pi^2/12)}{\tilde{t}}} \right)
$$
(2.92a)

- $n_{\max}^{(1)}$ is the maximum number of terms to make negligible the "tail" of the single n-summation in Eq. (2.55) at a level of $E = 10^{-A}/6$. It is

$$
n_{\max}^{(1)} = \mathrm{ceil}\left(\frac{\tilde{W}}{\pi} \sqrt{\frac{A \ln(10) - \ln\left[\pi^3/\left(12\tilde{W}^2\right)\right]}{\tilde{t}}} \right)
$$
(2.92b)

- $\left[m_{\max}^{(2)} n_{\max}^{(2)}(m) \right]_{\max}$ is the maximum number of terms to make negligible the "tail" of the m by n double summation appearing in Eq. (2.55) at a level of $E = 10^{-A}/6$. It is given by

$$\left[m_{\max}^{(2)} n_{\max}^{(2)}(m)\right]_{\max} = \text{ceil}\left(\frac{A \ln(10) - \ln\left[\pi^3\left(1 + \tilde{W}^2\right)/\left(24\tilde{W}^2\right)\right]}{4\pi} \frac{\tilde{W}}{\tilde{t}}\right) \tag{2.92c}$$

For computational efficiency purposes, however, the maximum numbers of terms to be used in either of the two summations of the double sum are

$$m_{\max}^{(2)} = \text{ceil}\left(\frac{1}{\pi}\sqrt{\frac{A \ln(10) - \ln\left[\pi^3\left(1 + \tilde{W}^2\right)/\left(24\tilde{W}^2\right)\right]}{\tilde{t}}}\right) \tag{2.93a}$$

$$n_{\max}^{(2)}(m) = \text{ceil}\left(\tilde{W}\sqrt{\left(m_{\max}^{(2)}\right)^2 - m^2}\right), \quad m = 1,2,\ldots,m_{\max}^{(2)} \tag{2.93b}$$

Equations (2.92a) and (2.92b) as well as Eqs. (2.93a) and (2.93b) are less conservative than the original equations given by Woodbury et al. (2017). Similarly, Eq. (2.92c) is less conservative than the original equation defined by Beck et al. (2004).

By using the equations listed before, that are valid for any location within the rectangular body, the maximum number of terms for the three summations appearing in the complementary transient part of the computational solution can be computed for different times of interest and accuracies. As an example, the results are shown by Table 2.2 for an aspect ratio of $\tilde{W} = 2$ at two different times, $\tilde{t} = 0.01$ and $\tilde{t} = 0.1$, and several accuracies.

It is evident that the double summation requires many terms, and this number increases linearly with the aspect ratio $\tilde{W}$ and inversely with the time $\tilde{t}$. Also, the accuracy plays an important role. The last column of the table gives $n_{\max}^{(2)}$ only for $m = 1$, when it exhibits the maximum number of terms.

In this case, the relative error E defined by Eq. (2.57) becomes

$$E = \frac{\left|\tilde{T}_e(\tilde{x}, \tilde{y}, \tilde{t}) - \left[\tilde{T}_{e,qs}(\tilde{x}, \tilde{y}, \tilde{t}) + \tilde{T}_{ct}\left(\tilde{x}, \tilde{y}, \tilde{t}, S_{ct}^{(A)}\right)\right]\right|}{\tilde{T}_{e,qs}(0, 0, \tilde{t}) + \tilde{T}_{ct}\left(0, 0, \tilde{t}, S_{ct}^{(A)}\right)} = \frac{10^{-A}}{2} \tag{2.94}$$

where $\tilde{T}_{e,qs}(\tilde{x}, \tilde{y}, \tilde{t})$ is the exact quasi-steady part of the solution as shown in Section 2.4.3.3.

Therefore, the exact numerical value of temperature for $\tilde{t}_d < \tilde{t} < \tilde{t}_{qs}$ (most general case) is

$$\tilde{T}_e(\tilde{x}, \tilde{y}, \tilde{t}) = \left[\tilde{T}_{e,qs}(\tilde{x}, \tilde{y}, \tilde{t}) + \tilde{T}_{ct}\left(\tilde{x}, \tilde{y}, \tilde{t}, S_{ct}^{(A)}\right)\right]$$
$$\pm \frac{10^{-A}}{2}\left[\tilde{T}_{e,qs}(0, 0, \tilde{t}) + \tilde{T}_{ct}\left(0, 0, \tilde{t}, S_{ct}^{(A)}\right)\right] \tag{2.95}$$

Table 2.2 Number of terms required from the three different summations appearing in the complementary transient part of the computational solution when $\tilde{W} = 2$ to get an error of $E = 10^{-A}/2$.

			$\tilde{t} = 0.01$			
A	$m_{\max}^{(1)}$	$n_{\max}^{(1)}$	$(m^{(2)}n^{(2)})_{\max}$	$S_{ct}^{(A)}$	$m_{\max}^{(2)}$	$n_{\max}^{(2)}(m = 1)$
2	7	15	66	**88**	7	14
3	9	18	103	**130**	9	18
5	11	23	176	**210**	11	22
10	16	31	359	**406**	16	32
			$\tilde{t} = 0.1$			
A	$m_{\max}^{(1)}$	$n_{\max}^{(1)}$	$(m^{(2)}n^{(2)})_{\max}$	$S_{ct}^{(A)}$	$m_{\max}^{(2)}$	$n_{\max}^{(2)}(m = 1)$
2	3	5	7	**15**	3	6
3	3	6	11	**20**	3	6
5	4	7	18	**29**	4	8
10	5	10	36	**51**	5	10

Substituting Eq. (2.78) in the above equation yields

$$
\tilde{T}_e(\tilde{x}, \tilde{y}, \tilde{t}) = \underbrace{\tilde{T}_{qs}\left(\tilde{x}, \tilde{y}, \tilde{t}, S_{qs}^{(A)}\right) + \tilde{T}_{ct}\left(\tilde{x}, \tilde{y}, \tilde{t}, S_{ct}^{(A)}\right)}_{\tilde{T}_c\left(\tilde{x}, \tilde{y}, \tilde{t}, S^{(A)}\right)}
$$
$$
\pm 10^{-A}\underbrace{\left[\tilde{T}_{qs}\left(0,\ 0,\ \tilde{t},\ S_{qs}^{(A)}\right)\left(1 + \frac{10^{-A}}{4}\right) + \frac{1}{2}\tilde{T}_{ct}\left(0,0,\tilde{t},S_{ct}^{(A)}\right)\right]}_{\approx \tilde{T}_{qs}\left(0, 0, \tilde{t}, S_{qs}^{(A)}\right) + \tilde{T}_{ct}\left(0, 0, \tilde{t}, S_{ct}^{(A)}\right) = \tilde{T}_c\left(0, 0, \tilde{t}, S^{(A)}\right)}
\tag{2.96}
$$

where $S^{(A)} = S_{qs}^{(A)} + S_{ct}^{(A)}$ and the approximation performed to the second term is conservative.

2.4.3.9 Computer Code and Plots

A MATLAB function called `fdX22By1pt10Y22B00T0.m` allows the dimensionless temperature to be computed for the current case. It is available for ease of use on the web site related to this book.

By using the above function, temperature plots can be obtained for any accuracy though $A = 3$ is enough for visual comparison. They are shown in Figure 2.11 versus the space coordinate $\tilde{x}$ with $\tilde{y}$ as a parameter at two different times for a rectangular body having $\tilde{W} = 1$ as aspect ratio and $\tilde{W}_0 = \tilde{W}/2$ as heated or cooled region.

Both figures indicate that, for an interior point close to the back side $\tilde{x} = 1$ of the rectangle and in correspondence of the heated or cooled region, that is, $0 \leq \tilde{y} \leq \tilde{W}_0$, the response is slow, being both damped and lagged. Also, they show that the maximum temperature rise occurs at the corner point $(\tilde{x}, \tilde{y}) = (0, 0)$. The heating/cooling edge $(\tilde{x}, \tilde{y}) = (0, \tilde{W}_0 = 1/2)$ is also interesting as a step change in the heat flux occurs at that surface point. The thermal response is also slow at a surface point of the boundary $\tilde{x} = 0$ which is located outside the active region of this boundary, that is, for $\tilde{W}_0 \leq \tilde{y} \leq \tilde{W}$, being both damped and lagged.

As an example, when $\tilde{W} = 1$ and $\tilde{W}_0 = \tilde{W}/2$, and the time of interest is $\tilde{t} = 0.05$, $\tilde{T}(0, 0, \tilde{t}) = 0.2469$, whereas $\tilde{T}(0, \tilde{y} = \tilde{W} = 1, \tilde{t}) = 0.00542$, a factor of nearly 50 smaller. (These numerical values can be derived by using the `fdX22By1pt10Y22B00T0.m` function.) This factor increases as $\tilde{t}$ becomes smaller. In fact, for $\tilde{t} = 0.01$, it is almost 40000. On the other hand, for sufficiently large times (much greater than the quasi-steady time of $\tilde{t}_{qs} \approx 0.84$ for $A = 3$; see Section 2.4.3.3), the factor approaches unity. In fact, for $\tilde{t} = 5$, $\tilde{T}(0, 0, \tilde{t}) = 2.853$, whereas $\tilde{T}(0, \tilde{y} = \tilde{W} = 1, \tilde{t}) = 2.480$, a factor of 1.15 smaller. On the back side $\tilde{x} = 1$ of the plate and at $\tilde{y} = 0$, it results in $\tilde{T}(1, 0, \tilde{t}) = 0.000245$, that is about 1000 time smaller than $\tilde{T}(0, 0, \tilde{t}) = 0.2469$ at the dimensionless time of 0.05.

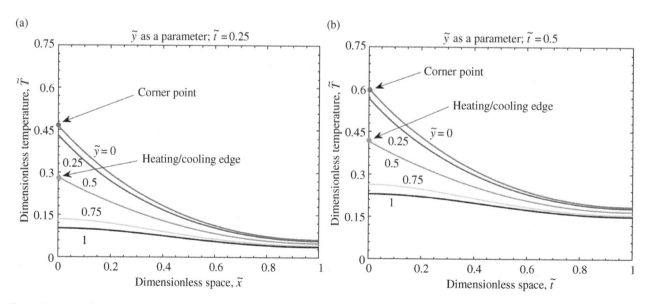

Figure 2.11 Dimensionless temperature of the X22B(y1pt1)0Y22B00T0 problem versus $\tilde{x}$ with $\tilde{y}$ as a parameter when $\tilde{W} = 1$ and $\tilde{W}_0 = \tilde{W}/2$: (a) at time $\tilde{t} = 0.25$; (b) at time $\tilde{t} = 0.5$.

For the special case of $\tilde{y} = \tilde{W}_0 = \tilde{W}/2 = 1/2$ and $\tilde{t} \geq \tilde{t}_{qs} \approx 0.84$ for $A = 3$, the exact solution reduces to Eq. (B.8) apart from a factor of 2. In this special case, the temperature ratio has an algebraic expression as

$$\frac{\tilde{T}_c(0, 1/2, \tilde{t})}{\tilde{T}_c(1, 1/2, \tilde{t})} = \frac{\tilde{t} + 1/3}{\tilde{t} - 1/6} = 1 + \frac{1}{2\tilde{t} - 1/3} \tag{2.97}$$

which simplifies to $1 + 1/(2\tilde{t})$ for $\tilde{t} \gg 1$. When $\tilde{t} = \tilde{t}_{qs} \approx 0.84$, the ratio defined by Eq. (2.97) is of about 1.75. But when $\tilde{t} = 5$, the ratio is of about 1.1. Also, note that Eq. (2.97) is the same as Eq. (2.40) of the X22B10T0 problem.

A comparison with the X22B10T0 case of Section 2.3.8 indicates that $\tilde{T}(0, 0, \tilde{t}) = 0.2469$ vs. $\tilde{T}(0, \tilde{t}) = 0.2523$ for $\tilde{t} = 0.05$ as well as $\tilde{T}(1, 0, \tilde{t}) = 0.000245$ vs. $\tilde{T}(1, \tilde{t}) = 0.000269$ for the same time $\tilde{t} = 0.05$. Therefore, the numerical values of temperature of the 2D problem are not very different from the ones of the 1D problem. This is because, at early times, the 2D deviation effects due to the heating edge at $\tilde{y} = \tilde{W}_0$ are negligible and, hence, the thermal field may be considered 1D, as shown in Section 2.4.3.2.

2.5 Chapter Summary

Exact analytical solutions to DHCPs can be used as fundamental building blocks to solve IHCPs when thermal properties can be assumed constant. A standard solution numbering scheme is useful to concisely identify these solutions. Specific 1D solutions presented in this chapter are:

- X12B10T0 – rectangular coordinates with first and second kind boundaries ("X12") with constant value at first boundary and homogeneous condition at second boundary ("B10") in a transient problem with homogeneous initial condition ("T0").
- X12B20T0 – rectangular coordinates with first and second kind boundaries ("X12") with linearly increasing in time value at first boundary and homogeneous condition at second boundary ("B20") in a transient problem with homogeneous initial condition ("T0").
- X22B10T0 – rectangular coordinates with two second kind boundaries ("X22") with constant value at first boundary and homogeneous condition at second boundary ("B10") in a transient problem with homogeneous initial condition ("T0").
- X22B20T0 – rectangular coordinates with two second kind boundaries ("X12") with linearly increasing in time value at first boundary and homogeneous condition at second boundary ("B20") in a transient problem with homogeneous initial condition ("T0").

Differing mathematical expressions result from various solution techniques. In particular, short time solutions result from LT approach, and large time solutions result from separation of variables evaluation. The best expression to use in a particular application depends on both the location and the time of evaluation. This is because of the presence of infinite series in the solutions and the fact that some of these converge faster for small values of time, while others converge faster for larger values of time.

These considerations give rise to the concept of a computational analytical solution. Such a solution uses the most appropriate form of the exact solution at each time and location; and uses the correct number of terms in the series to guarantee a desired approximation accuracy.

To guide the selection of appropriate form of solution, the deviation time concepts are used. The first deviation time, t_d, establishes a bound on the time during which the solution at a point is the same as that for a semi-infinite domain. For times less than t_d, the semi-infinite solution, which involves no infinite series, can be used. The second deviation time, t_δ, is a bound on the time when a nearby homogeneous boundary affects the solution at a point. Up until t_δ, the solution from LT can be used with only one term in the infinite series. For all other times, the SOV solutions are most efficient and considerably simplify for times greater than the quasi-steady (or steady-state but also unsteady) time.

Only one 2D solution is needed in this text as a building block for determining spatially varying heat fluxes in the IHCP (Chapter 8). This problem is a rectangular region insulated on all sides but with a constant magnitude heating or cooling on a portion of one face on the corner. The solution number for this problem is X22B(y1pt1)0Y22B00T0. Considerations for short- and long-term evaluation for 2D solutions are much more complex than those for 1D cases and depend on the location of the point of interest in proximity to both the heating/cooling source and nearby homogeneous boundaries.

MATLAB computer codes for evaluation of the computational analytical solutions presented in this chapter are available on the companion website for the book.

Problems

2.1 Write the mathematical formulation of the heat conduction problems denoted by X23B11T2 and X23B02T1.

2.2 Give the numbering system designation for the following 2D transient heat conduction problem in rectangular coordinates

$$\frac{\partial^2 T}{\partial x^2} + \frac{\partial^2 T}{\partial y^2} = \frac{1}{\alpha}\frac{\partial T}{\partial t} \quad (0 < x < L; 0 < y < W; t > 0)$$

$$T(0, y, t) = 0 \quad (0 < y < W; t > 0)$$

$$-k\left(\frac{\partial T}{\partial x}\right)_{x=L} = h[T(L, y, t) - T_\infty] \quad (0 < y < W; t > 0)$$

$$-k\left(\frac{\partial T}{\partial y}\right)_{y=0} = q_0(t) \quad (0 < x < L; t > 0)$$

$$T(x, W, t) = T_W \quad (0 < x < L; t > 0)$$

$$T(x, y, 0) = 0 \quad (0 < x < L; 0 < y < W)$$

where $q_0(t) = 5t$ W/m^2. Also, draw a schematic of the problem.

2.3 Write the governing equations of the heat conduction problem denoted by X21B00Y20B(x1pt1)T(x2y1). Also, draw a schematic of the problem.

2.4 A concrete wall ($\alpha = 7 \times 10^{-7}$ m^2/s) of thickness 30 cm is initially at a uniform temperature of 20 °C. Suddenly one surface is raised to 120 °C and maintained at that temperature, while the other surface is kept insulated. By treating the body as a 1D semi-infinite solid, calculate the temperatures at 5, 10, and 15 cm from the heated boundary 30 min after the raising of the surface temperature. Then, compare these approximate values of temperature with the exact ones (when the body is considered finite using the fdX12B10T0.m MATLAB function with $A = 15$ as accuracy) and calculate the relative error with respect to the maximum temperature rise. Lastly, calculate the values of deviation time, second deviation time, and steady state time.

2.5 A wall 0.12 m thick having a thermal diffusivity of 1.5×10^{-6} m^2/s is initially at a uniform temperature of 85 °C. Suddenly one face is lowered to a temperature of 20 °C, while the other face is perfectly insulated. Plot, on T–x coordinates, the temperature distributions: initial, steady-state, and at two intermediate times, 30 and 60 min, by using the fdX12B10T0.m MATLAB function. Also, calculate the steady state time.

2.6 The 150-mm-thick wall of a gas-fired furnace is constructed of fireclay brick having $k = 1.5$ W/(m °C), $\rho = 2600$ kg/m^3, $c = 1470$ J/(kg K) and is well insulated at its outer surface. The wall is at a uniform initial temperature of 20 °C, when the burners are fired and the inner surface is exposed to products of combustion for which its temperature increases linearly with time as $10t$ °C (with t in seconds). Calculate the time that it takes for the outer surface of the wall to reach a temperature of 200 °C by using the fdX12B20T0.m MATLAB function. Plot, on T–x coordinates, the temperature distribution at the foregoing time by means of the same tool.

2.7 A slab of copper ($k = 401$ W/(m °C) and $\alpha = 117 \times 10^{-6}$ m^2/s) of thickness 15 cm is initially at a uniform temperature of 20 °C. It is suddenly exposed to radiation at one surface such that the next heat flux is maintained at a constant value of 3×10^5 W/m^2, while the other surface is kept insulated. Using the approximation of 1D semi-infinite solid, determine the temperature at the irradiated surface and at the back side after 2 min have elapsed. Compare the results with those obtained from an exact analytical solution (when the body is considered finite using the fdX22B10T0.m Matlab function with $A = 15$ as accuracy) and calculate the relative error with respect to the maximum temperature rise. Lastly, calculate the values of deviation time, second deviation time, and quasi-steady time.

2.8 Standards for firewalls may be based on their thermal response to a prescribed radiant heat flux. Consider a 0.25-m-thick concrete wall, of $k = 1.4\ \text{W/(m\,°C)}$, $\rho = 2300\ \text{kg/m}^3$, $c = 880\ \text{J/(kg K)}$, which is at an initial temperature of 25 °C and irradiated at one surface by lamps that provide a uniform linearly-in-time dependent heat flux of $5t\ \text{W/m}^2$ (with t in seconds). The absorptivity of the surface to the irradiation is nearly 1. If building code requirements dictate that the temperatures of the irradiated and back (insulated) surfaces must not exceed 325 and 25 °C, respectively, after 30 min of heating, will the requirements be met? Use the approximation of 1D semi-infinite solid and compare the results with those obtainable from the application of the exact analytical solution (when the body is considered finite using the `fdX22B20T0.m` MATLAB function with $A = 15$ as accuracy). Also, calculate the values of deviation time, second deviation time, and unsteady time.

2.9 Consider a 6-mm-thick ceramic plate of thermal conductivity 2 W/(m °C) and thermal diffusivity $1.5 \times 10^{-6}\ \text{m}^2/\text{s}$. The width is of 50 mm and 50% of it (starting from one corner, as shown in Figure 2.10) is heated by means of electrical heating elements dissipating $2\ \text{kW/m}^2$, while the other boundary surfaces of the plate are well insulated. Initially, the ceramic solid is at a uniform temperature of 30 °C, and suddenly the heating elements are energized. Using the `fdX22By1pt10Y22B00T0.m` MATLAB function, estimate the time required for the difference between the heated surface corner and initial temperatures to reach 90% of the difference for quasi-steady conditions. Also, calculate the quasi-steady time.

2.10 Consider the system of Problem 2.9. By using the `fdX22By1pt10Y22B00T0.m` and `fdX22B10T0.m` functions, demonstrate numerically that

$$T_{\text{X22B(y1pt1)0Y22B00T1}}(0, y, t) + T_{\text{X22B(y1pt1)0Y22B00T1}}(0, W - y, t) = T_{\text{X22B10T1}}(0, t)$$

for $0 \leq y \leq W$ and $t > 0$. Then, plot the three temperatures on the same graph versus time for an arbitrary value of y. Why do the temperatures satisfy the above condition? Can the above condition be valid for any x?

References

ASME V V 20-2009 R (2021) Standard for verification and validation in computational fluid dynamics and heat transfer, *ASME V & V*, p. 20.

Beck, J. V. and Litkouhi, B. (1988) Heat conduction number system, *International Journal of Heat and Mass Transfer*, 31(3), pp. 505–515. https://doi.org/10.1016/0017-9310(88)90032-4.

Beck, J. V., Haji-Sheikh, A., Amos, D. E. and Yen, D.H.Y. (2004) Verification solution for partial heating of rectangular solid, *International Journal of Heat and Mass Transfer*, 47(19–20), pp. 4243–4255. https://doi.org/10.1016/j.ijheatmasstransfer.2004.04.021.

Beck, J. V., McMasters, R. L., Dowding, K. J. and Amos, D. E. (2006) Intrinsic verification methods in linear heat conduction, *International Journal of Heat and Mass Transfer*, 49(17–18), pp. 2984–2994. https://doi.org/10.1016/j.ijheatmasstransfer.2006.01.045.

Beck, J. V., Wright, N.T. and Haji-Sheikh, A. (2008) Transient power variation in surface conditions in heat conduction for plates, *International Journal of Heat and Mass Transfer*, 51(9–10), pp. 2553–2565. https://doi.org/10.1016/j.ijheatmasstransfer.2007.07.043.

Carslaw, H. S. and Jaeger, C. J. (1959) *Conduction of Heat in Solids*. 2nd edn. Oxford, UK: Oxford University Press.

Cole, K. D., Beck, J. V., Haji-Sheikh, A. and Litkouhi, B. (2011) *Heat Conduction using Green's Functions*. 2nd edn. Boca Raton, FL: CRC Press.

Cole, K. D., Beck, J. V., Woodbury, K. A. and de Monte, F. (2014) Intrinsic verification and a heat conduction database, *International Journal of Thermal Sciences*, 78, pp. 36–47. http://dx.doi.org/10.1016/j.ijthermalsci.2013.11.002.

Cole, K. D., de Monte, F., McMasters, R. L., Woodbury, K. A., Beck, J. V. and Haji-Sheikh, A. (2016) Steady heat conduction with generalized boundary conditions, in *Proceedings of the ASME 2016 International Mechanical Engineering Congress and Exposition (IMECE2016)*. Phoenix, AZ, USA.

D'Alessandro, G. and de Monte, F. (2018) Intrinsic verification of an exact analytical solution in transient heat conduction, *Computational Thermal Sciences*, 10(3), pp. 251–272. http://dx.doi.org/10.1615/ComputThermalScien.2017021201

de Monte, F. and Beck, J. V. (2009) Eigen-periodic-in-space surface heating in conduction with application to conductivity measurement of thin films, *International Journal of Heat and Mass Transfer*, 52(23–24), pp. 5567–5576. https://doi.org/10.1016/j.ijheatmasstransfer.2009.06.024.

de Monte, F., Beck, J. V. and Amos, D. E. (2008) Diffusion of thermal disturbances in two-dimensional Cartesian transient heat conduction, *International Journal of Heat and Mass Transfer*, 51(25), pp. 5931–5941. Available at: http://10.0.3.248/j.ijheatmasstransfer.2008.05.015.

de Monte, F., Beck, J. V. and Amos, D. E. (2012) Solving two-dimensional Cartesian unsteady heat conduction problems for small values of the time, *International Journal of Thermal Sciences*, 60, pp. 106–113. https://doi.org/10.1016/j.ijthermalsci.2012.05.002.

de Monte, F., Woodbury, K. A. and Najafi, A. (2023) Construction of short-time solutions in heat conduction, *ASME Summer Heat Transfer Conference (SHTC 2023)*. Washington D.C., USA, July 10–12.

Fernandes, A. P., Sousa, P.F.B., Borges, V.L. and Guimaraes, G. (2010) Use of 3D transient analytical solution based on Green's function to reduce computational time in inverse heat conduction problems, *Applied Mathematical Modelling*, 34(12), pp. 4040–4049. https://doi.org/10.1016/j.apm.2010.04.006.

Lienhard IV, J. H. and Lienhard, J. H. V. (2012) *A Heat Transfer Textbook*. 4th edn. Cambridge, MA: Phlogiston Press.

Luikov, A. V. (1968) *Analytical Heat Diffusion Theory*. New York: Academic Press.

McMasters, R. L., de Monte, F., Beck, J. V. and Amos, D.E. (2018) Transient two-dimensional heat conduction problem with partial heating near corners, *ASME Journal of Heat Transfer*, 140(2), pp. 021301–1–021301–10. https://doi.org/10.1115/1.4037542.

McMasters, R. L., de Monte, F. and Beck, J. V. (2019) Generalized solution for two-dimensional transient heat conduction problems with partial heating near a corner, *ASME Journal of Heat Transfer*, 141(7), pp. 071301-1071301-8. https://doi.org/10.1115/1.4043568.

McMasters, R. L., de Monte, F., D'Alessandro, G. and Beck, J. V. (2021) Verification of ansys and matlab heat conduction results using an "intrinsically" verified exact analytical solution, *Journal of Verification, Validation and Uncertainty Quantification*, 6(2), pp. 021005–1–021005–9. https://doi.org/10.1115/1.4050610.

Mishra, D. K., Dolan, K. D., Beck, J. V. and Ozadali, F. (2017) Use of scaled sensitivity coefficient relations for intrinsic verification of numerical codes and parameter estimation for heat conduction, *Journal of Verification, Validation and Uncertainty Quantification*, 2(3), p. 31005. https://doi.org/10.1115/1.4038494.

Oldham, K., Myland, J. and Spanier, J. (2009) *An Atlas of Functions*. 2nd edn. New York: Springer.

Özişik, M. N. (1993) *Heat Conduction*. 2nd edn. New York: Wiley.

Polyanin, A. D. (2002) *Handbook of Linear Partial Differential Equations for Engineers and Scientists*. Boca Raton, FL: Chapman & Hall/CRC Press.

Thambynayagam, R. K. M. (2011) *The Diffusion Handbook: Applied Solutions for Engineers*. New York: McGraw-Hill.

Toptan, A., Porter, N.W., Hales, J.D., Spencer, B.W., Pilch, M. and Williamson, R.L. (2020) Construction of a code verification matrix for heat conduction with finite element code applications, *Journal of Verification, Validation and Uncertainty Quantification*, 5(4), p. 41002. https://doi.org/10.1115/1.4049037.

Yen, D.H.Y., Beck, J. V., McMasters, R.L. and Amos, D. E., (2002). Solution of an initial-boundary value problem for heat conduction in a parallelepiped by time partitioning, *International Journal of Heat and Mass Transfer*, 45 (21), pp. 4267–4279. https://doi.org/10.1016/S0017-9310(02)00145-X.

Woodbury, K. A. and Beck, J. V. (2013) Heat conduction in a planar slab with power-law and polynomial heat flux input, *International Journal of Thermal Sciences*, 71, pp. 237–248. https://doi.org/10.1016/j.ijthermalsci.2013.04.009.

Woodbury, K. A., Najafi, H. and Beck, J. V. (2017) Exact analytical solution for 2-D transient heat conduction in a rectangle with partial heating on one edge, *International Journal of Thermal Sciences*, 112, pp. 252–262. https://doi.org/10.1016/j.ijthermalsci.2016.10.014.

3

Approximate Methods for Direct Heat Conduction Problems

3.1 Introduction

In this chapter an approximate method based on superposition of linear solutions is given for solving direct transient heat conduction problems with arbitrary boundary conditions. This concept can also be interpreted as a convolution integral or finite sum derived using Green's functions or Duhamel's theorem. This direct technique is the first stage of solution procedures for the inverse heat conduction problems. The second stage, which involves specific algorithms for the IHCP, is developed in Chapter 4. The approximate heat conduction solution procedures of this chapter are combined in later chapters with the IHCP algorithms of Chapter 4.

3.1.1 Various Numerical Approaches

Numerical methods are useful for solving heat conduction problems when dealing with boundary conditions arbitrary either in time, or in space or in both time and space. They treat as their basic unknowns the values of the temperature at discrete points of the domain (called the "grid" points) providing a set of algebraic equations for these unknowns. There are two types of numerical procedures. One is based on a differential formulation of the transient heat conduction equation (linear or nonlinear) and involves finite difference (FD) and finite element (FE) methods (Patankar 1980; Özişik 1993). An alternative method is the finite control-volume formulation. The other is based on an integral form of convolutive-type of the same equation and employs Duhamel's theorem and Green's functions (GF) both of which are only valid for linear cases (Özişik 1993; Cole et al. 2011). The advantage of the integral form is that the space dependence can be represented exactly leaving only the time dependence to be approximated when dealing not only with one-dimensional transient problems but also with multi-dimensional ones.

Superposition of linear solutions can be interpreted using the numerical approximate form of Duhamel's theorem (not always applicable) or Green's functions method. Duhamel's theorem can in fact be thought of as a special case of the more general GF solution equation. (Recall that Duhamel's integral is applicable only to bodies initially at zero temperature and subject to time-dependent but space-independent boundary conditions and/or heat generation term.) For this reason, only a Green's function method interpretation is given here in a numerical approximate form (de Monte et al. 2011).

The superposition of linear solutions proposed in this chapter can be relevant in investigation of the IHCPs because it gives a convenient expression for the temperature in terms of the unknown boundary condition components. Also, the linear solutions used in the superposition are the "*computational*" analytical solutions presented in Chapter 2.

3.1.2 Scope of Chapter

Development in this chapter builds heavily on the concepts of superposition of solutions. Recall that superposition is valid only for linear differential equations and their solutions. This chapter opens with a review of superposition concepts.

The bulk of this chapter gives a superposition-based numerical approximation of the temperature solution when the body is subject to a surface temperature (Section 3.3) and a surface heat flux (Section 3.4). This can be interpreted as a numerical approximation of the 1D transient Green's functions solution equation based on temperature and heat-flux formulations, respectively (Appendix C). Both formulations deal with a piecewise-constant approximation in time as well as a linear one. The former approximation employs only one basic "building block" solution, which is the computational analytical solution of a direct problem (Chapter 2); while the latter utilizes two basic "building block" solutions, still computational analytical solutions of a direct problem (Chapter 2).

Inverse Heat Conduction: Ill-Posed Problems, Second Edition. Keith A. Woodbury, Hamidreza Najafi, Filippo de Monte, and James V. Beck.
© 2023 John Wiley & Sons, Inc. Published 2023 by John Wiley & Sons, Inc.

As the heat transfer may be two- and three-dimensional, the single forcing input (surface heat flux) is considered space dependent in Section 3.5 and a superposition-based numerical piecewise-uniform approximation of the temperature solution is presented. This can still be interpreted as a numerical approximation in space of the 2D Green's functions solution equation based on a heat-flux formulation. Lastly, in Section 3.6, the single forcing input (surface heat flux) is assumed not only space dependent but also time dependent. Concerning this, a superposition-based numerical piecewise-uniform and -constant approximation of the thermal field is proposed, whose basic "building block" is the 2D transient temperature solution of Section 2.4 of the previous chapter. This can be considered as a numerical approximation in both space and time of the 2D Green's functions solution equation based on a heat-flux formulation (Appendix C).

3.2 Superposition Principles

Consider the X22B-0T1 problem, where "B-" indicates an arbitrary function of time at the first boundary (refer to Chapter 2 and Appendix A for full description of the numbering system). In brief, this is a flat plate, initially at a uniform temperature T_{in}, subject to a time-dependent surface heat flux $q_0(t)$ at $x = 0$ and thermally insulated at $x = L$.

Its mathematical formulation is

$$\frac{\partial^2 T}{\partial x^2} = \frac{1}{\alpha}\frac{\partial T}{\partial t} \qquad (0 < x < L; t > 0) \tag{3.1a}$$

$$-k\left(\frac{\partial T}{\partial x}\right)_{x=0} = q_0(t) \qquad (t > 0) \tag{3.1b}$$

$$\left(\frac{\partial T}{\partial x}\right)_{x=L} = 0 \qquad (t > 0) \tag{3.1c}$$

$$T(x, t = 0) = T_{in} \qquad (0 < x < L) \tag{3.1d}$$

If $q_0(t) = q_{0,1}(t) + q_{0,2}(t)$, the problem can be solved by breaking it into two separate subproblems (Woodbury and Beck 2013). Let $T(x,t) = T_1(x,t) + T_2(x,t)$. Substituting it into Eqs. (3.1a)–(3.1d) gives

$$\frac{\partial^2 T_1}{\partial x^2} + \frac{\partial^2 T_2}{\partial x^2} = \frac{1}{\alpha}\frac{\partial T_1}{\partial t} + \frac{1}{\alpha}\frac{\partial T_2}{\partial t} \qquad (0 < x < L; t > 0) \tag{3.2a}$$

$$-k\left(\frac{\partial T_1}{\partial x} + \frac{\partial T_2}{\partial x}\right)_{x=0} = q_{0,1}(t) + q_{0,2}(t) \qquad (t > 0) \tag{3.2b}$$

$$\left(\frac{\partial T_1}{\partial x} + \frac{\partial T_2}{\partial x}\right)_{x=L} = 0 \qquad (t > 0) \tag{3.2c}$$

$$T_1(x, t = 0) + T_2(x, t = 0) = T_{in} \qquad (0 < x < L) \tag{3.2d}$$

These equations can be satisfied by requiring

$$\frac{\partial^2 T_1}{\partial x^2} = \frac{1}{\alpha}\frac{\partial T_1}{\partial t} \qquad (0 < x < L; t > 0) \tag{3.3a}$$

$$-k\left(\frac{\partial T_1}{\partial x}\right)_{x=0} = q_{0,1}(t) \qquad (t > 0) \tag{3.3b}$$

$$\left(\frac{\partial T_1}{\partial x}\right)_{x=L} = 0 \qquad (t > 0) \tag{3.3c}$$

$$T_1(x, t = 0) = T_{in} \qquad (0 < x < L) \tag{3.3d}$$

and

$$\frac{\partial^2 T_2}{\partial x^2} = \frac{1}{\alpha}\frac{\partial T_2}{\partial t} \qquad\qquad (0 < x < L; t > 0) \tag{3.4a}$$

$$-k\left(\frac{\partial T_2}{\partial x}\right)_{x=0} = q_{0,2}(t) \qquad (t > 0) \tag{3.4b}$$

$$\left(\frac{\partial T_2}{\partial x}\right)_{x=L} = 0 \qquad\qquad (t > 0) \tag{3.4c}$$

$$T_2(x, t = 0) = 0 \qquad\quad (0 < x < L) \tag{3.4d}$$

So, solution to the original problem Eqs. (3.1a)–(3.1d) can be constructed from separate solutions of similar problems, that is, Eqs. (3.3a)–(3.3d) and (3.4a)–(3.4d). While on the surface this appears to be more work (solving two problems instead of one), this idea becomes a powerful principle to take advantage of existing analytical solutions to create solutions to new problems. The only requirement is that the non-homogeneity in the problem (in this case, the heat flux) can be considered as a linear combination (simple sum or difference with possible scaling factors) and the governing differential equations of the problem be linear (temperature-independent properties of the body).

Similarly, when dealing with the companion X12B-0T1 problem, that is, the plate is subject to an arbitrary boundary condition of the first kind. If the assigned surface temperature $T_0(t) = T_{0,1}(t) + T_{0,2}(t)$, the problem can still be solved by breaking it into two separate subproblems.

3.2.1 Green's Function Solution Interpretation

The superposition principle can be seen in application of Green's functions to solution of a heat conduction problem. Consider the X22B-0T1 problem described in Eqs. (3.1a)–(3.1d). The temperature solution using Green's functions can be written as (Cole et al. 2011; see also Eq. (C.19) in Appendix C of this book)

$$T(x, t) = T_{in} + \frac{\alpha}{k}\int_{\tau=0}^{t} q_0(\tau)G_{X22}(x, x' = 0, t - \tau)d\tau \tag{3.5a}$$

where $G_{X22}(x, x', t - \tau)$ is given in Appendix C, Section C.3. Clearly, when $q_0(t) = q_{0,1}(t) + q_{0,2}(t)$, the integral on the right splits into two parts, and each represents solution to the subproblems in Eqs. (3.3a)–(3.3d) and (3.4a)–(3.4d).

Similarly, the temperature solution using Green's functions to the X12B-0T1 case is (Cole et al. 2011; see also Eq. (C.3) in Appendix C of the current book)

$$T(x, t) = T_{in} + \alpha\int_{\tau=0}^{t} \theta_0(\tau)\frac{\partial G_{X12}}{\partial x'}(x, x', t - \tau)\Big|_{x'=0} d\tau \tag{3.5b}$$

where $G_{X12}(x, x', t - \tau)$ is given in Appendix C, Section C.2. If $\theta_0(t) = T_0(t) - T_{in} = \theta_{0,1}(t) + \theta_{0,2}(t)$, where $\theta_0(t)$ is the surface temperature rise, the integral on the right-hand side can be split into two parts, and each represents solution to the corresponding subproblem.

3.2.2 Superposition Example – Step Pulse Heating

The solution to the X22B10T0 problem is known and described fully in Chapter 2, Section 2.3.8. The solution is in a dimensionless form with $\tilde{T} = (T - T_{in})/(q_0L/k)$. Suppose the solution to a new problem with step changes in heat flux is needed:

$$q_0(t) = \begin{cases} 0 & 0 \le t < t_{i-1} \\ q_{0,i} & t_{i-1} \le t < t_i \\ 0 & t \ge t_i \end{cases} \tag{3.6}$$

This function can also be expressed using the Heaviside unit step function $H(t)$ as

$$q_0(t) = q_{0,i}H(t - t_{i-1}) - q_{0,i}H(t - t_i) \tag{3.7}$$

The solution $\tilde{T}(\tilde{x}, \tilde{t}_M)$ at time $\tilde{t}_M > \tilde{t}_i$ can be constructed by combining multiples of the basic $\tilde{T}_{\text{X22B10T0}}$ solution:

$$\tilde{T}(\tilde{x}, \tilde{t}_M) = \underbrace{\frac{q_{0,i}}{q_0} \tilde{T}_{\text{X22B10T0}}(\tilde{x}, \tilde{t} - \tilde{t}_{i-1})}_{\substack{\text{effect of initial step activated} \\ \text{at time } t_{i-1}}} - \underbrace{\frac{q_{0,i}}{q_0} \tilde{T}_{\text{X22B10T0}}(\tilde{x}, \tilde{t} - \tilde{t}_i)}_{\substack{\text{"turn off" effect of initial} \\ \text{step at } t_i}} \qquad t_M > t_i \tag{3.8}$$

where the dimensionless variables are defined in Section 2.3.3. Notice the similarity in the arguments to the Heaviside functions in Eq. (3.7) and the time dependence in the building block functions in Eq. (3.8).

3.3 One-Dimensional Problem with Time-Dependent Surface Temperature

Consider a flat plate, initially at a uniform temperature T_{in}, subject to a time-dependent surface temperature at $x = 0$ and thermally insulated at $x = L$. The surface temperature $T_0(t)$ is an arbitrary function of the time. Its value at the initial time, $T_0(t = 0)$, can be different from the uniform initial temperature T_{in} of the plate. This problem may be denoted by X12B-0T1.

In this section, approximate solutions to the above problem are outlined based on superposition principles. Both X12B10T0 and X12B20T0 building blocks from Chapter 2 are utilized in constructing the approximate solutions.

One simple way to treat this problem is to divide the forcing function $T_0(t)$ into a number M of equally spaced intervals, Δt, up to the time t of interest (denoted by $t_M = M\Delta t$), and to substitute a constant temperature within each of these intervals for the real $T_0(t)$. This gives the "*piecewise-constant*" approximation that will be analyzed in Section 3.3.1.

Another type of approximation of $T_0(t)$ is the "*piecewise-linear*" approximation which can in general represent a $T_0(t)$ profile more accurately than the constant elements approximation (see Example 3.3 of Section 3.3.2.4). The piecewise linear approximation is described in Section 3.3.2.

3.3.1 Piecewise-Constant Approximation

This approximation gives the piecewise profile of Figure 3.1, where $\theta_0(t) = T_0(t) - T_{in}$ is the arbitrary surface temperature rise (in general, variation as $T_0(t)$ can be greater than or less than T_{in} at any time). The piecewise profile is a sequence of constant segments where $\theta_{0,i}$ (ith surface temperature rise component) is an approximation for $\theta_0(t)$ between $t_{i-1} = (i-1)\Delta t$ and $t_i = i\Delta t$ (with $i = 1, 2, ..., M$, being $t_M = M\Delta t$ the time of interest). The ith surface temperature component, $\theta_{0,i}$, is constant and may be identified with the time coordinate $t_{i-1/2} = (i-1/2)\Delta t$, that is, $\theta_0(t = t_{i-1/2}) = \theta_{0,i}$.

The temperature solution $T_i(x, t_M)$ at time t_M within the flat plate due only to the ith surface temperature rise component $\theta_{0,i}$ applied at $x = 0$ for the finite period of time between t_{i-1} and t_i, with $i \leq M$, as shown in Figure 3.2, when the plate is initially at zero temperature, may be calculated by applying superposition using the X12B10T0 building block:

$$T_i(x, t_M) = \theta_{0,i}\left[\tilde{T}_{\text{X12B10T0}}(\tilde{x}, \tilde{t}_M - \tilde{t}_{i-1}) - \tilde{T}_{\text{X12B10T0}}(\tilde{x}, \tilde{t}_M - \tilde{t}_i)\right] \tag{3.9}$$

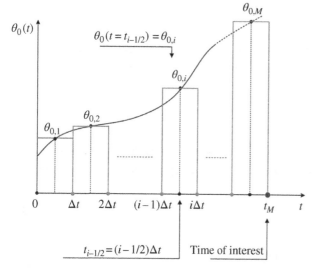

where, according to the dimensionless variables defined by Eqs. (2.4) and (2.5) in Chapter 2, $\tilde{x} = x/L$, $\tilde{t} = \alpha t/L^2$ and $\tilde{T} = (T - T_{in})/(T_0 - T_{in})$. Also, $\tilde{t}_M = M\Delta\tilde{t}$, $\tilde{t}_{i-1} = (i-1)\Delta\tilde{t}$ and $\tilde{t}_i = i\Delta\tilde{t}$, with $\Delta\tilde{t} = \alpha\Delta t/L^2$. Note that $\tilde{T}_{\text{X12B10T0}}(\tilde{x}, \tilde{t}_M - \tilde{t}_{i-1})$ and $\tilde{T}_{\text{X12B10T0}}(\tilde{x}, \tilde{t}_M - \tilde{t}_i)$ are the temperature rises at time $\tilde{t} = \tilde{t}_M$ due to unit step changes in surface ($\tilde{x} = 0$) temperature applied at times $\tilde{t} = \tilde{t}_{i-1}$ and $\tilde{t} = \tilde{t}_i$, respectively. (See also the last paragraph of Section 2.3.4.)

For brevity, let

$$\tilde{T}_{\text{X12B10T0}}(\tilde{x}, \tilde{t}_M - \tilde{t}_{i-1}) = \varphi(\tilde{x}, \tilde{t}_{M-i+1}) = \varphi_{x,M-i+1} \tag{3.10a}$$

where $\tilde{t}_{M-i+1} = (M - i + 1)\Delta\tilde{t}$. Similarly, $\tilde{T}_{\text{X12B10T0}}(\tilde{x}, \tilde{t}_M - \tilde{t}_i) = \varphi_{x,M-i}$ with $\tilde{t}_M - \tilde{t}_i = \tilde{t}_{M-i} = (M - i)\Delta\tilde{t}$. Also, let

$$T_i(x, t_M) = T_{i,M}(x) \tag{3.10b}$$

Figure 3.1 Piecewise-constant approximation for $\theta_0(t)$, where the ith component, $\theta_{0,i}$, is time-independent.

Then, Eq. (3.9) becomes

$$T_{i,M}(x) = \theta_{0,i}\left(\varphi_{x,M-i+1} - \varphi_{x,M-i}\right) = \theta_{0,i}\Delta\varphi_{x,M-i} \qquad (3.11)$$

where $\varphi_{x,M-i} = 0$ for $i = M$ as the plate is initially at zero dimensionless temperature. Also, the symbol "Δ" indicates a first difference or derivative in time. In the current case, it denotes a forward difference (FD) of the φ_x temperature at time $t_{M-i} = (M-i)\Delta t$. A different derivation of Eq. (3.11) is given in Appendix C, Section C.2.1, using Green's functions.

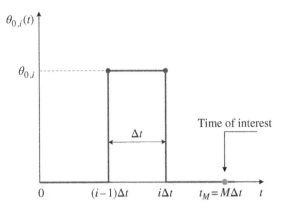

Figure 3.2 Component "i" of the piecewise-constant approximation for the surface temperature rise over the time interval $t = (i-1)\Delta t \rightarrow i\Delta t$.

3.3.1.1 Superposition-Based Numerical Approximation of the Solution

To compute the temperature $T(x, t)$ at time $t = t_M$ due to $\theta_0(t)$ applied up to $t = t_M$, the principle of superposition can be used. The temperature solution $T(x, t_M)$ may be calculated summing up all the contributions $T_{i,M}(x)$ due to the corresponding surface temperature rise components $\theta_{0,i}(t) = \theta_{0,i}$, each applied over its own time range $[(i-1)\Delta t, i\Delta t]$, with $i \leq M$. Bearing in mind that the plate is initially at uniform temperature T_{in} that is in general other than zero, it results in

$$T(x, t_M) = T_M(x) = T_{in} + \sum_{i=1}^{M} T_{i,M}(x) = T_{in} + \sum_{i=1}^{M} \theta_{0,i}\Delta\varphi_{x,M-i} \qquad (3.12)$$

where $T_{i,M}(x)$ defined by Eq. (3.11) has been used. In addition, the subscript "M" for $T_M(x)$ indicates that this temperature is calculated at the time step M as well as its ith contribution $T_{i,M}(x)$.

Equation (3.12) may be considered as the superposition-based numerical approximation of the temperature solution when the 1D body is subject to an arbitrary surface temperature and a piecewise-constant approximation for it is used. Also, it can be interpreted as the numerical form of the "*temperature-based*" Green's function 1D solution when dealing with a "*piecewise-constant*" approximate-in-time profile of the surface temperature, as shown in Appendix C, Section C.2.1. Equation (3.12) is very important in investigation of transient 1D inverse heat conduction problems (IHCPs) because it gives a convenient expression for the temperature in terms of the surface temperature rise components, i.e. $\theta_{0,1}, \theta_{0,2}, ..., \theta_{0,i}, ..., \theta_{0,M}$, as will be discussed in Chapter 4.

At the $x = 0$ boundary surface, Eq. (3.12) reduces rightly to $T_M(0) = T_{in} + \theta_{0,M} = T_{0,M}$ as $\Delta\varphi_{0,M-i} = 0$ for $i = 1, 2, ..., M-1$ and $\Delta\varphi_{0,M-i} = \varphi_{0,1} = 1$ for $i = M$ (see Section 2.3.6). This is an application of "symbolic" intrinsic verification of Eq. (3.12) (Cole et al. 2014; D'Alessandro and de Monte 2018).

3.3.1.2 Sequential-in-time Nature and Sensitivity Coefficients

A sequential-in-time nature of Eq. (3.12) can also be observed, one segment after another can be estimated, starting with the earliest times, and then moving successively to larger times. As an example, the first three time-steps as well as the Mth time-step are now considered:

- *First time step.* For $M = 1$, that is, when the time of interest is $t_1 = \Delta t$, Eq. (3.12) reduces to

$$T_1(x) = T_{in} + \theta_{0,1}\Delta\varphi_{x,0} = T_{in} + \theta_{0,1}\left(\varphi_{x,1} - \varphi_{x,0}\right) = T_{in} + \theta_{0,1}\varphi_{x,1} \qquad (3.13a)$$

where the surface temperature rise component $\theta_{0,1}$ refers to the midpoint $t = \Delta t/2$ of the time interval $t \in [0, \Delta t]$, that is, $\theta_{0,1} = \theta_0(t = \Delta t/2)$; while $T_1(x)$ refers to the endpoint of the same interval. Also, the first derivative of the temperature $T_1(x)$ with respect to the surface temperature component $\theta_{0,1}$ has a physical significance. In fact, it gives the sensitivity of $T_1(x)$ with respect to $\theta_{0,1}$ at the location of interest. For this reason, it is called temperature-based "*sensitivity coefficient*" for $T_1(x)$ and, in general, is denoted by $\Psi_{1,i}(x)$, with $i = 1, 2, ..., M$. It results in

$$\frac{\partial T_1(x)}{\partial \theta_{0,i}} = \Psi_{1,i}(x) = \begin{cases} \Delta\varphi_{x,0}, & i = 1 \\ 0, & i > 1 \end{cases} \qquad (3.13b)$$

- *Second time step.* For $M = 2$, that is, when the time of interest is $t_2 = 2\Delta t$, Eq. (3.12) reduces to

$$T_2(x) = T_{in} + \theta_{0,1}\Delta\varphi_{x,1} + \theta_{0,2}\Delta\varphi_{x,0} = T_{in} + \theta_{0,1}\left(\varphi_{x,2} - \varphi_{x,1}\right) + \theta_{0,2}\varphi_{x,1} \qquad (3.14a)$$

where the surface temperature rise component $\theta_{0,2}$ refers to the midpoint $t = 3\Delta t/2$ of the time interval $t \in [\Delta t, 2\Delta t]$, that is, $\theta_{0,2} = \theta_0(t = 3\Delta t/2)$; while $T_2(x)$ refers to the endpoint of the same interval. In the current case, the temperature-based sensitivity coefficient $\Psi_{2,i}(x)$ for $T_2(x)$ may be taken as

$$\frac{\partial T_2(x)}{\partial \theta_{0,i}} = \Psi_{2,i}(x) = \begin{cases} \Delta \varphi_{x,1-(i-1)}, & i \leq 2 \\ 0, & i > 2 \end{cases} \tag{3.14b}$$

- *Third time step.* For $M = 3$, that is, when the time of interest is $t_3 = 3\Delta t$, Eq. (3.12) becomes

$$T_3(x) = T_{in} + \theta_{0,1}\Delta \varphi_{x,2} + \theta_{0,2}\Delta \varphi_{x,1} + \theta_{0,3}\Delta \varphi_{x,0}$$
$$= T_{in} + \theta_{0,1}(\varphi_{x,3} - \varphi_{x,2}) + \theta_{0,2}(\varphi_{x,2} - \varphi_{x,1}) + \theta_{0,3}\varphi_{x,1} \tag{3.15a}$$

where the surface temperature rise component $\theta_{0,3}$ refers to the midpoint $t = 5\Delta t/2$ of the time interval $t \in [2\Delta t, 3\Delta t]$, that is, $\theta_{0,3} = \theta_0(t = 5\Delta t/2)$; while $T_3(x)$ refers to the endpoint of the same interval. The temperature-based sensitivity coefficient $\Psi_{3,i}(x)$ for $T_3(x)$ is given by

$$\frac{\partial T_3(x)}{\partial \theta_{0,i}} = \Psi_{3,i}(x) = \begin{cases} \Delta \varphi_{x,2-(i-1)}, & i \leq 3 \\ 0, & i > 3 \end{cases} \tag{3.15b}$$

- *Mth time step.* In the current case, Eq. (3.12) applies, and the sensitivity coefficient $\Psi_{M,i}(x)$ for $T_M(x)$ is given by

$$\frac{\partial T_M(x)}{\partial \theta_{0,i}} = \Psi_{M,i}(x) = \begin{cases} \Delta \varphi_{x,M-i}, & i \leq M \\ 0, & i > M \end{cases} \tag{3.16a}$$

Therefore, when using the above sensitivity coefficients, Eq. (3.12) may be rewritten as

$$T_M(x) = T_{in} + \sum_{i=1}^{M} \theta_{0,i}\Psi_{M,i}(x) \tag{3.16b}$$

The above sensitivity coefficients are very important when solving an inverse problem. In fact, they give insight on the location of temperature sensors for estimating the unknown surface temperature.

3.3.1.3 Basic "Building Block" Solution

The required building block solution for the approximation in Eq. (3.12) is the $\tilde{T}_{\text{X12B10T0}}(\tilde{x}, \tilde{t}) = \varphi(\tilde{x}, \tilde{t})$ temperature solution of the X12B10T0 case given in Section 2.3.6. The exact and computational analytical solutions for the X12B10T0 problem are described fully in this section of Chapter 2.

The computational analytical solution can be utilized choosing the accuracy desired (e.g., one part in 10^{15}) without excessive computational costs. This is of great concern as smaller time steps in the piecewise approximation will require higher accuracy of the building block solution to accommodate small differences in time, $\Delta \varphi_{x,M-i}$, that appear in the sensitivity matrix shown in Section 3.3.1.5. If the X12B10T0 fundamental solution is derived numerically (e.g., using finite element method), it might be not very accurate and numerical difficulties might arise when computing the sensitivity matrix. On the other side, the numerical solution of the building block might be very accurate but with an additional computational effort that might slow down remarkably the temperature computation.

3.3.1.4 Computer Code and Example

A MATLAB function implementing Eq. (3.12) and called fdX12B_0T0_pca.m (where "pca" denotes a piecewise-constant approximation) allows the dimensionless temperature to be computed at time $t = t_M = M\Delta t$ when the surface temperature at $x = 0$ varies with time as

$$T_0(t) = T_{in} + (T_0 - T_{in})\left(\frac{t}{t_{ref}}\right)^p = T_{in} + s_T t^p \quad (t \geq 0) \tag{3.17}$$

where p is a real number (positive or negative though zero is possible too) and $s_T = (T_0 - T_{in})/t_{ref}^p$.

Note that, for $p > 0$, the above equation gives $T_0(t = 0) = T_{0,0} = T_{in}$, that is, the surface temperature value at the initial time, $T_0(t = 0)$, is equal to the uniform initial temperature T_{in} of the plate. This choice does not lead to any loss of generality. In fact, if $T_{0,0} \neq T_{in}$, the principle of superposition shown in Section 3.2 and used throughout Section 3.3.1 can be applied. In other words, the problem can still be solved by breaking it into two separate subproblems, one subject to $T_{0,0}$ (constant) and, hence, denoted by X12B10T0 (see Section 2.3.6), and the other to Eq. (3.17); and then summing up the resulting temperature distributions. For $p < 0$, the initial ($t = 0$) surface temperature is infinitely large. However, this result is consistent with the current piecewise-constant approximation as its first component, $\theta_{0,1}$, is calculated at $t = \Delta t/2$, as shown in Figure 3.1. For the special case of $p = 0$, the current X12B-0T0 problem reduces to the X12B10T0 building block treated in Section 2.3.6 for which a MATLAB function is already available.

If the surface temperature rise at $x = 0$ is considered, that is, $\theta_0(t) = T_0(t) - T_{in}$, Eq. (3.17) may also be rewritten in dimensionless form as

$$\tilde{\theta}_0(\tilde{t}) = \frac{\theta_0(t)}{s_T(L^2/\alpha)^p} = \tilde{t}^p \quad (\tilde{t} \geq 0) \tag{3.18}$$

The fXd12B_0T0_pca.m function calculates the dimensionless temperature, $\tilde{T} = (T - T_{in})/[s_T(L^2/\alpha)^p]$, when the flat plate is subject to the dimensionless surface temperature listed above, and requires five dimensionless variables as input data. Two of these are the location $\tilde{x} = x/L$ and time $\tilde{t} = \alpha t/L^2$ of interest; one is related to the exponent of the input forcing, that is, p; lastly, A and M to the accuracy desired, where the former is related to the building block precision (one part in 10^4). Also, this function is available for ease of use on the web site related to this book and calls one basic "building block" function, that is, fdX12B10T0.m.

Example 3.1 A steel plate ($\alpha = 10^{-5}\,\mathrm{m}^2/\mathrm{s}$), initially at 30 °C, is subject at $x = 0$ to a surface temperature as given by Eq. (3.17), starting at time zero, with $p = 1$ (linear-in-time increase) and a heating rate of $s_T = (T_0 - T_{in})/t_{ref} = 20°\,\mathrm{C/s}$. The plate is insulated on the back side $x = L = 5$ cm. Calculate the temperature at 1 cm inside the flat plate at times 1, 2, and 3 seconds by using Eq. (3.12). Then, compare it with the exact values coming from the exact analytical solution defined in Section 2.3.7.

Solution
The temperature at the interior point of interest $x_P = 1$ cm is $T(x_P,t)$ and increases with time. If the piecewise-constant approximation of the applied surface temperature is based on a time step $\Delta t = 1$ s, by using Eq. (3.12) the temperatures at $t = 1$, 2, and 3 seconds are given, respectively, by

$$T_1(x_P) = T_{in} + \theta_{0,1}\varphi_1(\tilde{x}_P) = 30.253473\,°\mathrm{C} \tag{3.19a}$$

$$T_2(x_P) = T_{in} + \theta_{0,1}[\varphi_2(\tilde{x}_P) - \varphi_1(\tilde{x}_P)] + \theta_{0,2}\varphi_1(\tilde{x}_P) = 31.645409\,°\mathrm{C} \tag{3.19b}$$

$$T_3(x_P) = T_{in} + \theta_{0,1}[\varphi_3(\tilde{x}_P) - \varphi_2(\tilde{x}_P)]$$
$$+ \theta_{0,2}[\varphi_2(\tilde{x}_P) - \varphi_1(\tilde{x}_P)] + \theta_{0,3}\varphi_1(\tilde{x}_P) = 34.750928\,°\mathrm{C} \tag{3.19c}$$

where $\varphi_M(\tilde{x}_P) = \tilde{T}_{X12B10T0}(\tilde{x}_P, M\Delta\tilde{t})$, with $M = 1$, 2 or 3. Also, $\tilde{x}_P = x_P/L = 0.2$, $\Delta\tilde{t} = \alpha\Delta t/L^2 = 0.004$, and the first six decimal-places have been considered for the building block solution ($A = 6$).

The exact temperatures can be obtained from the fdX12B20T0.m MATLAB function of Section 2.3.7 as the current problem can be denoted by X12B20T1, where "2" in B20 indicates a linear-in-time variation of the surface temperature. They are

$$T(x_P = 1\,\mathrm{cm}, \Delta t = 1\,\mathrm{s}) = T_{in} + \underbrace{\tilde{t}_{ref}\tilde{T}\left(\tilde{x}_P = 0.2, \Delta\tilde{t} = 0.004, \tilde{t}_{ref} = 1, A = 15\right)}_{= \text{fdX12B20T0}(0.2,\ 0.004,\ 1,\ 15)}$$
$$\times \underbrace{\frac{(T_0 - T_{in})}{t_{ref}}}_{s_T = 20°\mathrm{C}\ \mathrm{s}^{-1}}\left(\frac{L^2}{\alpha}\right) = 30.112681°\mathrm{C} \tag{3.20a}$$

$$T(x_P = 1\,\mathrm{cm}, 2\Delta t = 2\,\mathrm{s}) = T_{in} + \underbrace{\tilde{t}_{ref}\tilde{T}\left(\tilde{x}_P = 0.2, 2\Delta\tilde{t} = 0.008, \tilde{t}_{ref} = 1, A = 15\right)}_{= \text{fdX12B20T0}(0.2,\ 0.008,\ 1,\ 15)}$$
$$\times \frac{(T_0 - T_{in})}{t_{ref}}\left(\frac{L^2}{\alpha}\right) = 31.480690°\mathrm{C} \tag{3.20b}$$

$$T(x_P = 1\,\mathrm{cm}, 3\Delta t = 3\,\mathrm{s}) = T_{in} + \underbrace{\tilde{t}_{ref}\tilde{T}\left(\tilde{x}_P = 0.2, 3\Delta\tilde{t} = 0.012, \tilde{t}_{ref} = 1, A = 15\right)}_{= \text{fdX12B20T0}(0.2,\ 0.012,\ 1,\ 15)}$$
$$\times \frac{(T_0 - T_{in})}{t_{ref}}\left(\frac{L^2}{\alpha}\right) = 34.613044°\mathrm{C} \tag{3.20c}$$

where $\tilde{t}_{ref}\tilde{T}$ is independent of $\tilde{t}_{ref}$ (see Section 2.3.7). For this reason, the simple numerical value of $\tilde{t}_{ref} = 1$ can be chosen. Also, an accuracy of one part in 10^{15} is chosen ($A=15$) in such a way as the computational analytical solution can be considered in every practical respect as exact (see Section 2.3.5.1).

Table 3.1 Comparison of exact and approximate temperatures for three different time steps using the piecewise-constant approximation (fdX12B_0T0_pca.m) at $x_p = 1$ cm.

t, s	Exact temperature, °C	Approximate temperature, °C					
		$\Delta t = 1$ s	M	$\Delta t = 0.25$ s	M	$\Delta t = 0.1$ s	M
1	30.112681	30.253473	1	30.120350	4	30.113902	10
2	31.480690	31.645409	2	31.490126	8	31.482196	20
3	34.613044	34.750928	3	34.620836	12	34.614287	30

Hence, the absolute errors in the approximate temperatures are +0.140792, +0.164719, and +0.137884 °C, respectively. If these errors are normalized relatively to the maximum temperature *rises* that occur at the boundary surface $x = 0$ up to and including the time of evaluation (i.e. 20, 40 and 60 °C, at $t = 1, 2$ and 3 seconds, respectively), they are: 0.7, 0.41, and 0.23%, respectively.

The above errors can be reduced if shorter time steps are chosen, as shown in Table 3.1, where the temperatures (°C) are obtained by using the fdX12B_0T0_pca.m function and the following equation that relates the dimensional and dimensionless temperatures:

$$T(x_P, t_M) = T_{in} + \underbrace{\tilde{T}(\tilde{x}_P, \tilde{t}_M, p, A, M)}_{= \text{fdX12B\_0T0\_pca}} \times s_T \left(\frac{L^2}{\alpha}\right)^p \tag{3.21}$$

where $\tilde{x}_P = x_P/L = 0.2$, $\tilde{t}_M = M\Delta\tilde{t}$, $\Delta\tilde{t} = \alpha\Delta t/L^2 = 0.004$, $p = 1$, and $A = 6$.

Note that, when using $\Delta t = 0.1$ s, the first two decimal places of the approximate temperatures are the same as the ones of the exact temperatures being the underlined digits different, as shown by Table 3.1.

3.3.1.5 Matrix Form of the Superposition-Based Numerical Approximation

It is frequently advantageous to perform algebraic manipulations utilizing a matrix form of the model. This section displays a matrix form for the superposition-based numerical statement of the solution Eq. (3.12), that is, the numerical statement of the temperature-based GFSE when dealing with the piecewise-constant approximation.

An expansion of the numerical form of Eq. (3.12), with M replaced by 1 to M, is

$$T_1(x) = T_{in} + \theta_{0,1}\Delta\varphi_{x,0} \tag{3.22a}$$

$$T_2(x) = T_{in} + \theta_{0,1}\Delta\varphi_{x,1} + \theta_{0,2}\Delta\varphi_{x,0} \tag{3.22b}$$

$$T_3(x) = T_{in} + \theta_{0,1}\Delta\varphi_{x,2} + \theta_{0,2}\Delta\varphi_{x,1} + \theta_{0,3}\Delta\varphi_{x,0} \tag{3.22c}$$

$$T_M(x) = T_{in} + \theta_{0,1}\Delta\varphi_{x,M-1} + \theta_{0,2}\Delta\varphi_{x,M-2} + \cdots + \theta_{0,M-1}\Delta\varphi_{x,1} + \theta_{0,M}\Delta\varphi_{x,0} \tag{3.22d}$$

The equations listed before can be written in a more compact form by using an appropriate matrix notation, resulting in

$$\mathbf{T} = \mathbf{\Psi}\mathbf{\theta}_0 + \mathbf{1}T_{in} \tag{3.22e}$$

where $\mathbf{T}$ and $\mathbf{\theta}_0$ are the vectors (M by 1) of temperature and surface temperature-rise components, respectively; $\mathbf{\Psi}$ is a square matrix M by M containing the forward differences $\Delta\varphi_{x,i}$ of the X12B10T0 basic building block solution, $\tilde{T}_{\text{X12B10T0}} = \varphi$, at time steps $i = 0, 1, 2, ..., M-1$; and $\mathbf{1}$ is a vector (M by 1) of ones. In detail,

$$\mathbf{T} = \begin{bmatrix} T_1(x) \\ T_2(x) \\ T_3(x) \\ \vdots \\ T_M(x) \end{bmatrix}, \quad \mathbf{\theta}_0 = \begin{bmatrix} \theta_{0,1} \\ \theta_{0,2} \\ \theta_{0,3} \\ \vdots \\ \theta_{0,M} \end{bmatrix} \tag{3.23}$$

$$\mathbf{\Psi} = \begin{bmatrix} \Delta\varphi_{x,0} & 0 & 0 & 0 & \cdots & 0 & 0 \\ \Delta\varphi_{x,1} & \Delta\varphi_{x,0} & 0 & 0 & \cdots & 0 & 0 \\ \Delta\varphi_{x,2} & \Delta\varphi_{x,1} & \Delta\varphi_{x,0} & 0 & \cdots & 0 & 0 \\ \vdots & \vdots & \vdots & \vdots & \vdots & \vdots & \vdots \\ \Delta\varphi_{x,M-1} & \Delta\varphi_{x,M-2} & \Delta\varphi_{x,M-3} & \cdots & \Delta\varphi_{x,2} & \Delta\varphi_{x,1} & \Delta\varphi_{x,0} \end{bmatrix} \qquad (3.24)$$

The $\mathbf{\Psi}$ matrix is called the sensitivity coefficient matrix for $\boldsymbol{\theta}_0$ as its elements or entries are the sensitivity coefficients of the temperature with respect to the surface temperature components according to Eq. (3.16a). Its structure is "lower triangular" as the off-diagonal entries lying above the main diagonal are zero (Potter et al. 2016). Also, the entries along the main diagonal are $\Delta\varphi_{x,0} = \varphi_{x,1}$, the elements along the diagonal just below the main one are $\Delta\varphi_{x,1}$, and so on.

3.3.2 Piecewise-Linear Approximation

This approximation gives the piecewise profile of Figure 3.3, that is a sequence of linear segments where $\theta_{0,i}$ (ith surface temperature-rise component) is a linear approximation for $\theta_0(t)$ between $t_{i-1} = (i-1)\Delta t$ and $t_i = i\Delta t$ (with $i = 1, 2, ..., M$, being $t_M = M\Delta t$ the time of interest).

Therefore, contrary to the piecewise-constant approximation of Section 3.3.1, now the ith surface temperature rise component, $\theta_{0,i}$, depends linearly on the time as

$$\theta_{0,i}(t) = \theta_{0,i-1} + \Delta\theta_{0,i}\frac{t-(i-1)\Delta t}{\Delta t}, \quad t \in [(i-1)\Delta t, i\Delta t] \qquad (3.25)$$

where

$$\theta_{0,i-1} = \theta_0[t = (i-1)\Delta t] \qquad (3.26a)$$

$$\Delta\theta_{0,i} = \theta_{0,i} - \theta_{0,i-1} \qquad (3.26b)$$

$$\theta_{0,i} = \theta_0(t = i\Delta t) \qquad (3.26c)$$

The temperature solution $T_i(x, t)$ at time $t = t_M$, that is, $T_i(x, t_M) = T_{i,M}(x)$ due only to the surface temperature-rise component $\theta_{0,i}(t)$ (with $i \leq M$) given by Eqs. (3.25)–(3.26c) and shown in Figure 3.4, when the plate is initially at zero temperature, may be calculated by applying the principles of superposition using the X12B10T0 and X12B20T0 building blocks:

$$T_i(\tilde{x},\tilde{t}_M) = T_{i,M}(\tilde{x}) = \theta_{0,i-1}\left[\tilde{T}_{X12B10T0}(\tilde{x},\tilde{t}_M - \tilde{t}_{i-1}) - \tilde{T}_{X12B10T0}(\tilde{x},\tilde{t}_M - \tilde{t}_i)\right]$$
$$+ \Delta\theta_{0,i}\left[\tilde{T}_{X12B20T0}\left(\tilde{x},\tilde{t}_M - \tilde{t}_{i-1}, \tilde{t}_{ref} = \Delta\tilde{t}\right) - \tilde{T}_{X12B20T0}(\tilde{x},\tilde{t}_M - \tilde{t}_i,\tilde{t}_{ref} = \Delta\tilde{t})\right] \qquad (3.27)$$
$$- \Delta\theta_{0,i}\tilde{T}_{X12B10T0}(\tilde{x},\tilde{t}_M - \tilde{t}_i)$$

where, according to the dimensionless variables defined by Eqs. (2.4) and (2.5) in Chapter 2, $\tilde{x} = x/L$, $\tilde{t} = \alpha t/L^2$ and $\tilde{T} = (T - T_{in})/(T_0 - T_{in})$. Also, $\tilde{t}_{ref} = \alpha t_{ref}/L^2$ (where t_{ref} is defined through Eq. (2.2)), $\tilde{t}_M = M\Delta\tilde{t}$, $\tilde{t}_{i-1} = (i-1)\Delta\tilde{t}$ and $\tilde{t}_i = i\Delta\tilde{t}$, with $\Delta\tilde{t} = \alpha\Delta t/L^2$. Note that $\tilde{T}_{X12B20T0}(\tilde{x}, \tilde{t}_M - \tilde{t}_{i-1}, \tilde{t}_{ref} = \Delta\tilde{t})$ and $\tilde{T}_{X12B20T0}(\tilde{x}, \tilde{t}_M - \tilde{t}_i, \tilde{t}_{ref} = \Delta\tilde{t})$ are the temperature rises at time $\tilde{t} = \tilde{t}_M$ due to unit linear-in-time increases in surface ($\tilde{x} = 0$) temperature (over the time $\tilde{t}_{ref} = \Delta\tilde{t}$) applied at times $\tilde{t} = \tilde{t}_{i-1}$ and $\tilde{t} = \tilde{t}_i$, respectively. (See also the last paragraph of Section 2.3.4.)

Eq. (3.27) is a little complicated and deserves some explanation. The term on the right-hand side in the first line accounts for the "step" pulse of magnitude $\theta_{0,i-1}$ that is under the triangle in Figure 3.4. This is the same as the expression obtained in the piecewise-constant approximation. The second line is the effect of the ramped increase $\Delta\theta_{0,i}$: the first term in the brackets activates this component, and the second term

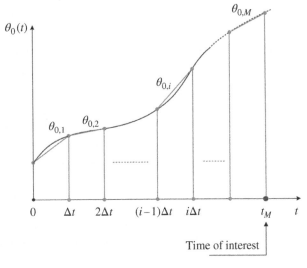

Figure 3.3 Piecewise-linear approximation for $\theta_0(t)$ where the ith component, $\theta_{0,i}$, varies linearly with time.

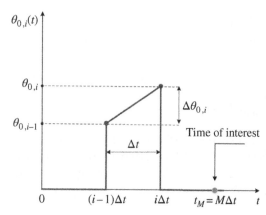

Figure 3.4 Component "*i*" of the piecewise-linear approximation for the surface temperature rise defined by Eq. (3.25)–(3.26c).

deactivates the ramped increase. The last term is needed to suppress the remnant of the difference of the two ramps.[1]

For brevity, let

$$\tilde{T}_{\text{X12B20T0}}\big(\tilde{x}, \tilde{t}_M - \tilde{t}_{i-1}, \tilde{t}_{ref} = \Delta\tilde{t}\big) = \mu\big(\tilde{x}, \tilde{t}_{M-i+1}\big) = \mu_{x,M-i+1} \qquad (3.28)$$

where $\tilde{t}_{M-i+1} = (M - i + 1)\Delta\tilde{t}$. Similarly, $\tilde{T}_{\text{X12B20T0}}\big(\tilde{x}, \tilde{t}_M - \tilde{t}_i, \tilde{t}_{ref} = \Delta\tilde{t}\big)$
$= \mu_{x,M-i}$ with $\tilde{t}_M - \tilde{t}_i = \tilde{t}_{M-i} = (M - i)\Delta\tilde{t}$.

Recalling also the definition of $\varphi_{x,M-i}$ given in Section 3.3.1, Eq. (3.27) can be written as

$$T_{i,M}(x) = \theta_{0,i-1}\big(\varphi_{x,M-i+1} - \varphi_{x,M-i}\big)$$
$$+ \Delta\theta_{0,i}\big(\mu_{x,M-i+1} - \mu_{x,M-i}\big) - \Delta\theta_{0,i}\varphi_{x,M-i} \qquad (3.29a)$$

Alternatively, because $\Delta\theta_i = \theta_i - \theta_{i-1}$, Eq. (3.29a) may be written as

$$T_{i,M}(x) = \theta_{0,i-1}\big(\varphi_{x,M-i+1} - \Delta\mu_{x,M-i}\big) + \theta_{0,i}\big(\Delta\mu_{x,M-i} - \varphi_{x,M-i}\big) \qquad (3.29b)$$

where

$$\Delta\mu_{x,M-i} = \mu_{x,M-i+1} - \mu_{x,M-i} \qquad (3.29c)$$

indicates a forward difference of the μ_x temperature at time $t = (M - i)\Delta t$. Also, $\mu_{x,M-i} = 0$ for $i = M$ as the plate is initially at zero dimensionless temperature. Therefore, $\Delta\mu_{x,0} = \mu_{x,1}$. Another explanation of Eq. (3.29b) is given in Appendix C, Section C.2.2, using Green's functions.

3.3.2.1 Superposition-Based Numerical Approximation of the Solution

To compute the temperature $T(x,t)$ at time $t = t_M$ due to $\theta_0(t)$ applied up to $t = t_M$, the principle of superposition can be utilized. The temperature solution $T(x, t_M) = T_M(x)$ may be calculated summing up all the contributions $T_{i,M}(x)$ given by Eq. (3.29b) and due to the corresponding surface temperature rise components $\theta_{0,i}(t)$, defined by Eqs. (3.25)-(3.26c), each applied over its own time range $[(i - 1)\Delta t, i\Delta t]$, with $i \le M$. Bearing also in mind that the plate is initially at the T_{in} uniform temperature, the result is

$$T_M(x) = T_{in} + \sum_{i=1}^{M} T_{i,M}(x)$$

$$= T_{in} + \sum_{i=1}^{M} \theta_{0,i-1}\big(\varphi_{x,M-i+1} - \Delta\mu_{x,M-i}\big) + \sum_{i=1}^{M} \theta_{0,i}\big(\Delta\mu_{x,M-i} - \varphi_{x,M-i}\big) \qquad (3.30)$$

An alternative form and more convenient expression of Eq. (3.30) is as follows:

$$T_M(x) = T_{in} + \theta_{0,0}\big(\varphi_{x,M} - \Delta\mu_{x,M-1}\big) + \sum_{i=1}^{M-1} \theta_{0,i}\big[\Delta(\Delta\mu_x)_{M-i}\big] + \theta_{0,M}\Delta\mu_{x,0} \qquad (3.31)$$

where $\theta_{0,0} = T_0(t = 0) - T_{in} = T_{0,0} - T_{in}$ can in general be different from zero, and

$$\Delta(\Delta\mu_x)_{M-i} = \Delta\mu_{x,M-i} - \Delta\mu_{x,M-i-1} = \mu_{x,M-i+1} - 2\mu_{x,M-i} + \mu_{x,M-i-1} \qquad (3.32)$$

Note that the quantity $T_{in} + \theta_{0,0}\varphi_{x,M}$ appearing in Eq. (3.31) represents the solution to a constant surface temperature rise equal to $\theta_{0,0}$. Equation (3.31) may be derived by observing the sequential-in-time nature of Eq. (3.30), as shown in the next section. Also, $\Delta(\Delta\mu_x)_{M-i}$ represents a second difference or derivative in time. It indicates a second central difference (CD) of the $\mu(\tilde{x}, \tilde{t})$ temperature at time $t = (M - i)\Delta t$. Also, at the $x = 0$ boundary surface, Eq. (3.31) reduces rightly to $T_M(0) = T_{in} + \theta_{0,M}$ as $\varphi_{0,M} = \Delta\mu_{0,M-1} = 1$, $\Delta(\Delta\mu_0)_{M-i} = 0$ for $i = 1, 2, ..., M - 1$ and $\Delta\mu_{0,0} = 1$ (see Sections 2.3.6 and 2.3.7). This is an application of "symbolic" intrinsic verification of Eqs. (3.31)–(3.32) (Cole et al. 2014; D'Alessandro and de Monte 2018). Another application of symbolic intrinsic verification is when the surface temperature rise is time-independent, that is, $\theta_{0,0} = \theta_{0,1} = ... = \theta_{0,i} = ... = \theta_{0,M}$. In such a case, in fact, Eq. (3.31) reduces properly to $T_M(x) = T_{in} + \theta_{0,0}\varphi_{x,M}$, as expected.

Equation (3.31) may be considered as the superposition-based numerical approximation of the temperature solution when the 1D body is subject to an arbitrary surface temperature and a piecewise-linear approximation for it is utilized.

---

1 For example, consider $f(t)=t$. The difference $f(t)-f(t-1)=t-(t-1)=1$. The increase with time is eliminated but a constant remains.

Also, it can be interpreted as the numerical form of the "*temperature-based*" Green's function 1D solution when dealing with a *piecewise-linear* approximate-in-time profile of the applied surface temperature, as shown in Appendix C, Section C.2.2. Equation (3.31) is very important in investigation of transient 1D IHCPs because it gives a convenient expression for the temperature in terms of the surface temperature values at different times, that is, $\theta_{0,0}$, $\theta_{0,1}$, $\theta_{0,2}$, ..., $\theta_{0,i}$ ..., $\theta_{0,M}$, as discussed in Chapter 4.

3.3.2.2 Sequential-in-time Nature and Sensitivity Coefficients

The sequential-in-time nature of Eq. (3.30) indicates that one segment after another can be estimated, starting with the earliest times, and then moving successively to larger times. As an example, the first three time-steps, as well as the Mth one, are now considered:

- *First time step.* For $M = 1$, that is, when the time of interest is $t = \Delta t$, bearing in mind that $\varphi_{x,0} = 0$, and $\mu_{x,0} = 0$, Eq. (3.30) reduces to

$$T_1(x) = T_{in} + \theta_{0,0}\left(\varphi_{x,1} - \mu_{x,1}\right) + \theta_{0,1}\mu_{x,1} \tag{3.33a}$$

where $T_1(x)$ refers to the time $t = \Delta t$. In addition, the first derivative of the temperature $T_1(x)$ with respect to the surface temperature rise $\theta_{0,i}$ (with $i = 0, 1, 2, ..., M$) gives the sensitivity of $T_1(x)$ with respect to $\theta_{0,i}$ at the location of interest. It is the so-called temperature-based "sensitivity coefficient" for $T_1(x)$ (already defined in Section 3.3.1.2) and, in general, is denoted by $\Psi_{1,i}(x)$. It results in

$$\frac{\partial T_1(x)}{\partial \theta_{0,i}} = \Psi_{1,i}(x) = \begin{cases} \varphi_{x,1} - \mu_{x,1}, & i = 0 \\ \mu_{x,0}, & i = 1 \\ 0, & i = 2, 3, ... \end{cases} \tag{3.33b}$$

- *Second time step.* For $M = 2$, that is, when the time of evaluation is $t = 2\Delta t$, bearing in mind that $\varphi_{x,0} = 0$ and $\mu_{x,0} = 0$, Eq. (3.30) becomes

$$T_2(x) = T_{in} + \theta_{0,0}\left(\varphi_{x,2} - \Delta\mu_{x,1}\right) + \theta_{0,1}\left[\Delta(\Delta\mu_x)_1\right] + \theta_{0,2}\mu_{x,1} \tag{3.34a}$$

where $T_2(x)$ refers to $t = 2\Delta t$ and, according to Eq. (3.32),

$$\Delta(\Delta\mu_x)_1 = \Delta\mu_{x,1} - \Delta\mu_{x,0} = \mu_{x,2} - 2\mu_{x,1} + \mu_{x,0} = \mu_{x,2} - 2\mu_{x,1} \tag{3.34b}$$

Also, in the current case, the temperature-based sensitivity coefficient $\Psi_{2,i}(x)$ for $T_2(x)$ may be taken as

$$\frac{\partial T_2(x)}{\partial \theta_{0,i}} = \Psi_{2,i}(x) = \begin{cases} \varphi_{x,2} - \Delta\mu_{x,1}, & i = 0 \\ \Delta(\Delta\mu_x)_1, & i = 1 \\ \mu_{x,1}, & i = 2 \\ 0, & i = 3, 4, ... \end{cases} \tag{3.34c}$$

- *Third time step.* For $M = 3$, that is, when the time of interest is $t = 3\Delta t$, Eq. (3.30) becomes

$$T_3(x) = T_{in} + \theta_{0,0}\left(\varphi_{x,3} - \Delta\mu_{x,2}\right) + \theta_{0,1}\left[\Delta(\Delta\mu_x)_2\right] + \theta_{0,2}\left[\Delta(\Delta\mu_x)_1\right] \\ + \theta_{0,3}\mu_{x,1} \tag{3.35a}$$

where $T_3(x)$ refers to $t = 3\Delta t$, and

$$\Delta(\Delta\mu_x)_2 = \Delta\mu_{x,2} - \Delta\mu_{x,1} = \mu_{x,3} - 2\mu_{x,2} + \mu_{x,1} \tag{3.35b}$$

Then, the temperature-based sensitivity coefficient $\Psi_{3,i}(x)$ for $T_3(x)$ is given by

$$\frac{\partial T_3(x)}{\partial \theta_{0,i}} = \Psi_{3,i}(x) = \begin{cases} \varphi_{x,3} - \Delta\mu_{x,2}, & i = 0 \\ \Delta(\Delta\mu_x)_{3-i}, & i = 1, 2 \\ \mu_{x,1}, & i = 3 \\ 0, & i = 4, 5, ... \end{cases} \tag{3.35c}$$

- *Mth time step.* In the current case, Eq. (3.31) with Eq. (3.32) is obtained. Then, the sensitivity coefficients $\Psi_{M,i}(x)$ for $T_M(x)$ are given by

$$\frac{\partial T_M(x)}{\partial \theta_{0,i}} = \Psi_{M,i}(x) = \begin{cases} \varphi_{x,M} - \Delta\mu_{x,M-1}, & i = 0 \\ \Delta(\Delta\mu_x)_{M-i}, & i = 1, 2, ..., M-1 \\ \mu_{x,1}, & i = M \\ 0, & i = M+1, M+2, ... \end{cases} \tag{3.36a}$$

Therefore, when using the above sensitivity coefficients, Eq. (3.31) may be rewritten as

$$T_M(x) = T_{in} + \theta_{0,0}\Psi_{M,0}(x) + \sum_{i=1}^{M-1} \theta_{0,i}\Psi_{M,i}(x) + \theta_{0,M}\Psi_{M,M}(x) \tag{3.36b}$$

The above sensitivity coefficients are very important when solving an inverse problem. In fact, they give insight on the location of temperature sensors for estimating the unknown surface temperature.

3.3.2.3 Basic "Building Block" Solutions

The required building block solutions for the approximation in Eq. (3.31) are: (i) the $\tilde{T}_{X12B10T0}(\tilde{x}, \tilde{t}) = \varphi(\tilde{x}, \tilde{t})$ temperature solution of the X12B10T0 case given in Section 2.3.6, and (ii) $\tilde{T}_{X12B20T0}(\tilde{x}, \tilde{t}, \tilde{t}_{ref} = \Delta\tilde{t}) = \mu(\tilde{x}, \tilde{t})$ that is the dimensionless temperature solution of the X12B20T0 case treated in Section 2.3.7. The corresponding exact and computational analytical solutions are described fully in Sections 2.3.6 and 2.3.7. However, when $\theta_{0,0} = 0$, Eq. (3.31) uses only one basic building block, that is, $\mu(\tilde{x}, \tilde{t})$.

The computational solution can be utilized choosing the accuracy desired, and smaller time steps in the piecewise approximation will require higher accuracy to accommodate small differences in time, that is, $\Delta\mu_{x,M-1}$ and $\Delta(\Delta\mu_x)_{M-i}$.

3.3.2.4 Computer Code and Examples

A MATLAB function implementing Eq. (3.31) and called fdX12B_0T0_pla.m allows the dimensionless temperature, $\tilde{T} = (T - T_{in})/[s_T(L^2/\alpha)^p]$, to be computed when the surface temperature varies with time following Eqs. (3.17) and (3.18), where p is now only a positive real number though zero is possible too (see Section 3.3.1.4). In fact, a negative value of p would lead to an infinite initial value of the surface temperature that is not consistent with the current approximate piecewise-linear profile whose first value, $\theta_{0,0}$, is calculated at time $t = 0$, as shown in Figure 3.3. Also, "pla" indicates a piecewise-linear approximation.

This function requires as input data five dimensionless variables, namely $\tilde{x}, \tilde{t}, p, A,$ and M, defined in Section 3.3.1.4. It is available for ease of use on the web site related to this book, and calls only one basic "building block" function, that is, fdX12B20T0.m, as $\theta_{0,0} = \theta_0(t = 0) = 0$ according to Eqs. (3.17) and (3.18).

Example 3.2   Calculate the temperature at 1 cm inside the steel plate of Example 3.1 of Section 3.3.1.4 at times 1, 2, and 3 seconds by using Eq. (3.31). Then, compare it with the exact values coming from the exact analytical solution defined in Section 2.3.7.

Solution

The temperature at the interior point of interest $x_P = 1$ cm is $T(x_P, t)$ and increases with time. If the piecewise-linear approximation of the applied surface temperature is based on a time step $\Delta t = 1$ s, by using Eq. (3.31) with $\theta_{0,0} = 0$ the temperatures at $t = 1$, 2 and 3 seconds are given, respectively, by

$$T_1(x_P) = T_{in} + \theta_{0,1}\mu_1(\tilde{x}_P) = 30.112681\,°C \tag{3.37a}$$

$$T_2(x_P) = T_{in} + \theta_{0,1}[\Delta(\Delta\mu(\tilde{x}_P))_1] + \theta_{0,2}\mu_1(\tilde{x}_P) = 31.480690\,°C \tag{3.37b}$$

$$T_3(x_P) = T_{in} + \theta_{0,1}[\Delta(\Delta\mu(\tilde{x}_P))_2] + \theta_{0,2}[\Delta(\Delta\mu(\tilde{x}_P))_1] + \theta_{0,3}\Delta\mu_1(\tilde{x}_P) = 34.613044\,°C \tag{3.37c}$$

where $\mu_M(\tilde{x}_P) = \tilde{T}_{X12B20T0}(\tilde{x}_P, M\Delta\tilde{t}, \tilde{t}_{ref} = \Delta\tilde{t})$, with $M = 1$, 2 or 3. Also, $\tilde{x}_P = x_P/L = 0.2$, $\Delta\tilde{t} = \alpha\Delta t/L^2 = 0.004$, and the first six decimal-places have been considered for the building block solution ($A = 6$).

The exact temperatures can be obtained from the fdX12B20T0.m MATLAB function of Section 2.3.7 with an accuracy of one part in 10^{15} ($A = 15$) in such a way as the computational analytical solution can be assumed exact (see Section 2.3.5.1). They are 30.112681, 31.480690, and 34.613044\,°C, respectively, as already given in Example 3.1.

Therefore, the approximate and exact temperatures are the same (as expected) as the applied surface temperature changes linearly with time and, hence, the piecewise-linear approximation cannot perform any approximation in the current case. This is an application of "numerical" intrinsic verification of Eq. (3.31) and, hence, of the fdX12B_0T0_pla.m MATLAB function (D'Alessandro and de Monte 2018).

Example 3.3 A steel plate ($\alpha = 10^{-5}$ m²/s), initially at 20 °C, is subject at $x = 0$ to a surface temperature as given by Eq. (3.17), starting at time zero, with $p = 2$ (quadratic-in-time increase) and $s_T = (T_0 - T_{in})/t_{ref}^2 = 2$ °C s^{-2}. The plate is insulated on the back side $x = L = 5$ cm. Calculate the temperature at 1.25 cm inside the flat plate at times 5 and 10 seconds by using the fdX12B_0T0_pla.m MATLAB function and by choosing five different time steps, that is, $\Delta t = 5, 1, 0.1, 0.01$ and 0.001 s. Then, compare it with the approximate values coming from the fdX12B_0T0_pca.m function of Section 3.3.1.4 (piecewise-constant approximation) and with the exact values coming from the exact analytical solution given by Beck et al. (2008; see Eq. (16c) and Tables 1 and 2).

Solution

The current problem can be denoted by X12B30T1, where "3" in B30 indicates a power-in-time variation of the surface temperature (see Table A.3 in Appendix A). The temperature at the interior point of interest $x_P = 1.25$ cm is $T(x_P, t)$ and increases with time. By using the fdX12B_0T0_pca.m and fdX12B_0T0_pla.m functions, approximate values of temperature are obtained at times 5 and 10 seconds and shown in Table 3.2, where the temperatures in units of °C are obtained by using Eq. (3.21) for $p = 2$ of Example 3.1.

The exact temperatures can be obtained by using the exact analytical solution given by Beck et al. (2008) in dimensionless form and setting an accuracy of one part in 10^{15} ($A = 15$) when truncating the infinite series-solution. In such a way, the computational analytical solution can be assumed exact (see Section 2.3.5.1). The solution is

$$
\tilde{T}(\tilde{x}, \tilde{t}) = \frac{1}{\tilde{t}_{ref}^2}\left[\tilde{t}^2 + \tilde{t}(\tilde{x}^2 - 2\tilde{x}) + \left(\frac{\tilde{x}^4}{12} - \frac{\tilde{x}^3}{3} + \frac{2}{3}\tilde{x}\right)\right]
$$

$$
- \frac{4}{\tilde{t}_{ref}^2}\sum_{m=1}^{\infty}\frac{e^{-[(m-1/2)\pi]^2\tilde{t}}}{[(m-1/2)\pi]^5}\sin[(m-1/2)\pi\tilde{x}]
$$

(3.38)

where $\tilde{x} = x/L$, $\tilde{t} = \alpha t/L^2$ and $\tilde{T} = (T - T_{in})/(T_0 - T_{in})$. Also, $\tilde{t}_{ref} = \alpha t_{ref}/L^2$.

Table 3.2 Comparison of exact and approximate temperatures (limited to the first eight decimal-places) at $x_p = 1.25$ cm and times 5 and 10 seconds for different time steps using the piecewise-constant and piecewise-linear approximations for the surface temperature rise.

			$t = 5$ s	
Δt, s	M	Approximate temperature (fdX12B_0T0_pca.m), °C	Approximate temperature (fdX12B_0T0_pla.m), °C	Exact temperature, °C
5	1	22.64124434	24.24161882	
1	5	22.26761678	22.30555709	22.23522294
0.1	50	22.23557509	22.23592727	
0.01	500	22.23522646	22.23522998	
0.001	5000	22.23522297	22.23522301	
$t = 10$ s				
5	2	45.83944371	47.87473453	
1	10	44.55980777	44.62897732	44.50351183
0.1	100	44.50413975	44.50476769	
0.01	1000	44.50351811	44.50352439	
0.001	10000	44.50351189	44.50351195	

Table 3.3 Exact and approximate variations of the surface temperature into the range $t \in [0, 5\text{ s}]$ when dealing with $\Delta t = 5$ s.

	$\theta_0(t)$, °C, with $t \in [0, 5\text{ s}]$		
t, s	Exact: quadratic $\theta_0(t) = 2t^2$	Approximate: constant, $\theta_{0,1}$	Approximate: linear, $10t$
0	0	0	0
1	2	12.5	10
2	8	12.5	20
3	18	12.5	30
4	32	12.5	40
5	50	12.5	50

The exact values of temperature are

$$T(x_P = 1.25\,\text{cm}, t = 5\,\text{s}) = T_{in} + \underbrace{\tilde{t}_{ref}^2 \tilde{T}(\tilde{x}_P = 0.25, \tilde{t} = 0.02, \tilde{t}_{ref} = 1, A = 15)}_{= 1.788178354316260e-05}$$

$$\times \underbrace{\frac{(T_0 - T_{in})}{t_{ref}^2}\left(\frac{L^2}{\alpha}\right)^2}_{s_T = 2°\text{C s}^{-2}} = 22.23522294\ °\text{C} \qquad (3.39\text{a})$$

$$T(x_P = 1.25\,\text{cm}, t = 10\,\text{s}) = T_{in} + \underbrace{\tilde{t}_{ref}^2 \tilde{T}(\tilde{x}_P = 0.25, \tilde{t} = 0.04, \tilde{t}_{ref} = 1, A = 15)}_{= 1.960280946552985e-04}$$

$$\times \frac{(T_0 - T_{in})}{t_{ref}^2}\left(\frac{L^2}{\alpha}\right)^2 = 44.50351183\ °\text{C} \qquad (3.39\text{b})$$

where $\tilde{t}_{ref}^2 \tilde{T}$ is independent of $\tilde{t}_{ref}$, as shown by Eq. (3.38). For this reason, the simple numerical value of $\tilde{t}_{ref} = 1$ can be chosen.

These temperatures are given in Table 3.2 for visual comparison where the different digits of the approximate temperatures with respect to the exact ones are underlined. Note that, when dealing with the piecewise-constant approximation and a time step of 0.001 seconds, the first seven decimal-places of the approximate temperature are the same as the ones of the exact temperature for both times 5 and 10 seconds. In this case, the absolute errors in the approximate temperatures are +0.00000003 °C and +0.00000006 °C at times 5 seconds and 10 seconds, respectively. By using the piecewise-linear approximation and $\Delta t = 0.001$ s, these errors are +0.00000007 °C and +0.00000012 °C, respectively, that is, they are doubled.

Therefore, in the current case of a quadratic-in-time variation of the surface temperature, the piecewise-constant approximation is just a bit more accurate than the piecewise-linear one. This is so, since the linear approximation of the quadratic-in-time increase of the surface temperature gives higher values than the constant approximation during the single time step apart from early times. As an example, for $t = 5$ seconds and with a time step of 5 seconds, the surface temperature variation into the range $t \in [0, 5\text{ s}]$ is shown in both Table 3.3 and Figure 3.5.

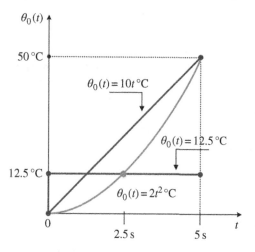

Figure 3.5 Plot of exact and approximate variations of the surface temperature into the range $t \in [0, 5\text{ s}]$ when dealing with $\Delta t = 5$ s.

3.3.2.5 Matrix Form of the Superposition-Based Numerical Approximation

It is frequently advantageous to perform algebraic manipulations utilizing a matrix form of the model. This section displays a matrix form for the superposition-based numerical statement of the solution Eq. (3.31), that is, the numerical statement of the temperature-based GFSE when dealing with the piecewise-linear approximation.

An expansion of the numerical form of Eq. (3.31), with M replaced by 1 to M, is (recalling that $\Delta\mu_{x,0} = \mu_{x,1}$)

$$T_1(x) = T_{in} + \theta_{0,0}(\varphi_{x,1} - \Delta\mu_{x,0}) + \theta_{0,1}\Delta\mu_{x,0} \tag{3.40a}$$

$$T_2(x) = T_{in} + \theta_{0,0}(\varphi_{x,2} - \Delta\mu_{x,1}) + \theta_{0,1}[\Delta(\Delta\mu_x)_1] + \theta_{0,2}\Delta\mu_{x,0} \tag{3.40b}$$

$$T_3(x) = T_{in} + \theta_{0,0}(\varphi_{x,3} - \Delta\mu_{x,2}) + \theta_{0,1}[\Delta(\Delta\mu_x)_2] + \theta_{0,2}[\Delta(\Delta\mu_x)_1] \\ + \theta_{0,3}\Delta\mu_{x,0} \tag{3.40c}$$

$$T_M(x) = T_{in} + \theta_{0,0}(\varphi_{x,M} - \Delta\mu_{x,M-1}) + \theta_{0,1}[\Delta(\Delta\mu_x)_{M-1}] + \theta_{0,2}[\Delta(\Delta\mu_x)_{M-2}] \\ + \cdots + \theta_{0,M-2}[\Delta(\Delta\mu_x)_2] + \theta_{0,M-1}[\Delta(\Delta\mu_x)_1] + \theta_{0,M}\Delta\mu_{x,0} \tag{3.40d}$$

The equations listed before can be written in a more compact form by using an appropriate matrix notation. It results in

$$\mathbf{T} = \mathbf{\Psi}\mathbf{\theta}_0 + \mathbf{\Psi}_0\theta_{0,0} + \mathbf{1}T_{in} \tag{3.41}$$

where $\mathbf{T}$ and $\mathbf{\theta}_0$ are the vectors (M by 1) of temperature $T_i(x)$ and surface temperature-rise values $\theta_{0,i}$, respectively, at time steps $i = 1, 2, ..., M$; $\mathbf{\Psi}_0$ is a vector (M by 1) containing the forward differences of the X12B20T0 basic building block solution at time steps $i = 0, 1, 2, ..., M-1$ and the values of the X12B10T0 solution at time steps $i = 1, 2, ..., M$; $\mathbf{\Psi}$ is a square matrix M by M containing the central differences of the X12B20T0 solution at time steps $i = 1, 2, ..., M-1$ with the exception of the main diagonal where the forward difference at $i = 0$ of the same building block appears; and, lastly, $\mathbf{1}$ is a vector (M by 1) of ones. In detail,

$$\mathbf{T} = \begin{bmatrix} T_1(x) \\ T_2(x) \\ T_3(x) \\ \vdots \\ T_M(x) \end{bmatrix}, \quad \mathbf{\theta}_0 = \begin{bmatrix} \theta_{0,1} \\ \theta_{0,2} \\ \theta_{0,3} \\ \vdots \\ \theta_{0,M} \end{bmatrix}, \quad \mathbf{\Psi}_0 = \begin{bmatrix} \varphi_{x,1} - \Delta\mu_{x,0} \\ \varphi_{x,2} - \Delta\mu_{x,1} \\ \varphi_{x,3} - \Delta\mu_{x,2} \\ \vdots \\ \varphi_{x,M} - \Delta\mu_{x,M-1} \end{bmatrix} \tag{3.42}$$

$$\mathbf{\Psi} = \begin{bmatrix} \Delta\mu_{x,0} & 0 & 0 & 0 & \cdots & 0 & 0 \\ \Delta(\Delta\mu_x)_1 & \Delta\mu_{x,0} & 0 & 0 & \cdots & 0 & 0 \\ \Delta(\Delta\mu_x)_2 & \Delta(\Delta\mu_x)_1 & \Delta\mu_{x,0} & 0 & \cdots & 0 & 0 \\ \vdots & \vdots & \vdots & \vdots & \vdots & \vdots & \vdots \\ \Delta(\Delta\mu_x)_{M-1} & \Delta(\Delta\mu_x)_{M-2} & \Delta(\Delta\mu_x)_{M-3} & \cdots & \Delta(\Delta\mu_x)_2 & \Delta(\Delta\mu_x)_1 & \Delta\mu_{x,0} \end{bmatrix} \tag{3.43}$$

The $\mathbf{\Psi}_0$ vector is called the sensitivity coefficient vector for $\theta_{0,0}$ as its entries are the $\Psi_{M,0}(x)$ sensitivity coefficients of the temperature $T_M(x)$ ($M = 1, 2, 3, ...$) with respect to the surface temperature rise value $\theta_{0,0}$ according to Eq. (3.36a) for $i = 0$.

The $\mathbf{\Psi}$ matrix is called the sensitivity coefficient matrix for $\mathbf{\theta}_0$ as its elements are the $\Psi_{M,i}(x)$ sensitivity coefficients of the temperature $T_M(x)$ ($M = 1, 2, 3, ...$) with respect to the surface temperature rise values $\theta_{0,i}$ (with $i = 1, 2, ..., M$) according to Eq. (3.36a) for $i \geq 1$. Its structure is "lower triangular" as the off-diagonal entries lying above the main diagonal are zero (Potter et al. 2016). Also, the entries along the main diagonal are $\Delta\mu_{x,0} = \mu_{x,1}$, the elements along the diagonal just below the main one are $\Delta(\Delta\mu_x)_1$, then $\Delta(\Delta\mu_x)_2$, and so on.

3.4 One-Dimensional Problem with Time-Dependent Surface Heat Flux

Consider a flat plate, initially at a uniform temperature T_{in}, subject to a time-dependent surface heat flux at $x = 0$ and thermally insulated at $x = L$. The surface heat flux $q_0(t)$ is an arbitrary function of time and can be positive (heating) or negative (cooling). Also, its value at the initial time, $q_0(t = 0)$, can be other than zero. This problem may be denoted by X22B-0T1, and its governing equations are given by Eqs. (3.1a)–(3.1d) in Section 3.2.

In this section, approximate solutions to the above problem are outlined based on superposition principles. Both X22B10T0 and X22B20T0 building blocks from Chapter 2 are utilized in constructing the approximate solutions.

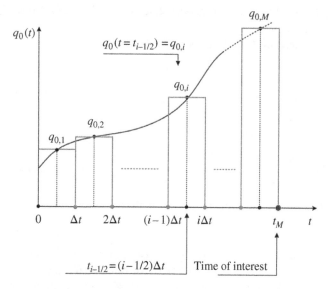

Figure 3.6 Piecewise-constant approximation for $q_0(t)$ where the *i*th component, $q_{0,i}$, is time-independent.

One simple way to treat this problem is to apply the procedure proposed by de Monte et al. (2011). The same function may be divided into a number M of equally spaced intervals, Δt, up to the time t of interest (denoted by $t_M = M\Delta t$), and to substitute a constant surface heat flux within each of these intervals for the real $q_0(t)$. This gives the "*piecewise-constant*" approximation that will be analyzed in Section 3.4.1.

Another type of approximation of $q_0(t)$ is the "*piecewise-linear*" approximation which can in general represent a $q_0(t)$ profile more accurately than the constant elements approximation (see Example 3.6 of Section 3.4.2.4). The piecewise linear approximation is described in Section 3.4.2.

3.4.1 Piecewise-Constant Approximation

This approximation gives the piecewise profile of Figure 3.6, that is a sequence of constant segments where $q_{0,i}$ (*i*th surface heat flux component) is an approximation for $q_0(t)$ between $t_{i-1} = (i-1)\Delta t$ and $t_i = i\Delta t$ (with $i = 1, 2, ..., M$, being $t_M = M\Delta t$ the time of evaluation). The *i*th surface heat flux component, $q_{0,i}$, is time-independent and may be identified with the time coordinate $t_{i-1/2} = (i-1/2)\Delta t$, that is, $q_0(t = t_{i-1/2}) = q_{0,i}$.

The temperature solution $T_i(x, t_M)$ at time t_M due only to the surface heat flux component $q_{0,i}$ of Figure 3.6, when the plate is initially at zero temperature, may be calculated by applying superposition using the X22B10T0 building block:

$$T_i(x, t_M) = q_{0,i}\frac{L}{k}\left[\tilde{T}_{\text{X22B10T0}}(\tilde{x}, \tilde{t}_M - \tilde{t}_{i-1}) - \tilde{T}_{\text{X22B10T0}}(\tilde{x}, \tilde{t}_M - \tilde{t}_i)\right] \tag{3.44}$$

where, according to the dimensionless variables defined by Eqs. (2.4) and (2.6) in Chapter 2, $\tilde{x} = x/L$, $\tilde{t} = \alpha t/L^2$ and $\tilde{T} = (T - T_{in})/(q_0 L/k)$. Also, $\tilde{t}_M = M\Delta\tilde{t}$, $\tilde{t}_{i-1} = (i-1)\Delta\tilde{t}$ and $\tilde{t}_i = i\Delta\tilde{t}$, with $\Delta\tilde{t} = \alpha\Delta t/L^2$. Note that $\tilde{T}_{\text{X22B10T0}}(\tilde{x}, \tilde{t}_M - \tilde{t}_{i-1}) \times L/k$ and $\tilde{T}_{\text{X22B10T0}}(\tilde{x}, \tilde{t}_M - \tilde{t}_i) \times L/k$ are the temperature rises at time $\tilde{t} = \tilde{t}_M$ due to unit step changes in surface ($\tilde{x} = 0$) heat flux applied at times $\tilde{t} = \tilde{t}_{i-1}$ and $\tilde{t} = \tilde{t}_i$, respectively. (See also the last paragraph of Section 2.3.4.)

For brevity, let

$$\frac{L}{k}\tilde{T}_{\text{X22B10T0}}(\tilde{x}, \tilde{t}_M - \tilde{t}_{i-1}) = \phi(\tilde{x}, \tilde{t}_{M-i+1}) = \phi_{x,M-i+1} \tag{3.45a}$$

where $\tilde{t}_{M-i+1} = (M-i+1)\Delta\tilde{t}$ and $\phi_{x, M-i+1}$ has units of °C-m^2/W. Similarly, $\tilde{T}_{\text{X22B10T0}}(\tilde{x}, \tilde{t}_M - \tilde{t}_i) \times L/k = \phi_{x,M-i}$ with $\tilde{t}_M - \tilde{t}_i = \tilde{t}_{M-i} = (M-i)\Delta\tilde{t}$. Also, let

$$T_i(x, t_M) = T_{i,M}(x) \tag{3.45b}$$

Then, Eq. (3.44) becomes

$$T_{i,M}(x) = q_{0,i}\left(\phi_{x,M-i+1} - \phi_{x,M-i}\right) = q_{0,i}\Delta\phi_{x,M-i} \tag{3.46}$$

where $\phi_{x, M-i} = 0$ for $i = M$ as the plate is initially at zero temperature, that is, $\phi_{x,0} = \phi(x, 0) = 0$; while $\Delta\phi_{x, M-i}$ denotes a forward difference (FD) of the ϕ_x temperature at time $t_{M-i} = (M-i)\Delta t$. Another way to obtain Eq. (3.46) is given in Appendix C, Section C.3.1, using Green's functions.

3.4.1.1 Superposition-Based Numerical Approximation of the Solution

To compute the temperature $T(x, t_M) = T_M(x)$ due to $q_0(t)$ applied up to $t_M = M\Delta t$, the principle of superposition valid for linear problems can now be used. Therefore, the temperature solution $T_M(x)$ may be calculated summing up all the contributions $T_{i,M}(x)$ given by Eq. (3.46) due to the corresponding surface heat flux components $q_{0,i}$, each applied over its own time range $[(i-1)\Delta t, i\Delta t]$, with $i \le M$. Bearing also in mind that the plate is initially at the T_{in} uniform temperature, it is found that

$$T_M(x) = T_{in} + \sum_{i=1}^{M} T_{i,M}(x) = T_{in} + \sum_{i=1}^{M} q_{0,i}\Delta\phi_{x,M-i} \tag{3.47}$$

where $T_{i,M}(x)$ defined by Eq. (3.46) has been used.

Equation (3.47) may be considered as the superposition-based numerical approximation of the temperature solution when the 1D body is subject to an arbitrary surface heat flux and a piecewise-constant approximation is used. Also, it can be interpreted as the numerical form of the "*heat flux-based*" Green's function 1D solution when dealing with a "*piecewise-constant*" approximate-in-time profile of the surface heat flux, as shown in Appendix C, Section C.3.1. Equation (3.47) is very important in investigation of transient 1D inverse heat conduction problems (IHCPs) because it gives a convenient expression for the temperature in terms of the applied surface heat flux components, that is, $q_{0,1}$, $q_{0,2}$, ..., $q_{0,i}$, ..., $q_{0,M}$, as discussed in Chapter 4. Also, it is very important for the "heat flux-based" unsteady surface element method (USEM) (Cole et al. 2011; see Chapter 12).

A sequential-in-time nature of Eq. (3.47) can also be observed. In fact, one segment after another can be estimated, starting with the earliest times, and then moving successively to larger times.

3.4.1.2 Heat Flux-Based Sensitivity Coefficients

The first derivative of the temperature $T_M(x)$ of Eq. (3.47) with respect to the surface heat flux component $q_{0,i}$ gives the sensitivity of $T_M(x)$ with respect to $q_{0,i}$ at the location of interest. For this reason, it is called heat flux-based "sensitivity coefficient" for $T_M(x)$ and is denoted by $X_{M,i}(x)$, with $i = 1, 2, ..., M$. It is found that

$$\frac{\partial T_M(x)}{\partial q_{0,i}} = X_{M,i}(x) = \begin{cases} \Delta\phi_{x,M-i}, & i \leq M \\ 0, & i > M \end{cases} \tag{3.48}$$

Therefore, when using the above sensitivity coefficients (units of °C m²/W), Eq. (3.47) may be rewritten as

$$T_M(x) = T_{in} + \sum_{i=1}^{M} q_{0,i} X_{M,i} \tag{3.49}$$

The above sensitivity coefficients are very important when solving an inverse problem. In fact, they give insight on the location of temperature sensors when estimating the unknown surface heat flux.

3.4.1.3 Basic "Building Block" Solution

The required building block solution for the approximation in Eq. (3.47) is the $\tilde{T}_{X22B10T0}(\tilde{x}, \tilde{t}) \times L/k = \phi(\tilde{x}, \tilde{t})$ related to the temperature solution of the X22B10T0 case given in Section 2.3.8.

The exact and computational analytical solutions for the X22B10T0 problem are described fully in the above section of Chapter 2. The computational solution discussed in Section 2.3.8.1 with a finite number of terms can be utilized choosing the accuracy desired (e.g., one part in 10^{15}) without excessive computational costs. This contrasts significantly with the additional computational effort required to reduce errors using the numerical finite element method (FEM).

3.4.1.4 Computer Code and Example

A MATLAB function implementing Eq. (3.47) and called `fdX22B_0T0_pca.m` (where "pca" denotes a piecewise-constant approximation) allows the dimensionless temperature to be computed when the surface heat flux at $x = 0$ varies with time as

$$q_0(t) = q_0\left(\frac{t}{t_{ref}}\right)^p = s_q t^p \quad (t \geq 0) \tag{3.50}$$

where p is a real number (positive or negative though zero is also possible) and $s_q = q_0/t_{ref}^p$. In addition, Eq. (3.50) may be taken in dimensionless form as $\tilde{q}_0(\tilde{t}) = \tilde{t}^p$, where $\tilde{q}_0(\tilde{t}) = q_0(t)/\left[s_q(L^2/\alpha)^p\right]$.

Note that, for $p > 0$, the above equation gives $q_0(t = 0) = q_{0,0} = 0$, that is, the surface heat flux value at the initial time, $q_0(t = 0)$, is equal to zero. This choice does not lead to any loss of generality. In fact, if $q_{0,0}$ is other than zero, the principle of superposition shown in Section 3.2 and used throughout Section 3.4.1 can be applied. In other words, the problem can still be solved by breaking it into two separate subproblems, one subject to $q_{0,0}$ (constant) and, hence, denoted by X22B10T0 (see Section 2.3.8), and the other to Eq. (3.50); and then summing up the resulting temperature distributions. For $p < 0$, the initial ($t = 0$) surface heat flux is infinitely large. However, this result is consistent with the current piecewise-constant approximation of the surface heat flux. In fact, its first component, $q_{0,1}$, is calculated at $t = \Delta t/2$, as shown in Figure 3.6. For the special case of $p = 0$, the current X22B-0T0 problem reduces to the building block X22B10T0 described in Section 2.3.8 for which a MATLAB function is already available.

The `fdX22B_0T0_pca.m` function calculates the dimensionless temperature, $\tilde{T} = (T - T_{in})/\left[s_q(L/k)(L^2/\alpha)^p\right]$, and requires five dimensionless variables as input data. Two of these are the location $\tilde{x} = x/L$ and time $\tilde{t} = \alpha t/L^2$ of interest;

one is related to the exponent of the input forcing, that is, p; lastly, A and M to the accuracy desired, where the former refers to the building block precision (one part in 10^4). Also, this function is available for ease of use on the web site related to this book, and calls one basic "building block" function, that is, $\texttt{fdX22B10T0.m}$.

Example 3.4 A steel plate ($\alpha = 10^{-5}$ m^2/s, $k = 40$ W/m/°C), initially at 30 °C, is subject at $x = 0$ to a surface heat flux as given by Eq. (3.50), starting at time zero, with $p = 1$ (linear-in-time increase) and a rate of $s_q = q_0/t_{ref} = 10^6$ W/m^2/s. The plate is insulated on the back side $x = L = 5$ cm. Calculate the temperature at 1 cm inside the flat plate at times 1, 2, and 3 seconds by using Eq. (3.47). Then, compare it with the exact values coming from the exact analytical solution defined in Section 2.3.9.

Solution

The temperature at the interior point of interest $x_P = 1$ cm is $T(x_P, t)$ and increases with time. If the piecewise-constant approximation of the applied surface heat flux is based on a time step $\Delta t = 1$ s, by using Eq. (3.47) the temperatures at $t = 1, 2,$ and 3 seconds are given, respectively, by

$$T_1(x_P) = T_{in} + q_{0,1}\phi_1(\tilde{x}_P) = 30.492831°C \tag{3.51a}$$

$$T_2(x_P) = T_{in} + q_{0,1}\Delta\phi_1(\tilde{x}_P) + q_{0,2}\phi_1(\tilde{x}_P) = 34.827114°C \tag{3.51b}$$

$$T_3(x_P) = T_{in} + q_{0,1}\Delta\phi_2(\tilde{x}_P) + q_{0,2}\Delta\phi_1(\tilde{x}_P) + q_{0,3}\phi_1(\tilde{x}_P) = 47.655181°C \tag{3.51c}$$

where $\phi_M(\tilde{x}_P) = \tilde{T}_{X22B10T0}(\tilde{x}_P, M\Delta\tilde{t}) \times L/k$, with $M = 1, 2$ or 3. Also, $\tilde{x}_P = x_P/L = 0.2$, $\Delta\tilde{t} = \alpha\Delta t/L^2 = 0.004$, and the first six decimal-places have been considered for the building block solution ($A = 6$).

The exact temperatures can be obtained from the $\texttt{fdX22B20T0.m}$ MATLAB function of Section 2.3.9 as the current problem can be denoted by X22B20T1, where "2" in B20 indicates a linear-in-time increase of the surface heat-flux. They are

$$T(x_P = 1 \text{ cm}, \Delta t = 1 \text{ s}) = T_{in} + \frac{L}{k}\underbrace{\tilde{t}_{ref}\tilde{T}\left(\tilde{x}_P = 0.2, \Delta\tilde{t} = 0.004, \tilde{t}_{ref} = 1, A = 15\right)}_{= \text{fdX22B20T0}(0.2,\ 0.004,\ 1,15)}$$
$$\times \underbrace{\left(\frac{q_0}{t_{ref}}\right)}_{s_q = 10^6 \text{ W m}^{-2}\text{ s}^{-1}} \left(\frac{L^2}{\alpha}\right) = 30.187600 \text{ °C} \tag{3.52a}$$

$$T(x_P = 1 \text{ cm}, 2\Delta t = 2 \text{ s}) = T_{in} + \frac{L}{k}\underbrace{\tilde{t}_{ref}\tilde{T}\left(\tilde{x}_P = 0.2, 2\Delta\tilde{t} = 0.008, \tilde{t}_{ref} = 1, A = 15\right)}_{= \text{fdX22B20T0}(0.2,\ 0.008,\ 1,15)}$$
$$\times \frac{q_0}{t_{ref}}\left(\frac{L^2}{\alpha}\right) = 34.074329 \text{ °C} \tag{3.52b}$$

$$T(x_P = 1 \text{ cm}, 3\Delta t = 3 \text{ s}) = T_{in} + \frac{L}{k}\underbrace{\tilde{t}_{ref}\tilde{T}\left(\tilde{x}_P = 0.2, 3\Delta\tilde{t} = 0.012, \tilde{t}_{ref} = 1, A = 15\right)}_{= \text{fdX22B20T0}(0.2,\ 0.012,\ 1,15)}$$
$$\times \frac{q_0}{t_{ref}}\left(\frac{L^2}{\alpha}\right) = 46.725442 \text{ °C} \tag{3.52c}$$

where $\tilde{t}_{ref}\tilde{T}$ is independent of $\tilde{t}_{ref}$ (see Section 2.3.9). For this reason, the simple numerical value of $\tilde{t}_{ref} = 1$ can be chosen. Also, an accuracy of one part in 10^{15} is chosen ($A = 15$) in such a way as the computational analytical solution can be assumed exact (see Section 2.3.5.1).

Hence, the absolute errors in the approximate temperatures are +0.305231, +0.752785, and +0.929739 °C, respectively. If these errors are normalized relatively to the maximum exact temperature *rises* that occur at the boundary surface $x = 0$ up to and including the time of interest (i.e. $T(0, t) - T_{in} = 59.470803, 168.208835$ and 309.019362 ° C, at $t = 1, 2,$ and 3 seconds, respectively), they are: 0.51, 0.45, and 0.30%, respectively. The exact temperature rises at $x = 0$ can still be obtained from the $\texttt{fdX22B20T0.m}$ MATLAB function of Section 2.3.9.

The above errors are very small. However, if the approximate temperatures are computed at the heated surface $x = 0$ of the plate by still choosing a time step of $\Delta t = 1$ s, the absolute and relative errors can be very large. In such a case, to get more accurate results, it is convenient to use smaller time steps, as shown in Table 3.4, where the temperatures (°C) are obtained

Table 3.4 Comparison of exact and approximate temperatures for three different time steps using the piecewise-constant approximation (fdX22B_0T0_pca.m) at $x = 0$.

		Approximate temperature (°C)					
t, s	Exact temperature (°C)	$\Delta t = 1$ s	M	$\Delta t = 0.1$ s	M	$\Delta t = 0.01$ s	M
1	89.470803	74.603103	1	88.921532	10	89.452631	100
2	198.208835	182.284519	2	197.648680	20	198.190553	200
3	339.019362	322.617672	3	338.454385	30	339.001031	300

by using the fdX22B_0T0_pca.m function and the following equation that relates the dimensional and dimensionless temperatures:

$$T(x_P, t_M) = T_{in} + \underbrace{\tilde{T}(\tilde{x}_P, \tilde{t}_M, p, A, M)}_{= \text{ fdX22B\_0T0\_pca}} \times s_q \left(\frac{L}{k}\right) \left(\frac{L^2}{\alpha}\right)^p \tag{3.53}$$

where $\tilde{x}_P = x_P/L = 0$, $\tilde{t}_M = M\Delta\tilde{t}$, $\Delta\tilde{t} = \alpha\Delta t/L^2$, $p = 1$, and $A = 6$.

Note that, when using $\Delta t = 0.01$ s, the first digits of the approximate temperatures are the same as the ones of the exact temperatures, being the underlined digits different, as shown by Table 3.4.

3.4.1.5 Matrix Form of the Superposition-Based Numerical Approximation

This section displays a matrix form for the superposition-based numerical statement of the solution Eq. (3.47), that is, the numerical statement of the heat flux-based GFSE when dealing with the piecewise-constant approximation. It results in

$$\mathbf{T} = \mathbf{X}\mathbf{q}_0 + \mathbf{1}T_{in} \tag{3.54}$$

where $\mathbf{T}$ and $\mathbf{q}_0$ are the vectors (M by 1) of temperature and surface heat-flux components, respectively; $\mathbf{X}$ is a square matrix M by M containing the forward differences $\Delta\phi_{x,i}$ of the X22B10T0 basic building block solution, $\phi = \tilde{T}_{\text{X22B10T0}} \times (L/k)$, at time steps $i = 0, 1, 2, ..., M-1$; and $\mathbf{1}$ is a vector (M by 1) of ones. In detail,

$$\mathbf{T} = \begin{bmatrix} T_1(x) \\ T_2(x) \\ T_3(x) \\ \vdots \\ T_M(x) \end{bmatrix}, \quad \mathbf{q}_0 = \begin{bmatrix} q_{0,1} \\ q_{0,2} \\ q_{0,3} \\ \vdots \\ q_{0,M} \end{bmatrix} \tag{3.55}$$

$$\mathbf{X} = \begin{bmatrix} \Delta\phi_{x,0} & 0 & 0 & 0 & \cdots & 0 & 0 \\ \Delta\phi_{x,1} & \Delta\phi_{x,0} & 0 & 0 & \cdots & 0 & 0 \\ \Delta\phi_{x,2} & \Delta\phi_{x,1} & \Delta\phi_{x,0} & 0 & \cdots & 0 & 0 \\ \vdots & \vdots & \vdots & \vdots & \vdots & \vdots & \vdots \\ \Delta\phi_{x,M-1} & \Delta\phi_{x,M-2} & \Delta\phi_{x,M-3} & \cdots & \Delta\phi_{x,2} & \Delta\phi_{x,1} & \Delta\phi_{x,0} \end{bmatrix} \tag{3.56}$$

The $\mathbf{X}$ matrix is called the sensitivity coefficient matrix for $\mathbf{q}_0$ as its entries are the sensitivity coefficients of the temperature response with respect to the surface heat-flux components according to Eq. (3.48). Its structure is "lower triangular" as the off-diagonal elements lying above the main diagonal are zero. Also, the entries along the main diagonal are $\Delta\phi_{x,0} = \phi_{x,1}$, the entries along the diagonal just below the main one are $\Delta\phi_{x,1}$, and so on.

3.4.2 Piecewise-Linear Approximation

This approximation gives the piecewise profile of Figure 3.7, where $q_{0,i}$ (ith surface heat flux component) is a linear approximation for $q_0(t)$ between $t_{i-1} = (i-1)\Delta t$ and $t_i = i\Delta t$ (with $i = 1, 2, ..., M$, being $t_M = M\Delta t$ the time of interest).

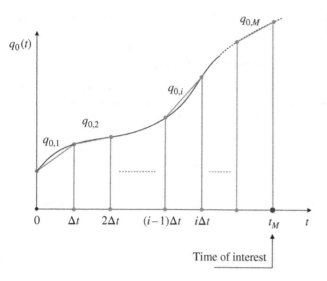

Therefore, contrary to the piecewise-constant approximation of Section 3.4.1, now $q_{0,i}$ depends linearly on the time as

$$q_{0,i}(t) = q_{0,i-1} + \Delta q_{0,i} \frac{t - t_{i-1}}{\Delta t}, \ t \in [t_{i-1}, t_i] \tag{3.57}$$

where $q_{0,i-1} = q_0(t = t_{i-1})$, $\Delta q_{0,i} = q_{0,i} - q_{0,i-1}$, and $q_{0,i} = q_0(t = t_i)$.

The temperature solution $T_i(x, t)$ at time $t = t_M$, that is, $T_i(x, t_M) = T_{i,M}(x)$ due only to the surface heat flux component $q_{0,i}(t)$ (with $i \le M$) given by Eq. (3.57) and shown in Figure 3.7, when the plate is initially at zero temperature, may be derived applying the principles of superposition using the X22B10T0 and X22B20T0 building blocks (See Section 3.3.2 for the companion problem of surface temperature assigned):

$$T_i(\tilde{x}, \tilde{t}_M) = T_{i,M}(\tilde{x}) = q_{0,i-1} \frac{L}{k} \left[\tilde{T}_{\text{X22B10T0}}(\tilde{x}, \tilde{t}_M - \tilde{t}_{i-1}) - \tilde{T}_{\text{X22B10T0}}(\tilde{x}, \tilde{t}_M - \tilde{t}_i) \right]$$

$$+ \Delta q_{0,i} \frac{L}{k} \left[\tilde{T}_{\text{X22B20T0}}(\tilde{x}, \tilde{t}_M - \tilde{t}_{i-1}, \ \tilde{t}_{ref} = \Delta \tilde{t}) - \tilde{T}_{\text{X22B20T0}}(\tilde{x}, \tilde{t}_M - \tilde{t}_i, \ \tilde{t}_{ref} = \Delta \tilde{t}) \right] \tag{3.58}$$

$$- \Delta q_{0,i} \frac{L}{k} \tilde{T}_{\text{X22B10T0}}(\tilde{x}, \tilde{t}_M - \tilde{t}_i)$$

where, according to the dimensionless variables defined by Eqs. (2.4) and (2.6) in Chapter 2, $\tilde{x} = x/L$, $\tilde{t} = \alpha t/L^2$ and $\tilde{T} = (T - T_{in})/(q_0 L/k)$. Also, $\tilde{t}_{ref} = \alpha t_{ref}/L^2$ (where t_{ref} is defined through Eq. (2.3)), $\tilde{t}_M = M \Delta \tilde{t}$, $\tilde{t}_{i-1} = (i-1)\Delta \tilde{t}$ and $\tilde{t}_i = i \Delta \tilde{t}$, with $\Delta \tilde{t} = \alpha \Delta t/L^2$. Note that $\tilde{T}_{\text{X22B20T0}}(\tilde{x}, \tilde{t}_M - \tilde{t}_{i-1}, \tilde{t}_{ref} = \Delta \tilde{t}) \times L/k$ and $\tilde{T}_{\text{X22B20T0}}(\tilde{x}, \tilde{t}_M - \tilde{t}_i, \tilde{t}_{ref} = \Delta \tilde{t}) \times L/k$ are the temperature rises at time $\tilde{t} = \tilde{t}_M$ due to unit linear-in-time increases in surface ($\tilde{x} = 0$) heat flux (over the time $\tilde{t}_{ref} = \Delta \tilde{t}$) applied at times $\tilde{t} = \tilde{t}_{i-1}$ and $\tilde{t} = \tilde{t}_i$, respectively. (See also the last paragraph of Section 2.3.4.)

This expression is parallel to that of Eq. (3.27) concerning the X12 problem, and the terms have similar meaning in consideration of the linear change in heat flux from $q_{0,i-1}$ to $q_{0,i}$ over the time interval Δt.

For brevity, let

$$\frac{L}{k} \tilde{T}_{\text{X22B20T0}}(\tilde{x}, \tilde{t}_M - \tilde{t}_{i-1}, \tilde{t}_{ref} = \Delta \tilde{t}) = \gamma(\tilde{x}, \tilde{t}_{M-i+1}) = \gamma_{x,M-i+1} \tag{3.59}$$

where $\tilde{t}_{M-i+1} = (M - i + 1)\Delta \tilde{t}$ and $\gamma_{x,M-i+1}$ has units of °C·m²/W. Similarly, $\tilde{T}_{\text{X22B20T0}}(\tilde{x}, \tilde{t}_M - \tilde{t}_i, \tilde{t}_{ref} = \Delta \tilde{t}) \times L/k = \gamma_{x,M-i}$ with $\tilde{t}_M - \tilde{t}_i = \tilde{t}_{M-i} = (M - i)\Delta \tilde{t}$.

Recalling also the definition given by Eq. (3.45a) of $\phi_{x,M-i}$, Eq. (3.58) can be written as

$$T_{i,M}(x) = q_{0,i-1}(\phi_{x,M-i+1} - \phi_{x,M-i}) + \Delta q_{0,i}(\gamma_{x,M-i+1} - \gamma_{x,M-i}) - \Delta q_{0,i}\phi_{x,M-i} \tag{3.60a}$$

Alternatively, because $\Delta q_i = q_i - q_{i-1}$, Eq. (3.60a) may be written as

$$T_{i,M}(x) = q_{0,i-1}(\phi_{x,M-i+1} - \Delta \gamma_{x,M-i}) + q_{0,i}(\Delta \gamma_{x,M-i} - \phi_{x,M-i}) \tag{3.60b}$$

where

$$\Delta\gamma_{x,M-i} = \gamma_{x,M-i+1} - \gamma_{x,M-i} \tag{3.60c}$$

indicates a forward difference of the γ_x temperature at time $t = (M - i)\Delta t$. Also, $\gamma_{x, M-i} = 0$ for $i = M$ as the plate is initially at zero dimensionless temperature. Therefore, $\Delta\gamma_{x,0} = \gamma_{x,1}$. A different procedure to derive Eq. (3.60b) is given in Appendix C, Section C.3.2, using Green's functions.

3.4.2.1 Superposition-Based Numerical Approximation of the Solution

To compute the temperature $T(x, t)$ at time $t = t_M$ due to $q_0(t)$ applied up to $t = t_M$, the principle of superposition can be utilized. Bearing in mind that the plate is initially at T_{in} temperature, the solution $T(x, t_M) = T_M(x)$ due to $q_0(t)$ applied up to $t_M = M\Delta t$ may be calculated summing up all the contributions $T_{i,M}(x)$ given by Eq. (3.60b):

$$T_M(x) = T_{in} + \sum_{i=1}^{M} T_{i,M}(x) = \sum_{i=1}^{M} q_{0,i-1}\left(\phi_{x,M-i+1} - \Delta\gamma_{x,M-i}\right)$$
$$+ \sum_{i=1}^{M} q_{0,i}\left(\Delta\gamma_{x,M-i} - \phi_{x,M-i}\right) \tag{3.61}$$

An alternative form and more convenient expression of Eq. (3.61) is as follows:

$$T_M(x) = T_{in} + q_{0,0}\left(\phi_{x,M} - \Delta\gamma_{x,M-1}\right) + \sum_{i=1}^{M-1} q_{0,i}\left[\Delta(\Delta\gamma_x)_{M-i}\right] + q_{0,M}\Delta\gamma_{x,0} \tag{3.62}$$

where $q_{0,0} = q_0(t = 0)$ can in general be different from zero; and $\Delta(\Delta\gamma_x)_{M-i}$ represents a central difference (CD) of the $\gamma(x, t)$ temperature at time $t = (M - i)\Delta t$, that is,

$$\Delta(\Delta\gamma_x)_{M-i} = \Delta\gamma_{x,M-i} - \Delta\gamma_{x,M-i-1} = \gamma_{x,M-i+1} - 2\gamma_{x,M-i} + \gamma_{x,M-i-1} \tag{3.63}$$

Note that the quantity $T_{in} + q_{0,0}\phi_{x,M}$ appearing in Eq. (3.62) represents the solution to a constant surface heat flux equal to $q_{0,0}$. Also, when the surface heat flux is time-independent, that is, $q_{0,0} = q_{0,1} = ... = q_{0,i} = ... = q_{0,M}$, Eq. (3.62) reduces properly to $T_M(x) = T_{in} + q_{0,0}\phi_{x,M}$, as expected. This is an application of symbolic intrinsic verification (Cole et al. 2014; D'Alessandro and de Monte 2018). Equation (3.62) may be derived by observing the sequential-in-time nature of Eq. (3.61), as was done in Section 3.3.2.2 for the companion X12B-0T1 problem of arbitrary surface temperature assigned. Also, Eq. (3.62) may be considered as the superposition-based numerical approximation of the temperature solution when the 1D body is subject to an arbitrary surface heat flux and a piecewise-linear approximation for it is used. In addition, it can be interpreted as the numerical form of the "*heat flux-based*" Green's function 1D solution when dealing with a "*piecewise-linear*" approximate-in-time profile of the surface heat flux, as shown in Appendix C, Section C.3.2.

It is very important in investigation of transient 1D IHCPs because it gives a convenient expression for the temperature in terms of the surface heat-flux values at different times, that is, $q_{0,0}, q_{0,1}, q_{0,2}, ..., q_{0,i}, ..., q_{0,M}$, as discussed in Chapter 4.

3.4.2.2 Heat Flux-Based Sensitivity Coefficients

The first derivative of the temperature $T_M(x)$ of Eq. (3.62) with respect to the surface heat flux value $q_{0,i}$ at time $t_M = M\Delta t$ gives the sensitivity of $T_M(x)$ with respect to $q_{0,i}$ at the location of interest. For this reason, it is called heat flux-based "sensitivity coefficient" for $T_M(x)$ (already defined in Section 3.4.1.2) and is denoted by $X_{M,i}$, with $i = 1, 2, ..., M$. It is found that

$$\frac{\partial T_M(x)}{\partial q_{0,i}} = X_{M,i}(x) = \begin{cases} \phi_{x,M} - \Delta\gamma_{x,M-1}, & i = 0 \\ \Delta(\Delta\gamma_x)_{M-i}, & i = 1, 2, ..., M-1 \\ \Delta\gamma_{x,0}, & i = M \\ 0, & i = M+1, M+2, ... \end{cases} \tag{3.64}$$

where $\Delta\gamma_{x,0} = \gamma_{x,1}$. Therefore, when using the above sensitivity coefficients (units of $°C\ m^2/W$), Eq. (3.62) may be rewritten as

$$T_M(x) = T_{in} + q_{0,0}X_{M,0}(x) + \sum_{i=1}^{M-1} q_{0,i}X_{M,i}(x) + q_{0,M}X_{M,M}(x) \tag{3.65}$$

The above sensitivity coefficients are very important when solving an inverse problem. In fact, they give insight on the location of temperature sensors when estimating the unknown surface heat flux.

3.4.2.3 Basic "Building Block" Solutions

The required building block solutions for Eq. (3.62) are: (i) $\phi(\tilde{x}, \tilde{t}) = \tilde{T}_{X22B10T0}(\tilde{x}, \tilde{t}) \times L/k$, that is related to the temperature solution of the X22B10T0 case given in Section 2.3.8; and (ii) $\gamma(\tilde{x}, \tilde{t}) = \tilde{T}_{X22B20T0}(\tilde{x}, \tilde{t}, \tilde{t}_{ref} = \Delta\tilde{t}) \times L/k$ that is associate to the X22B20T0 case of Section 2.3.9. The corresponding exact and computational analytical solutions are described fully in Sections 2.3.8 and 2.3.9. However, when $q_{0,0} = 0$, Eq. (3.62) uses only one basic building block, that is, $\gamma(\tilde{x}, \tilde{t})$.

The computational solution can be utilized choosing the accuracy desired, and smaller time steps in the piecewise approximation will require higher accuracy to accommodate small forward and central differences in time, that is, $\Delta\gamma_{x,M-1}$ and $\Delta(\Delta\gamma_x)_{M-i}$, respectively.

3.4.2.4 Computer Code and Examples

A MATLAB function implementing Eq. (3.62) and called `fdX22B_0T0_pla.m` allows the dimensionless temperature, $\tilde{T} = (T - T_{in})/[s_q(L/k)(L^2/\alpha)^p]$, to be computed when the surface heat flux at $x = 0$ varies with time according to Eq. (3.50), where p is now only a positive real number though zero is also possible (see Section 3.4.1.4). In fact, a negative value of p would lead to $q_0(t = 0) = q_{0,0} \to \infty$ that is not consistent with the current approximate piecewise-linear profile whose first value, $q_{0,0}$, is calculated at time $t = 0$, as shown in Figure 3.7. Also, "pla" indicates a piecewise-linear approximation.

This function requires five dimensionless variables defined in Section 3.4.1.4 as input data, namely $\tilde{x}, \tilde{t}, p, A$, and M. Also, this function is available for ease of use on the web site related to this book, and calls only one basic "building block" function, that is, `fdX22B20T0.m`, as $q_{0,0} = q_0(t = 0) = 0$ according to Eq. (3.50).

Example 3.5 Calculate the temperature at 1 cm inside the steel plate of the X22B20T1 case of Example 3.4 of Section 3.4.1.4 at times 1, 2, and 3 seconds by using Eq. (3.62). Then, compare it with the exact values coming from the exact analytical solution defined in Section 2.3.9 of Chapter 2.

Solution

The temperature at the interior point of interest $x_P = 1$ cm is $T(x_P, t)$ and increases with time. If the piecewise-linear approximation of the applied surface heat flux is based on a time step of $\Delta t = 1$ s, by using Eq. (3.62) with $q_{0,0} = 0$ the temperatures at $t = 1, 2$, and 3 seconds are given, respectively, by

$$T_1(x_P) = T_{in} + q_{0,1}\gamma_1(\tilde{x}_P) = 30.187600 \,°C \tag{3.66a}$$

$$T_2(x_P) = T_{in} + q_{0,1}[\Delta(\Delta\gamma(\tilde{x}_P))_1] + q_{0,2}\gamma_1(\tilde{x}_P) = 34.074329°C \tag{3.66b}$$

$$T_3(x_P) = T_{in} + q_{0,1}[\Delta(\Delta\gamma(\tilde{x}_P))_2] + q_{0,2}[\Delta(\Delta\gamma(\tilde{x}_P))_1] + q_{0,3}\gamma_1(\tilde{x}_P) = 46.725442 \,°C \tag{3.66c}$$

where $\gamma_M(\tilde{x}_P) = \tilde{T}_{X22B20T0}(\tilde{x}_P, M\Delta\tilde{t}, \tilde{t}_{ref} = \Delta\tilde{t}) \times L/k$, with $M = 1, 2$ or 3. Also, $\tilde{x}_P = x_P/L = 0.2$, $\Delta\tilde{t} = \alpha\Delta t/L^2 = 0.004$, and the first six decimal-places have been considered for the building block solution ($A = 6$).

The exact temperatures can be obtained from the `fdX22B20T0.m` MATLAB function of Section 2.3.9 with an accuracy of one part in 10^{15} ($A = 15$) as the current problem is denoted by X22B20T1. They are 30.187600, 34.074329, and 46.725442 °C, respectively, as already given in Example 3.4. Note that the accuracy of $A = 15$ allows the computational analytical solution to be considered as exact.

Therefore, the approximate and exact temperatures are the same (as expected) as the applied surface heat-flux changes linearly with time and, hence, the piecewise-linear approximation cannot perform any approximation in the current case. This is an application of "numerical" intrinsic verification of Eq. (3.62) and, hence, of the `fdX22B_0T0_pla.m` MATLAB function (D'Alessandro and de Monte 2018).

Example 3.6 A steel plate ($\alpha = 10^{-5}$ m²/s), initially at 20 °C, is subject at $x = 0$ to a surface heat flux as given by Eq. (3.50), starting at time zero, with $p = 2$ (quadratic-in-time increase) and $s_q = q_0/t_{ref}^2 = 10^5$ W/m²/s². The plate is insulated on the back side $x = L = 5$ cm. Calculate the temperature at 1.25 cm inside the flat plate at times 5 and 10 seconds by using the `fdX22B_0T0_pla.m` MATLAB function and choosing five different time steps, that is, $\Delta t = 5, 1, 0.1, 0.01$ and 0.001 s.

Table 3.5 Comparison of exact and approximate temperatures (limited to the first eight decimal-places) at $x_p = 1.25$ cm and times 5 and 10 seconds for different time steps using the piecewise-constant and -linear approximations for the surface heat flux.

		t = 5 s		
Δt, s	M	Approximate temperature (fdX22B_0T0_pca.m), (°C)	Approximate temperature (fdX22B_0T0_pla.m), (°C)	Exact temperature, (°C)
5	1	35.80839635	40.06395886	
1	5	29.26946481	29.48749397	
0.1	50	29.06820322	29.07031095	29.06609539
0.01	500	29.06611647	29.06613754	
0.001	5000	29.06609560	29.06609581	
		t = 10 s		
5	2	192.17820805	202.92273691	
1	10	172.49088307	173.11720901	
0.1	100	171.90433103	171.91042584	171.89823623
0.01	1000	171.89829718	171.89835813	
0.001	10000	171.89823684	171.89823745	

Then, compare it with the approximate values coming from the `fdX22B_0T0_pca.m` function of Section 3.4.1.4 (piecewise-constant approximation) and with the exact values coming from the exact analytical solution given by Beck et al. (2008; see Eq. (25c), p. 2560).

Solution
The current problem can be denoted by X22B30T1, where "3" in B30 indicates a power-in-time variation of the surface temperature (see Table A.3 in Appendix A). The temperature at the interior point of interest $x_P = 1.25$ cm is $T(x_P, t)$ and increases with time. By using the `fdX22B_0T0_pca.m` and `fdX22B_0T0_pla.m` functions, approximate values of temperature are obtained at times 5 and 10 seconds and shown in Table 3.5. where the temperatures in units of °C are obtained by using Eq. (3.53) for $p = 2$ of Example 3.4.

The exact temperatures can be obtained by using the exact analytical solution given by Beck et al. (2008) in dimensionless form and setting an accuracy of one part in 10^{15} ($A = 15$) when truncating the infinite series-solution. This accuracy allows the computational analytical solution to be considered as exact. The solution is

$$\tilde{T}\left(\tilde{x}, \tilde{t}, \tilde{t}_{ref}\right) = \frac{1}{\tilde{t}_{ref}^2}\left[\frac{\tilde{t}^3}{3} + \left(\frac{\tilde{x}^2}{2} - \tilde{x} + \frac{1}{3}\right)\tilde{t}^2 + \left(\frac{\tilde{x}^4}{12} - \frac{\tilde{x}^3}{3} + \frac{\tilde{x}^2}{3} - \frac{2}{45}\right)\tilde{t}\right.$$
$$\left. + \left(\frac{\tilde{x}^6}{360} - \frac{\tilde{x}^5}{60} + \frac{\tilde{x}^4}{36} - \frac{\tilde{x}^2}{45} + \frac{4}{945}\right)\right] - \frac{4}{\tilde{t}_{ref}^2}\sum_{m=1}^{\infty}\frac{e^{-(m\pi)^2\tilde{t}}}{(m\pi)^6}\cos\left(m\pi\tilde{x}\right) \tag{3.67}$$

where $\tilde{x} = x/L$, $\tilde{t} = \alpha t/L^2$ and $\tilde{T} = (T - T_{in})/(q_0 L/k)$. Also, $\tilde{t}_{ref} = \alpha t_{ref}/L^2$.

The exact values of temperature are

$$T(x_P = 1.25 \text{ cm}, t = 5 \text{ s}) = T_{in} + \frac{L}{k}\underbrace{\tilde{t}_{ref}^2 \tilde{T}\left(\tilde{x}_P = 0.25, \tilde{t} = 0.02, \tilde{t}_{ref}, A = 15\right)}_{= 1.160460209707701e - 06}$$

$$\times \underbrace{\left(\frac{q_0}{t_{ref}^2}\right)}_{s_q = 10^5 \text{ W m}^{-2} \text{ s}^{-2}}\left(\frac{L^2}{\alpha}\right)^2 = 29.06609539°\text{C} \tag{3.68a}$$

Table 3.6 Exact and approximate variation of the surface heat flux into the range $t \in [0, 5\ \mathrm{s}]$ when dealing with $\Delta t = 5$ seconds.

	$q_0(t)$ [W/m²], with $t \in [0, 5\ \mathrm{s}]$		
t [s]	Exact: quadratic $q_0(t) = 10^5 t^2$	Approximate: constant, $q_{0,1}$	Approximate: linear, $5 \cdot 10^5 t$
0	0	0	0
1	10^5	$6.25 \cdot 10^5$	$5 \cdot 10^5$
2	$4 \cdot 10^5$	$6.25 \cdot 10^5$	$10 \cdot 10^5$
3	$9 \cdot 10^5$	$6.25 \cdot 10^5$	$15 \cdot 10^5$
4	$16 \cdot 10^5$	$6.25 \cdot 10^5$	$20 \cdot 10^5$
5	$25 \cdot 10^5$	$6.25 \cdot 10^5$	$25 \cdot 10^5$

$$T(x_P = 1.25\ \mathrm{cm}, t = 10\ \mathrm{s}) = T_{in} + \frac{L}{k}\underbrace{\tilde{t}_{ref}^2 \tilde{T}\left(\tilde{x}_P = 0.25, \tilde{t} = 0.04, \tilde{t}_{ref}, A = 15\right)}_{= 1.160460209707701e-06}$$

$$\times \underbrace{\left(\frac{q_0}{t_{ref}^2}\right)}_{s_q = 10^5\ \mathrm{W\ m^{-2}\ s^{-2}}} \left(\frac{L^2}{\alpha}\right)^2 = 171.89823623°\mathrm{C} \tag{3.68b}$$

where $\tilde{t}_{ref}^2 \tilde{T}$ is independent of $\tilde{t}_{ref}$, as shown by Eq. (3.67). These temperatures are given in Table 3.5 for visual comparison where the different digits of the approximate temperatures with respect to the exact ones are underlined. Note that, when dealing with the piecewise-constant approximation and a time step of 0.001 seconds, the first six decimal-places of the approximate temperature are the same as the ones of the exact temperature for both times 5 and 10 seconds. In this case, the absolute errors in the approximate temperatures are +0.00000021 °C and +0.00000061°C at times 5 and 10 seconds, respectively. By using the piecewise-linear approximation and $\Delta t = 0.001$ s, these errors are +0.00000042 °C and +0.00000122 °C, respectively, that is, they are doubled.

Therefore, in the current case of a quadratic-in-time variation of the surface heat flux, the piecewise-constant approximation is just a bit more accurate than the piecewise-linear one. This is since the linear approximation of the quadratic-in-time increase of the surface heat flux gives higher values than the constant approximation during the single time step apart from early times. As an example, for $t = 5$ seconds and with a time step of 5 seconds, the surface heat-flux variation into the range $t \in [0, 5\ \mathrm{s}]$ is shown in both Table 3.6 and Figure 3.8.

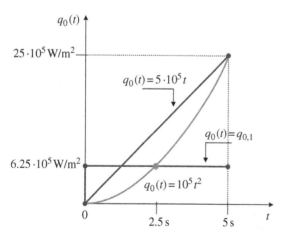

Figure 3.8 Plot of exact and approximate variations of the surface heat flux into the range $t \in [0, 5\ \mathrm{s}]$ when dealing with $\Delta t = 5$ s.

3.4.2.5 Matrix Form of the Superposition-Based Numerical Approximation

This section gives a matrix form for the superposition-based numerical statement of the solution Eq. (3.62), that is, the numerical statement of the heat flux-based GFSE when dealing with the piecewise-linear approximation. It results in

$$\mathbf{T} = \mathbf{X}\mathbf{q}_0 + \mathbf{X}_0 q_{0,0} + \mathbf{1}T_{in} \tag{3.69}$$

where $\mathbf{T}$ and $\mathbf{q}_0$ are the vectors (M by 1) of temperature and surface heat-flux values, respectively, at time steps $i = 1, 2, ..., M$; $\mathbf{X}_0$ is a vector (M by 1) containing the forward differences $\Delta \gamma_{x,i}$ of the X22B20T0 building block solution, $\gamma = (L/k)\tilde{T}_{X22B20T0}$, at time steps $i = 0, 1, 2, ..., M - 1$, and the values of the X22B10T0 building block solution, $\phi = (L/k)\tilde{T}_{X22B10T0}$, at time steps $i = 1, 2, ..., M$; $\mathbf{X}$ is a square matrix M by M containing the central differences of the γ

solution at time steps $i = 1, 2, ..., M - 1$ with the exception of the main diagonal where the forward difference at $i = 0$ of the same building block solution appears; and, lastly, $\mathbf{1}$ is a vector (M by 1) of ones. In detail,

$$\mathbf{T} = \begin{bmatrix} T_1(x) \\ T_2(x) \\ T_3(x) \\ \vdots \\ T_M(x) \end{bmatrix}, \quad \mathbf{q_0} = \begin{bmatrix} q_{0,1} \\ q_{0,2} \\ q_{0,3} \\ \vdots \\ q_{0,M} \end{bmatrix}, \quad \mathbf{X_0} = \begin{bmatrix} \phi_{x,1} - \Delta\gamma_{x,0} \\ \phi_{x,2} - \Delta\gamma_{x,1} \\ \phi_{x,3} - \Delta\gamma_{x,2} \\ \vdots \\ \phi_{x,M} - \Delta\gamma_{x,M-1} \end{bmatrix} \tag{3.70}$$

$$\mathbf{X} = \begin{bmatrix} \Delta\gamma_{x,0} & 0 & 0 & 0 & \cdots & 0 & 0 \\ \Delta(\Delta\gamma_x)_1 & \Delta\gamma_{x,0} & 0 & 0 & \cdots & 0 & 0 \\ \Delta(\Delta\gamma_x)_2 & \Delta(\Delta\gamma_x)_1 & \Delta\gamma_{x,0} & 0 & \cdots & 0 & 0 \\ \vdots & \vdots & \vdots & \vdots & \vdots & \vdots & \vdots \\ \Delta(\Delta\gamma_x)_{M-1} & \Delta(\Delta\gamma_x)_{M-2} & \Delta(\Delta\gamma_x)_{M-3} & \cdots & \Delta(\Delta\gamma_x)_2 & \Delta(\Delta\gamma_x)_1 & \Delta\gamma_{x,0} \end{bmatrix} \tag{3.71}$$

The $\mathbf{X_0}$ vector is called the sensitivity coefficient vector for $q_{0,0}$ as its entries are the sensitivity coefficients of the temperature rise $T_M(x)$ ($M = 1, 2, 3, ...$) with respect to the surface heat flux value $q_{0,0}$ according to Eq. (3.64) for $i = 0$.

The $\mathbf{X}$ matrix is called the sensitivity coefficient matrix for $\mathbf{q_0}$ as its elements are the sensitivity coefficients of the temperature rise $T_M(x)$ ($M = 1, 2, 3, ...$) with respect to the surface heat flux values $q_{0,i}$ (with $i = 1, 2, ..., M$) according to Eq. (3.64) for $i \geq 1$. Its structure is "lower triangular" as the off-diagonal entries lying above the main diagonal are zero. Also, the entries along the main diagonal are $\Delta\gamma_{x,0} = \gamma_{x,1}$, the elements along the diagonal just below the main one are $\Delta(\Delta\gamma_x)_1$, then $\Delta(\Delta\gamma_x)_2$, and so on.

3.5 Two-Dimensional Problem with Space-Dependent and Constant Surface Heat Flux

Consider the X22B(y-t1)0Y22B00T1 problem, where "B(y-)" indicates an arbitrary space variation along y of the applied surface heat flux (refer to Chapter 2 and Appendix A for full description of the numbering system). In brief, this is a flat plate of length L and width W, initially at a uniform temperature T_{in}, subject to a space-dependent but constant surface heat flux at $x = 0$, that is, $q_0(y,t) = q_0(y)$, and is thermally insulated at $x = L$. Adiabatic conditions exist over the remainder of the boundaries, say $y = 0$ and $y = W$, which make the problem quasi-steady. The surface heat flux $q_0(y)$ is an arbitrary function of space. Also, its value at the corner point, $q_0(y = 0)$, can be other than zero. This 2D problem is similar to the one of Figure 2.9 where, however, the surface heat flux is uniform in space but with a step change at $y = W_0$.

In this section, an approximate solution to the X22B(y-t1)0Y22B00T1 problem is outlined based on superposition-in-space principles. The X22B(y1pt1)0Y22B00T0 transient building block from Chapter 2 (Section 2.4) is utilized in constructing the approximate numerical solution.

One simple way to treat this problem is to apply the procedure proposed by de Monte et al. (2011) and, then, by Najafi et al. (2015). The arbitrary $q_0(y)$ curve may be divided into a number of N equally spaced intervals, Δy, up to the $y = W$ corner (denoted by $y_N = N\Delta y$), and to substitute a uniform heat flux within each of these intervals for the real $q_0(y)$. This gives the "*piecewise-uniform*" approximation that will be analyzed in the following.

Another type of approximation of $q_0(y)$ is the linear elements ("*piecewise-linear*" space profile) which can in general represent a $q_0(y)$ curve more accurately than the uniform elements. However, this approximation is not analyzed in this book.

3.5.1 Piecewise-Uniform Approximation

This approximation gives the piecewise profile of Figure 3.9, that is a sequence of uniform segments where $q_{0,j}$ (jth surface heat flux component) is an approximation for $q_0(y)$ between $y_{j-1} = (j-1)\Delta y$ and $y_j = j\Delta y$ (with $j = 1, 2, ..., N$, being $y_N = N\Delta y = W$). The heat flux $q_{0,j}$ may be identified with the space coordinate $y_{j-1/2} = (j-1/2)\Delta y$, that is, $q_{0,j} = q_0(y = y_{j-1/2})$, where $y_{j-1/2}$ is the coordinate of the grid point $(j-1/2)$ shown in Fig. 3.10.

The temperature solution $T_j(x, y, t)$ at location (x, y) and time t due only to the surface heat flux uniform and constant component $q_{0,j}$ of Figure 3.9, when the plate is initially at zero temperature, may be calculated by applying superposition-in-space (see Figure 3.10) using the X22B(y1pt1)0Y22B00T0 building block:

$$T_j(x, y, t) = q_{0,j} \frac{L}{k} \tilde{T}_{\text{X22B(y1pt1)0Y22B00T0}}\left(\tilde{x}, \tilde{y}, \tilde{t}, \tilde{W}_0 = j\Delta\tilde{y}\right)$$

$$- q_{0,j} \frac{L}{k} \tilde{T}_{\text{X22B(y1pt1)0Y22B00T0}}\left[\tilde{x}, \tilde{y}, \tilde{t}, \tilde{W}_0 = (j-1)\Delta\tilde{y}\right] \tag{3.72}$$

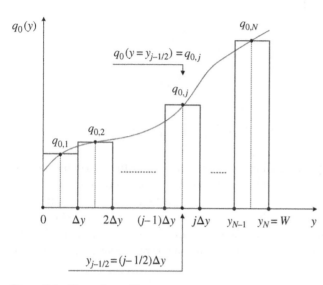

Figure 3.9 Piecewise-uniform approximation for the surface heat flux where the jth component, $q_{0,j}$, is space-independent.

where the former is the temperature due to a uniform and constant heat flux having $q_{0,j}$ as a magnitude and applied to the region $y \in [0, j\Delta y]$; while the latter is the temperature due to a uniform and constant heat flux still having $q_{0,j}$ as a magnitude but applied to the region $y \in [0, (j-1)\Delta y]$. Also, according to the dimensionless variables defined by Eq. (2.49) in Chapter 2, $\tilde{x} = x/L$, $\tilde{y} = y/L$, $\tilde{t} = \alpha t/L^2$ and $\tilde{T} = (T - T_{in})/(q_0 L/k)$, with the dimensionless space step taken as $\Delta\tilde{y} = \Delta y/L$.

For the sake of brevity, let

$$\frac{L}{k} \tilde{T}_{\text{X22B(y1pt1)0Y22B00T0}}\left(\tilde{x}, \tilde{y}, \tilde{t}, \tilde{W}_0 = j\Delta\tilde{y}\right) = \lambda_j(\tilde{x}, \tilde{y}, \tilde{t}) = \lambda_{j,xyt} \tag{3.73}$$

where λ_j (°C-m^2/W) is the temperature rise per unit of surface heat flux of the X22B(y1pt1)0Y22B00T0 heat conduction case when $\tilde{W}_0 = j\Delta\tilde{y}$, that is, when the active surface region is between the corner point $(0,0)$ and the jth grid point $\left(0, \tilde{y}_j = j\Delta\tilde{y}\right)$. Similar considerations when dealing with λ_{j-1}.

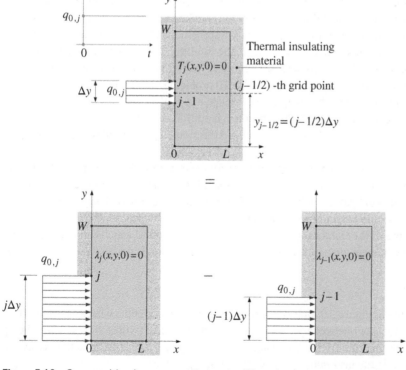

Figure 3.10 Superposition in space applied to the $T_j(x, y, t)$ temperature due to the $q_{0,j}$ heat flux component.

Then, Eq. (3.72) becomes

$$T_j(x, y, t) = q_{0,j}\left(\lambda_{j,xyt} - \lambda_{j-1,xyt}\right) = q_{0,j}\Delta_j\lambda_{j,xyt} \tag{3.74}$$

where $\Delta_j\lambda_{j,xyt}$ denotes a backward difference (BD) in space (Δ_j) at the jth grid point of the $\lambda_{j,xyt}$ temperature variation per unit of surface heat flux.

3.5.1.1 Superposition-Based Numerical Approximation of the Solution

To compute the temperature $T(x, y, t)$ due to $q_0(y)$ applied at $x = 0$, the principle of superposition valid for linear problems can still be utilized. Therefore, the temperature solution may be derived summing up all the contributions $T_j(x, y, t)$ given by Eq. (3.74) due to the corresponding surface heat flux components $q_{0,j}$, each applied over its own space range $[0, j\Delta y]$, with $j \leq N$ as $y_N = W$. Bearing also in mind that the plate is initially at the T_{in} uniform temperature, it is found that

$$T(x, y, t) = T_{in} + \sum_{j=1}^{N} T_j(x, y, t) = T_{in} + \sum_{j=1}^{N} q_{0,j}\Delta_j\lambda_{j,xyt} \tag{3.75}$$

where $T_j(x, y, t)$ defined by Eq. (3.74) has been used. Also, for $j = 1$, it results in $\Delta_1\lambda_{1,xyt} = \lambda_{1,xyt}$.

Equation (3.75) may be considered as the superposition-based numerical approximation of the temperature solution when the 2D body is subject to an arbitrary-in-space variation of the surface heat flux. Also, it can be interpreted as the approximate-in-space numerical form of the *"heat flux-based"* 2D transient Green's function solution when dealing with a *"piecewise-uniform"* profile of the surface heat flux. In fact, the solution using Green's functions can be written as (Cole et al. 2011)

$$T(x, y, t) = T_{in} + \frac{\alpha}{k}\int_{\tau=0}^{t} G_{X22}(x, x' = 0, t - \tau)\int_{y'=0}^{W} q(y')G_{Y22}(y, y', t - \tau)dy'd\tau \tag{3.76}$$

where $G_{X22}(x, x', t - \tau)$ is given in Appendix C, Section C.3. As regards $G_{Y22}(y, y', t - \tau)$, it is simply obtained by replacing (x, x') in $G_{X22}(x, x', t - \tau)$ with (y, y').

Applying the piecewise-uniform heat flux approximation of Figure 3.9 to Eq. (3.76) gives

$$T(x, y, t) = T_{in} + \sum_{j=1}^{N} q_{0,j}\underbrace{\left[\frac{\alpha}{k}\int_{\tau=0}^{t} G_{X22}(x, 0, t-\tau)\int_{y'=(j-1)\Delta y}^{j\Delta y} G_{Y22}(y, y', t-\tau)dy'd\tau\right]}_{=\,\Delta_j\lambda_{j,xyt}} \tag{3.77}$$

where, bearing in mind Eq. (3.75), the quantity between square brackets is recognized to be $\Delta_j\lambda_{j,xyt}$ defined in Eq. (3.74).

Equation (3.75) is very important in investigation of transient 2D inverse heat conduction problems (IHCPs) when the unknown surface heat flux is space-dependent and constant because it gives a convenient expression for the temperature in terms of the applied surface heat flux space components, that is, $q_{0,1}, q_{0,2}, ..., q_{0,j}, ..., q_{0,N}$.

3.5.1.2 Heat Flux-Based Sensitivity Coefficients

The first derivative of the temperature $T(x, y, t)$ of Eq. (3.75) with respect to the surface heat flux component $q_{0,j}$ gives the sensitivity of $T(x, y, t)$ with respect to $q_{0,j}$ at the location and time of interest. For this reason, it is called heat flux-based *"sensitivity coefficient"* for $T(x, y, t)$ and is denoted by $\Lambda_j(x, y, t)$, with $j = 1, 2, ..., N$. It is found that

$$\frac{\partial T(x, y, t)}{\partial q_{0,j}} = \Lambda_j(x, y, t) = \Lambda_{j,xyt} = \Delta_j\lambda_{j,xyt} \tag{3.78}$$

Therefore, when using the above sensitivity coefficients (units of °C-m²/W), Eq. (3.75) may be rewritten as

$$T(x, y, t) = T_{in} + \sum_{j=1}^{N} q_{0,j}\Lambda_{j,xyt} \tag{3.79}$$

The above sensitivity coefficients are very important when solving an inverse problem. In fact, they give insight on the location of temperature sensors for estimating the unknown space-dependent surface heat flux.

3.5.1.3 Basic "Building Block" Solution

The required building block solution for the approximation in Eq. (3.75) is the $\tilde{T}_{X22B(y1pt1)0Y22B00T0}(\tilde{x}, \tilde{y}, \tilde{t}) \times L/k = \lambda(\tilde{x}, \tilde{y}, \tilde{t})$ related to the temperature solution of the X22B(y1pt1)0Y22B00T0 case given in Section 2.4.

The exact solution and computational analytical solution for the X22B(y1pt1)0Y22B00T0 problem are described fully in Chapter 2, Section 2.4, and Appendix B. The computational analytical solution discussed in Section 2.4.3 with a finite number of terms can be utilized choosing the accuracy desired.

3.5.1.4 Computer Code and Examples

A MATLAB function fdX22By_t10Y22B00T0_pua.m (where "pua" denotes a piecewise underline{u}niform underline{a}pproximation) implementing Eq. (3.75) allows the dimensionless temperature to be computed when the surface heat flux at $x = 0$ varies with space as

$$q_0(y) = q_{y0}\left(\frac{y}{W}\right)^p = \sigma_q y^p \quad (0 \le y \le W) \tag{3.80}$$

where p can be a positive or negative real number (the zero value is possible too), and $\sigma_q = q_{y0}/W^p$. Also, Eq. (3.80) may be taken in dimensionless form as $\tilde{q}_0(\tilde{y}) = \tilde{y}^p$, where $\tilde{q}_0(\tilde{y}) = q_0(y)/(\sigma_q L^p)$. Also, $q_{y0} > 0$ when heating, while $q_{y0} < 0$ when cooling.

Note that, for $p > 0$, the above equation gives $q_0(y = 0) = q_{y0,0} = 0$, that is, the surface heat flux value at the heated/cooled corner point $(x, y) = (0, 0)$ is equal to zero. This choice does not lead to any loss of generality. In fact, if $q_{y0,0}$ is other than zero, the principle of superposition used throughout the current section can be applied. In other words, the problem can still be solved by breaking it into two separate subproblems, one subject to $q_{y0,0}$ (uniform) and, hence, 1D and denoted by X22B10T0 (see Section 2.3.8), and the other to Eq. (3.80); and then summing up the resulting temperature distributions. For $p < 0$, Eq. (3.80) states that the surface heat flux value at $y = 0$ tends to infinity. This results is however consistent with the current piecewise-uniform approximation as its first space component, $q_{0,1}$, is calculated at the grid point $y_{1/2} = \Delta y/2$. For the special case of $p = 0$, the current X22B(y-t1)0Y22B00T0 problem reduces to the X22B10T0 building block of Section 2.3.8 for which a MATLAB function is already available.

The fdX22By_t10Y22B00T0_pua.m function calculates the dimensionless temperature, $\tilde{T} = (T - T_{in})/[\sigma_q (L/k)L^p]$, and requires seven dimensionless variables as input data. Three of these are the location $\tilde{x} = x/L$, $\tilde{y} = \tilde{y}/L$, and time $\tilde{t} = \alpha t/L^2$ of interest; one is the aspect ratio $\tilde{W} = W/L$ of the rectangular body; one is related to the input forcing, that is, p; lastly, A and N to the accuracy desired, where the former refers to the building block precision (one part in 10^4). Also, the above function calls the basic "building block" solution, that is, fdX22By1pt10Y22B00T0.m. As the building block is based on a computational analytical solution, the running time of the fdX22By_t10Y22B00T0_pua.m function is very short, usually a fraction of second and up to 3 or 4 seconds for several thousands of space steps. It is available for ease of use on the web site related to this book.

Example 3.7   A steel plate ($\alpha = 10^{-5}$ m^2/s, $k = 40$ W/m/°C), initially at 30 °C, is subject at $x = 0$ to a surface heat flux as given by Eq. (3.80), starting at $y = 0$, with $p = 1$ (linear-in-space variation), and ending at $y = W$, with a slope of $\sigma_q = q_{y0}/W = 10^7$ (W/m^2)/m. The plate is insulated on the back side $x = L = 5$ cm as well as on the lateral sides $y = 0$ and $y = W = 20$ cm. Calculate the temperature of the rectangular body at the middle point of its heated surface and at time 12.5 seconds by using Eq. (3.75) with a space step of 5 cm.

Solution

The temperature at the heated surface point of interest $y_P = 10$ cm and time $t_P = 12.5$ s is $T(0, y_P, t_P)$. If the piecewise-uniform approximation of the applied surface heat flux is based on a space step $\Delta y = 5$ cm (i.e., $N = 4$), by using Eq. (3.75) the temperature is given by

$$
\begin{aligned}
T(0, y_P, t_P) &= T_{in} + \sum_{j=1}^{4} q_{0,j} \Delta_j \lambda_j(0, \tilde{y}_P, \tilde{t}_P) \\
&= T_{in} + q_{0,1} \lambda_1(0, \tilde{y}_P, \tilde{t}_P) \\
&\quad + q_{0,2}[\lambda_2(0, \tilde{y}_P, \tilde{t}_P) - \lambda_1(0, \tilde{y}_P, \tilde{t}_P)] \\
&\quad + q_{0,3}[\lambda_3(0, \tilde{y}_P, \tilde{t}_P) - \lambda_2(0, \tilde{y}_P, \tilde{t}_P)] \\
&\quad + q_{0,4}[\lambda_4(0, \tilde{y}_P, \tilde{t}_P) - \lambda_3(0, \tilde{y}_P, \tilde{t}_P)]
\end{aligned}
\tag{3.81}
$$

where $\lambda_j(0, \tilde{y}_P, \tilde{t}_P)$ may be computed by using Eq. (3.73) with $\tilde{y}_P = 2$, $\tilde{t}_P = 0.05$, $\tilde{W} = 4$ and $\tilde{W}_0 = j\Delta\tilde{y} = j$ as $\Delta\tilde{y} = \Delta y/L = 1$ and $\Delta y = W/N$.

Table 3.7 Surface heat flux components and building block temperatures at the surface point y_p = 10 cm and time t_p = 12.5 seconds for a space step Δy = 5 cm using the piecewise-uniform approximation for the surface heat flux, that is, $q_{0,j} = \sigma_q(j - 1/2)\Delta y$.

j	$q_{0,j}$, W/m^2	$\lambda_j(0, \tilde{y}_P, \tilde{t}_P)$, °C m^2/W	$\Delta_j\lambda_j(0, \tilde{y}_P, \tilde{t}_P)$, °C m^2/W	$q_{0,j}\Delta_j\lambda_j(0, \tilde{y}_P, \tilde{t}_P)$ °C
1	0.25×10^6	0.0000000184	0.0000000184	0.0046027962
2	0.75×10^6	0.0001576958	0.0001576774	118.2580500000
3	1.25×10^6	0.0003153732	0.0001576774	197.0967500000
4	1.75×10^6	0.0003153916	0.0000000184	0.0322195734

Temperature rise: $T(0, y_P, t_P) - T_{in} = \sum_{j=1}^{4} q_{0,j}\Delta_j\lambda_j(0, \tilde{y}_P, \tilde{t}_P)$ 315.4 °C

Also, the various terms appearing in Eq. (3.81) are shown in Table 3.7 where, for the numerical values of the "building block" temperature, $\lambda_j(0, \tilde{y}_P, \tilde{t}_P)$, the first ten decimal-places have been considered ($A = 10$). The temperature rise at the location and time of interest is of about 315 °C.

Table 3.7 also shows that $\Delta_1\lambda_1 = \Delta_4\lambda_4$ and $\Delta_2\lambda_2 = \Delta_3\lambda_3$ according to the fact that the surface point of interest is located at the middle of the heated region and, hence, there exists a thermal symmetry. This is an application of "numerical" intrinsic verification of Eq. (3.75) and, hence, of the fdX22By_t10Y22B00T0_pua.m function (Cole et al. 2014; D'Alessandro and de Monte 2018).

Example 3.8 Calculate the temperature of the rectangular body of Example 3.7 at three different locations of its heated surface, say 1, 10 and 19 cm, at time 12.5 seconds by using the fdX22By_t10Y22B00T0_pua.m MATLAB function for different space steps.

Solution
The temperatures of the rectangle at the surface points and time of interest may be derived by using the fdX22By_-t10Y22B00T0_pua.m function. The results are given in Table 3.8 where, for the numerical values of the "building block" temperature, $\lambda_j(0, \tilde{y}_P, \tilde{t}_P)$, the first ten decimal-places have been considered ($A = 10$). In addition, the numerical values of temperature in °C given in the table are obtained by using the following equation that relates the dimensional and dimensionless temperatures:

$$T(x_P, y_P, t_P) = T_{in} + \underbrace{\tilde{T}(\tilde{x}_P, \tilde{y}_P, \tilde{t}_P, \tilde{W}, p, A, N)}_{= \text{ fdX22By\_t10Y22B00T0\_pua}} \times \sigma_q\left(\frac{L}{k}\right)L^p \tag{3.82}$$

where $\tilde{x}_P = 0$, $\tilde{t}_P = 0.05$, $\tilde{W} = 4$, and $p = 1$.

Table 3.8 Comparison of approximate temperatures for six different space steps using the piecewise-uniform approximation of the surface heat flux (fdX22By_t10Y22B00T0_pua.m) at three different points of the heated boundary and time 12.5 seconds.

Δy, (cm)	N	$T(0, 1$ cm, 12.5 s), (°C)	$T(0, 10$ cm, 12.5 s), (°C)	$T(0, 19$ cm, 12.5 s), (°C)
20	1	345.3915652828	345.3915652828	345.3915652828
10	2	187.6957826962	345.3915652828	503.0873478695
5	4	108.9419981851	345.3915652828	581.8411323806
2.5	8	73.7866265325	345.3915652828	616.9965040332
1	20	66.8558205589	345.3915652828	623.9273100068
0.1	200	66.0760874784	345.3915652828	624.7067897970

Discussion

It is interesting to observe that the numerical value of temperature does not change at the middle point of the heated surface (10 cm) when reducing the space step (as expected) according to (i) linear-in-space variation of the surface heat flux, and (ii) middle point of the boundary. This is an application of "numerical" intrinsic verification of the fdX22By_t10Y22B00T0_pua.m function (Cole et al. 2014; D'Alessandro and de Monte 2018). Another observation concerns the same numerical value of temperature at 1, 10, and 19 cm when dealing with only one space step, that is, $N = 1$. In such a case, in fact, the heating is uniform and the 2D problem reduces to a 1D one, that is, X22B10T1 (in other words, the temperature is independent on y, as expected). This is another application of "numerical" intrinsic verification of the MATLAB function.

3.5.1.5 Matrix Form of the Superposition-Based Numerical Approximation

Contrary to the superposition-based numerical form of the temperature solution derived in Sections 3.3 and 3.4, where the surface temperature and heat flux were varying arbitrarily with the time, Eq. (3.75) here derived allows the temperature to be computed at any time within the rectangular body as the heat flux depends only on space. For this reason, a matrix form of it might be not necessary. However, for the sake of consistency with the previous sections, a matrix form of Eq. (3.75) is given as

$$T(x, y, t) = \mathbf{\Lambda}\mathbf{q}_0 + T_{in} \tag{3.83}$$

where $\mathbf{\Lambda}$ is a vector (1 by N) containing the backward differences $\Delta_j \lambda_{j,\,xyt}$ of the X22B(y1pt1)0Y22B00T0 basic building block solution defined by Eq. (3.73); while $\mathbf{q}_0$ is a vector (N by 1) of surface heat-flux components in space. In detail,

$$\mathbf{\Lambda} = \begin{bmatrix} \Delta_1 \lambda_{1,xyt} & \Delta_2 \lambda_{2,xyt} & \Delta_3 \lambda_{3,xyt} & \cdots & \Delta_N \lambda_{N,xyt} \end{bmatrix}, \qquad \mathbf{q}_0 = \begin{bmatrix} q_{0,1} \\ q_{0,2} \\ q_{0,3} \\ \vdots \\ q_{0,N} \end{bmatrix} \tag{3.84}$$

The $\mathbf{\Lambda}$ vector is called the sensitivity coefficient vector for $\mathbf{q}_0$ as its entries are the sensitivity coefficients of the temperature with respect to the surface heat-flux components in space according to Eq. (3.78).

3.6 Two-Dimensional Problem with Space- and Time-Dependent Surface Heat Flux

Consider the same rectangular domain L by W of Section 3.5, initially at a uniform temperature T_{in}, but now subject to an arbitrary space- and time-dependent surface heat flux at $x = 0$, that is, $q_0(y, t)$. Adiabatic conditions exist over the remainder of the boundaries. This case may be denoted by X22B(y-t-)0Y22B00T1, where "B(y-t-)" indicates an arbitrary space (along y) and time variation of the boundary condition (refer to Chapter 2 and Appendix A for full description of the numbering system).

In this section, an approximate solution to the X22B(y-t-)0Y22B00T1 problem is outlined based on both superposition-in-time and superposition-in-space principles. The former has already been applied in Section 3.4.1 where a piecewise-constant approximation was used, while the latter in Section 3.5.1 where a piecewise-uniform approximation was applied. The X22B(y1pt1)0Y22B00T0 building block from Chapter 2 is utilized in constructing the approximate numerical solution.

One simple way to treat this problem is to apply the procedure proposed in Section 3.5.1. The arbitrary $q_0(y, t)$ curve may be divided into a number N of equally spaced intervals, Δy, up to the $y = W$ (denoted by $y_N = N\Delta y$), and to substitute a time-dependent uniform heat flux, $q_{0,j}(t)$, with $j = 1, 2, ..., N$, within each of these space intervals for the real $q_0(y, t)$. This gives the "*piecewise-uniform*" approximation that will be analyzed in Section 3.6.1.

Then, as the surface heat flux component in space, $q_{0,j}(t)$, is an arbitrary function of time, a simple way to treat this problem is to apply the procedure proposed in Section 3.4.1. The same $q_{0,j}(t)$ function may be divided into a number M of equally spaced intervals, Δt, up to the time t of interest (denoted by $t_M = M\Delta t$), and to substitute a constant heat flux, $q_{0,i,j}$, with $i = 1, 2, ..., M$, within each of these time intervals for the real $q_{0,j}(t)$. This gives the "*piecewise-constant*" approximation that will be analyzed in Section 3.6.2.

3.6.1 Piecewise-Uniform Approximation

This approximation applied to $q_0(y, t)$ gives the piecewise profile of Figure 3.9, that is a sequence of uniform segments where $q_{0,j}$ (jth surface heat flux component in space) is now an arbitrary function of time, that is, $q_{0,j} = q_{0,j}(t)$. Also, it is an approximation for $q_0(y, t)$ between $y_{j-1} = (j-1)\Delta y$ and $y_j = j\Delta y$ (with $j = 1, 2, ..., N$, being $y_N = N\Delta y = W$). The heat flux-component $q_{0,j}(t)$ may be identified with the space coordinate $y_{j-1/2} = (j-1/2)\Delta y$, that is, $q_{0,j}(t) = q(y = y_{j-1/2}, t)$, where $y_{j-1/2}$ is the coordinate of the grid point $(j-1/2)$ shown in Fig. 3.10.

The temperature solution $T_j(x, y, t)$ at location (x, y) and time t due only to the surface heat flux uniform component $q_{0,j}(t)$, when the plate is initially at zero temperature, may be calculated by applying superposition-in-space similar to the one shown by Figure 3.10:

$$T_j(x, y, t) = \vartheta_j(x, y, t) - \vartheta_{j-1}(x, y, t) = \Delta_j \vartheta_j(x, y, t) \tag{3.85}$$

where the former, ϑ_j, is the temperature due to a uniform and time-dependent heat flux having $q_{0,j}(t)$ as a magnitude and applied to the region $y \in [0, j\Delta y]$; while the latter, ϑ_{j-1}, is the temperature due to a uniform and time variable heat flux still having $q_{0,j}(t)$ as a magnitude but applied to the region $y \in [0, (j-1)\Delta y]$. Also, $\Delta_j \vartheta_j$ denotes a backward difference (BD) in space (Δ_j) at the jth grid point of the ϑ_j temperature. Note that Eq. (3.85) can also be derived using Green's functions as shown in Appendix C, Section C.4.

3.6.1.1 Numerical Approximation in Space

To compute the temperature $T(x, y, t)$ due to $q_0(y, t)$ applied at $x = 0$, the principle of superposition in space valid for linear problems can still be utilized. Therefore, the temperature solution may be derived summing up all the contributions $T_j(x, y, t)$ given by Eq. (3.85) due to the corresponding surface heat flux components $q_{0,j}(t)$, each applied over its own space range $[0, j\Delta y]$, with $j \le N$ as $y_N = W$. Bearing also in mind that the plate is initially at the T_{in} uniform temperature, it is found that

$$T(x, y, t) = T_{in} + \sum_{j=1}^{N} T_j(x, y, t) = T_{in} + \sum_{j=1}^{N} \Delta_j \vartheta_j(x, y, t)$$

$$= T_{in} + \sum_{j=1}^{N} \left[\vartheta_j(x, y, t) - \vartheta_{j-1}(x, y, t) \right] \tag{3.86}$$

where $T_j(x, y, t)$ defined by Eq. (3.85) has been used. Also, for $j = 1$, it results in $\Delta_1 \vartheta_1 = \vartheta_1$.

3.6.2 Piecewise-Constant Approximation

This approximation applied to $q_{0,j}(t)$ gives the piecewise profile of Figure 3.6, that is a sequence of constant segments where $q_{0,i,j}$ (ith surface heat flux component in time of the jth surface heat flux component in space) is an approximation for $q_{0,j}(t)$ between $t_{i-1} = (i-1)\Delta t$ and $t_i = i\Delta t$ (with $i = 1, 2, ..., M$, being $t_M = M\Delta t$ the time of interest). The surface heat flux component, $q_{0,i,j}$, is time-independent and may be identified with the time coordinate $t_{i-1/2} = (i-1/2)\Delta t$, that is, $q_{0,j}(t = t_{i-1/2}) = q_{0,i,j}$.

The temperature solution $\vartheta_{i,j,M}(x, y, t_M) = \vartheta_{i,j,M}(x, y)$ at time t_M within the flat plate due only to the surface heat flux component $q_{0,i,j}$ applied over the region $y \in [0, j\Delta y]$ between t_{i-1} and t_i, when the plate is initially at zero temperature, may be calculated by applying superposition in time (as described in Section 3.4.1) using here the X22B(y1pt1)0Y22B00T0 building block. It results in

$$\vartheta_{i,j,M}(x, y) = q_{0,i,j} \frac{L}{k} \tilde{T}_{\text{X22B(y1pt1)0Y22B00T0}}\left(\tilde{x}, \tilde{y}, \tilde{t}_M - \tilde{t}_{i-1}, \tilde{W}_0 = j\Delta\tilde{y}\right)$$

$$- q_{0,i,j} \frac{L}{k} \tilde{T}_{\text{X22B(y1pt1)0Y22B00T0}}\left(\tilde{x}, \tilde{y}, \tilde{t}_M - \tilde{t}_i, \tilde{W}_0 = j\Delta\tilde{y}\right) \tag{3.87}$$

where, according to the dimensionless variables defined by Eq. (2.49) in Chapter 2, $\tilde{x} = x/L$, $\tilde{y} = y/L$, $\tilde{t} = \alpha t/L^2$ and $\tilde{T} = (T - T_{in})/(q_0 L/k)$, with the dimensionless space step taken as $\Delta\tilde{y} = \Delta y/L$. Note that $\tilde{T}_{\text{X22B(y1pt1)0Y22B00T0}}(\tilde{x}, \tilde{y}, \tilde{t}_M - \tilde{t}_{i-1}, \tilde{W}_0 = j\Delta\tilde{y}) \times L/k$ and $\tilde{T}_{\text{X22B(y1pt1)0Y22B00T0}}(\tilde{x}, \tilde{y}, \tilde{t}_M - \tilde{t}_i, \tilde{W}_0 = j\Delta\tilde{y}) \times L/k$ are the temperature rises at time $\tilde{t} = \tilde{t}_M$ due to unit step changes in surface $(\tilde{x} = 0)$ heat flux applied at times $\tilde{t} = \tilde{t}_{i-1}$ and $\tilde{t} = \tilde{t}_i$, respectively. (See also the last paragraph of Sections B.1 and B.4 of App. B.) In addition, $\tilde{t}_M - \tilde{t}_{i-1} = (M - i + 1)\Delta\tilde{t} = \tilde{t}_{M-i+1}$ and $\tilde{t}_M - \tilde{t}_i = (M - i)\Delta\tilde{t} = \tilde{t}_{M-i}$, with $\Delta\tilde{t} = \alpha\Delta t/L^2$.

For the sake of brevity, let

$$\frac{L}{k} \tilde{T}_{\text{X22B(y1pt1)0Y22B00T0}}\left(\tilde{x}, \tilde{y}, \tilde{t}_{M-i+1}, \tilde{W}_0 = j\Delta\tilde{y}\right) = \lambda_j(\tilde{x}, \tilde{y}, \tilde{t}_{M-i+1}) = \lambda_{j,xy,M-i+1} \tag{3.88}$$

where $\lambda_{j,xy,M-i+1}$ (°C-m²/W) is the temperature rise per unit of surface heat flux of the X22B(y1pt1)0Y22B00T0 heat conduction case at a generic location $(\tilde{x}, \tilde{y})$ and time $\tilde{t}_{M-i+1}$ when $\tilde{W}_0 = j\Delta\tilde{y}$, that is, the heated or cooled surface region is between the corner point (0, 0) and the jth grid point $\left(0, \tilde{y}_j = j\Delta\tilde{y}\right)$. Similarly, $\lambda_{j,xy,M-i}$.

Then, Eq. (3.87) becomes

$$\vartheta_{i,j,M}(x, y) = q_{0,i,j}\left(\lambda_{j,xy,M-i+1} - \lambda_{j,xy,M-i}\right) = q_{0,i,j}\Delta_{M-i}\left(\lambda_{j,xy,M-i}\right) \tag{3.89}$$

where $\lambda_{j,xy,M-i} = 0$ for $i = M$ as the plate is initially at zero temperature, that is, $\lambda_{j,xy,0} = \lambda_j(x, y, 0) = 0$; while $\Delta_{M-i}\left(\lambda_{j,xy,M-i}\right)$ denotes a forward difference (FD) at time $t_{M-i} = (M-i)\Delta t$ of the $\lambda_j(x, y, t_{M-i})$ temperature. An alternative way to derive Eq. (3.89) is given in Appendix C, Section C.4, using Green's functions.

By applying the same methodology, the temperature solution $\vartheta_{i,j-1}(x, y, t_M) = \vartheta_{i,j-1,M}(x, y)$ at time t_M within the flat plate due only to the surface heat flux component $q_{0,i,j}$ applied to the region $y \in [0, (j-1)\Delta y]$ between t_{i-1} and t_i, when the plate is initially at zero temperature, is

$$\vartheta_{i,j-1,M}(x, y) = q_{0,i,j}\left(\lambda_{j-1,xy,M-i+1} - \lambda_{j-1,xy,M-i}\right) = q_{0,i,j}\Delta_{M-i}\left(\lambda_{j-1,xy,M-i}\right) \tag{3.90}$$

where, according to Eq. (3.88),

$$\lambda_{j-1,xy,M-i} = \lambda_{j-1}(\tilde{x}, \tilde{y}, \tilde{t}_{M-i}) = \frac{L}{k}\tilde{T}_{\text{X22B(y1pt1)0Y22B00T0}}\left[\tilde{x}, \tilde{y}, \tilde{t}_{M-i}, \tilde{W}_0 = (j-1)\Delta\tilde{y}\right] \tag{3.91}$$

Similarly, when dealing with $\lambda_{j-1,xy,M-i+1}$.

3.6.2.1 Numerical Approximation in Time

To compute the temperature $\vartheta_j(x, y, t_M) = \vartheta_{j,M}(x, y)$ due to $q_{0,j}(t)$ applied over the region $y \in [0, j\Delta y]$ between $t = 0$ up to $t = t_M = M\Delta t$, the principle of superposition in time valid for linear problems can now be used. Therefore, the temperature solution may be calculated summing up all the contributions $\vartheta_{i,j,M}(x, y)$ given by Eq. (3.89) due to the corresponding surface heat flux components $q_{0,i,j}$, each applied over its own time range $[(i-1)\Delta t, i\Delta t]$, with $i \leq M$. It is found that

$$\vartheta_{j,M}(x, y) = \sum_{i=1}^{M}\vartheta_{i,j,M}(x, y) = \sum_{i=1}^{M}q_{0,i,j}\Delta_{M-i}\left(\lambda_{j,xy,M-i}\right) \tag{3.92}$$

Similarly, the temperature $\vartheta_{j-1}(x, y, t_M) = \vartheta_{j-1,M}(x, y)$ due to $q_{0,j}(t)$ applied over the region $y \in [0, (j-1)\Delta y]$, between $t = 0$ up to $t = t_M = M\Delta t$, is

$$\vartheta_{j-1,M}(x, y) = \sum_{i=1}^{M}\vartheta_{i,j-1,M}(x, y) = \sum_{i=1}^{M}q_{0,i,j}\Delta_{M-i}\left(\lambda_{j-1,xy,M-i}\right) \tag{3.93}$$

where $\vartheta_{i,j-1,M}(x, y)$ defined by Eq. (3.90) has been used.

3.6.3 Superposition-Based Numerical Approximation of the Solution

Substituting Eqs. (3.92) and (3.93) in Eq. (3.86) for $t = t_M$ yields the temperature $T(x, y, t_M) = T_M(x, y)$ within the rectangular body due to the arbitrary surface heat flux, $q_0(y, t)$, applied at $x = 0$ over the region $y \in [0, W]$ and between $t = 0$ up to $t = t_M = M\Delta t$. Therefore,

$$T_M(x, y) = T_{in} + \sum_{j=1}^{N}\sum_{i=1}^{M}q_{0,i,j}\left[\Delta_{M-i}\left(\lambda_{j,xy,M-i}\right) - \Delta_{M-i}\left(\lambda_{j-1,xy,M-i}\right)\right]$$

$$= T_{in} + \sum_{j=1}^{N}\sum_{i=1}^{M}q_{0,i,j}\left[\Delta_j\Delta_{M-i}\left(\lambda_{j,xy,M-i}\right)\right] \tag{3.94}$$

(a)

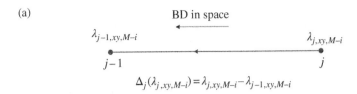

(b)

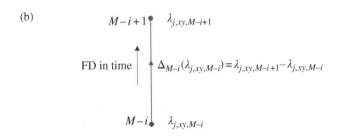

(c)

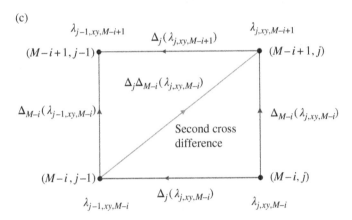

Figure 3.11 Building blocks differences: (a) backward difference (BD) in space; (b) forward difference (FD) in time; (c) second cross difference in both space and time.

where $q_{0,i,j} = q_0(y = y_{j-1/2}, t = t_{i-1/2})$, with $y_{j-1/2} = (j-1/2)\Delta y$ and $t_{i-1/2} = (i-1/2)\Delta t$; while $\Delta_j \Delta_{M-i}(\lambda_{j,xy,M-i})$ is a second cross difference in space (Δ_j) and time (Δ_{M-i}) of $\lambda_{j,xy,M-i}$ that is visually shown in Figure 3.11. Also, $\Delta_j \Delta_{M-i}(\lambda_{j,xy,M-i}) = \Delta_{M-i}\Delta_j(\lambda_{j,xy,M-i})$ according to Schwartz's theorem or equality of mixed partials (Sheldon 2020). In fact,

$$
\begin{aligned}
\Delta_j \Delta_{M-i}(\lambda_{j,xy,M-i}) &= \Delta_{M-i}(\lambda_{j,xy,M-i}) - \Delta_{M-i}(\lambda_{j-1,xy,M-i}) \\
&= (\lambda_{j,xy,M-i+1} - \lambda_{j,xy,M-i}) - (\lambda_{j-1,xy,M-i+1} - \lambda_{j-1,xy,M-i}) \\
&= (\lambda_{j,xy,M-i+1} - \lambda_{j-1,xy,M-i+1}) - (\lambda_{j,xy,M-i} - \lambda_{j-1,xy,M-i}) \\
&= \Delta_j(\lambda_{j,xy,M-i+1}) - \Delta_j(\lambda_{j,xy,M-i}) = \Delta_{M-i}\Delta_j(\lambda_{j,xy,M-i})
\end{aligned}
\tag{3.95}
$$

Equation (3.94) may be considered as the superposition-based numerical approximation of the temperature solution. It can also be interpreted as the approximate form in both space and time of the "*heat flux-based*" 2D transient Green's function solution equation when dealing with a "*piecewise-uniform*" and "*piecewise-constant*" profile of the surface heat flux, as shown in Appendix C, Section C.4. This equation is very important in investigation of 2D transient, rectangular, inverse heat conduction problems (IHCPs) when the unknown surface heat flux is both space- and time-dependent because it gives a convenient expression for the temperature in terms of the applied surface heat flux components in both space and time, that is, $q_{0,i,1}, q_{0,i,2}, ..., q_{0,i,j}, ..., q_{0,i,N}$, with $i = 1, 2, ..., M$, as discussed in Chapter 4.

3.6.3.1 Sequential-in-time Nature and Sensitivity Coefficients

A sequential-in-time nature of Eq. (3.94) can also be observed, one segment after another can be estimated, starting with the earliest times, and then moving successively to larger times. As an example, the first three time-steps as well as the Mth time-step are now considered:

- *First time step*. For $M = 1$, that is, when the time of interest is $t_1 = \Delta t$, bearing in mind that $\lambda_{j,xy,0} = \lambda_{j-1,xy,0} = 0$, Eq. (3.94) reduces to

$$
T_1(x, y) = T_{in} + \sum_{j=1}^{N} q_{0,1,j} \left[\Delta_j \Delta_0 \left(\lambda_{j,xy,0} \right) \right]
$$

$$
= T_{in} + \sum_{j=1}^{N} q_{0,1,j} \left(\lambda_{j,xy,1} - \lambda_{j-1,xy,1} \right) = T_{in} + \sum_{j=1}^{N} q_{0,1,j} \left[\Delta_j \left(\lambda_{j,xy,1} \right) \right]
$$

(3.96a)

where the jth surface heat flux component $q_{0,1,j}$ refers to the midpoint $t = \Delta t/2$ of the time interval $t \in [0, \Delta t]$, that is, $q_{0,1,j}$ equals to $q_{0,j}(t = \Delta t/2)$; while $T_1(x, y)$ refers to the endpoint of the same interval. Also, the first derivative of the temperature $T_1(x, y)$ with respect to the component $q_{0,1,j}$ gives the sensitivity of $T_1(x, y)$ with respect to $q_{0,1,j}$ at the location of interest. For this reason, it is the so-called heat flux-based "*sensitivity coefficient*" for $T_1(x, y)$ and, in general, is denoted by $X_{1,i,j}(x, y)$, with $i = 1, 2, ..., M$ and $j = 1, 2, ..., N$. It results in

$$
\frac{\partial T_1(x, y)}{\partial q_{0,i,j}} = X_{1,i,j}(x, y) = \begin{cases} \Delta_j \Delta_{1-i} \left(\lambda_{j,xy,1-i} \right), & i = 1 \\ 0, & i > 1 \end{cases}
$$

(3.96b)

- *Second time step*. For $M = 2$, that is, when the time of interest is $t_2 = 2\Delta t$, Eq. (3.94) reduces to

$$
T_2(x, y) = T_{in} + \sum_{j=1}^{N} q_{0,1,j} \left[\Delta_j \Delta_1 \left(\lambda_{j,xy,1} \right) \right] + \sum_{j=1}^{N} q_{0,2,j} \left[\Delta_j \Delta_0 \left(\lambda_{j,xy,0} \right) \right]
$$

(3.97a)

where the jth surface heat flux component $q_{0,2,j}$ refers to the midpoint $t = 3\Delta t/2$ of the time interval $t \in [\Delta t, 2\Delta t]$, that is, $q_{0,2,j} = q_{0,j}(t = 3\Delta t/2)$; while $T_2(x, y)$ refers to the endpoint of the same interval. In the current case, the heat flux-based sensitivity coefficient $X_{2,i,j}(x, y)$ for $T_2(x, y)$, with $j = 1, 2, ..., N$, may be taken as

$$
\frac{\partial T_2(x, y)}{\partial q_{0,i,j}} = X_{2,i,j}(x, y) = \begin{cases} \Delta_j \Delta_{2-i} \left(\lambda_{j,xy,2-i} \right), & i \leq 2 \\ 0, & i > 2 \end{cases}
$$

(3.97b)

- *Third time step*. For $M = 3$, that is, when the time of interest is $t_3 = 3\Delta t$, Eq. (3.94) becomes

$$
T_3(x, y) = T_{in} + \sum_{j=1}^{N} q_{0,1,j} \left[\Delta_j \Delta_2 \left(\lambda_{j,xy,2} \right) \right] + \sum_{j=1}^{N} q_{0,2,j} \left[\Delta_j \Delta_1 \left(\lambda_{j,xy,1} \right) \right]
$$
$$
+ \sum_{j=1}^{N} q_{0,3,j} \left[\Delta_j \Delta_0 \left(\lambda_{j,xy,0} \right) \right]
$$

(3.98a)

where the jth surface heat flux component $q_{0,3,j}$ refers to the midpoint $t = 5\Delta t/2$ of the time interval $t \in [2\Delta t, 3\Delta t]$, that is, $q_{0,3,j} = q_{0,j}(t = 5\Delta t/2)$; while $T_3(x, y)$ refers to the endpoint of the same interval. The heat flux-based sensitivity coefficient $X_{3,i,j}(x, y)$ for $T_3(x, y)$, with $j = 1, 2, ..., N$, is given by

$$
\frac{\partial T_3(x, y)}{\partial q_{0,i,j}} = X_{3,i,j}(x, y) = \begin{cases} \Delta_j \Delta_{3-i} \left(\lambda_{j,xy,3-i} \right), & i \leq 3 \\ 0, & i > 3 \end{cases}
$$

(3.98b)

- *Mth time step*. In the current case, Eq. (3.94) applies, and the sensitivity coefficient $X_{M,i,j}(x, y)$ for $T_M(x, y)$ is given by

$$
\frac{\partial T_M(x, y)}{\partial q_{0,i,j}} = X_{M,i,j}(x, y) = \begin{cases} \Delta_j \Delta_{M-i} \left(\lambda_{j,xy,M-i} \right), & i \leq M \\ 0, & i > M \end{cases}
$$

(3.99a)

Therefore, when using the above sensitivity coefficients, Eq. (3.94) may be rewritten as

$$T_M(x, y) = T_{in} + \sum_{j=1}^{N} \sum_{i=1}^{M} q_{0,i,j} X_{M,i,j}(x, y) \tag{3.99b}$$

The above sensitivity coefficients are very important when solving an inverse problem. In fact, they give insight on the location of temperature sensors for estimating the unknown surface heat flux.

3.6.3.2 Basic "Building Block" Solution

The required building block solution for the approximation in Eq. (3.94) is the $\tilde{T}_{X22B(y1pt1)0Y22B00T0}(\tilde{x}, \tilde{y}, \tilde{t}) \times L/k = \lambda(\tilde{x}, \tilde{y}, \tilde{t})$ related to the temperature solution of the X22B(y1pt1)0Y22B00T0 case given in Section 2.4.

The exact and computational analytical solutions for the X22B(y1pt1)0Y22B00T0 problem are described fully in Chapter 2, Section 2.4 and Appendix B. The computational solution discussed in Section 2.4.3 with a finite number of terms can be utilized choosing the accuracy desired.

3.6.3.3 Computer Code and Example

A MATLAB function `fdX22By_t_0Y22B00T0_puca.m` implementing Eq. (3.94) allows the dimensionless temperature to be computed when the surface heat flux at $x = 0$ varies with space ($0 \leq y \leq W$) and time ($t \geq 0$) as

$$q_0(y, t) = q_{y0t0} \left(\frac{y}{W}\right)^{p_y} \left(\frac{t}{t_{ref}}\right)^{p_t} = r_q y^{p_y} t^{p_t} \tag{3.100a}$$

where p_y and p_t are real numbers (positive or negative) in general other than zero though the zero value is also applicable (see afterwards), and $r_q = q_{y0t0}/\left(W^{p_y} t_{ref}^{p_t}\right)$. In addition, $q_{y0t0} > 0$ when heating, while $q_{y0t0} < 0$ when cooling. Also, the above equation may be taken in dimensionless form as $\tilde{q}_0(\tilde{y}, \tilde{t}) = \tilde{y}^{p_y} \tilde{t}^{p_t}$, where $\tilde{q}_0(\tilde{y}, \tilde{t}) = q_0(y, t)/\left[r_q L^{p_y}(L^2/\alpha)^{p_t}\right]$.

Note that, for $p_y > 0$ and $p_t > 0$, the above equation gives a zero surface heat flux at both $y = 0$ and $t = 0$, that is, $q_0(y = 0, t = 0) = q_{y0t0,0} = 0$. This choice does not lead to any loss of generality. In fact, if $q_{y0t0,0}$ is other than zero, the principle of superposition used throughout the current section can be applied. In other words, the problem can still be solved by breaking it into two separate subproblems, one subject to $q_{y0t0,0}$ (uniform and constant) and, hence, 1D and denoted by X22B10T0 (see Section 2.3.8), and the other to Eq. (3.100a); and then summing up the resulting temperature solutions. For $p_y < 0$, Eq. (3.100a) states that the surface heat flux value at $y = 0$ tends to infinity. This results is however consistent with the current piecewise-uniform approximation as its first space component is calculated at the grid point $y_{1/2} = \Delta y/2$. Similarly, for $p_t < 0$, the surface heat flux value at $t = 0$ becomes infinitely large. This is still consistent with the current piecewise-constant approximation as its first time component is computed at time $t_{1/2} = \Delta t/2$. For the special case of $p_y = p_t = 0$, the current X22B(y-t-)0Y22B00T0 problem (2D) reduces to the X22B10T0 problem (1D) of Section 2.3.8 for which a MATLAB function is already available. For $p_y = 0$ and/or $p_t = 0$, see Discussion at the end of Example 3.9.

The `fdX22By_t_0Y22B00T0_puca.m` function (where "puca" denotes a piecewise underline{u}niform and underline{c}onstant approximation of the surface heat flux) calculates the dimensionless temperature, $\tilde{T} = (T - T_{in})/\left[r_q(L/k)L^{p_y}(L^2/\alpha)^{p_t}\right]$, and requires nine dimensionless variables as input data:

- location and time of interest: $\tilde{x} = x/L$, $\tilde{y} = \tilde{y}/L$, $\tilde{t} = \alpha t/L^2$;
- aspect ratio: $\tilde{W} = W/L$;
- data related to the input forcing: p_y and p_t;
- data related to the accuracy desired: A (building block precision, one part in 10^4), M (number of time steps), and N (number of space steps).

Also, it calls the basic "building block" function, that is, `fdX22By1pt10Y22B00T0.m`. As the building block is based on a computational analytical solution, the running time of the `fdX22By_t_0Y22B00T0_puca.m` function is very short, right a few seconds for several hundreds of space and time steps. It is available for ease of use on the web site related to this book.

Example 3.9 A steel plate ($\alpha = 10^{-5}$ m²/s, $k = 40$ W/m-°C), initially at 30 °C, is subject at $x = 0$ to a surface heat flux as

$$q_0(y, t) = q_{y0} \left(\frac{y}{W}\right)^{p_y} + q_{t0} \left(\frac{t}{t_{ref}}\right)^{p_t} = \sigma_q y^{p_y} + s_q t^{p_t} \quad (0 \leq y \leq W; \ t \geq 0) \tag{3.100b}$$

with $p_y = 1$ (linear-in-space increase), a slope of $\sigma_q = q_0/W = 10^6$ (W/m^2)/m, $p_t = 1$ (linear-in-time increase) and a rate of $s_q = q_0/t_{ref} = 10^4$ W/m^2·s. The plate is insulated on the back side $x = L = 5$ cm as well as on the lateral sides $y = 0$ and $y = W = 20$ cm. Calculate the temperature of the rectangular body at the middle point of its heated surface at time 20 seconds by using Eq. (3.94) with a space step of 5 cm and a time step of 10 seconds.

Solution

The temperature at the surface point of interest $y_P = 10$ cm and time $t_P = 20$ s is $T(0, y_P, t_P)$. If the piecewise-uniform and -constant approximations of the applied surface heat flux are based on a space step $\Delta y = 5$ cm (i.e., $N = 4$) and a time step $\Delta t = 10$ s (hence, $M = 2$), by using Eq. (3.94) the temperature $T(0, y_P, t_P = 2\Delta t) = T_2(0, y_P)$ is given by

$$T_2(0, y_P) = T_{in} + \sum_{i=1}^{2}\sum_{j=1}^{4} q_{0,i,j}\left[\Delta_j\Delta_{2-i}\left(\lambda_{j,2-i}(0,\tilde{y}_P)\right)\right] \tag{3.101}$$

where the space and time heat flux component, $q_{0,i,j}$, may be taken as

$$q_{0,i,j} = \sigma_q(j - 1/2)\Delta y + s_q(i - 1/2)\Delta t \quad (i = 1, 2; j = 1, 2, 3, 4) \tag{3.102}$$

Then, bearing in mind Eq. (3.95), for $j = 1, 2, 3, 4$ it results in

$$\Delta_j\Delta_1\left(\lambda_{j,1}(0,\ \tilde{y}_P)\right) = \Delta_1\lambda_{j,1}(0,\tilde{y}_P) - \Delta_1\lambda_{j-1,1}(0,\tilde{y}_P) \tag{3.103a}$$

$$\Delta_j\Delta_0\left(\lambda_{j,0}(0,\ \tilde{y}_P)\right) = \Delta_0\lambda_{j,0}(0,\tilde{y}_P) - \Delta_0\lambda_{j-1,0}(0,\tilde{y}_P) \tag{3.103b}$$

where

$$\Delta_1\lambda_{j,1}(0,\tilde{y}_P) = \lambda_{j,2}(0,\tilde{y}_P) - \lambda_{j,1}(0,\tilde{y}_P) \tag{3.104a}$$

$$\Delta_1\lambda_{j-1,1}(0,\tilde{y}_P) = \lambda_{j-1,2}(0,\tilde{y}_P) - \lambda_{j-1,1}(0,\tilde{y}_P) \tag{3.104b}$$

$$\Delta_0\lambda_{j,0}(0,\tilde{y}_P) = \lambda_{j,1}(0,\tilde{y}_P) \tag{3.104c}$$

$$\Delta_0\lambda_{j-1,0}(0,\tilde{y}_P) = \lambda_{j-1,1}(0,\tilde{y}_P) \tag{3.104d}$$

The $\lambda_{j,2-i}(0,\tilde{y}_P)$ "building block" solutions in Eq. (3.101) may be computed by using Eqs. (3.88) and (3.91) with $\tilde{x}_P = 0$, $\tilde{y}_P = 2$, $\Delta\tilde{t} = 0.04$, $\tilde{W} = 4$, and $\tilde{W}_0 = j\Delta\tilde{y} = j$ as $\Delta\tilde{y} = \Delta y/L = 1$ and $\Delta y = W/N$. The numerical values are shown in Table 3.9, where an accuracy of ten decimal-places ($A = 10$) has been considered for the building block solution.

Also, the various terms appearing in Eq. (3.101) are shown in Table 3.10. The temperature rise at the location and time of interest is of about 88 °C.

Table 3.10 also shows that $\Delta_1\Delta_1(\lambda_{1,1}) = \Delta_4\Delta_1(\lambda_{4,1})$, $\Delta_1\Delta_0(\lambda_{1,0}) = \Delta_4\Delta_0(\lambda_{4,0})$, $\Delta_2\Delta_1(\lambda_{2,1}) = \Delta_3\Delta_1(\lambda_{3,1})$, and $\Delta_2\Delta_0(\lambda_{2,0}) = \Delta_3\Delta_0(\lambda_{3,0})$ according to the fact that the surface point of interest is located at the middle of the heated region and the surface heat flux varies linearly with space. Hence, there exists a thermal symmetry. This is an application of "numerical" intrinsic verification of Eq. (3.94) and, hence, of the `fdX22By_t_0Y22B00T0_puca.m` function (Cole et al. 2014; D'Alessandro and de Monte 2018).

Table 3.9 Building block temperatures at the surface point $y_P = 10$ cm for a space step $\Delta y = 5$ cm and a time step $\Delta t = 10$ s.

j	i	$\lambda_{j-1,2-i}(0,\tilde{y}_P)$, (°C-m^2/W)	$\lambda_{j,2-i}(0,\tilde{y}_P)$, (°C-m^2/W)	$\lambda_{j-1,2-i+1}(0,\tilde{y}_P)$, (°C-m^2/W)	$\lambda_{j,2-i+1}(0,\tilde{y}_P)$, (°C-m^2/W)
1	1	0	0.0000000036	0	0.0000002617
	2	0	0	0	0.0000000036
2	1	0.0000000036	0.0001410474	0.0000002617	0.0001994712
	2	0	0	0.0000000036	0.0001410474
3	1	0.0001410474	0.0002820912	0.0001994712	0.0003986807
	2	0	0	0.0001410474	0.0002820912
4	1	0.0002820912	0.0002820948	0.0003986807	0.0003989424
	2	0	0	0.0002820912	0.0002820948

Table 3.10 Surface heat flux components and sensitivity coefficients at the surface point $y_P = 10$ cm for a space step $\Delta y = 5$ cm and a time step $\Delta t = 10$ s using the piecewise-uniform and -constant approximations for the surface heat flux.

j	i	$\Delta_j \Delta_{2-i}(\lambda_{j,2-i}(0,\ \tilde{y}_P)) = X_{2,ij}$ (°C-m^2/W)	$q_{0,ij}$ – Eq. (3.102) (W/m^2)	$q_{0,i,j}X_{2,i,j}$ (°C)
1	1	$\Delta_1\Delta_1(\lambda_{1,\ 1}) = 0.0000002581$	0.75×10^5	0.0193575000
	2	$\Delta_1\Delta_0(\lambda_{1,\ 0}) = 0.0000000036$	1.75×10^5	0.0006300000
2	1	$\Delta_2\Delta_1(\lambda_{2,\ 1}) = 0.0000581657$	1.25×10^5	7.2707125000
	2	$\Delta_2\Delta_0(\lambda_{2,\ 0}) = 0.0001410438$	2.25×10^5	31.7348550000
3	1	$\Delta_3\Delta_1(\lambda_{3,\ 1}) = 0.0000581657$	1.75×10^5	10.1789975000
	2	$\Delta_3\Delta_0(\lambda_{3,\ 0}) = 0.0001410438$	2.75×10^5	38.7870450000
4	1	$\Delta_4\Delta_1(\lambda_{4,\ 1}) = 0.0000002581$	2.25×10^5	0.0580725000
	2	$\Delta_4\Delta_0(\lambda_{4,\ 0}) = 0.0000000036$	3.25×10^5	0.0011700000
				88.05 °C

Temperature rise: $T_2(0, y_P) - T_{in} = \sum\limits_{i=1}^{2} \sum\limits_{j=1}^{4} q_{0,i,j}X_{2,i,j}$

Discussion

Note that the assigned heat flux, $q_0(y, t)$, of this example given by Eq. (3.100b) only apparently is different from the one defined through Eq. (3.100a). In fact, it can be seen as a linear combination of two heat fluxes, either following equation Eq. (3.100a). The former is in fact

$$q_0(y, t) = q_{0y}(y) = q_{y0}\left(\frac{y}{W}\right)^{p_y} = \sigma_q y^{p_y} = r_q y^{p_y} t^{p_t} \tag{3.105a}$$

with $r_q = \sigma_q$ and $p_t = 0$; while the latter is

$$q_0(y, t) = q_{0t}(t) = q_{t0}\left(\frac{t}{t_{ref}}\right)^{p_t} = s_q t^{p_t} = r_q y^{p_y} t^{p_t} \tag{3.105b}$$

with $r_q = s_q$ and $p_y = 0$. Therefore, the dimensional temperature in °C may be obtained by using superposition principles as

$$
\begin{aligned}
T(x, y, t) = T_{in} + &\underbrace{\tilde{T}\left(\tilde{x}, \tilde{y}, \tilde{t}, \tilde{W}, p_y, p_t = 0, A, M = 1, N\right)}_{=\ \text{fdX22By\_t\_0Y22B00T0}} \times \underbrace{\left[r_q\left(\frac{L}{k}\right)L^{p_y}\left(\frac{L^2}{\alpha}\right)^{p_t}\right]}_{(r_q = \sigma_q;\, p_t = 0)} \\
+ &\underbrace{\tilde{T}\left(\tilde{x}, \tilde{y}, \tilde{t}, \tilde{W}, p_y = 0, p_t, A, M, N = 1\right)}_{=\ \text{fdX22By\_t\_0Y22B00T0}} \times \underbrace{\left[r_q\left(\frac{L}{k}\right)L^{p_y}\left(\frac{L^2}{\alpha}\right)^{p_t}\right]}_{(r_q = s_q;\, p_y = 0)}
\end{aligned}
\tag{3.106}
$$

Lastly, note that the former dimensionless temperature in Eq. (3.106) could also be derived by using the `fdX22By_-t10Y22B00T0_pua.m` function given in Section 3.5.1.4 as the heat flux defined by Eq. (3.105a) is time-independent. Similarly, the latter could be calculated by the `fdX22B_0T0_pca.m` function of Section 3.4.1.4 as the subproblem subject to Eq. (3.105b) reduces to a 1D problem.

3.6.3.4 Matrix Form of the Superposition-Based Numerical Approximation

It is frequently advantageous to perform algebraic manipulations utilizing a matrix form of the model. This section displays a matrix form for the superposition-based numerical statement of the solution Eq. (3.94), that is, the numerical statement of the 2D transient heat flux-based GFSE when dealing with piecewise-uniform and -constant approximations of the arbitrary (in both space and time) surface heat flux.

An expansion of the numerical form of the solution equation as given by Eq. (3.94), with M replaced by 1 to M, is

$$T_1(x, y) = T_{in} + \mathbf{X}_{1,1}\mathbf{q}_{0,1} \tag{3.107a}$$

$$T_2(x, y) = T_{in} + \mathbf{X}_{2,1}\mathbf{q}_{0,1} + \mathbf{X}_{2,2}\mathbf{q}_{0,2} \tag{3.107b}$$

$$T_3(x, y) = T_{in} + \mathbf{X}_{3,1}\mathbf{q}_{0,1} + \mathbf{X}_{3,2}\mathbf{q}_{0,2} + \mathbf{X}_{3,3}\mathbf{q}_{0,3} \tag{3.107c}$$

$$T_M(x, y) = T_{in} + \mathbf{X}_{M,1}\mathbf{q}_{0,1} + \mathbf{X}_{M,2}\mathbf{q}_{0,2} + \cdots + \mathbf{X}_{M,M-1}\mathbf{q}_{0,M-1} + \mathbf{X}_{M,M}\mathbf{q}_{0,M} \tag{3.107d}$$

where $\mathbf{X}_{M,i}$ is a vector (1 by N) at time step $t_i = i\Delta t$ (with $i = 1, 2, ..., M$) containing the second cross differences (or sensitivity coefficients) $\Delta_j\Delta_{M-i}(\lambda_{j,xy,M-i}) = X_{M,i,j}$ of the X22B(y1pt1)0Y22B00T0 basic building block solution at different space steps $y_j = j\Delta y$ (with $j = 1, 2, ..., N$); while $\mathbf{q}_{0,i}$ is a vector (N by 1) at the same time step $t_i = i\Delta t$ of surface heat-flux components in space. In detail,

$$\mathbf{X}_{M,i} = [X_{M,i,1} \quad X_{M,i,2} \quad \cdots \quad X_{M,i,j} \quad \cdots \quad X_{M,i,N}], \quad \mathbf{q}_{0,i} = \begin{bmatrix} q_{0,i,1} \\ q_{0,i,2} \\ q_{0,i,3} \\ \vdots \\ q_{0,i,N} \end{bmatrix} \tag{3.108}$$

Equations (3.107a)–(3.107d) can be written in a more compact form by using an appropriate matrix notation, which results in

$$\mathbf{T} = \mathbf{X}\mathbf{q}_0 + \mathbf{1}T_{in} \tag{3.109}$$

where $\mathbf{T}$ and $\mathbf{q}_0$ are the vectors (M by 1) of temperature and surface heat-flux components $\mathbf{q}_{0,i}$, respectively; $\mathbf{X}$ is a square matrix M by M containing the $\mathbf{X}_{M,i}$ vectors and, hence, the sensitivity coefficients $\Delta_j\Delta_{M-i}(\lambda_{j,xy,M-i}) = X_{M,i,j}$; and $\mathbf{1}$ is a vector (M by 1) of ones. In detail,

$$\mathbf{T} = \begin{bmatrix} T_1(x, \ y) \\ T_2(x, \ y) \\ T_3(x, \ y) \\ \vdots \\ T_M(x, \ y) \end{bmatrix}, \quad \mathbf{q}_0 = \begin{bmatrix} \mathbf{q}_{0,1} \\ \mathbf{q}_{0,2} \\ \mathbf{q}_{0,3} \\ \vdots \\ \mathbf{q}_{0,M} \end{bmatrix} \tag{3.110}$$

$$\mathbf{X} = \begin{bmatrix} \mathbf{X}_{1,1} & 0 & 0 & 0 & \cdots & 0 & 0 \\ \mathbf{X}_{2,1} & \mathbf{X}_{2,2} & 0 & 0 & \cdots & 0 & 0 \\ \mathbf{X}_{3,1} & \mathbf{X}_{3,2} & \mathbf{X}_{3,3} & 0 & \cdots & 0 & 0 \\ \vdots & \vdots & \vdots & \vdots & \vdots & \vdots & \vdots \\ \mathbf{X}_{M,1} & \mathbf{X}_{M,2} & \mathbf{X}_{M,3} & \cdots & \mathbf{X}_{M,M-2} & \mathbf{X}_{M,M-1} & \mathbf{X}_{M,M} \end{bmatrix} \tag{3.111}$$

The $\mathbf{X}$ matrix is called the sensitivity coefficient matrix for $\mathbf{q}_0$ as its entries are the sensitivity coefficients of the temperature with respect to the surface heat-flux components $q_{0,i,j}$ according to Eq. (3.99a). Its structure is "lower triangular" as the off-diagonal elements (vectors) lying above the main diagonal are zero. Also, the entries along the main diagonal are the same, that is, $\mathbf{X}_{1,1} = \mathbf{X}_{2,2} = \cdots = \mathbf{X}_{M,M}$. In fact, $X_{1,1,j} = X_{2,2,j} = \cdots = X_{M,M,j}$ as $X_{1,1,j} = \Delta_j\Delta_0(\lambda_{j,xy,0})$, $X_{2,2,j} = \Delta_j\Delta_0(\lambda_{j,xy,0})$ and so on. Then, the entries along the diagonal just below the main one are still the same, that is, $\mathbf{X}_{2,1} = \mathbf{X}_{3,2} = \cdots = \mathbf{X}_{M,\,M-1}$. In fact, $X_{2,1,j} = X_{3,2,j} = \cdots = X_{M,M-1,j}$ as $X_{2,1,j} = \Delta_j\Delta_1(\lambda_{j,xy,1})$, $X_{3,2,j} = \Delta_j\Delta_1(\lambda_{j,xy,1})$, and so on.

3.7 Chapter Summary

Superposition of exact solution creates a pathway to construct solutions with external heating (or cooling) that varies with time and/or space. These techniques are restricted to linear problems (in heat conduction, those with constant thermophysical properties).

One-dimensional heat conduction problems with time-varying surface heating/cooling can be solved by modeling the surface effect as either piecewise constant (stair step variation) or piecewise linear (connected line segments). The surface disturbance analyzed in this text can be either a known temperature variation with time or a known heat flux variation in time. Solutions constructed using the piecewise constant assumption make use of the XI2B10T0 building block solutions (here I = 1 for temperature and I = 2 for heat flux), and those using the piecewise linear assumption make use of the XI2B10T0 and XI2B20T0 building block solutions (I = 1 or 2). The computational analytical solutions from Chapter 2 can be used to evaluate the building block solutions to any desired degree of accuracy.

Two-dimensional problems with spatially varying or space- and time-varying boundary conditions can also be approximated using superposition of exact solutions. Only piecewise uniform spatial variations are considered in this text, but piecewise linear spatial approximations are also possible. Likewise, only piecewise constant time-wise variations are considered in this text, but other approximations (such as piecewise linear) can be developed.

The only approximation involved in the solutions in this chapter is in the character of the surface disturbance over space and time. It means that, if the actual disturbance is piecewise constant, then the solutions obtained through superposition will be exact. Similarly, if the actual disturbance is piecewise linear or, when dealing with 2D problems, piecewise uniform. Accordingly, if space and time steps are chosen small so that the piecewise uniform and constant approximation is an excellent assumption, then the resulting approximate solutions will be highly accurate and computationally very efficient. The numerical solution of the building block might be very accurate too, but with an additional computational effort that might slow down remarkably the temperature computation.

Problems

3.1 A "step" heat flux pulse can be created by superposition of two X22B10T0 solutions. This $q(t)$ is given by

$$q(t) = \begin{cases} 0 & 0 \leq t < t_1 \\ q_0 & t_1 \leq t \leq t_2 \\ 0 & t > t_2 \end{cases}$$

This function can also be expressed using the Heaviside unit step functions as

$$q(t) = q_0 H(t - t_1) - q_0 H(t - t_2)$$

which gives a more direct indication of how the X22B10T0 should be combined.

Use superposition of two X22B10T0 solutions to plot the response to this function. For simplicity use dimensionless variables (let $\alpha = 1$ m²/s, $k = 1$ W/m-°C, $L = 1$ m, $q_0 = 1$ W/m²). Plot the temperature response at $x = 0$ and at $x = L$ for $0 \leq t \leq 10$ seconds, if $t_1 = 2$ seconds and $t_2 = 6$ seconds.

3.2 A triangular heat flux pulse can be created by superposition of X22B20T0 solutions. This $q(t)$ is given by

$$q(t) = \begin{cases} 0 & 0 \leq t < t_1 \\ q_0(t - t_1)/t_{ref} & t_1 \leq t \leq t_2 \\ q_0(t_3 - t)/t_{ref} & t_2 \leq t \leq t_3 \\ 0 & t > t_3 \end{cases}$$

Time "t_2" is midway between t_1 and $t_3 = (2t_2 - t_1)$, which are the start and end times of the "pulse." This function can be expressed using the Heaviside functions as

$$q(t) = \frac{q_0}{t_{ref}} \times (t - t_1)H(t - t_1) - 2\frac{q_0}{t_{ref}} \times (t - t_2)H(t - t_2) + \frac{q_0}{t_{ref}} \times (t - t_3)H(t - t_3)$$

which gives a more direct indication of how the X22B20T0 should be combined.

Use superposition of three X22B20T0 solutions to plot the response to this function. For simplicity use dimensionless variables (let $\alpha = 1$ m²/s, $k = 1$ W/m-°C, $L = 1$ m, $q_0 = 1$ W/m², $t_{ref} = 1$ s). Plot the temperature response at $x = 0$ and at $x = L$ for $0 \leq t \leq 10$ seconds, if $t_1 = 1$ second and $t_2 = 4$ seconds.

3.3 Obtain analytical expressions for the surface temperature $T(x = L, t)$ of a plate (1D rectangular body) that is initially at the uniform temperature of T_{in}, insulated on the back side $x = L$ and exposed at $x = 0$ to the temperature history given by

$$T_0(t) = \begin{cases} T_{in} + (T_0 - T_{in}), & 0 < t < t_0 \\ T_{in} + 2(T_0 - T_{in}), & t_0 < t < 2t_0 \\ T_{in}, & t < 0 \text{ and } t > 2t_0 \end{cases}$$

A different expression is needed for each time interval. Plot the dimensionless temperature, $\tilde{T} = (T - T_{in})/(T_0 - T_{in})$, at the plate back side $\tilde{x} = x/L = 1$ versus $\tilde{t} = \alpha t/L^2$ for $0 \leq \tilde{t} \leq 4\tilde{t}_0$ with $\tilde{t}_0 = 1$. (Superposition of X12B10T0 solutions is needed.)

3.4 Derive mathematical expressions for the surface temperature $T(x = 0, t)$ of a 1D rectangular body that is initially at the uniform temperature of T_{in}, insulated on the back side $x = L$ and exposed at $x = 0$ to the heat flux history given by

$$q_0(t) = \begin{cases} q_0, & 0 < t < t_0 \\ -q_0, & t_0 < t < 2t_0 \\ 0, & t < 0 \text{ and } t > 2t_0 \end{cases}$$

A different expression is needed for each time interval. Plot the dimensionless temperature, $\tilde{T} = (T - T_{in})/(q_0 L/k)$, at the heated boundary $\tilde{x} = 0$ versus $\tilde{t} = \alpha t/L^2$ for $0 \leq \tilde{t} \leq 4\tilde{t}_0$ with $\tilde{t}_0 = 1$. (Superposition of X22B10T0 solutions is needed.) Why does the surface temperature change after $\tilde{t} = 2\tilde{t}_0 = 2$ even though the net energy added is zero? Also, compare the plot obtained with the one derived for Problem 3.3.

3.5 Find symbolic expressions for the surface temperature $T(x = L, t)$ of a slab that is initially at the uniform temperature of T_{in}, insulated on the back side $x = L$ and that is subjected at $x = 0$ to a temperature that varies in a triangular fashion with time as

$$T_0(t) = \begin{cases} T_{in} + (T_0 - T_{in})(t/t_{ref}), & 0 < t < t_0 \\ T_{in} + (T_0 - T_{in})(2t_0 - t)/t_{ref}, & t_0 < t < 2t_0 \\ T_{in}, & t < 0 \text{ and } t > 2t_0 \end{cases}$$

where $(T_0 - T_{in})/t_{ref} = s_T$ is the constant heating $(T_0 > T_{in})$ rate.

Give expressions for the three intervals of time: (i) $0 < t < t_0$; (ii) $t_0 < t < 2t_0$; and (iii) $t > 2t_0$. Plot the dimensionless temperature, $\tilde{T} = (T - T_{in})/[s_T(L^2/\alpha)]$, at the slab back side $\tilde{x} = x/L = 1$ versus $\tilde{t} = \alpha t/L^2$ for $0 \leq \tilde{t} \leq 3\tilde{t}_0$ with $\tilde{t}_0 = 1$. (Superposition of X12B20T0 solutions is needed.)

3.6 Find exact expressions for the surface temperature $T(x = 0, t)$ of a 1D rectangular body that is initially at the uniform temperature of T_{in}, insulated on the back side $x = L$ and that is subjected at $x = 0$ to a heat flux that varies in a triangular fashion with time as given by

$$q_0(t) = \begin{cases} q_0(t/t_{ref}), & 0 < t < t_0 \\ q_0(2t_0 - t)/t_{ref}, & t_0 < t < 2t_0 \\ 0, & t < 0 \text{ and } t > 2t_0 \end{cases}$$

where $q_0/t_{ref} = s_q$ is a positive constant (heating).

Give expressions for the three intervals of time: (i) $0 < t < t_0$; (ii) $t_0 < t < 2t_0$; and (iii) $t > 2t_0$. Plot the dimensionless temperature, $\tilde{T} = (T - T_{in})/[s_q(L/k)(L^2/\alpha)]$, at the heated boundary $\tilde{x} = 0$ versus $\tilde{t} = \alpha t/L^2$ for $0 \leq \tilde{t} \leq 3\tilde{t}_0$ with $\tilde{t}_0 = 1$ (Superposition of X22B20T0 solutions is needed.) Also, compare the plot obtained with the one derived for Problem 3.5.

3.7 A slab of copper ($k = 401$ W/(m °C) and $\alpha = 117 \times 10^{-6}$ m²/s) of thickness 15 cm is initially at a uniform temperature of 20 °C. It is suddenly exposed to radiation at one surface (while the other surface is kept insulated) such that the next heat flux is given by

$$q_0(t) = \begin{cases} q_0, & 0 < t < t_0 \\ q_0 + q_0(t - t_0)/t_{ref}, & t_0 < t < 2t_0 \\ 0, & t < 0 \text{ and } t > 2t_0 \end{cases}$$

where $q_0 = 3 \times 10^5$ W m^{-2}, $s_q = q_0/t_{ref} = 10^5$ W m^{-2} s^{-1}, and $t_0 = 1$ min. Also, the heat transfer by convection at the irradiated side with the surrounding ambient at 20 °C is negligible.

Using the MATLAB functions of Section 3.4, determine the temperature at the irradiated surface and at the back side of the slab after 2 and 4 minutes have elapsed.

3.8 A slab of thickness L initially at the uniform temperature of zero is exposed at $t = 0$ and at $x = 0$ to the heat flux

$$q_0(t) = s_q \times t^{-1/2}$$

The back side of the plate is adiabatic. Let $L = 10$, $\alpha = 1$, $k = 1$, $s_q = \pi^{-1/2}$. Calculate the surface temperature at $x = 0$ and time $t = 1$ using the MATLAB function of Section 3.4.1.4. Note that the exact temperature is 1 ± 10^{-8} up to time $t = 5$. Comment on the accuracy of the procedure by changing the time step (starting with a time step of 1) as well as the accuracy A of the building block solution. Can the MATLAB function of Section 3.4.2.4 be used to solve this problem?

3.9 A two-dimensional rectangular body of steel ($k = 40$ W/(m °C) and $\alpha = 10^{-5}$ m²/s) of thickness 1 cm and width 5 cm is initially at a uniform temperature of 20 °C. It is suddenly exposed at the larger boundary surface $x = 0$ (while the other surfaces are kept insulated) to a heat flux variable in space and time as

$$q_0(y, t) = q_{y0t0}\left(\frac{y}{W}\right) \cdot \left(\frac{t}{t_{ref}}\right) = r_q y t$$

where $r_q = q_{y0t0}/(t_{ref}W) = 10^9$ W/(m²s m).

Using the MATLAB function of Section 3.6.3.3, determine the temperature at the corner points of the heated surface of the body after 2 and 4 seconds.

3.10 A 2D rectangular body of thickness L and width W, initially at the uniform temperature of zero, is exposed at $t = 0$ and at $x = 0$ to the heat flux

$$q_0(y, t) = r_q \times (yt)^{-1/2}$$

The other boundaries are adiabatic. Let $L = 1$, $W = 5$, $\alpha = 1$, $k = 1$, $r_q = \pi^{-1/2}$. Calculate the surface temperature at the corner points of the heated surface of the body at time $t = 1$ using the MATLAB function of Section 3.6.3.3. Comment on the accuracy of the procedure by changing the time and space steps as well as the accuracy A of the building block solution.

References

Beck, J. V., Wright, N. T. and Haji-Sheikh, A. (2008) Transient power variation in surface conditions in heat conduction for plates, *International Journal of Heat and Mass Transfer*, 51(9–10), p. 2553. https://doi.org/10.1016/j.ijheatmasstransfer.2007.07.043.

Cole, K. D., Beck, J. V., Haji-Sheikh, A. and Litkouhi, B. (2011) *Heat Conduction using Green's Functions*. 2nd edn. Boca Raton, FL: CRC Press.

Cole, K. D., Beck, J. V., Woodbury, K. A. and de Monte, F. (2014) Intrinsic verification and a heat conduction database, *International Journal of Thermal Sciences*, 78, pp. 36–47. https://doi.org/10.1016/j.ijthermalsci.2013.11.002.

D'Alessandro, G. and de Monte, F. (2018) Intrinsic verification of an exact analytical solution in transient heat conduction, *Computational Thermal Sciences*, 10(3), pp. 251–272. https://doi.org/10.1615/ComputThermalScien.2017021201.

de Monte, F., Beck, J. V. and Amos, D. E. (2011) A heat-flux based building block approach for solving heat conduction problems, *International Journal of Heat and Mass Transfer*, 54(13–14), pp. 2789–2800. https://doi.org/10.1016/j. ijheatmasstransfer.2011.02.060.

Najafi, H., Woodbury, K. A. and Beck, J. V. and Keltner, N. R. (2015) Real-time heat flux measurement using directional flame thermometer, *Applied Thermal Engineering*, 86, pp. 229–237. https://doi.org/10.1016/j.applthermaleng.2015.04.053.

Özişik, M. N. (1993) *Heat Conduction*. 2nd edn. New York: Wiley.

Patankar, S. V. (1980) *Numerical Heat Transfer and Fluid Flow*. Boca Raton: Hemisphere Publishing Corporation.

Potter, M. C., Goldberg, J. L. and Aboufadel, E. F. (2016) *Advanced Engineering Mathematics*. 3rd edn. Wildwood, MO, USA: Meridian Press, Inc.

Sheldon, A. (2020) *Measure, Integration & Real Analysis, Graduate Texts in Mathematics*. Springer.

Woodbury, K. A. and Beck, J. V. (2013) Heat conduction in a planar slab with power-law and polynomial heat flux input, *International Journal of Thermal Sciences*, **71**, pp. 237–248. https://doi.org/10.1016/j.ijthermalsci.2013.04.009.

4

Inverse Heat Conduction Estimation Procedures

4.1 Introduction

The inverse heat conduction problem (IHCP) is one of many ill-posed problems. The notion of a well-posed or "correctly set" mathematical problem made its debut in the 1923 discussions in Hadamard's work (Hadamard 1923). Hadamard famously set well-posed problems apart from those that are not, arguing that the latter problems are not suitable for mathematical analysis. The conditions for "well-posedness" require sufficiently general properties of *existence, uniqueness*, and (by implication) *stability* of solutions. In Section 4.3, it is shown that the IHCP is not well-posed and hence it is called ill-posed.

Ill-posed problems, according to Tikhonov and Arsenin (1977), can be divided into two subclasses: those involving estimation using data and those involving design of automatic controls. The IHCP is in the first subclass. In general, this subclass includes problems of mathematical processing and interpretation of data in varied fields where functions are to be estimated. Several of these areas, including manufacturing, aerophysics, biological, and electronics, were reviewed extensively in Chapter 1. Other fields include nuclear physics, radiophysics, electronics, interpretation of geophysical observations, mineral and petroleum exploration, and nuclear reactor engineering. The second subclass includes many optimization problems such as optimal control and optimal economic planning. In this book, only problems involving estimation with data are considered with the emphasis on the IHCP, but the techniques can be applied to many other problems.

Ill-posed problems include the mathematical problems of solution of singular or ill-conditioned systems of linear algebraic equations, differentiation of functions known only approximately, solution of partial differential equations using "interior" measurements and solution of integral equations of the first kind utilizing measurements. Even though there are many such problems that are of engineering importance, the ill-posed nature of the problems not only defies easy solution but has served to discourage the type of massive study that has accompanied direct or well-posed problems.

One of the difficulties of ill-posed problems is in defining what is meant by a "solution" because the solution does not satisfy general conditions of existence, uniqueness, and stability. This difficulty is addressed in this chapter.

The plan of this chapter is first to outline the difficulties is solving the IHCP in Section 4.2 and then further explore the concept of ill-posed problems in Section 4.3. Section 4.4 presents a broad overview of IHCP solution methodology, and Section 4.5 defines and illustrates sensitivity coefficients which are necessary for IHCP solutions. Sections 4.6–4.11 present a number of IHCP algorithms.

To better relate the various algorithms given in Sections 4.6–4.11, the contents are outlined next. These sections are devoted to estimating a single heat flux history for the linear IHCP; however, one section addresses nonlinear problems directly.

Section 4.6 presents the classical exact matching sequential method. For a single temperature sensor, the heat flux component at any given time can be found by setting the calculated temperature equal to the measured temperature at that time. This is called "exact" matching because the calculated temperature is made equal to the measured temperature. It is called a "sequential" procedure because q_M is estimated after q_{M-1}, for $M = 2,3,....$ If superposition of exact solution is used to obtain the calculated temperatures, the exact matching algorithm is called the Stolz (1960) algorithm. The paper by Stolz in 1960 was one of the earliest on the IHCP in Western literature.

Section 4.7 develops the function specification method (FSM) in which a functional form for the unknown heat flux is assumed. The functional form contains a number of unknown parameters that are estimated utilizing the method of least squares. Section 4.7.2 gives the sequential function specification method (SFSM) which was first discussed by Beck in 1961 (Beck 1961, 1962, 1968, 1970; Blackwell 1981).

Inverse Heat Conduction: Ill-Posed Problems, Second Edition. Keith A. Woodbury, Hamidreza Najafi, Filippo de Monte, and James V. Beck.
© 2023 John Wiley & Sons, Inc. Published 2023 by John Wiley & Sons, Inc.

Section 4.8 presents the Tikhonov regularization (TR) method. A number of authors have championed this whole domain regularization procedure, including Tikhonov and Arsenin (1977), Alifanov (1974, 1975, 1983, 1994), Alifanov and Artyukhin (1975), and Bell and Wardlaw (1981a, 1981b). The zeroth order TR is related to ridge regression (Marquardt 1963, 1970; Hoerl and Kennard 1970a, b; Lawson and Hanson 1974). A sequential regularization method is given in Section 4.8.4 which was developed by Lamm et al. (1997), Lamm (2005), and Brooks et al. (2010).

Iterative methods for whole domain solution are described in Section 4.9. These methods are based on gradient methods for minimization. Section 4.9.1 presents the general framework which is restricted to linear problems. Section 4.9.2 outlines the procedures for adapting these methods for nonlinear problems and this approach is often called the adjoint method (AM).

Section 4.10 describes truncated singular value decomposition (TSVD or SVD). This method uses features of the SVD approach to computing the matrix inverse to directly solve the linear IHCP.

Section 4.11 presents application of Kalman filtering (KF) to solution of the IHCP. Kalman filters are used extensively in estimation and control applications and can be adapted to solving the IHCP.

4.2 Why is the IHCP Difficult?

The IHCP is inherently *ill-posed*, and the concept of ill-posedness is explained in Section 4.3. However, the difficulties in solving the IHCP arise from two physical sources: sensitivity to measurement errors and damping and lagging of information. These ideas are explored in the next two sections.

4.2.1 Sensitivity to Errors

The IHCP is difficult because it is extremely sensitive to measurement errors. This sensitivity to errors is a manifestation of the ill-posedness of the IHCP, which is discussed in Section 4.3. These difficulties are more pronounced as maximal information is sought in the solution. For the one-dimensional IHCP, when discrete values of the q curve are estimated, maximizing the amount of information implies using small time steps between q_i values (see Figure 1.2). However, as will be seen, the use of small time steps amplifies instabilities in the solution of the IHCP.

The instability can be countered by incorporating some restriction on the solution, a process known as *regularization*. Several types of regularization for IHCP solutions will be explored later in this chapter.

Notice the condition of *small* time steps has the opposite effect in the IHCP compared to that in the numerical solution of the heat conduction equation. In the latter, stability problems often can be corrected by reducing the size of the time steps.

4.2.2 Damping and Lagging

The transient temperature response of an internal point in an opaque, heat-conducting body is quite different from that of a point at the surface. The internal temperature excursions are much smaller as compared to the surface temperature changes. This is a damping effect. A large time delay or lag in the internal response can also be noted. These damping and lagging effects for the direct problem are important to study because they provide engineering insight into the difficulties encountered in the inverse problem.

The damping and lagging of the thermal response can be visualized using Figure 2.7b for the X22 case and Figure 2.3b for the X12 case. In both figures, the response at the surface is instantaneous, as the temperature rises immediately upon excitation. However, for any subsurface location, the temperature rise is always less than that at the surface (damped relative to the surface), and the response begins later than the surface disturbance (lagged relative to the excitation).

4.2.2.1 Penetration Time

The lagging effect is well described through the concept of *penetration time* (de Monte et al. 2008), which is the time required for the temperature response at a subsurface location to reach a specified fraction of the temperature at the surface. This time is also tied to the magnitude specified for a "noticeable" temperature rise at the location. For a sensor located a distance d from the active surface, the time, t_p, required for the thermal response at that location, *relative to the surface response,* to be at least 10^{-A}, where A is a positive integer, is at least

$$t_p^+ = \frac{\alpha t_p}{d^2} \simeq \frac{0.1}{A} \tag{4.1}$$

Equation (4.1) is derived considering a step change in surface temperature, which provides the fastest internal response; therefore, the resulting t_p time is a lower bound for other types of disturbances – the amount of time needed is *at least* that specified by Eq. (4.1). The important distance dimension in Eq. (4.1) is the distance from the active surface to the sensor.

For a sensor is located at a dimensionless distance from the surface $\tilde{d} = x/L = 0.5$, Eq. (4.1) predicts the dimensionless penetration time for the response to rise to 0.01 ($A = 2$), based on the distance d, as $0.1/2 = 0.05$. This is related to the dimensionless time, *based on length scale L*, as

$$t_p^+ = \frac{\alpha t_p}{d^2} = \frac{\alpha t_p}{(0.5L)^2} = 4\tilde{t}_p \simeq 0.05 \quad \Rightarrow \quad \tilde{t}_p = \frac{0.1}{4A} = 0.0125 \tag{4.2}$$

Consider the case of a flat plate insulated at $x = L$ and exposed to a constant heat flux q_c at $x = 0$, which is denoted X22B10T0. The computational analytical solution for this case from Section 2.3.8.1 is used to find the response at $\tilde{x} = 0.5$ relative to the corresponding surface temperature. For a step change in heat flux at the surface, at time $\tilde{t} = 0.0125$, this ratio is less than 0.01; in fact $\tilde{T}(0.5, 0.0125)/\tilde{T}(0, 0.0125) = 0.000534$. The minimum penetration time given by Eq. (4.1) can be computed more accurately using power law curve fit equations to the actual penetration times.

$$t_{p,X22}^+ = 0.2662A^{-1.4168}, \qquad 1 \le A \le 6; \quad 0 \le \tilde{x} \le 0.75 \tag{4.3}$$

$$t_{p,X12}^+ = 0.1826A^{-1.2474}, \qquad 1 \le A \le 6; \quad 0 \le \tilde{x} \le 0.75 \tag{4.4}$$

Equations (4.3) and (4.4) predict the penetration times within 3.5% for $A < 4$, and within 9% for $4 < A < 5.75$. The depth is limited to about $\tilde{x} = 0.75$ because the effect of the insulated boundary begins to decrease the time required for the temperature to rise. For the example with $A = 2$, the time required is found from Eq. (4.3) as $t_p^+ = 0.0997 \cong 0.10$, which corresponds to a dimensionless time based on L of $\tilde{t}_p = t_p^+/4 = 0.025$. From Eq. (2.36), the ratio of the response at $\tilde{x} = 0.5$ to that at $\tilde{x} = 0$ at time $t_p^+ = 0.025$ is, in fact, 0.0110.

4.2.2.2 Importance of the Penetration Time

The time required for the temperature rise to reach the depth of a sensor is important in selecting sensor locations and data sampling rate. An IHCP solution cannot produce any estimation of the surface action until data are obtained in response to the changed condition at the surface. In the example from the preceding paragraph, the dimensionless time required for the proposed sensor located at $\tilde{x} = 0.5$ to achieve 1% of the temperature change at the surface is $\tilde{t}_p = 0.025$. It means it will be very difficult to estimate changes in surface conditions at dimensionless time intervals less than 0.025 using this sensor location. Smaller penetration times will require more *regularization* (introduced in Section 4.3.3), which increases the bias in the estimation for the surface disturbance.

4.3 Ill-Posed Problems

The IHCP is shown to be ill-posed in this section. This problem will be viewed first from the perspective of a solution to a partial differential equation and then as set of linear algebraic equations.

4.3.1 An Exact Solution

Burggraf (1964) developed an exact solution which is applicable to an IHCP in a one-dimensional slab. Figure 1.3 illustrates the overall problem where the slab is heated with an unknown flux $q(t)$ at the active surface at $x = 0$ and a conventional boundary condition is known at the inactive surface at $x = L$.

Temperature measurements are made at a location x_1 and are used to approximate a specified temperature history $T(x_1, t) = Y$. With known boundary conditions for the "direct" region, the temperature in region (2) can be determined, and the derivative of the temperature at x_1 can be used to determine the heat flux at that location, $q_{x1}(t) = q(x_1, t)$.

Burggraf's solution makes use of the resulting overspecified information at $x = x_1$ to solve for the temperature in the indirect region. The solution is achieved by assuming the temperature field in the indirect region $0 \leq x \leq x_1$ is an infinite series of temperature gradients at the sensor location. For the planar geometry, the temperature distribution is written as

$$\tilde{T}(\tilde{x}, \tilde{t}) = \tilde{Y}(\tilde{t}) + \sum_{n=1}^{\infty} \frac{1}{(2n)!} (\tilde{x}_1 - \tilde{x})^{2n} \frac{d^n \tilde{Y}}{d\tilde{t}^n} - (\tilde{x}_1 - \tilde{x}) \left[\tilde{q}_{x_1}(t) + \sum_{n=1}^{\infty} \frac{1}{(2n+1)!} (\tilde{x}_1 - \tilde{x})^{2n} \frac{d^n \tilde{q}_{x_1}}{d\tilde{t}^n} \right] \tag{4.5}$$

where

$$\tilde{x} = \frac{x}{L}; \quad \tilde{t} = \frac{\alpha t}{L^2}; \quad \tilde{T} = \frac{T}{q_{\text{ref}} L / k}; \quad \tilde{q} = \frac{q}{q_{\text{ref}}} \tag{4.6}$$

The planar geometry solution is similar in appearance to a Taylor series expansion about the temperature sensor depth.

For the problem to be well-posed, it is necessary that the solution (i) exist, (ii) be unique, and (iii) be continuously dependent on the data or, equivalently, be stable. Each of these aspects is now considered for the foregoing solution.

The *existence* of this solution was proved by Widder (1975) (see also Weber (1981)). The series given by Eq. (4.5) converges uniformly for bounded t, provided Y and q_{x1} satisfy the absolute value conditions,

$$\left| \frac{d^n Y(t)}{dt^n} \right| \leq M \frac{(2n)!}{x_1} \tag{4.7}$$

$$\left| \frac{d^n q_{x_1}(t)}{dt^n} \right| \leq P \frac{(2n)!}{x_1}, \quad n = 0, 1, 2, \dots \tag{4.8}$$

for some constants M and P and where $x = L$ is the heated surface. Hence, the solution exists for these conditions for the nth time derivatives of Y and q_{x1} growing with n less than the factorial of $2n$.

The *uniqueness* of the solution given by Eq. (4.5) is proved by noting that any two solutions must be equal to the right side of Eq. (4.5) and therefore to each other.

The question of *stability* is now considered. This requires measures of the differences between corresponding solutions. For that reason, for the time and space intervals of $0 \leq t \leq t_f$ and $0 \leq x \leq L$, the following norms are defined:

$$\|Y\| \equiv \max_{0 \leq t \leq t_f} |Y(t)| \tag{4.9a}$$

$$\|q_{x_1}\| \equiv \max_{0 \leq t \leq t_f} |q_{x_1}(t)| \tag{4.9b}$$

$$\|T\| \equiv \max_{\substack{0 \leq t \leq t_f \\ 0 \leq x \leq x_1}} |T(x, t)| \tag{4.9c}$$

The single bars are for absolute values and the double bars are for the norms which are defined in Eqs. (4.9a)–(4.9c). Consider two solutions of Eq. (4.5), denoted $T_i(x, t)$, resulting from $Y_i(t)$ and $q_{x1,i}(t)$ where $i = 1$ and 2.

A necessary condition for stability is that the norm $\|T_1 - T_2\|$ can be made arbitrarily small by choosing Y_1, Y_2, $q_{x1,1}$, and $q_{x1,2}$ so that the norms $\|Y_1 - Y_2\|$ and $\|q_{x1,1} - q_{x1,2}\|$ are sufficiently small. This is impossible, in general, as shown next.

To prove that the solution is not always stable, we need display only one exceptional case; namely

$$q_{x1,1} = q_{x1,2} = 0 \tag{4.10}$$

$$Y_2(t) = Y_1(t) + \frac{1}{\beta} \cos (\beta^2 t), \quad \beta > 0 \tag{4.11}$$

and $Y_1(t)$ is an arbitrary analytic function. The error term, $\beta^{-1} \cos \beta^2 t$, can be made small by making β large. This forces Y_1 and Y_2 to be arbitrarily close as measured by the norm,

$$\|Y_1 - Y_2\| = \max_{0 \leq t \leq t_f} \left| \frac{1}{\beta} \cos \beta^2 t \right| = \frac{1}{\beta} \tag{4.12}$$

The corresponding solutions for T_1 and T_2 may not be close at all. The difference between T_1 and T_2 is

$$T_1 - T_2 = \sum_{n=1}^{\infty} \frac{1}{(2n)!} \left[\frac{(x_1 - x)^2}{\alpha} \right]^n \frac{d^n}{dt^n} \left(\frac{1}{\beta} \cos \beta^2 t \right)$$

$$= \frac{1}{\beta} \sum_{n=1}^{\infty} \frac{(-1)^n}{(4n)!} \left(\frac{(x_1 - x)^2}{\alpha} \right)^{2n} \beta^{4n} \cos \beta^2 t \qquad (4.13a, b)$$

$$+ \frac{1}{\beta} \sum_{n=0}^{\infty} \frac{(-1)^{n+1}}{(4n+2)!} \left(\frac{(x_1 - x)^2}{\alpha} \right)^{2n+1} \beta^{4n+2} \sin \beta^2 t$$

For the special location x where $(x_1 - x)^2/\alpha = 1\,\text{s}$ and for time $t = 0$, Eq. (4.13b) gives

$$T_1 - T_2 = \frac{1}{\beta} \sum_{n=0}^{\infty} \frac{(-1)^n}{(4n)!} \beta^{2n}$$

$$= \frac{1}{\beta} \cosh\left(\frac{\beta}{\sqrt{2}} \right) \cos\left(\frac{\beta}{\sqrt{2}} \right) \qquad (4.14a, b)$$

The equality between Eqs. (4.14a) and (4.14b) is proved in Weber (1981). Using Eq. (4.14b) results in

$$\|T_1 - T_2\| \geq \frac{1}{\beta} \left| \cosh\left(\frac{\beta}{\sqrt{2}} \right) \cos\left(\frac{\beta}{\sqrt{2}} \right) \right| \qquad (4.15)$$

which as $\beta \to \infty$ gives $(2\beta)^{-1} \exp\left(2^{-1/2} \beta \right) \exp\left(\beta/\sqrt{2} \right)/(2\beta)$ which is unbounded. In other words, for the special case considered, arbitrarily small differences in the input temperature can result in arbitrarily large differences in the surface temperature.

The conclusion is that the solution given by Eq. (4.5) is unstable. This is observed to be true although Eq. (4.5) is the *exact* solution for an IHCP with continuous data. Thus, even with existence and uniqueness, stability is not present. Hence, it is proved that the IHCP solution coming from a partial differential equation is ill-posed because it does not satisfy the three conditions for a well-posed problem.

4.3.2 Discrete System of Equations

The Burggraf solution provides an excellent mathematical platform to prove the inherent ill-posedness of the IHCP; however, it is not a practical method for solution of inverse problems due to the requirement of infinite differentiability of the temperature and heat flux at the measurement location.

A solution of the transient heat conduction equation can be obtained by using superposition of linear solutions (Chapter 3), or by numerical methods (finite differences or finite elements). These approaches provide sets of linear algebraic equations that can be used to obtain a system of equations in the form

$$\mathbf{T} = \mathbf{Xq} \qquad (4.16)$$

Most techniques for solving the IHCP considered in this textbook are based on the discrete form of Eq. (4.16).

For a given set of measurement observations, $\mathbf{Y}$, the solution for the heat flux may naively be attempted using Eq. (4.16) through matrix inversion:

$$\hat{\mathbf{q}} = \mathbf{X}^{-1}\mathbf{Y} \qquad (4.17)$$

For very limited cases, a solution for $\hat{\mathbf{q}}$ will result. However, the nature of the structure of $\mathbf{X}$ results in a matrix with a high *condition number*. The inverse $\mathbf{X}^{-1}$ invariably has very large elements, which will amplify small perturbations in the measured data $\mathbf{Y}$.

A very simple example of ill-posedness in algebraic systems of the form of Eq. (4.16) is given by Hensel (1990). Consider the following system of equations:

$$\begin{bmatrix} 1 & 1 \\ 1 & 1.001 \end{bmatrix} \begin{Bmatrix} a \\ b \end{Bmatrix} = \begin{Bmatrix} 1 \\ 1 \end{Bmatrix} \qquad (4.18)$$

The coefficient matrix in Eq. (4.18) is very nearly singular but can be inverted to yield the solution $a = 1$, $b = 0$. If a very small error is introduced in the right side, as

$$\begin{bmatrix} 1 & 1 \\ 1 & 1.001 \end{bmatrix} \begin{Bmatrix} a \\ b \end{Bmatrix} = \begin{Bmatrix} 1 \\ 1.001 \end{Bmatrix} \tag{4.19}$$

the solution becomes $a = 0$ and $b = 1$.

The notion of the condition number helps inform these results. The condition number of a matrix is connected to its SVD and is defined as the ratio of the largest eigenvalue to the smallest eigenvalue. When the condition number is infinite, the matrix is singular and cannot be inverted. When the condition number is "large," then the matrix is *ill-conditioned*. Wolfram Mathworld (Lichtblau and Weisstein 2002) define that a "large" condition number exists when the logarithm of the condition number is greater than the precision of the matrix entries. For the coefficient matrix of Eqs. (4.18) and (4.19), the condition number is 4002, and $\ln(4002) = 8.3$, which is much larger than the precision of the matrix entries (0.001).

For an interior temperature measurement location and "small" time steps, the solution of Eq. (4.17) for $\hat{\mathbf{q}}$ is ill-conditioned. That is, small changes in the $\mathbf{Y}$ vector will make large changes in the $\hat{\mathbf{q}}$ vector.

4.3.3 The Need for Regularization

The IHCP is inherently ill-posed, and the effects of ill-posedness are exacerbated by noisy data and/or small calculation time steps. These effects are combatted through *regularization* of the IHCP solution technique.

Regularization can be defined as any modification to an original, ill-posed problem, to convert it to a "mathematically nearby" well-posed one.[1] Various procedures for estimating the surface disturbance are presented in Sections 4.6–4.11. The mechanism for regularization in each technique will be described.

Regularization, by its very nature, necessarily alters the underlying ill-posed problem, and therefore introduces bias to the estimates. The trade-off is increased stability of the calculation, and an obvious challenge is the task of selecting the proper amount of regularization to stabilize the calculations without overly biasing the results. The topic of optimal regularization is discussed in Chapter 6.

4.4 IHCP Solution Methodology

Several methods for solving the IHCP will be presented later in this chapter. Generic components necessary for solution of an inverse problem are:

1) A "direct solver," such as Eq. (4.16), is needed to evaluate the solution to the heat conduction equation for a known surface disturbance. The direct solver can be based on exact analytical solutions (for linear heat conduction problems) or any suitable numerical solver. This textbook emphasizes use of analytical solutions as direct solvers.
2) Experimental observations or measurements of temperatures in the domain are needed.
3) A measure of goodness or objective function is needed to match the measurements to the solution from the direct solver. This measure could be as simple as exactly matching the measurement to the computed values, but a sum of squared errors between the measured and computed temperature values is often used. A suitable technique is used to solve the IHCP for the unknown surface disturbance by optimizing the measure of goodness, for example, by minimizing the sum of the squared errors.
4) Some form of regularization should be present to counteract the inherent ill-posedness of the IHCP.
5) Once the IHCP is solved for the unknown action at the boundary, the direct solver can be used to evaluate the temperature at any location in the domain.

4.5 Sensitivity Coefficients

4.5.1 Definition of Sensitivity Coefficients and Linearity

In function estimation, as in parameter estimation, a detailed examination of the sensitivity coefficients can provide considerable insight into the estimation problem. These coefficients can show possible areas of difficulty and lead to improved experimental design.

---

1 Paraphrased from a definition given by Professor Patti Lamm of Michigan State University at a tutorial lecture delivered at the Inverse Problems Symposium around 1991.

The sensitivity coefficient is defined as the first derivative of a dependent variable, such as temperature, with respect to an unknown parameter, such as a heat flux component. If the sensitivity coefficients are either small or correlated with one another, the estimation problem is difficult and very sensitive to measurement errors.

For the IHCP, the sensitivity coefficients of interest are those of the first derivatives of temperature T at location x_j and time t_i with respect to a heat flux component, q_M, and are defined by

$$X_{jM}(x_j, t_i) \equiv \frac{\partial T(x_j, t_i)}{\partial q_M} \qquad (4.20a)$$

for $j = 1,...,J$ measurement locations, $i = 1,2,...,N$ time steps, and $M = 1,2,...,N$ heat flux components. Here, q_M is the component of the heat flux associated with time t_M. Note that the number of times t_i equals the number of these heat flux components. For the *piecewise constant* approximation to the heat flux (Figure 4.1, dashed lines), q_M is the constant heat flux

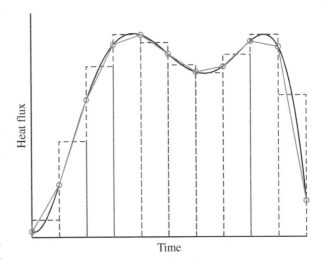

Figure 4.1 Piecewise constant (dashed lines) and piecewise linear approximations (solid lines and circles) to heat flux history

between times t_{M-1} and t_M. For the *piecewise linear* approximation to the heat flux (Figure 4.1, solid lines), q_M is the value of the heat flux at time t_M. If there is only one interior location, that is, $J = 1$, the sensitivity coefficient is simply given by

$$X_M(t_i) \equiv \frac{\partial T_i}{\partial q_M} \qquad (4.20b)$$

For the transient problems considered in the IHCP, the sensitivity coefficients are zero for $M > i$. In other words, the temperature at time t_i is independent of a yet-to-occur future heat flux component of q_M, $M > i$.

One of the important characteristics of the IHCP is that it is a linear problem if the heat conduction equation is linear and the boundary conditions are linear. The thermal properties (k, ρ, and c) can be functions of position and not affect the linearity. They cannot, however, be functions of temperature without causing the IHCP to be nonlinear. Linearity, if it exists, is an important property because it allows superposition in various ways and it generally eliminates the need for iteration in the solution. Techniques illustrated in this textbook emphasize linearity and superposition.

One way to determine the linearity of an estimation problem is to inspect the sensitivity coefficients. If the sensitivity coefficients are *not* functions of the parameters, then the estimation problem is linear. If they are, then the problem is nonlinear. To illustrate this, consider the response of a semi-infinite body to a constant surface heat flux q_c. The temperature response at the surface ($x = 0$) is

$$T(0, t) = \frac{q_c}{k}\left[\frac{4\alpha t}{\pi}\right]^{1/2} \qquad (4.21)$$

Differentiating T with respect to q_c gives

$$\frac{\partial T(0, t)}{\partial q_c} = \frac{1}{k}2\left(\frac{\alpha t}{\pi}\right)^{1/2} \qquad (4.22)$$

which is independent of q_c. Similarly, for finite domains, the X22B10T0 solution can be used to arrive at the same conclusion.

An example of a nonlinear estimation problem is that of estimating α. Taking the derivative of T in Eq. (4.21) with respect to α yields

$$\frac{\partial T(0, t)}{\partial \alpha} = \frac{q_c}{k}\left(\frac{t}{\pi \alpha}\right)^{1/2} \qquad (4.23)$$

The right side of Eq. (4.23) is a function of α; thus the estimation of α from transient temperature measurements is a nonlinear problem. (The same conclusion is reached in consideration of the X22B10T0 case for finite geometry.)

This principle of the sensitivity coefficients being independent of the parameter to be estimated can be employed in cases when the solution is not explicitly known. The equations for a flat plate with temperature-independent properties are given as an example:

$$\frac{\partial}{\partial x}\left[k(x)\frac{\partial T}{\partial x}\right] = \rho c(x)\frac{\partial T}{\partial t} \tag{4.24a}$$

$$-k\frac{\partial T}{\partial x}\Big|_{x=0} = \left\{\begin{array}{ll} q_M = \text{constant} & t_{M-1} < t < t_M \\ q(t) & t > t_M \end{array}\right\} \tag{4.24b}$$

$$-k\frac{\partial T}{\partial x}\Big|_{x=L} = q_{\text{loss}} \tag{4.24c}$$

$$T(x, t_{M-1}) = T_{M-1}(x) \tag{4.24d}$$

where $T_{M-1}(x)$ denotes the temperature distribution at time t_{M-1} and q_{loss} is a heat flux history due to losses that are independent of q_M. The heat flux, $q(t)$, for $t > t_M$ is an arbitrary function of time. The thermal properties can be functions of x. And the parameter of interest is the heat flux, q_M, which is a constant between times t_{M-1} and t_M. This is the piecewise constant assumption for heat flux variation as indicated by the dashed lines in Figure 4.1. The temperature distribution at time t_{M-1} is known and is specified through Eq. (4.24d).

The differential equation and boundary conditions for the q_M sensitivity coefficient defined by

$$X_M(x, t) \equiv \frac{\partial T(x, t)}{\partial q_M} \tag{4.25}$$

are to be found. For $t < t_{M-1}$ the solution is $X_M(x, t) = 0$; that is, the body has not yet been exposed to q_M. For times greater than t_{M-1}, Eqs. (4.24a)–(4.24d) are differentiated with respect to q_M to obtain,

$$\frac{\partial}{\partial x}\left[k(x)\frac{\partial X_M}{\partial x}\right] = \rho c(x)\frac{\partial X_M}{\partial t} \tag{4.26a}$$

$$-k\frac{\partial X_M}{\partial x}\Big|_{x=0} = \left\{\begin{array}{ll} 1 & t_{M-1} < t < t_M \\ 0 & t > t_M \end{array}\right. \tag{4.26b}$$

$$\frac{\partial X_M}{\partial x}\Big|_{x=L} = 0 \tag{4.26c}$$

$$X_M(x, t_{M-1}) = 0 \tag{4.26d}$$

Equations (4.26a)–(4.26d) describe the mathematical problem for the sensitivity coefficient X_M which can be explicitly found if the functions $k(x)$ and $\rho c(x)$ are known. Notice that it is not necessary to know q_M, $q(t)$, or even $T_{M-1}(x)$ to obtain a solution for the sensitivity coefficient $X_M(x, t)$ because Eqs. (4.26a)–(4.26d) are not functions of q_M, $q(t)$, or $T_{M-1}(x)$.

An important conclusion that can be drawn from Eqs. (4.26a)–(4.26d) is that the estimation problem for q_M is linear, a consequence of $X_M(x, t)$ not being a function of q_M. This means that the (unknown) value of q_M is not needed to find its sensitivity coefficient. It is also significant that the same differential equation is given for $X_M(x, t)$, Eq. (4.26a), as for $T(x, t)$, Eq. (4.24a). Also the boundary conditions are of the same type; that is, the gradient conditions given by Eqs. (4.24b) and (4.24c) for $T(x, t)$ are still gradient conditions, Eqs. (4.26b) and (4.26c) for $X_M(x, t)$. The main differences are that the $X_M(x, t)$ boundary conditions are simpler and the initial condition is zero. Due to this similarity between the $T(x, t)$ and $X_M(x, t)$ problems, the same solution procedure or computer program can be used for the $X_M(x, t)$ solution as for the $T(x, t)$, which can result in considerable programming and computational efficiency.

4.5.2 One-Dimensional Sensitivity Coefficient Examples

4.5.2.1 X22 Plate Insulated on One Side

The heat flux sensitivity coefficients for a flat plate heated on one side and insulated on the other (the X22B-0T0 case) are discussed in this section.

4.5.2.1.1 Piecewise Constant Heat Flux

For a constant heat flux of infinite duration (X22B10T0), the temperature distribution is given in equation form by Eq. (2.34a) or Eq. (2.35) and in graphical form by Figure 2.7. These results can also be interpreted as being equal to the dimensionless heat flux sensitivity coefficient:

$$T = \frac{q_{\text{ref}}L}{k}\tilde{T} + T_0; \quad \frac{\partial T}{\partial q_{\text{ref}}} = \frac{L}{k}\tilde{T}; \quad \tilde{X} = q_{\text{ref}} \times \frac{\partial T}{\partial q_{\text{ref}}} \times \frac{k}{q_{\text{ref}}L} = \tilde{T} \tag{4.27a}$$

$$\tilde{X}(\tilde{x}, \tilde{t}) \equiv \frac{k\partial T}{L\partial q_{\text{ref}}} = \tilde{T}(\tilde{x}, \tilde{t}) \tag{4.27b}$$

The sensitivity coefficient for a finite-width constant heat flux pulse, q_1, given by

$$q(t) = \begin{cases} q_1, & 0 < t < \Delta t \\ 0 & t > \Delta t \end{cases} \tag{4.28}$$

was examined in Section 3.4.1.2. All the sensitivity coefficients for $q_2, q_3, ..., q_n$ have exactly the same shape but are displaced Δt apart, so it is only necessary to examine $\tilde{X}_1$, which is given by

$$\tilde{X}_1 \equiv \frac{k}{L}\frac{\partial T}{\partial q_1} = \begin{cases} \tilde{T}(\tilde{x}, \tilde{t}), & 0 < \tilde{t} < \Delta\tilde{t} \\ \tilde{T}(\tilde{x}, \tilde{t}) - \tilde{T}(\tilde{x}, \tilde{t} - \Delta\tilde{t}), & \tilde{t} > \Delta\tilde{t} \end{cases} \tag{4.29}$$

where

$$\Delta\tilde{t} \equiv \frac{\alpha\Delta t}{L^2} \tag{4.30}$$

and Δt is the duration of the heating pulse. For the large dimensionless times of

$$\tilde{t} - \Delta\tilde{t} > 0.5 \tag{4.31}$$

the second expression of Eq. (4.29) goes to the time- and space-independent value of

$$\tilde{X}_{1,\text{max}} = \Delta\tilde{t} \tag{4.32}$$

This expression shows the dependence of $\tilde{X}_1$ on the duration of heating; hence, as more components of q are estimated using smaller time intervals over a fixed time period, the sensitivity coefficients become smaller and thus the $q(t)$ curve becomes more difficult to estimate. Notice that as $\tilde{t}$ increases, the temperatures increase as shown by Figure 2.7, but the sensitivity coefficients do not.

The sensitivity coefficient $\tilde{X}_1$ normalized with respect to $\tilde{X}_{1,\text{max}}$ is plotted in Figure 4.2. To present these results compactly, the time has been made dimensionless by normalization with respect to the heating duration, Δt.

4.5.2.1.2 Piecewise Linear Heat Flux

An alternative to the piecewise constant discretization of the continuous heat flux is the *piecewise linear approximation* seen as the open circles in Figure 4.1. This assumption arguably can represent a smoothly varying heat flux more accurately, and this is visualized in Figure 4.1 through comparison of the solid linear segments connected with nodes against the dashed-line piecewise constant segments. In the piecewise-linear approximation, the heat flux values characterizing the underlying function are associated directly with the time steps ((t_1,q_1), (t_2,q_2), etc.).

The sensitivity coefficients for this case can be computed from the X22B20T0 solution for a linear in time heat flux function at the surface (Section 3.4.2.2). For this case, the solution in Eq. (2.45a) can be rewritten slightly as:

$$\tilde{T}_{\text{X22B20T0}}(\tilde{x}, \tilde{t}) = \frac{T(\tilde{x}, \tilde{t})}{q_{\text{ref}}L/k} = \frac{1}{\tilde{t}_0}\left[\begin{array}{l} \frac{\tilde{t}^2}{2} + \left(\frac{1}{2}\tilde{x}^2 - \tilde{x} + \frac{1}{3}\right)\tilde{t} + \frac{1}{24}\tilde{x}^4 - \frac{1}{6}\tilde{x}^3 + \frac{1}{6}\tilde{x}^2 - \frac{1}{45} \\ + 2\sum_{m=1}^{\infty}\frac{1}{(m\pi)^4}\cos(m\pi\tilde{x})e^{-(m\pi)^2\tilde{t}} \end{array}\right] \tag{4.33}$$

This form provides some flexibility in scaling the magnitude of the increasing ramp heat flux since the variable $\tilde{t}_0$ is the dimensionless time at which $q_{\text{ref}} = 1$.

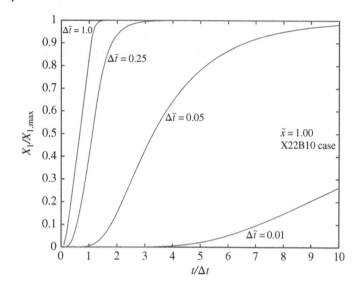

Figure 4.2 Heat flux sensitivity coefficients at insulated surface of flat plate for piecewise constant heat flux model.

Sensitivity coefficients for the first heat flux value, q_1, in Figure 4.1 can be found using superposition (all subsequent sensitivity coefficients will be identical). In this case, the dimensionless sensitivity coefficient will be

$$\tilde{X}_{1,\text{X22B20}} \equiv \frac{k}{L}\frac{\partial T}{\partial q_1} = \begin{cases} \tilde{T}_{\text{X22B20}}(\tilde{x}, \tilde{t}), & 0 < \tilde{t} < \Delta\tilde{t} \\ \tilde{T}_{\text{X22B20}}(\tilde{x}, \tilde{t}) - 2\tilde{T}_{\text{X22B20}}(\tilde{x}, \tilde{t} - \Delta\tilde{t}), & \Delta\tilde{t} < \tilde{t} < 2\Delta\tilde{t} \\ \tilde{T}_{\text{X22B20}}(\tilde{x}, \tilde{t}) - 2\tilde{T}_{\text{X22B20}}(\tilde{x}, \tilde{t} - \Delta\tilde{t}) + \tilde{T}_{\text{X22B20}}(\tilde{x}, \tilde{t} - 2\Delta\tilde{t}), & \tilde{t} > 2\Delta\tilde{t} \end{cases} \tag{4.34}$$

which is equivalent to Eq. (3.64) when $I = 1$ and M is varied.

For the large dimensionless times, the second expression of Eq. (4.34) goes to the time- and space-independent value of

$$\tilde{X}_{1,\text{X22B20T0}}(\tilde{x}, \tilde{t} \to \infty) = \Delta\tilde{t} \tag{4.35}$$

which is the same as for the piecewise constant approximation. However, the values at smaller times are less than those for the piecewise constant approximation. This, in turn, suggests the IHCP solution for the linear assumption may be more ill-posed than for the piecewise constant assumption for the same time step size.

The sensitivity coefficient $\tilde{X}_{1,\text{X22B20}}$ normalized with respect to $\tilde{X}_{1,\max}$ is plotted in Figure 4.3. To present these results compactly, the time has been made dimensionless by normalization with respect to the heating duration, Δt.

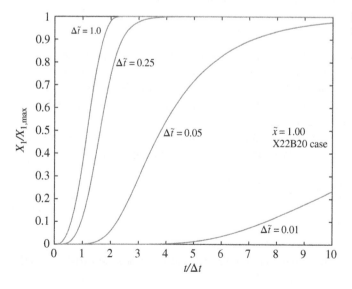

Figure 4.3 Heat flux sensitivity coefficients at insulated surface of flat plate for piecewise linear heat flux model.

4.5.2.2 X12 Plate Insulated on One Side, Fixed Boundary Temperature

A final example of a one-dimensional sensitivity coefficient is given based on assuming the unknown surface disturbance is a temperature function of time. The basic solution for a step change in surface temperature is used to generate coefficients by assuming a piecewise constant time variation of the unknown surface heat flux:

$$\tilde{T}_{X12}(\tilde{x}, \tilde{t}) = 1 - 2\sum_{m=1}^{\infty} \frac{1}{(m-0.5)\pi} e^{-((m-0.5)\pi)^2\tilde{t}} \sin((m-0.5)\pi\tilde{x}) \tag{4.36}$$

The X12B10T0 solution provides the dimensionless sensitivity to the constant surface temperature T_0 as

$$\tilde{X}_{X12} = \frac{\partial T_{X12}(\tilde{x}, \tilde{t})}{\partial T_0} = \frac{T_{X12}(\tilde{x}, \tilde{t})}{T_0} = \tilde{T}_{X12}(\tilde{x}, \tilde{t}) \tag{4.37}$$

As before, the sensitivity to the first of the piecewise constant surface temperature changes over each time interval Δt, $X_{X12,1}$, can be found by superposition, and each of the subsequent pulse sensitivity coefficients ($X_{X12,2}$, $X_{X12,3}$, etc.) will be the same.

$$\tilde{X}_{1,X12} \equiv T_1 \frac{\partial T}{\partial T_1} = \begin{cases} \tilde{T}_{X12}(\tilde{x}, \tilde{t}), & 0 < \tilde{t} < \Delta\tilde{t} \\ \tilde{T}_{X12}(\tilde{x}, \tilde{t}) - \tilde{T}_{X12}(\tilde{x}, \tilde{t} - \Delta\tilde{t}), & \tilde{t} > \Delta\tilde{t} \end{cases} \tag{4.38}$$

Note that, because the T_{X12} solution has a steady-state ($T = T_0$), the final value of $X_{1,X12}$ is zero. In contrast, the T_{X22} solution has no steady state, and the quasi-steady state solution increases linearly in time, so the $X_{1,X22}$ coefficients reach a constant value.

Figure 4.4 shows the X_1 values for the surface temperature history assumption (X12) along the those from the heat flux history assumption (X22, from Figure 4.1). In Figure 4.4, the values of X_1 for all curves are normalized by the corresponding steady-state value for the piecewise constant heat flux assumption (as in Figure 4.1). This scaling shows that, for smaller values of Δt, the magnitudes of $X_{1,X12}$ from the X12 model are approximately 1.8 times the maximum value $X_{1,X22max}$ from the X22 model. Figure 4.4 also shows that for any Δt, the magnitude of the X12 components always rise faster than those for X22.

Because the $X_{1,X12}$ magnitudes are larger for small Δt values, the surface temperature disturbance model offers some advantage for these cases. However, because of the persistence of the $X_{1,X22}$ coefficients (they attain a constant value of Δt) the surface heat flux disturbance model is generally preferred.

4.5.2.3 X32 Plate Insulated on One Side, Fixed Heat Transfer Coefficient

Of course, the excitation at the active boundary may also be modeled as that due to a constant (fixed) heat transfer coefficient driven by a known environmental temperature: the Type 3 boundary condition. If the heat transfer coefficient is considered unknown, the sensitivity coefficients for h (or, equivalently, the Biot number hL/k) will depend on the value of h itself, resulting in a nonlinear estimation problem. See Problem 4.1. If h is considered time-invariant, there is a single unknown, and the procedure for estimating h is a parameter estimation problem, not a function estimation problem; see Chapter 9. (Note that superposition principles cannot be used to combine X32 solutions to describe a time-varying heat transfer coefficient, $h(t)$.)

Figure 4.4 Sensitivity coefficients for temperature (X12) compared to those for heat flux (X22).

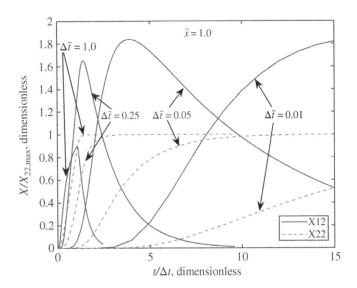

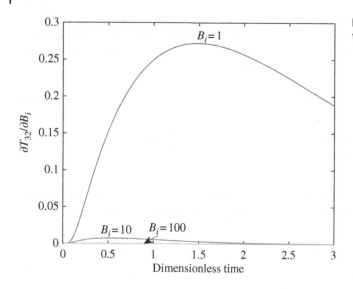

Figure 4.5 Sensitivity coefficients for estimating a constant Bi for Bi = 1, 10, and 100. $x_s = L$.

Figure 4.5 displays the sensitivity coefficients at the insulated surface for constant Bi = 1, 10, 100. The sensitivities are highest for small values of Bi, and become nearly zero for large values of Bi. When Bi is very large, the boundary action is indistinguishable from a sudden change in temperature, and the magnitude of Bi required to produce this result covers a wide range. In other words, the temperature response at the remote boundary becomes insensitive to the magnitude of Bi.

4.5.3 Two-Dimensional Sensitivity Coefficient Example

In this section, sensitivities for the estimation of two-dimensional heat flux histories are investigated. The heat flux is a function of time, as previously discussed, but also is a function of position over the surface. Herein, the surface heat flux is modeled as piecewise uniform spatially and piecewise constant in time. In other words, the surface heat flux distribution at any time is subdivided into a number of simple uniform building blocks over small spatial regions (as depicted in Figure 4.6), and each building block level changes with time. Other functional variations such as linear, parabolic, sinusoidal, and so on, are possible but the basic ideas can be more easily presented for the constant approximation.

The two-dimensional and time-dependent heat flux, $q(x, t)$, is approximated by

$$q(y, t) = \sum_i \sum_j f_{ji}(y, t) \tag{4.39}$$

where i refers to time and j to position. The function $f_{ji}(y, t)$ is given by

$$f_{ji}(y, t) = \begin{cases} q_{ji} & t_{i-1} \leq t < t_1 \text{ and } y_{j+1/2} \\ 0 & \text{otherwise} \end{cases} \tag{4.40}$$

for constant heat flux "building blocks." The sensitivity coefficient for q_{ji} is then

$$X_{ji}(x, y, t) \equiv \frac{\partial T(x, y, t)}{\partial q_{ji}} \tag{4.41}$$

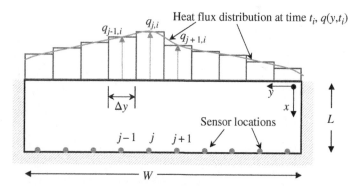

Figure 4.6 Finite two-dimensional slab with piecewise constant heat flux pulses at time t_i.

An example of a two-dimensional body is a finite body heated with a space-variable flux as shown in Figure 4.6. The heat flux on the upper surface (along $x = 0$) is a function of space and time, and all other surfaces are insulated. The heat flux $q(y,t)$ is generally a continuous function in both time and position, but it is approximated by a series of uniform pulses of time-varying intensity, as discussed previously.

Consider the arrangement of pulses on the surface of a finite slab as shown in Figure 4.6, where several equally spaced heat flux components for time t_i are considered. Specifically, consider the number of pulses $n_p = 10$, and let the heated regions be equal to $\Delta y = W/n_p$. For pulses in the center of the plate (where insulated lateral boundaries are remote), the sensitivity coefficients for these locations will all have the same character. A prototype sensitivity coefficient is the one for q_{j1} pulse centered at location y_j which is for heating over the region $(y_j - \Delta x/2) < y < (y_j + \Delta y/2)$ and for a heating duration of $0 < t < \Delta t$. The dimensionless sensitivity coefficient for the locations $y = y_j + s\Delta y, (s = 0,1,...)$ at the remote (insulated) boundary $x = L$ and time $t = n\Delta t (n = 1,2,...)$ is desired. The notation is $\tilde{X}_{j1,sn}$, where the first two subscripts refer to the location of the application of the heat flux component (i.e., j) and to the time of the applied heat flux component ($i = 1$), and the last two subscripts are for the location and time of evaluation. The first two subscripts are for the cause (i.e. the impulse) and the last two are for the effect (i.e. response).

These sensitivity coefficients can be developed from the $T(x, y, t)$ solution for a finite body heated continuously over a portion of the upper surface (Woodbury et al. 2017). This solution is presented in Chapter 2 as X22B(y1pt1)0Y22B00T0 solution. The solution also depends on the aspect ratio of the rectangle, $\tilde{W} = W/L$, and the fraction of the surface that is heated, $\tilde{W}_1 = W_1/W$. If this dependence is represented for convenient shorthand as $\tilde{T}_{X22Y22}(\tilde{x}, \tilde{y}, \tilde{t}, \tilde{W}, \tilde{W}_1)$, then the response in the domain due to heating at the surface of a constant unit pulse of width $\Delta\tilde{y}$, centered at location $\tilde{y}_j$, for a specified aspect ratio $\tilde{W}$ is found using superposition as

$$\tilde{T}_{response}(\tilde{x}, \tilde{y}, \tilde{t}) = \tilde{T}_{X22Y22}\left(\tilde{x}, \tilde{y}, \tilde{t}, \tilde{W}, \tilde{y}_j + \frac{\Delta\tilde{y}}{2}\right) - \tilde{T}_{X22Y22}\left(\tilde{x}, \tilde{y}, \tilde{t}, \tilde{L}, \tilde{y}_j - \frac{\Delta\tilde{y}}{2}\right) \tag{4.42}$$

The sensitivity of a sensor at any location $(\tilde{x}, \tilde{y})$ to a constant pulse of duration $\Delta\tilde{t}$, can then be found by superposition of Eq. (4.42):

$$\tilde{X}_{j1,sn}(\tilde{x}, \tilde{y}_s, n\Delta\tilde{t}) = \tilde{T}_{response}(\tilde{x}, \tilde{y}_s + s\Delta\tilde{y}, n\Delta\tilde{t}) - \tilde{T}_{response}(\tilde{x}, \tilde{y}_s + s\Delta\tilde{y}, (n-1)\Delta\tilde{t}) \tag{4.43}$$

Figure 4.7 gives plots of $\tilde{X}_{j1,in}$ for the sensors at the remote surface ($x = L$) due to a unit impulse at $\tilde{y}_j = 0.45$ ($j = 5$ with $n_p = 10$) for $s = 0, 1, ... 4$. These curves are for $\Delta\tilde{t} = 0.05$, which is a relatively small value. Figure 4.7a is for an aspect ratio $W/L = 4$, which is consistent with the proportions illustrated in Figure 4.6, and Figure 4.7b is for $W/L = 16$, which is a much

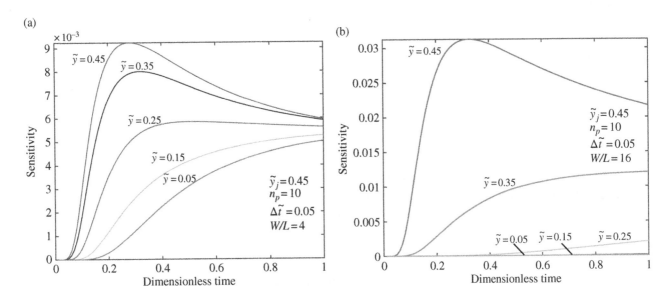

Figure 4.7 Two-dimensional sensitivity coefficients (a) $W/L = 4$ and (b) $W/L = 16$.

thinner plate. Note that all the curves will attain a steady-state value $\Delta t/n_p = 0.005$ due to conservation of energy in the insulated domain. Figure 4.7a shows that the maximum sensitivity coefficients are not quite two times the steady-state value and occur at the sensor directly below the heating location. All other sensor locations have sensitivity coefficients that are damped in magnitude and displaced in time relative to this dominant result. Figure 4.7b for a much thinner plate shows that the sensitivity of the sensor beneath the excitation is significantly increased, as are the effects of the damping and lagging of the adjacent sensors. There is little correlation (or cross-coupling) of the individual heat pulses.

These results have implications for experimental design. To resolve surface heat fluxes with remote measurements, thinner plates (larger aspect ratios) are desired. Larger values of W/L result in larger sensitivity coefficients and reduce the correlation (cross-coupling) of the surface fluxes.

4.6 Stolz Method: Single Future Time Step Method

4.6.1 Introduction

One of the earliest methods developed in western scientific literature to address the IHCP is due to Stolz (1960), and often bears his name. This method is based on exact matching of the solutions computed from a direct solver to experimental data using only a single future time step. A single heat flux component is estimated for that time step before advancing to the next time step. Consequently, a *sequential* algorithm is the result. Early applications of the Stolz method were restricted to data from a single temperature sensor.

The Stolz method is not very powerful or robust, so is not used frequently. However, it is presented here for historical perspective and as segue to the function specification algorithm. The case of a single future time step and a single temperature sensor is presented in Section 4.6.2, and an example follows. Only the linear IHCP is discussed.

4.6.2 Exact Matching of Measured Temperatures

When discrete transient temperatures from a single temperature sensor are used to estimate the surface heat flux as a function of time, the calculated temperatures are made equal to the measured values. (This is called "exact matching" to distinguish it from approximate matching obtained by using least squares.) For the piecewise-constant heat flux approximation (Eq. (3.47)), the temperature at the sensor location and at time t_M can be written as

$$T_M = T_M|_{q_M = 0} + \Delta\phi_0 q_M \tag{4.44}$$

where $\Delta\phi_0$ is the temperature rise at the sensor location at time t_1 for a unit step rise in the surface heat flux at time $t = 0$; $\Delta\phi_0$ is also the sensitivity coefficient of T_M for q_M,

$$\Delta\phi_0 = \frac{\partial T_M}{\partial q_M} \tag{4.45}$$

In exact matching, T_M is made equal to the measured temperature at time t_M, which is denoted Y_M. Then, Eq. (4.44) is solved for the estimated heat flux component, which is denoted $\hat{q}_M$,

$$\hat{q}_M = \frac{Y_M - \hat{T}_M|_{q_M = 0}}{\phi_1}, \text{ since } \Delta\phi_0 = \phi_1 - \phi_0 = \phi_1 \tag{4.46}$$

Here, $\hat{T}_M|_{q_M = 0}$ is the temperature computed from the direct problem using all the heat flux components estimated prior to time t_M and may include a nonzero initial temperature at $t = 0$ (see Eq. (3.47)).

An advantage of Eq. (4.46) is its simplicity which permits easy understanding of the method and the development of relatively simple and computationally efficient algorithms. The major weakness is the absence of any regularization mechanism to combat noisy data and its extreme sensitivity as time steps are made small. Due to the sensitivity of Eq. (4.46) to measurement errors and even instability for small Δt values, it is rarely recommended for the IHCP.

Using superposition of linear solutions, the $\hat{T}_M|_{qM=0}$ term in Eq. (4.46) can be inferred in consideration of the matrix equation Eqs. (3.54)–(3.56). For the Mth row of the matrix, all the heat fluxes up through $M - 1$ have been determined, so

$$\hat{T}_M\Big|_{q_M=0} = \sum_{i=1}^{M-1} \hat{q}_i \Delta\phi_{M-i} + T_0 \tag{4.47}$$

and thus Eq. (4.46) becomes

$$\hat{q}_M = \frac{Y_M - \sum_{i=1}^{M-1} \hat{q}_i \Delta\phi_{M-i} + T_0}{\phi_1} \tag{4.48}$$

This equation is called the Stolz algorithm because Stolz was the first to apply superposition of linear solutions through Duhamel's theorem to the IHCP for the single sensor and single future temperature case. An important characteristic of Eq. (4.48) is its sequential nature; that is, q_M depends on Y_M and the previous q's (q_1, q_2,..., q_{M-1}) and M is sequentially increased by one after each time step. This contrasts with estimating all q components simultaneously, which is the whole domain procedure.

Another important characteristic of Eq. (4.48) is the linearity in the measurements, Y_1, Y_2,.... To demonstrate this, write Eq. (4.48) for q_1, q_2, and q_3 to obtain

$$\hat{q}_1 = \frac{Y_1 - T_0}{\phi_1} \tag{4.49}$$

$$\hat{q}_2 = \frac{Y_2 - \hat{q}_1 \Delta\phi_1 - T_0}{\phi_1} \tag{4.50}$$

$$\hat{q}_3 = \frac{Y_3 - \hat{q}_1 \Delta\phi_2 - \hat{q}_2 \Delta\phi_1 - T_0}{\phi_1} \tag{4.51}$$

Equation (4.49) is a linear function of Y_1. Use Eq. (4.49) in Eq. (4.50) to get

$$\hat{q}_2 = \frac{Y_2 - (Y_1 - T_0)\Delta\phi_1/\phi_1 - T_0}{\phi_1} \tag{4.52}$$

which shows that q_2 is a linear function of Y_1 and Y_2. If Eqs. (4.49) and (4.52) are introduced into Eq. (4.51), q_3 is revealed as a function of Y_1, Y_2, and Y_3, although the expression is somewhat complicated. This procedure can be extended to show that q_M is a linear function of Y_1, Y_2,..., Y_M. This linearity leads to several important concepts that are related to insight into the IHCP, stability considerations, and the development of important filter algorithms; see Chapter 5.

Example 4.1 The temperatures at 10 mm inside a 100 mm steel plate ($k = 40$ W/m-C, $\alpha = 10^{-5}$ m^2/s) initially at 30 °C are 35.706, 62.419, 109.741, and 175.387 °C for $t = 5$, 10, 15, and 20 s, respectively. The plate is insulated on the rear surface. Calculate the surface heat flux estimates, q_1, q_2, q_3, and q_4 using the Stolz method.

Solution
The q_i values are obtained using Eq. (4.48) or Eqs. (4.49)–(4.51). The ϕ_i values are obtained from the solution of a finite body problem subjected to a step change in surface heat flux. The dimensionless time steps are

$$\Delta\tilde{t} = \frac{\alpha\Delta t}{L^2} = \left(10^{-5}\text{m}^2/\text{s}\right)(5\text{s})/(0.1\,\text{m})^2 = 0.005 \tag{4.53}$$

and the times of $\tilde{t}_1 = 0.005$, $\tilde{t}_2 = 0.01$, $\tilde{t}_3 = 0.015$, and $\tilde{t}_4 = 0.020$ are used in Eq. (2.35) to get the associated $\tilde{T}$ values by setting $q_o = 1$ W/m^2 and using

$$\phi_1 = \tilde{T}(\tilde{t}_1)\frac{L}{k} = 0.016663\frac{0.1}{40} = 4.16577\text{E-5}°\text{C-m}^2/\text{W} \tag{4.54}$$

$$\phi_2 = \tilde{T}(\tilde{t}_2)\frac{L}{k} = 0.0399282\frac{0.1}{40} = 9.98206\text{E-5}°\text{C-m}^2/\text{W} \tag{4.55}$$

$$\phi_3 = \tilde{T}(\tilde{t}_3)\frac{L}{k} = 0.060612\frac{0.1}{40} = 1.51529\text{E-4}°\text{C-m}^2/\text{W} \tag{4.56}$$

$$\phi_4 = \tilde{T}(\tilde{t}_4)\frac{L}{k} = 0.079119\frac{0.1}{40} = 1.97797\text{E-4}°\text{C-m}^2/\text{W} \tag{4.57}$$

$$\begin{aligned}
\Delta\phi_1 &= \phi_2 - \phi_1 = 5.8163\text{E-5}\\
\Delta\phi_2 &= \phi_3 - \phi_2 = 5.1708\text{E-5}\\
\Delta\phi_3 &= \phi_4 - \phi_3 = 4.6268\text{E-5}
\end{aligned} \tag{4.58a–c}$$

For $\hat{q}_1$, Eq. (4.49) is used to get

$$\hat{q}_1 = \frac{Y_1 - Y_0}{\phi_1} = \frac{35.706 - 30}{4.1658\text{E-5}} = 136\ 973\ \text{W/m}^2 \tag{4.59}$$

For $\hat{q}_2$, Eq. (4.50) gives

$$\begin{aligned}
\hat{q}_2 &= \frac{Y_2 - \hat{q}_1\Delta\phi_1 - T_0}{\phi_1} = \frac{62.419 - 136\ 973(5.8163\text{E-5}) - 30}{4.1658\text{E-5}}\\
&= 586\ 980\ \text{W/m}^2
\end{aligned} \tag{4.60}$$

For $\hat{q}_3$,

$$\begin{aligned}
\hat{q}_3 &= \frac{Y_3 - \hat{q}_1\Delta\phi_2 - \hat{q}_2\Delta\phi_1 - T_0}{\phi_1}\\
&= \frac{109.741 - 136\ 973(5.1708\text{E-5}) - 586\ 980(5.8163\text{E-5}) - 30}{4.1658\text{E-5}}\\
&= 924\ 629\ \text{W/m}^2
\end{aligned} \tag{4.61}$$

And $\hat{q}_4$ is

$$\hat{q}_4 = 1\ 318\ 335\ \text{W/m}^2 \tag{4.62}$$

Discussion

These $\hat{q}_i$ values can be compared with the exact input values, which were generated considering a ramp heat flux input at the surface (the X22B20T0 case). The exact heat flux input is linearly increasing with time and is given by

$$q(t) = 75\ 000 \times t, \ t \text{ in s}$$

For the piecewise constant heat flux assumption, the calculated heat fluxes are best associated with the values at the times $t - \Delta t/2$ (see Figure 4.1) with $t = 2.5, 7.5, 12.5,$ and 17.5 s, respectively, so that the exact heat flux components are $q_1 = 187\ 500; 562\ 500; 937\ 500,$ and $1\ 312\ 500\ \text{W/m}^2$. The errors in q_i, $i = 1, 2, 3, 4$ are 26.9, −4.4, 1.4, and −0.4%, respectively. The first value is quite approximate, but the errors decrease as time goes on.

If the time step is too small, or the temperature data too approximate, the Stolz algorithm becomes unstable, resulting in totally nonphysical predictions for the heat fluxes. This is due to the absence of any regularization mechanism in the method.

4.7 Function Specification Method

4.7.1 Introduction

One way to treat the IHCP is to assume a functional form of the surface heat flux variation with time. This is called the FSM. The function can be a sequence of constant segments, straightline segments, or it can be one of many other forms such as parabolas, cubics (Samadi et al. 2018), exponentials, etc.

Two possible variations in this method are (i) to estimate simultaneously all the parameters for the total time interval and (ii) to estimate the parameters sequentially. In the first scheme, called whole domain estimation, the complete functional form is found as a unit. In the sequential procedure, one segment after another is estimated, starting with the earliest time and then moving successively to larger times. The following discussion considers various functional forms and focuses on the sequential procedures. The sequential method is more efficient in terms of computational resources than the whole domain procedure for the IHCP.

This sequential procedure can be viewed as an extension or generalization of the Stolz method. The method adds regularization to the procedure by including data from future time steps. The algorithm is referred to as the SFSM and sometimes as Beck's Method.

4.7.2 Sequential Function Specification Method

Several observations based on the characteristics of heat flux sensitivity coefficients were made earlier in this chapter. One is that the nature of transient heat conduction is "diffusive"; that is, the effect of a surface heat input at any time on an interior location is both lagged and damped. Another aspect of the diffusive character of transient heat conduction is that for two time-adjacent heat flux pulses of similar magnitude the long-time effects on the temperature are virtually indistinguishable. Because of these characteristics, the most computationally efficient estimation procedure for the IHCP does not simultaneously estimate all the heat flux components (q_M, $M = 1,2...,N$) but instead uses a sequential procedure such as given in this subsection.

The purpose of this subsection is to present sequential methods for the function specification procedure. The basic principles are developed in a form that can utilize direct problem solutions based on superposition of linear solutions and numerical solutions for nonlinear problems.

The basic concepts in the function specification sequential procedure are:

1) A functional form for $q(t)$ is assumed for times t_M, t_{M+}, t_{M+r-1} (the heat flux is *known* for $t < t_{M-1}$). Typical assumptions are piecewise constant and piecewise linear (Figure 4.1).
2) A sum of squares function is used for these times and involves squares of differences between the measured and the corresponding calculated temperatures. Measured data from several future time steps are included in the sum:

$$S = \sum_{i=1}^{r} \left(Y_{M+i-1} - T_{M+i-1}(q_M,\ q_{M+1},\ ...,\ q_{M+r-1}) \right)^2 \tag{4.63a}$$

or, in matrix form

$$S = (\mathbf{Y} - \mathbf{T})^T (\mathbf{Y} - \mathbf{T}) \tag{4.63b}$$

3) *Temporarily* hold the assumed form of the heat flux unchanged over the r future time steps.
4) The heat flux components are estimated for the assumed functional form.
5) Only the first heat flux component, q_M, is retained.
6) M is increased by one and the procedure is repeated.

This is the general procedure for the *linear* IHCP. Some iteration may be necessary for the nonlinear case. The FSMs for both the linear and nonlinear cases were originally proposed by Beck (1961, 1962, 1968, 1970).

4.7.2.1 Piecewise Constant Functional Form

The simplest sequential procedure results from the piecewise constant functional form (dashed lines in Figure 4.1). Consistent with step 3 in the outline above, the value of q_M is then temporarily assumed constant with time over the next r time steps. Hence, r "future" heat flux components are *temporarily* made equal

$$q_{M+1} = q_{M+2} = \cdots = q_{M+r-1} = q_M \tag{4.64}$$

All previously determined heat flux components $\hat{q}_1, \hat{q}_2,...,\hat{q}_{M-1}$ are assumed known and the objective is to estimate q_M. If $r = 1$, no "additional" information is introduced for a single interior temperature sensor, and the method is identical to the Stolz method.

4.7.2.1.1 *X22 – Unknown Surface Heat Flux*

Consider the matrix form of the direct solution of the linear heat conduction equation for piecewise constant heat flux over the time interval $t_M \leq t \leq t_{M+r-1}$:

$$
\begin{Bmatrix} T_M \\ T_{M+1} \\ \vdots \\ T_{M+r-1} \end{Bmatrix} = \begin{bmatrix} \Delta\phi_1 & 0 & 0 & 0 \\ \Delta\phi_2 & \Delta\phi_1 & 0 & 0 \\ \vdots & & \ddots & \vdots \\ \Delta\phi_{M+r-1} & \cdots & \Delta\phi_2 & \Delta\phi_1 \end{bmatrix} \begin{Bmatrix} q_M \\ q_{M+1} \\ \vdots \\ q_{M+r-1} \end{Bmatrix} + \begin{Bmatrix} \hat{T}_M \\ \hat{T}_{M+1} \\ \vdots \\ \hat{T}_{M+r-1} \end{Bmatrix}
\tag{4.65a}
$$

which can be represented in matrix form as:

$$
\mathbf{T} = \mathbf{Xq} + \hat{\mathbf{T}}\big|_{\mathbf{q}=0}
\tag{4.65b}
$$

With Eq. (4.65b), the sum of squares function (Eq. 4.63b) can be written as

$$
S = \left(\mathbf{Y} - \mathbf{Xq} - \hat{\mathbf{T}}\big|_{\mathbf{q}=0}\right)^T \left(\mathbf{Y} - \mathbf{Xq} - \hat{\mathbf{T}}\big|_{\mathbf{q}=0}\right)
\tag{4.66}
$$

Minimization of Eq. (4.64) with respect to the unknown vector $\mathbf{q}$ leads to

$$
\hat{\mathbf{q}} = \left(\mathbf{X}^T\mathbf{X}\right)^{-1}\mathbf{X}^T\left(\mathbf{Y} - \hat{\mathbf{T}}\big|_{\mathbf{q}=0}\right)
\tag{4.67}
$$

By making the *temporary* (but incorrect!) assumption that $q_M = q_{M+1} = \ldots = q_{M+r-1}$, Eq. (4.65a) degenerates into a function of the single unknown q_M.

$$
\begin{Bmatrix} T_M \\ T_{M+1} \\ \vdots \\ T_{M+r-1} \end{Bmatrix} = \cdot \begin{bmatrix} \phi_1 & 0 & 0 & 0 \\ 0 & \phi_2 & 0 & 0 \\ \vdots & & \ddots & \vdots \\ 0 & \cdots & 0 & \phi_r \end{bmatrix} q_M + \begin{Bmatrix} \hat{T}_M \\ \hat{T}_{M+1} \\ \vdots \\ \hat{T}_{M+r-1} \end{Bmatrix}
\tag{4.68}
$$

The coefficient matrix, $\mathbf{X}$, is simplified considerably. Note that $\sum_{i=0}^{j-1} \Delta\phi_i = \phi_j$ has been used in simplification of Eq. (4.65a). The structure of Eq. (4.68) allows for a simple equation for computing the only unknown heat flux q_M. Since,

$$
\mathbf{X}^T\mathbf{X} = \sum_{i=1}^{r} \phi_i^2
\tag{4.69a}
$$

the inverse needed to solve Eq. (4.68) is

$$
\left(\mathbf{X}^T\mathbf{X}\right)^{-1} = \frac{1}{\sum\limits_{i=1}^{r} \phi_i^2}
\tag{4.69b}
$$

and q_M can be computed directly as

$$
\hat{q}_M = \frac{\sum\limits_{i=1}^{r} \phi_i \left(Y_{M+i-1} - \hat{T}_{M+i-1}\big|_{q=0}\right)}{\sum\limits_{i=1}^{r} \phi_i^2}
\tag{4.70a}
$$

This equation provides an algorithm that is used in a sequential manner by increasing M by one for each time step. The value of r should be small (about 3 or 4) and should be selected in accord with some optimality criterion (see Chapter 6).

In Eq. (4.70a), $\hat{T}_{M+i-1}\big|_{q=0}$ represents the effect of the initial uniform temperature distribution and all the previously estimated heat flux components $\hat{q}_1, \hat{q}_2, \ldots, \hat{q}_{M-1}$:

$$
\hat{T}_{M+i-1}\big|_{q=0} = \sum_{j=1}^{M-1} \hat{q}_j \Delta\phi_{M+i-1-j}, \quad i = 1,2,\ldots,r
\tag{4.70b}
$$

and $\Delta\phi_i = \phi_{i+1} - \phi_i$.

Regularization in the SFSM is afforded by the number of future times r considered. If $r = 1$, the method is identical to the Stoltz method, since exact matching of the data, point by point, will result. Of course, the regularization introduces a bias, as the (incorrect) assumption that $q_M = q_{M+1} = \ldots = q_{M+r-1}$ is made, and so the heat flux function $q(t)$ is biased toward a constant value.

Equation (4.70a) can be written as

$$\hat{q}_M = \sum_{i=1}^{r} K_i \left(Y_{M+i-1} - \hat{T}_{M+i-1} \big|_{q_M = \cdots = 0} \right) \tag{4.71a}$$

where K_i is called a gain coefficient and is defined by

$$K_i \equiv \frac{\phi_i}{\sum\limits_{j=1}^{r} \phi_j^2} \tag{4.71b}$$

Note that K_i has units of reciprocal ϕ. For $r = 2$, $\hat{q}_M$ is given by

$$\hat{q}_M = K_1 \left(Y_M - \hat{T}_M \big|_{q_M = 0} \right) + K_2 \left(Y_{M+1} - \hat{T}_{M+1} \big|_{q_M = q_{M+1} = 0} \right) \tag{4.71c}$$

$$K_1 \equiv \frac{\phi_1}{\phi_1^2 + \phi_2^2}; \quad K_2 \equiv \frac{\phi_2}{\phi_1^2 + \phi_2^2} \tag{4.71d}$$

Table 4.1 displays the dimensionless gain coefficients, $\tilde{K}_i$ for a sensor at $x = L$ in a finite insulated plate. The ϕ_i's are numerically equal to the dimensionless temperatures calculated from Eq. (2.35) for dimensionless time steps of 0.05, 0.2, and 0.5. Notice that the $\tilde{K}_i$ values[2] decrease with increasing $\Delta \tilde{t}$ values, indicating decreased sensitivity to measurement errors with increasing $\Delta \tilde{t}$. A large reduction in sensitivity to measurement errors is also obtained for a given $\Delta \tilde{t}$ by increasing r, the number of future time steps.

Table 4.1 Dimensionless gain coefficients for Eqs. (4.71a)–(4.71d) for sensor at $x = L$ in a finite insulated plate.

Gain coefficient	$r = 1$	$r = 2$	$r = 3$	$r = 4$	$r = 5$
		$\Delta \tilde{t} = 0.05$			
$\tilde{K}_1$	3720	4.33	0.292	0.057	0.018
$\tilde{K}_2$		127	8.56	1.68	0.533
$\tilde{K}_3$			31.8	6.24	1.98
$\tilde{K}_4$				13.08	4.15
$\tilde{K}_5$					6.79
		$\Delta \tilde{t} = 0.2$			
$\tilde{K}_1$	16.3	1.02	0.248	0.095	0.046
$\tilde{K}_2$		3.95	0.955	0.365	0.176
$\tilde{K}_3$			1.75	0.668	0.323
$\tilde{K}_4$				0.975	0.471
$\tilde{K}_5$					0.620
		$\Delta \tilde{t} = 0.5$			
$\tilde{K}_1$	2.9869	0.41509	0.129	0.056	0.029
$\tilde{K}_2$		1.0332	0.322	0.140	0.073
$\tilde{K}_3$			0.516	0.224	0.117
$\tilde{K}_4$				0.308	0.161
$\tilde{K}_5$					0.205

Function specification method, q_M assumed constant.

---

2 The dimensionless values of K_i can be denoted $\tilde{K}_i$ and are related to K_i (with dimensions) by $\tilde{K}_i = (q_c L/k) K_i$ where $q_c = 1 \text{ W/m}^2$.

4.7.2.1.2 Multiple Sensors

In some experiments more than one temperature sensor is used because the additional information can aid in more accurately estimating the surface conditions. In this case, the sum of squares function,

$$S = \sum_{i=1}^{r} \sum_{j=1}^{J} \left(Y_{j,M+i-1} - T_{j,M+i-1} \right)^2 \tag{4.72}$$

is minimized with respect to q_M. The first subscript refers to space (sensor number) and the second to time. There are r future times and J temperature sensors. For the temporary assumption of $q_M = $ const, $T_{j,M+i-1}$ can be written as

$$T_{j,M+i-1} = \hat{T}_{j,M+i-1}|_{q_M = \cdots = 0} + \phi_{ji} q_M \tag{4.73}$$

Introducing Eq. (4.73) into Eq. (4.72) and performing the usual operations of differentiating with respect to q_M, setting equal to zero and replacing $\hat{q}_M$ by gives

$$\hat{q}_M = \frac{\sum_{i=1}^{r} \sum_{j=1}^{J} \left(Y_{j,M+i-1} - \hat{T}_{j,M+i-1}|_{q_M = \cdots = 0} \right) \phi_{ji}}{\sum_{i=1}^{r} \sum_{j=1}^{J} \phi_{ji}^2} \tag{4.74}$$

Using the gain coefficient concept for this equation yields

$$\hat{q}_M = \sum_{i=1}^{r} \sum_{j=1}^{J} K_{ji} \left(Y_{j,M+i-1} - \hat{T}_{j,M+i-1}|_{q_M = \cdots = 0} \right) \tag{4.75a}$$

$$K_{ji} \equiv \frac{\phi_{ji}}{\sum_{i=1}^{r} \sum_{j=1}^{J} \phi_{ji}^2} \tag{4.75b}$$

The foregoing expressions reduce to the single-sensor and r-future times equations, Eqs. (4.71a) and (4.71b), for $J = 1$. As in the equations just mentioned, Eqs. (4.75a) and (4.75b) are also linear functions of the measured temperatures.

Some gain coefficients are displayed in Table 4.2 for the case of two future times ($r = 2$) and two sensors ($J = 2$). Dimensionless time steps of 0.05 and 0.5 are used. The geometry is a flat plate heated at $x = 0$ and insulated at $x = L$, for which the ϕ values are given by Eq. (2.35). Three sets of sensor locations are considered: $x = 0$ and L, $x = 0.5L$ and L, and $x = L$ and L. The last set of x values is similar to the case of single x (which is $x = L$) used in Table 4.1; the $\tilde{K}_{ji}$ values in Table 4.2 for $x = L$ and L are one-half of the corresponding values in Table 4.1.

The most striking difference between Tables 4.1 and 4.2 is the great reduction in the gain coefficients for small time steps when a sensor is located at or near the heated surface. This means that the heat flux component, q_M, is much less sensitive to measurement errors when sensors are used near the heated surface. Also note for $\Delta \tilde{t} = 0.05$ that the gain coefficients are quite small for $x = L$ (for the sensors at $x = 0$ and L); as a consequence, the sensor at $x = L$ makes little contribution to the

Table 4.2 Dimensionless gain coefficients for Eqs. (4.75a) and (4.75b) for two sensors in a finite plate insulated at $x = L$.

	$x = 0$	$x = L$	$x = 0.5L$	$x = L$	$x = L$	$x = L$
i	$(J = 1)$	$(J = 2)$	$(J = 1)$	$(J = 2)$	$(J = 1)$	$(J = 2)$
			$\tilde{K}_{ji}$ values for $\Delta \tilde{t} = 0.05$			
1	1.32	0.001	4.03	0.071	2.16	2.16
2	1.87	0.041	15.5	2.07	63.3	63.3
			$\tilde{K}_{ji}$ values for $\Delta \tilde{t} = 0.5$			
1	0.254	0.102	0.237	0.173	0.208	0.208
2	0.407	0.254	0.495	0.431	0.517	0.517

Function specification method, q_M assumed constant. Two future times. $r = 2$. ($\Delta \tilde{t} = \alpha \Delta t / L^2$).

estimation of the surface heat flux for this small $\Delta \tilde{t}$. For the relatively large time step of $\Delta \tilde{t} = 0.5$, the locations of the temperature sensors are much less important. The gain coefficients in Table 4.2 illustrate that, whenever practical, sensors should be located as near the heated surface as possible for small dimensionless times. (As pointed out in Section 4.2.2, these "small dimensionless times" should be based on the distance d from the heated surface to the temperature sensor nearest the heated surface. Note, however, that $\alpha \Delta t / d^2$ is always large if d approaches zero.)

Example 4.2 For the same plate and data as in Example 4.1, calculate the surface heat flux estimates, q_1 and q_2, using the FSM with the $q = $ constant temporary assumption and two future temperatures.

Solution

Equation (4.71a,b) is used with $r = 2$. The ϕ_i values are the same as for Example 4.1 and

$$K_1 = \frac{\phi_1}{\phi_1^2 + \phi_2^2} = 3560.6 \, \text{W/m}^2\text{-°C} \tag{4.76}$$

$$K_2 = \frac{\phi_2}{\phi_1^2 + \phi_2^2} = 8532.0 \, \text{W/m}^2\text{-°C} \tag{4.77}$$

Using Eq. (4.71a) with $M = 1$ gives

$$\hat{q}_1 = (Y_1 - T_0)K_1 + (Y_2 - T_0)K_2 = 296 \ 917 \, \text{W/m}^2 \tag{4.78}$$

and repeating with $M = 2$ gives

$$\hat{q}_2 = (Y_2 - \hat{q}_1 \Delta \phi_1 - T_0)K_1 + (Y_3 - \hat{q}_1 \Delta \phi_2 - T_0)K_2 = 603\,302 \, \text{W/m}^2 \tag{4.79}$$

Discussion

These q values are larger than the exact values of 187 500 and 562 500 W/m², with errors of 58 and 7%, respectively. The errors are somewhat larger than those given in Example 4.1 for the Stolz method but the present algorithm is much less sensitive to measurement errors and can permit considerably smaller time steps.

Example 4.3 A finite plate with $\alpha = 1 \, \text{m}^2/\text{s}$, $k = 1 \, \text{W/m-°C}$, and $L = 1 \, \text{m}$ has measured temperatures at the insulated surface ($x = L$) of 16, 45, 99, and 179 °C for times 0.5, 1, 1.5, and 2 s, respectively. The initial temperature is 10 °C. Calculate q_1, q_2, and q_3 using Eqs. (4.71a, b) with $r = 1$ and 2.

Solution

The dimensionless time step is 0.5 and the ϕ_i values are given by Eq. (2.35), but in this case the K_i values are given in Table 4.1 for $\Delta \tilde{t} = 0.5$. The results are:

t_i (s)	$\hat{q}_i, r = 1$	$\hat{q}_i, r = 2$	q_i, exact
0.5	17.92	38.65	25
1	77.85	78.52	75
1.5	123.1	126.8	125

Discussion

The units are W/m². After the first time step both the $r = 1$ and $r = 2$ results are acceptable but only the $r = 2$ method can produce acceptable results if Δt is decreased to 0.25 s in this example. The errors in the calculated values are caused by inaccuracies of the algorithms and by approximate precision in Y_i values.

4.7.2.1.3 *X12 – Unknown Surface Temperature*

As described in Section 4.4, the resulting temperatures in the heat conducting body can be computed from the direct solver once the unknown surface heat fluxes are determined. This includes the surface temperature. However, it is possible to devise the inverse heat conduction algorithm so that the unknown surface temperatures are estimated directly. Furthermore, Section 4.5.2 indicates that the sensitivity coefficients for the temperature-based IHCP approach are greater than those for the heat flux based approach for early times, suggesting this approach may have some benefit. This section explores the IHCP based on a piecewise constant surface temperature disturbance, the X12 case.

The solution to the X12B10T0 case is given in Chapter 2, and using superposition of piecewise constant surface temperature disturbances, the temperature at any location x_j can be computed from known temperatures at discrete times at the surface, $T_0(i\Delta t)$ using the matrix relation:

$$\{T(x_j, i\Delta t)\} = [\Delta\varphi_{i-k}(x_j, i\Delta t)]\{T_0(i\Delta t)\} \qquad k = 1,2,...,i \tag{4.80}$$

Here, $\Delta\varphi$ is the pulse sensitivity response due to a unit surface temperature disturbance and plays the analogous role of $\Delta\phi$ in the heat flux-based (X22) solution.

$$\Delta\varphi_n = \varphi(x_j, (n+1)\Delta t) - \varphi(x_j, n\Delta t) \tag{4.81}$$

$\varphi(x,t)$ is the X12B10T0 solution for a constant unit temperature impulse at $x = 0$ and an insulated boundary at $x = L$.

Consequently, the function specification solution for the surface temperatures can be found by applying the same techniques as for the X22 IHCP after substituting $\Delta\phi \to \Delta\varphi$.

$$\hat{T}_{0,M} = \frac{\sum\limits_{i=1}^{r} \varphi_i\left(Y_{M+i-1} - \hat{T}_{M+i-1}\big|_{T_0=0}\right)}{\sum\limits_{i=1}^{r} \varphi_i^2} \tag{4.82}$$

and

$$\hat{T}_{M+i-1}\big|_{q=0} = \sum\limits_{j=1}^{M-1} \hat{T}_{0,j}\Delta\varphi_{M+i-1-j}, \qquad i = 1,2,...,r \tag{4.83}$$

Of course, the gain coefficient concept can also be used to perform the calculations in Eqs. (4.82) and (4.83).

Example 4.4   For the same plate and data as in Example 4.1, calculate the surface temperature estimates, $T_{0,1}$, $T_{0,2}$, and $T_{0,3}$, using the temperature-based (X12) FSM with the $T = $ constant temporary assumption for $r = 1$ and 2.

Solution

The φ values are obtained from the X12B10T0 solution using $\tilde{x} = 0.10$ and $\Delta\tilde{t} = 0.005$:

$$\varphi_1 = 0.317311 \quad \varphi_2 = 0.479500 \quad \varphi_3 = 0.563703 \quad \varphi_4 = 0.617075 \quad \varphi_5 = 0.654721$$

And the $\Delta\varphi$ values are found as the differences

$$\Delta\varphi_1 = 0.479500 - 0.317311 = 0.162190$$

$$\Delta\varphi_2 = 0.084203$$

$$\Delta\varphi_3 = 0.053372$$

$$\Delta\varphi_4 = 0.037646$$

For $r = 1$ (exact matching), the gain coefficient is

$$K_1 = \frac{\varphi_1}{\varphi_1^2} = \frac{1}{\varphi_1} = 3.1515°C/°C \tag{4.84}$$

For $M = 1$, the estimated surface temperature is

$$\hat{T}_{0,1} = (Y_1 - T_0)K_1 = (35.708 - 30)3.1515 = 48.0°C \tag{4.85}$$

and repeating with $M = 2$ gives

$$\hat{T}_{0,2} = (Y_2 - \hat{T}_{0,1}\Delta\varphi_1 - T_0)K_1 = (62.419 - 48.0 \times 0.162190 - 30)3.1515 = 107.6°C \tag{4.86}$$

And with $M = 3$,

$$\hat{T}_{0,3} = 213.5°C \tag{4.87}$$

When $r = 2$, the gain coefficients are

$$K_1 = \frac{\varphi_1}{\varphi_1^2 + \varphi_2^2} = 0.95978°C/°C \tag{4.88}$$

$$K_2 = \frac{\varphi_2}{\varphi_1^2 + \varphi_2^2} = 1.4504°\text{C}/°\text{C} \tag{4.89}$$

The surface temperature with $M = 1$ is

$$\hat{T}_{0,1} = (Y_1 - T_0)K_1 + (Y_2 - T_0)K_2 = 82.5°\text{C} \tag{4.90}$$

and repeating with $M = 2$ gives

$$\hat{T}_{0,2} = (Y_2 - \hat{T}_{0,1}\Delta\varphi_1 - T_0)K_1 + (Y_3 - \hat{T}_1\Delta\varphi_2 - T_0)K_2 = 153.9°\text{C} \tag{4.91}$$

And for $M = 3$,

$$\hat{T}_{0,3} = 261.6°\text{C} \tag{4.92}$$

Discussion

These exact values for the surface temperature are 47.6, 121.6, and 227.1 °C, so the relative errors for $r = 1$ are −0.7, 11.5, and 6.0%. For $r = 2$, these errors are −73.2, −26.5, and −15.2%.

4.7.2.2 Piecewise Linear Functional Form

4.7.2.2.1 *X22 – Unknown Surface Heat Flux*

Another algorithm for the SFSM uses the temporary assumption of a linear heat flux. Figure 4.1 illustrates the linearly connected segments. Note that the values at the time intervals $M\Delta t$ are connected linearly, which contrasts with the values at the times $M\Delta t/2$ that represent the piecewise constant approximation.

For constant thermal properties, the solution to an arbitrary sequence of heat flux inputs over uniform time increments Δt can be found through superposition of the X22B20T0 solution as described in Section 3.4.2.

When all the heat flux components up to, and including, q_{M-1} have been found, and the initial heat flux at time zero $q_{0,0} = 0$, the temperature at a time $M + r - 1$ can be written from Eq. (3.62) as

$$T_{M+r-1} = \sum_{i=1}^{M+r-2} q_i[\Delta(\Delta\gamma_{M+r-1-i})] + q_{M+r-1}\Delta\gamma_0 + \hat{T}\Big|_{q_M = q_{M+1} = \cdots = q_{M+r-1} = 0} \tag{4.93}$$

To determine the function specification algorithm, the heat flux is assumed to follow the assumed form for r future time steps, which in this case is linearly increasing in time. If q is constrained to be a linear function:

$$q_{M+r-1} = q_{M-1} + r(q_M - q_{M-1}) = rq_M - (r-1)q_{M-1} \tag{4.94}$$

and the $M + r - 1$ temperatures can be expressed as

$$T_M = \hat{T}_M\Big|_{q_M = 0} + q_M\Delta\gamma_0 = \hat{T}_M\Big|_{q_M = 0} + q_M\gamma_1 \tag{4.95a}$$

$$\begin{aligned}
T_{M+1} &= \hat{T}_{M+1}\Big|_{q_M = 0} + q_{M+1}\Delta\gamma_0 + q_M[\Delta(\Delta\gamma)_1] \\
&= \hat{T}_{M+1}\Big|_{q_M = 0} + (2q_M - q_{M-1})\Delta\gamma_0 + q_M[\gamma_2 - 2\gamma_1 + \gamma_0] \\
&= \hat{T}_{M+1}\Big|_{q_M = 0} + q_M(2\gamma_1 + \gamma_2 - 2\gamma_1 + \gamma_0) - q_{M-1}\gamma_1
\end{aligned} \tag{4.95b}$$

$$\begin{aligned}
&= \hat{T}_{M+1}\Big|_{q_M = 0} + q_M\gamma_2 - q_{M-1}\gamma_1 \\
T_{M+2} &= \hat{T}_{M+2}\Big|_{q_M = 0} + q_{M+2}\Delta\gamma_0 + q_{M+1}[\Delta(\Delta\gamma)_1] + q_M[\Delta(\Delta\gamma)_2] \\
&= \hat{T}_{M+2}\Big|_{q_M = 0} + (3q_M - 2q_{M-1})(\gamma_1 - \gamma_0) + (2q_M + q_{M-1})[\gamma_2 - 2\gamma_1 + \gamma_0] + q_M[\gamma_3 - 2\gamma_2 + \gamma_1] \\
&= \hat{T}_{M+2}\Big|_{q_M = 0} + q_M(3\gamma_1 + 2[\gamma_2 - 2\gamma_1 + \gamma_0] + [\gamma_3 - 2\gamma_2 + \gamma_1]) - q_{M-1}(2\gamma_1 + [\gamma_2 - 2\gamma_1 + \gamma_0]) \\
&= \hat{T}_{M+2}\Big|_{q_M = 0} + q_M(\gamma_3) - q_{M-1}(\gamma_2)
\end{aligned} \tag{4.95c}$$

Generalize these results to find

$$T_{M+r-1} = \hat{T}_{M+r-1}\Big|_{q_M = 0} + q_M(\gamma_r) - q_{M-1}\gamma_{r-1} \tag{4.95d}$$

The sum of the squares of the errors between the computed and measured temperatures is

$$S = \sum_{i=1}^{r} \left(Y_{M+i-1} - \hat{T}\Big|_{\hat{q}_M = \hat{q}_{M+1} = \cdots = \hat{q}_{M+r-1} = 0} - q_M\gamma_i\right)^2 \tag{4.96}$$

Minimizing S with respect to the heat flux component q_M yields

$$\frac{\partial S}{\partial q_M} = 0 = 2\sum_{i=1}^{r}\left[\left(Y_{M+i-1} - \hat{T}\Big|_{\hat{q}_M = \hat{q}_{M+1} = \cdots = \hat{q}_{M+r-1} = 0} - q_M\gamma_i\right)(-\gamma_i)\right] \tag{4.97a}$$

$$\sum_{i=1}^{r}\left[\left(Y_{M+i-1} - \hat{T}\Big|_{\hat{q}_M = \hat{q}_{M+1} = \cdots = \hat{q}_{M+r-1} = 0}\right)(\gamma_i)\right] + q_M\sum_{i=1}^{r}\left[(\gamma_i)^2\right] = 0 \tag{4.97b}$$

$$\hat{q}_M = \frac{\sum_{i=1}^{r}\left[\left(Y_{M+i-1} - \hat{T}\Big|_{\hat{q}_M = \hat{q}_{M+1} = \cdots = \hat{q}_{M+r-1} = 0}\right)\gamma_i\right]}{\sum_{i=1}^{r}\left[(\gamma_i)^2\right]} \tag{4.97c}$$

In terms of the gain coefficients:

$$K_i = \frac{\gamma_i}{\sum_{i=1}^{r}(\gamma_i)^2} \tag{4.98a}$$

$$\hat{q}_M = \sum_{i=1}^{r}\left[\left(Y_{M+i-1} - \hat{T}\Big|_{\hat{q}_M = \hat{q}_{M+1} = \cdots = \hat{q}_{M+r-1} = 0}\right)K_i\right] \tag{4.98b}$$

Here,

$$\hat{T}_{M+i-1}\Big|_{q_M = 0} = \sum_{j=1}^{M-1}\hat{q}_j\left[\Delta\left(\Delta\gamma_{M+i-1-j}\right)\right] - \hat{q}_{M-1}\gamma_{i-1} + T_0, \quad i = 1,2,\ldots,r \tag{4.98c}$$

Gain coefficients for Eq. (4.98a) for a plate, insulated at $x = L$ and heated at $x = 0$ are given in Table 4.3. The sensor is at $x = L$. The values can be compared with those in Table 4.1. For $r = 1$, the K_1 value is much larger than in Table 4.1 since the

Table 4.3 Dimensionless gain coefficients for Eq. (4.71a,b) for sensor at $x = L$ in a finite insulated plate.

Gain coefficient	$r = 1$	$r = 2$	$r = 3$	$r = 4$	$r = 5$
			$\Delta\tilde{t} = 0.05$		
$\tilde{K}_1$	29615	3.747	0.0785	0.0072	0.0013
$\tilde{K}_2$		333.1	6.980	0.6414	0.1158
$\tilde{K}_3$			47.71	4.385	0.7918
$\tilde{K}_4$				13.930	2.515
$\tilde{K}_5$					5.623
			$\Delta\tilde{t} = 0.2$		
$\tilde{K}_1$	61.36	0.6155	0.0596	0.0122	0.0037
$\tilde{K}_2$		6.114	0.5921	0.1212	0.0364
$\tilde{K}_3$			1.817	0.3721	0.1118
$\tilde{K}_4$				0.7717	0.2318
$\tilde{K}_5$					0.3967
			$\Delta\tilde{t} = 0.5$		
$\tilde{K}_1$	8.202	0.2378	0.0328	0.0081	0.0027
$\tilde{K}_2$		1.396	0.1900	0.0468	0.0157
$\tilde{K}_3$			0.4818	0.1186	0.0399
$\tilde{K}_4$				0.2236	0.0752
$\tilde{K}_5$					0.01216

Function specification method, q = linear assumption.

first sensitivity coefficient γ_1 is much smaller than ϕ_1. As r increases, the gain coefficients of the linear q values decrease more rapidly than for the $q = C$ coefficients.

Example 4.5   Repeat Example 4.3 with $r = 1,2,3$ using the linear-in-time heat flux gain coefficients.

Solution

The dimensionless time step is 0.5 and the K_i values are given in Table 4.3 for $\Delta \tilde{t} = 0.5$. The results are:

t_i (s)	$\hat{q}_{i,r=1}$	$\hat{q}_{i,r=2}$	q_i, exact
0.5	49.2	49.6	50
1	100.7	99.6	100
1.5	146.9	151.0	150
2.0	215.3	–	200

Discussion

The units are W/m^2. The exact heat flux is, again, linearly increasing with time, but the values for comparison are now evaluated at the times t_i, rather than $t_i - \Delta t/2$ as was required for the piecewise constant ($q = $ const) heat flux model. The predicted values are considerably closer to the exact values than in the $q = $ const model, but this can be attributed in part to the fact that the chosen representation (piecewise linear) is exactly the same as the true heat flux.

4.7.2.2.2   X12 – Unknown Surface Temperature

The same idea of superposition of solutions due to a linearly-increasing surface temperature can be used to model the surface disturbance as a piecewise linear function described by values coincident with the time steps. Superposition can be used to find the gain coefficients. See Problem 4.3.

4.7.3   General Remarks About Function Specification Method

The FSM is a generalization of the exact matching, or Stolz, method. The method requires an assumption about the functional form of the unknown surface disturbance in a forward solver model, and further assumes that this functional form remains unchanged for a few future time steps. A sum of squares errors between measured and calculated temperature values is minimized to determine the unknown heat flux components. In this way, additional data (measurements) are brought into the estimation of the heat flux component at the present time without adding additional unknown heat fluxes to the estimation problem.

Regularization is afforded in the FSM through these additional future times data. The number of future time steps considered, r, is the regularizing parameter. As will be discussed in Chapter 6, regularization combats noise in the data by introducing bias in the estimate.

The gain coefficients, K, in the FSM are indicative of the sensitivity of a particular FS formulation. On the one hand, large K coefficients are desirable because they indicate strong sensitivity of the measurement to the unknown surface disturbance. At the same time, large K coefficients will likewise be more sensitive to errors in the data.

4.8   Tikhonov Regularization Method

4.8.1   Introduction

Tikhonov Regularization (TR) is a whole domain procedure which modifies the least-squares approach by adding stabilizing terms to the objective function. These terms reduce excursions in the unknown surface excitation function (surface heat flux or surface temperature).

TR is named after Russian mathematician Tikhonov, who pioneered the method (Tikhonov 1943, 1963). He later coauthored the classic text *Solution of Ill-posed Problems* (Tikhonov and Arsenin 1977). The technique was adopted and widely employed in application to thermal problems by Alifanov (1975) and Alifanov and Artyukhin (1975). Early papers by western authors using TR includes Bell and Wardlaw (1981a, 1981b) and Beck and Murio (1986).

The TR method is related to some approximate least-squares procedures. One of these is called *ridge regression* (Hoerl and Kennard 1970a, b, 1976; Marquardt 1970) or *damped least squares* (Matz 1964). This method is also related to procedures that have been used for solving nonlinear least-squares problems; two of these authors are Levenberg, whose work dates back to 1944, and Marquardt (1963). Further discussion of ridge regression and nonlinear least-squares procedures is in Beck and Arnold (1977). An excellent book on least squares is authored by Lawson and Hanson (1974) and includes a discussion of ridge regression.

The plan of this section is to first discuss the physical significance of the regularization terms in Section 4.8.2, then the whole domain regularization is developed in Section 4.8.3, and finally Section 4.8.4 briefly covers the sequential regularization method. Section 4.8.5 summarizes the section with general remarks about TR.

4.8.2 Physical Significance of Regularization Terms

TR can be explained considering the continuous form of the modified sum of squared errors between the measured ($Y(x_j,t)$) and model-predicted ($T(x_j,t)$) temperatures. The basic idea is to add a penalty term to the objective function that imposes a constraint on the unknown function, $q(t)$.

4.8.2.1 Continuous Formulation

For a specific measurement location, x_j, in a problem to determine the unknown surface heat flux, $q(t)$, the continuous function is written as

$$S = \int_0^{t_f} \left(Y(x_j, \ \tau) - T(x_j, \ \tau : q(\tau))\right)^2 d\tau + \int_0^{t_f} \left[\alpha_0 (q(\tau))^2 + \alpha_1 \left(\frac{\partial q(\tau)}{\partial \tau}\right)^2 + \alpha_2 \left(\frac{\partial^2 q(\tau)}{\partial \tau^2}\right)^2\right] d\tau \quad (4.99)$$

This continuous objective function is properly termed a *functional*, as it depends on the unknown function $q(t)$. In Eq. (4.99), the last group of terms are the TR terms, and the 0, 1, 2, subscripts on the α coefficients denote zeroth-, first-, and second-order TR terms, respectively. These coefficients are called the *regularization parameters*. Higher order derivative terms can be used for regularization but are rarely used. Generally, only one of the three types of TR terms is included, with zeroth or first order used most commonly, and the magnitudes of the α_i coefficients are "small."

Zeroth-order regularization penalizes the function value directly. If the α_0 coefficient is made very large, the target sum of squared errors becomes insignificant in comparison to the regularization term. Minimizing the functional in Eq. (4.99) with a very large α_0 will bias the solution for $q(t)$ toward zero.

First-order regularization penalizes the first derivative (slope) of the unknown function. In the limit of a very large α_1 coefficient, $q(t)$ becomes a zero-slope, or constant-valued function.

Second-order regularization penalizes the second derivative (curvature) of the unknown function. Large α_2 values will produce $q(t)$ with a constant slope (straight line).

4.8.2.2 Discrete Formulation

The continuous functional in Eq. (4.99) can be cast into discrete form corresponding to observations at time intervals $t = i\Delta t$ using forward finite difference approximations for the first and second derivatives in the TR terms:

$$\left.\frac{\partial q}{\partial t}\right|_{t_i} \approx \frac{q_{i+1} - q_i}{\Delta t}; \qquad \left.\frac{\partial^2 q}{\partial t^2}\right|_{t_i} \approx \frac{q_{i+2} - 2q_{i+1} + q_i}{\Delta t^2} \quad (4.100)$$

Other discrete approximations for these derivatives can be used. The second expression in Eq. (4.100) is quite approximate, as it actually represents the centered second derivative approximation at time t_{i+1}. A more appropriate form for forward difference approximation at time t_i involves additional terms. This simpler approximation is effective because it still enforces a penalty on the second derivative of the heat flux at every time step up to time t_{n-2}.

The integrations in Eq. (4.99) reduce to discrete sums over the time interval of interest, so that equation becomes

$$S = \sum_{i=1}^{n} \left(Y_{j,i} - T_{j,i}\right)^2 + \left[\alpha_0 \sum_{i=1}^{n} q_i^2 + \alpha_1 \sum_{i=1}^{n-1} (q_{i+1} - q_i)^2 + \alpha_2 \sum_{i=1}^{n-2} (q_{i+2} - 2q_{i+1} + q_i)^2\right] \quad (4.101)$$

Here, the subscript "j" corresponds to sensor location x_j. Note that the denominator terms containing Δt in Eq. (4.100) are effectively absorbed into the TR coefficients (α terms) in Eq. (4.101). The resulting physical units of the α terms in Eq. (4.101) are $(°\text{C-m}^2/\text{W})^2$.

The whole domain zeroth-order regularization procedure for a single sensor involves minimizing

$$S = \sum_{i=1}^{n} (Y_i - T_i)^2 + \alpha_0 \sum_{i=1}^{n} q_i^2 \tag{4.102}$$

with respect to q_i, $i = 1, 2, ..., n$. If $\alpha_0 \to 0$, then exact matching of Y_i and T_i is approached but the sum of the q_i^2 terms becomes large for small time steps. For the opposite case of very large α_0, the magnitudes of the q_i's are reduced with the limit being

$$q_i = 0, \quad i = 1, 2, ..., n \tag{4.103}$$

The effect of a nonzero α_0 is to reduce the magnitude of the q_i values. For small time steps for an interior sensor, the case of $\alpha_0 \approx 0$ can result in an unstable procedure with the q_i values oscillating with changing signs and ever-increasing magnitudes. However, by properly selecting α_0, instability can be eliminated because the effect of the regularization term in Eq. (4.102) is to reduce the maximum magnitudes of estimated values of q_i.

The whole domain *first*-order regularization procedure for a single sensor involves minimizing

$$S = \sum_{i=1}^{n} (Y_i - T_i)^2 + \alpha_1 \sum_{i=1}^{n-1} (q_{i+1} - q_i)^2 \tag{4.104}$$

Again, if α_1 goes to zero, exact matching is obtained. On the other hand, an extremely large value of α_1 causes q_i to be

$$q_i = \text{constant} \quad i = 1, 2, ..., n \tag{4.105}$$

where the constant can be any positive or negative value. For moderate values of α_1, *differences* in the q_i values are reduced. Reducing the changes in the q_i values from one time step to the next during the minimizing of Eq. (4.104) has the effect of improving the stability of the q_i, $i = 1, 2, ..., n$ values.

Similar to Eqs. (4.102) and (4.104), the second-order regularization procedure minimizes

$$S = \sum_{i=1}^{n} (Y_i - T_i)^2 + \alpha_2 \sum_{i=1}^{n-2} (q_{i+2} - 2q_{i+1} + q_i)^2 \tag{4.106}$$

The effect of increasing α is to cause q_i to be a straight line or

$$q_i = q_0 + iq'\Delta t, \quad i = 1, 2, ..., n \tag{4.107}$$

where q_0 and q' are arbitrary; that is, large α_2 results in approximating $q(t)$ as a straight line with two unknowns, the intercept and slope. Moderate values of α_2 tend to reduce second differences of the estimated q_i's; in other words, the rate of change of $q(t)$ is reduced.

The foregoing discussion points out the different effects of the regularization terms of various orders. The zeroth order reduces the magnitude of q_i, the first order reduces the magnitude of changes in the q_i values from one i to the next, and the second-order regularization term tends to reduce rapid oscillations in the estimated heat flux.

This section also highlights the challenge of proper selection of the regularization parameter, α. If chosen too small, the degree of regularization is too little, and effects of ill-posedness and sensitivity to noise will dominate the solution. If α is chosen too large, then the unwanted bias effects of over-regularization become apparent. Proper selection of the amount of regularization to balance the competing effects of sensitivity to noise against bias in the solution is considered in Chapter 6.

4.8.3 Whole Domain TR Method

4.8.3.1 Matrix Formulation

A matrix formulation is used to develop the whole domain procedure for n times and n components of the heat flux. The sum of squares function becomes

$$S = (\mathbf{Y} - \mathbf{T})^T (\mathbf{Y} - \mathbf{T}) + \left[\alpha_0 (\mathbf{H}_0 \mathbf{q})^T \mathbf{H}_0 \mathbf{q} + \alpha_1 (\mathbf{H}_1 \mathbf{q})^T \mathbf{H}_1 \mathbf{q} + \alpha_2 (\mathbf{H}_2 \mathbf{q})^T \mathbf{H}_2 \mathbf{q} \right] \tag{4.108}$$

where, for the whole domain method with a single sensor,

$$\mathbf{Y} = \begin{bmatrix} Y_1 \\ \vdots \\ Y_n \end{bmatrix}, \quad \mathbf{T} = \begin{bmatrix} T_1 \\ \vdots \\ T_n \end{bmatrix}, \quad \mathbf{q} = \begin{bmatrix} q_1 \\ \vdots \\ q_n \end{bmatrix} \tag{4.109}$$

The α_i values in Eq. (4.108) are the regularizing coefficients for zeroth-, first-, and second-order regularization methods. The $\mathbf{H}_i$ matrices are square and are associated with the zeroth-, first-, and second-order regularization procedures. Corresponding to the algebraic expressions contained in Eq. (4.101), the $\mathbf{H}_i$ matrices are

$$\mathbf{H_0} = \mathbf{I} \tag{4.110a}$$

$$\mathbf{H_1} = \begin{bmatrix} -1 & 1 & 0 & 0 & 0 \\ 0 & -1 & 1 & 0 & 0 \\ 0 & 0 & -1 & 1 & 0 \\ 0 & 0 & 0 & -1 & 1 \\ 0 & 0 & 0 & 0 & 0 \end{bmatrix} \tag{4.110b}$$

$$\mathbf{H_2} = \begin{bmatrix} 1 & -2 & 1 & 0 & 0 \\ 0 & 1 & -2 & 1 & 0 \\ 0 & 0 & 1 & -2 & 1 \\ 0 & 0 & 0 & 0 & 0 \\ 0 & 0 & 0 & 0 & 0 \end{bmatrix} \tag{4.110c}$$

where $\mathbf{H}_1$ and $\mathbf{H}_2$ are written here for $n = 5$ as illustrations.

The $\mathbf{q}$ components in Eq. (4.108) are found by minimizing S, which is accomplished by taking the matrix first derivative with respect to the n components of the heat flux, replacing $\mathbf{q}$ by $\hat{\mathbf{q}}$, and setting the resulting equation equal to the zero vector,

$$\frac{\partial S}{\partial q} = \begin{bmatrix} \dfrac{\partial S}{\partial q_1} \\ \vdots \\ \dfrac{\partial S}{\partial q_n} \end{bmatrix}_{q=\hat{q}} = \frac{\partial S}{\partial \mathbf{q}} = \mathbf{0} \tag{4.111}$$

The model for the temperature $\mathbf{T}$ is given by Eq. (3.54) which is used in Eq. (4.108). Taking the matrix derivative as indicated by Eq. (4.111) gives

$$\left[\mathbf{X}^{\mathrm{T}}\mathbf{X} + \left(\alpha_0\mathbf{H_0^{\mathrm{T}}H_0} + \alpha_1\mathbf{H_1^{\mathrm{T}}H_1} + \alpha_2\mathbf{H_2^{\mathrm{T}}H_2}\right)\right]\hat{\mathbf{q}} = \mathbf{X}^{\mathrm{T}}(\mathbf{Y} - T_{in}\mathbf{1}) \tag{4.112}$$

because (see Beck and Arnold (1977), p. 221)

$$\frac{\partial}{\partial \mathbf{q}}\left[(\mathbf{Xq})^{\mathrm{T}}\mathbf{Xq}\right] = 2\mathbf{X}^{\mathrm{T}}\mathbf{Xq} \tag{4.113}$$

Note that $\mathbf{1}$ is a vector of ones.

The solution of Eq. (4.112) for the $\mathbf{q}$ vector can be symbolically written as

$$\hat{\mathbf{q}} = \left[\mathbf{X}^{\mathrm{T}}\mathbf{X} + \left(\alpha_0\mathbf{I} + \alpha_1\mathbf{H_1^{\mathrm{T}}H_1} + \alpha_2\mathbf{H_2^{\mathrm{T}}H_2}\right)\right]^{-1}\mathbf{X}^{\mathrm{T}}(\mathbf{Y} - T_{in}\mathbf{1}) \tag{4.114}$$

The regularization terms introduced by the procedure only appear as additive contributions to $\mathbf{X}^{\mathrm{T}}\mathbf{X}$. These terms reduce or eliminate the ill conditioning of the ill-posed IHCP.

The first-order regularization term $\mathbf{H_1^{\mathrm{T}}H_1}$ in Eqs. (4.112) and (4.114) is

$$\mathbf{H_1^{\mathrm{T}}H_1} = \begin{bmatrix} 1 & -1 & 0 & 0 & \cdots & 0 \\ -1 & 2 & -1 & 0 & \cdots & 0 \\ 0 & -1 & 2 & -1 & \ddots & \vdots \\ \vdots & \ddots & \ddots & \ddots & \ddots & 0 \\ 0 & \cdots & 0 & -1 & 2 & -1 \\ 0 & 0 & \cdots & 0 & -1 & 1 \end{bmatrix} \tag{4.115}$$

Notice that the coefficients of -1, 2, and -1 for all rows (except first and last) are the same as those for *second* differences.

Table 4.4 Condition number of coefficient matrix in Eq. (4.112) for piecewise constant (q = const) and piecewise linear (q = linear) heat flux assumptions with various regularization parameter values ($\Delta \tilde{t}$ = 0.05, n = 5 time steps).

	$\alpha_i \rightarrow$	0.0	10^{-6}	10^{-5}	10^{-4}	10^{-3}	10^{-2}
	x/L			Zeroth order			
q = const	0.0	5.2	5.2	5.2	5.2	5.1	4.3
	0.5	39 953	13 904	2025	213	22.3	3.1
	1.0	1.4781E + 16	5021	503	51.2	6.0	1.5
q = linear	0.0	21.5	21.5	21.5	21.3	19.2	10.0
	0.5	4.619E + 09	16 789	1680	169	17.8	2.7
	1.0	5.954E + 21	3337	335	34.4	4.3	1.3
				First order			
q = const	0.0	5.2	5.2	5.2	5.2	4.8	3.2
	0.5	39 953	7135	883	118	22.8	9.2
	1.0	1.4781E + 16	6102	802	118.2	20.2	22.8
q = *linear*	0.0	21.5	21.5	21.4	20.6	14.9	5.3
	0.5	4.619E + 09	12 636	1,337	172	29.1	12.1
	1.0	5.954E + 21	6304	791	112.6	17.5	31.9

Some insights regarding TR can be seen by considering the condition number of the coefficient matrix in Eq. (4.112) for various values of the regularization parameter. Recall (see Section 4.3.2) that the condition number relates to the "invertability" of the matrix: a very large condition number indicates a near singular matrix, leading to errors in numerical computation of the matrix inverse. Conversely, a small condition number indicates a matrix whose inverse is easy to compute accurately. Considering Eq. (4.112), if the condition number of the coefficient matrix is large, then errors in the matrix inversion will amplify any errors in the data **Y**.

Table 4.4 shows the condition number of a coefficient matrix with $n = 5$ time steps for different assumptions about the heat flux variation in time and for different values of the regularization parameter. The upper portion of the table is for zeroth-order TR, and the lower portion is for first-order TR. In each of these sections there are two assumptions about the heat flux variation: piecewise constant (q = const) and piecewise linear (q = linear). Three different sensor locations are considered: $x/L = 0, 0.5$, and 1.0. When $\alpha_i = 0$, there is no regularization, and the values of the condition numbers suggest that solutions with sensors at the active surface ($x = 0$) can produce stable results. However, at locations away from the active surface, the condition numbers become large, and solutions to Eq. (4.112) will produce unstable results with $\alpha_i = 0$. Note that the linear in time q = linear formulation gives higher condition numbers when $\alpha_i = 0$. This is consistent with observation made previously that the q = linear assumption results in *lower* sensitivity to the heat flux components, with smaller values in the **X** matrix, resulting in larger condition numbers.

Another observation regarding Table 4.4 can be made in comparison of the zeroth-order and first-order results. Although the regularization parameters in the two TR methods weight different measures in the objective function (q or Δq), for the same magnitude of α_i, the degree of regularization provided (as indicated by the reduction in the condition number of the matrix) is greater for q = linear over q = const. For example, consider zeroth-order TR with sensor at $x/L = 0.50$. For the q = const case, $\alpha_0 = 10^{-6}$ reduces the condition number from about 40 000 (with $\alpha_0 = 0$) to around 14 000 (about a factor of one-half), while for the q = linear case, $\alpha_0 = 10^{-6}$ reduces the condition number from approximately 4.5×10^9 to about 17 000 (five orders of magnitude). So, the magnitude of the regularization parameter required for q = linear will be smaller than for q = const. This can be interpreted in the context that the magnitude of the regularization coefficient needed to stabilize the solution is relative to the magnitude of the terms in the $\mathbf{X}^T\mathbf{X}$ matrix. The topic of optimal selection of the regularization parameter will be considered in Chapter 6.

Example 4.6   Calculate the coefficient matrix in Eq. (4.112) for zeroth-order TR using $\alpha_0 = 10^{-5}$ for a sensor at $x/L = 0.50$ using the piecewise-constant heat flux assumption. Use $\Delta \tilde{t} = 0.05$ and $n = 5$.

Solution
From the X22B10T0 solution (Eq. (2.35)), the values of ϕ_i are 0.0153660, 0.0593109, 0.1084691, 0.1583522, and 0.2083360. The values of $\Delta\phi_i$ are then

$$\Delta\phi_0 = \phi_1 - 0 = \phi_1 = 0.0153660 \tag{4.116}$$

$$\Delta\phi_1 = \phi_2 - \phi_1 = 0.0593109 - 0.0153660 = 0.0439449 \tag{4.117}$$

and $\Delta\phi_3 = 0.0491582$, $\Delta\phi_4 = 0.0498831$, and $\Delta\phi_5 = 0.0499838$. For these values, the $\mathbf{X}$ matrix is

$$\mathbf{X} = \begin{bmatrix} 0.0153659 & 0 & 0 & 0 & 0 \\ 0.0439450 & 0.0153659 & 0 & 0 & 0 \\ 0.0491582 & 0.0439450 & 0.0153659 & 0 & 0 \\ 0.0498831 & 0.0491582 & 0.0439450 & 0.0153659 & 0 \\ 0.0499838 & 0.0498831 & 0.0491582 & 0.0439450 & 0.0153659 \end{bmatrix} \tag{4.118}$$

The unregularized $\mathbf{X}^T\mathbf{X}$ matrix is

$$\mathbf{X}^T\mathbf{X} = \begin{bmatrix} 0.00957050 & 0.00778102 & 0.00540458 & 0.00296303 & 0.00076805 \\ 0.00778102 & 0.00707212 & 0.00528768 & 0.00294747 & 0.00076650 \\ 0.00540458 & 0.00528768 & 0.00458380 & 0.00283551 & 0.00075536 \\ 0.00296303 & 0.00294747 & 0.00283551 & 0.00216727 & 0.00067526 \\ 0.00076805 & 0.00076650 & 0.00075536 & 0.00067526 & 0.00023611 \end{bmatrix} \tag{4.119}$$

and the condition number of this matrix is 40 439, which differs from the entry in Table 4.4 only due to rounding of the matrix entries. By adding $\alpha_0\mathbf{I}$ to this matrix, the resulting coefficient matrix is

$$\mathbf{X}^T\mathbf{X} + \alpha_0\mathbf{I} = \begin{bmatrix} 0.00958050 & 0.00778102 & 0.00540458 & 0.00296303 & 0.00076805 \\ 0.00778102 & 0.00708212 & 0.00528768 & 0.00294747 & 0.00076650 \\ 0.00540458 & 0.00528768 & 0.00459380 & 0.00283551 & 0.00075536 \\ 0.00296303 & 0.00294747 & 0.00283551 & 0.00217727 & 0.00067526 \\ 0.00076805 & 0.00076650 & 0.00075536 & 0.00067526 & 0.00024611 \end{bmatrix} \tag{4.120}$$

The condition number of this matrix is 2027.

Discussion
Adding a "very small" term to the diagonal of the $\mathbf{X}^T\mathbf{X}$ matrix decreases its condition number by an order of magnitude.

Example 4.7 Repeat Example 4.6 using the piecewise linear heat flux assumption for heat flux discretization.

Solution
From the X22B20T0 solution (Eq. (2.45a)), the values of γ_i are 0.0040743, 0.0404770, 0.1242368, 0.2576293, and 0.4409709. The values of $\Delta\gamma_i$ are then

$$\Delta\gamma_0 = \gamma_1 - 0 = \gamma_1 = 0.0040743 \tag{4.121}$$

$$\Delta\gamma_1 = \gamma_2 - \gamma_1 = 0.0404770 - 0.0040743 = 0.0364026 \tag{4.122}$$

and $\Delta\gamma_3 = 0.0837598$, $\Delta\gamma_4 = 0.1333926$, and $\Delta\gamma_5 = 0.1833416$. The $\Delta(\Delta\gamma)_i$ values are

$$\Delta(\Delta\gamma_1) = \Delta\gamma_1 - \Delta\gamma_0 = 0.0364026 - 0.0040743 = 0.0323283 \tag{4.123}$$

and $\Delta(\Delta\gamma_2) = 0.0473571$, $\Delta(\Delta\gamma_3) = 0.0496328$, and $\Delta(\Delta\gamma_4) = 0.0499490$. The $\mathbf{X}$ matrix is

$$\mathbf{X} = \begin{bmatrix} \Delta\gamma_0 & 0 & 0 & 0 & 0 \\ \Delta(\Delta\gamma_1) & \Delta\gamma_0 & 0 & 0 & 0 \\ \Delta(\Delta\gamma_2) & \Delta(\Delta\gamma_1) & \Delta\gamma_0 & 0 & 0 \\ \Delta(\Delta\gamma_3) & \Delta(\Delta\gamma_2) & \Delta(\Delta\gamma_1) & \Delta\gamma_0 & 0 \\ \Delta(\Delta\gamma_4) & \Delta(\Delta\gamma_3) & \Delta(\Delta\gamma_2) & \Delta(\Delta\gamma_1) & \Delta\gamma_0 \end{bmatrix} \tag{4.124a}$$

$$\mathbf{X} = \begin{bmatrix} 0.0040743 & 0 & 0 & 0 & 0 \\ 0.0323283 & 0.0040743 & 0 & 0 & 0 \\ 0.0473571 & 0.0323283 & 0.0040743 & 0 & 0 \\ 0.0496328 & 0.0473571 & 0.0323283 & 0.0040743 & 0 \\ 0.0499490 & 0.0496328 & 0.0473571 & 0.0323283 & 0.0040743 \end{bmatrix} \tag{4.124b}$$

So the unregularized $\mathbf{X}^T\mathbf{X}$ matrix is

$$\mathbf{X}^T\mathbf{X} = \begin{bmatrix} 0.0082627 & 0.0064923 & 0.0041629 & 0.0018170 & 0.0002035 \\ 0.0064923 & 0.0057678 & 0.0040132 & 0.0017975 & 0.0002022 \\ 0.0041629 & 0.0040132 & 0.0033044 & 0.0016627 & 0.0001929 \\ 0.0018170 & 0.0017975 & 0.0016627 & 0.0010617 & 0.0001317 \\ 0.0002035 & 0.0002022 & 0.0001929 & 0.0001317 & 0.0000166 \end{bmatrix} \tag{4.125}$$

The condition number of this matrix is 1.3661×10^7. By adding $\alpha_0 \mathbf{I}$ to this matrix, the resulting coefficient matrix is

$$\mathbf{X}^T\mathbf{X} + \alpha_0 \mathbf{I} = \begin{bmatrix} 0.0082727 & 0.0064923 & 0.0041629 & 0.0018170 & 0.0002035 \\ 0.0064923 & 0.0057778 & 0.0040132 & 0.0017975 & 0.0002022 \\ 0.0041629 & 0.0040132 & 0.0033144 & 0.0016627 & 0.0001929 \\ 0.0018170 & 0.0017975 & 0.0016627 & 0.0010117 & 0.0001317 \\ 0.0002035 & 0.0002022 & 0.0001929 & 0.0001317 & 0.0000266 \end{bmatrix} \tag{4.126}$$

and the condition number of this matrix is 1680.

Discussion

As noted previously, for the same value of Δt, the entries in the $\mathbf{X}$ matrix for the q = linear assumption (Eq. 4.124b) are smaller than under the q = const assumption (Eq. 4.118). This results in much smaller values in the corresponding $\mathbf{X}^T\mathbf{X}$ matrices and a correspondingly larger condition number. As a result of the smaller values, the same value of regularization parameter, α_0, results in a more dramatic reduction in the condition number of the coefficient matrix.

4.8.4 Sequential TR Method

TR is inherently a whole domain solution technique – the complete time history of observations must be considered together. However, the method can be made sequential through at least two approaches. A sequential TR technique was developed extensively by Lamm et al. (1997), Lamm and Scofield (2001, Lamm (2005), Lamm and Dai (2005), Dai and Lamm (2008), Brooks et al. (2010), and this technique is described in this section. The filter coefficient concept can also be used to treat TR as a sequential method (see Chapter 5).

The sequential TR method can be envisioned as a combination of the future times concept from the FS method employing TR over only the next r future times. At the end of each time step, only the first heat flux component from the local TR is retained and the time window advanced to the next time step.

More explicitly, the method can be summarized as follows (Lamm et al. 1997). Assuming $q_1, q_2, ..., q_{M-1}$ have already been found, determine the vector $\boldsymbol{\beta}$ that solves the reduced dimension Tikhonov problem. That is, find $\boldsymbol{\beta}$ that minimizes

$$S = \left(\mathbf{Y}_r - \mathbf{X}_r\boldsymbol{\beta} - \hat{\mathbf{T}}_{q=0}\right)^T \left(\mathbf{Y}_r - \mathbf{X}_r\boldsymbol{\beta} - \hat{\mathbf{T}}_{q=0}\right) + \alpha_i (\mathbf{H}_i\boldsymbol{\beta})^T (\mathbf{H}_i\boldsymbol{\beta}) \tag{4.127}$$

Here, $\mathbf{Y}_r$ is the reduced set of data over times $t_M, t_{M+1}, ..., t_{M+r-1}$, and $\mathbf{X}_r$ is the $r \times r$ subset from the upper left of the whole-domain matrix $\mathbf{X}$. $\mathbf{H}_i$ and α_i result from the order (zeroth, first, second, etc.) of TR selected, and $\hat{\mathbf{T}}_{q=0}$ contains all the components of the computed value of temperature resulting from all the previously estimated q_i values (Eq. 4.70b). After solving the local TR problem, assign $q_M = \beta_1$, the first value in the $\boldsymbol{\beta}$ vector, discarding all the other entries in this vector. Next, advance to the next time step, and repeat the procedure.

Regarding this procedure, Lamm et al. (1997) comment: "It is important to note that this sequential approach is not simply a matter of performing a decomposition of the original matrix system into smaller subproblems, with standard TR then applied individually to each smaller problem. In contrast, the method performs TR sequentially on small over-lapping subproblems, updating the definition of each subproblem as new information about the solution is obtained."

4.8.5 General Comments About Tikhonov Regularization

TR results from adding a penalty term to the sum of squared errors between the measured and model-predicted temperatures. The *order* of TR corresponds to the order of the derivative of the unknown heat flux that is used in the penalty term of the resulting objective function. Typically, either zeroth- or first-order TR is used.

Regularization in the Tikhonov method is controlled by the magnitude of the regularization parameter, α_i. When $\alpha_i = 0$, no regularization occurs and exact matching of the data results (as in the Stolz method). For $\alpha_i \to \infty$, the penalty term in the objective function dominates the minimization, resulting in $q_i \approx 0$ for zeroth-order TR and $q_i \approx$ const for first-order TR.

Sequential TR, as described by Lamm, combines concepts of "future times" from function specification with TR over the specified r future times. This allows estimation of heat flux components one at a time. Regularization in this technique is now controlled through two parameters: the number of future times, r, and the regularization parameter α_i.

4.9 Gradient Methods

The conjugate gradient algorithm is a family of techniques for unconstrained minimization of an objective function. It can be applied to minimize the sum of squares objective function (e.g. Eq. (4.63b)), and the resulting IHCP solution technique is termed the conjugate gradient method (CG method or CGM). This is inherently a whole-domain method which makes successive approximations to an initial guess for the unknown parameters (in this case, entries of the vector $\mathbf{q}$) using gradient information (derivatives of the objective function). Since no numerical inversion of the matrix $\mathbf{X}$ (or $\mathbf{X}^T\mathbf{X}$) is required, issues of condition number and "invertability" discussed in the previous section are avoided in this method.

An excellent description of CGM is given by Jarny and Orlande (2011). Section 4.9.1 outlines CGM for linear problems, wherein the gradient of the objective function (S in Eq. (4.63b)) can be computed explicitly. Section 4.9.2 provides a summary of the *adjoint method*, which provides a framework for determination of the gradient of the objective function required for the CGM when the gradient is implicitly contained in the model equations and cannot be obtained directly. This section closes with some general remarks about CGM in Section 4.9.3.

4.9.1 Conjugate Gradient Method

The method assumes a computational model for the temperature at a measurement point given a known or assumed vector of heat flux components, such as

$$\mathbf{T} = \mathbf{X}\mathbf{q} \tag{4.128}$$

When Eq. (4.128) is inserted into S in Eq. (4.63b), the objective function to minimize becomes

$$S = (\mathbf{Y} - \mathbf{X}\mathbf{q})^T(\mathbf{Y} - \mathbf{X}\mathbf{q}) \tag{4.129}$$

An initial guess for the $\mathbf{q}$ vector ($\mathbf{q}^0$) must be made to begin the iterations. Updates to the initial guess are made at each iteration using

$$\mathbf{q}^k = \mathbf{q}^{k-1} + \Delta\mathbf{q}^k \tag{4.130}$$

where $\Delta\mathbf{q}^k$ is a finite increment to the $\mathbf{q}^{k-1}$ vector. To improve the initial guess for $\mathbf{q}^0$ that will minimize the objective function, adjustments should be made to move the current value of the objective function (S^0) "downhill," toward the minimum. Let

$$\Delta\mathbf{q}^k = \rho^k\mathbf{w}^k \tag{4.131}$$

where $\mathbf{w}^k$ is the *descent direction* at iteration step k and ρ^k is the *step size*. In order for $\mathbf{w}^k$ to produce an improvement in $\mathbf{q}^k$, it is necessary to satisfy the *descent condition*

$$\left(\nabla_{\mathbf{q}}S^k\right)^T\mathbf{w}^k < 0 \tag{4.132}$$

For the purely linear objective function (Eq. 4.129), the gradient of S^k at iteration k with respect to the vector $\mathbf{q}^k$ is

$$\nabla_{\mathbf{q}}S^k = -2\mathbf{X}^T\left(\mathbf{Y} - \mathbf{X}\mathbf{q}^k\right) \tag{4.133}$$

The simplest approach to moving the solution "downhill" from the initial S^0 is to let the descent direction be equal to the negative of the function gradient $(-\nabla S)$, which is known as the *steepest descent* (SD) algorithm. For SD, let

$$\mathbf{w}^k = -\nabla_{\mathbf{q}} S^k \tag{4.134}$$

The step size follows from a simple line search to move in the direction of $\mathbf{w}^k$ as far as possible before S increases again. Under the linear form (Eq. 4.129), this step size can be calculated in closed form:

$$\rho^k = \underset{\rho > 0}{\arg\min}\, S\left(\mathbf{q}^k + \rho\mathbf{w}^k\right) \tag{4.135a}$$

$$\rho^k = -\frac{1}{2}\frac{\left(\nabla_{\mathbf{q}} S^k\right)^T\left(\nabla_{\mathbf{q}} S^k\right)}{\left(\nabla_{\mathbf{q}} S^k\right)^T \mathbf{X}^T\mathbf{X}\left(\nabla_{\mathbf{q}} S^k\right)} = -\frac{1}{2}\frac{\left(\mathbf{Y}-\mathbf{Xq}^k\right)^T\mathbf{XX}^T\left(\mathbf{Y}-\mathbf{Xq}^k\right)}{\left(\mathbf{Y}-\mathbf{Xq}^k\right)^T\mathbf{XX}^T\mathbf{XX}^T\left(\mathbf{Y}-\mathbf{Xq}^k\right)} \tag{4.135b}$$

The SD method is direct but converges slowly. Faster convergence is possible using one of a family of CGMs, where the descent direction $\mathbf{w}^k$ is chosen *conjugate* to the previous direction, $\mathbf{w}^{k-1}$, with respect to the $\mathbf{X}^T\mathbf{X}$ matrix. That is, so that

$$\left(\mathbf{w}^k\right)^T\mathbf{X}^T\mathbf{X}\mathbf{w}^{k-1} = 0 \tag{4.136}$$

which results when

$$\mathbf{w}^k = -\nabla_{\mathbf{q}} S^k + \gamma^k \mathbf{w}^{k-1} \tag{4.137}$$

with

$$\mathbf{w}^0 = -\nabla_{\mathbf{q}} S^0 \tag{4.138}$$

Different CGMs follow from selection of the parameter γ in Eq. (4.137). Two popular methods are the Fletcher–Reeves algorithm (Section 4.9.1.1) and the Polak–Ribiere algorithm (Section 4.9.1.2).

The CG procedure for the linear IHCP can be summarized in the following steps:

1) Make an initial guess for $\mathbf{q}^0$ for step $k = 0$
   repeat steps 2–7 below until convergence
2) Evaluate the least-squares objective function S^k (Eq. 4.129)
3) Compute the gradient ∇S^k (Eq. 4.133) and determine the descent direction:
   a) if $k = 0$, $\mathbf{w}^0 = -\nabla S^0$ (Eq. 4.138)
   b) else, compute $\mathbf{w}^k$ from Eq. (4.137)

4) Compute the step size ρ^k from Eq. (4.135a)
5) Calculate the increment in the heat flux vector (Eq. 4.131)
6) Update the heat flux vector (Eq. 4.130)
7) Let $k = k + 1$. Return to step 2 until converged.

"Convergence" can be monitored in several ways, including by checking the magnitude of S or by checking the magnitude of the gradient. Both of these should be very small in the neighborhood of the minimum of S. One way to judge the magnitude of these changes is to consider the root mean square (RMS) of each and compare them to a suitably small value.

$$S_{\text{RMS}}^k = \sqrt{\frac{S^k}{n}} \leq \varepsilon \tag{4.139}$$

$$\nabla S_{\text{RMS}}^k = \sqrt{\frac{\left(\nabla S^k\right)^T \nabla S^k}{n}} \leq \varepsilon \tag{4.140}$$

Here, "n" is the number of entries in the $\mathbf{q}$ vector (and also the $\nabla_{\mathbf{q}} S$ vector). The RMS values represent the average value for the components of the vector. S_{RMS}^k is the average difference between the measured ($\mathbf{Y}$) and model-predicted ($\mathbf{Xq}$) values. ∇S_{RMS}^k, similarly, is the average of the components of the gradient vector.

For a quadratic objective function (such as the sum of squares in Eq. (4.129)), with n unknowns in the vector $\mathbf{q}$, the conjugate gradient algorithms will converge to the minimum in exactly n iterations for linear problems. In application to the IHCP, errors in the measurement data $\mathbf{Y}$ can lead to convergence issues, or even divergence of the iterative process. In essence, the number of iterations performed is the regularization parameter for CGM solution of the IHCP. Procedures for optimal regularization of the IHCP are the topic of Chapter 6.

4.9.1.1 Fletcher–Reeves CGM

The Fletcher–Reeves algorithm (Fletcher and Reeves 1964) satisfies the conjugacy requirement (Eq. (4.136)) with the parameter γ specified by

$$\gamma^k = \frac{\left(\nabla_{\mathbf{q}}S^k\right)^T \nabla_{\mathbf{q}}S^k}{\left(\nabla_{\mathbf{q}}S^{k-1}\right)\nabla_{\mathbf{q}}S^{k-1}} \tag{4.141}$$

The gradient of the quadratic function S is given in Eq. (4.133). The resulting values for $\mathbf{w}^k$ are

$$\mathbf{w}^k = \begin{cases} 2\mathbf{X}^T\left(\mathbf{Y}-\mathbf{Xq}^k\right) & k=1 \\ 2\mathbf{X}^T\left(\mathbf{Y}-\mathbf{Xq}^k\right) + \dfrac{\left\|2\mathbf{X}^T\left(\mathbf{Y}-\mathbf{Xq}^k\right)\right\|^2}{\left\|2\mathbf{X}^T\left(\mathbf{Y}-\mathbf{Xq}^{k-1}\right)\right\|^2}\mathbf{w}^{k-1} & k>1 \end{cases} \tag{4.142}$$

The step size at each iteration is

$$\rho^k = -\frac{1}{2}\frac{\left(\nabla_{\mathbf{q}}S^k\right)^T\mathbf{w}^k}{(\mathbf{w}^k)^T\mathbf{X}^T\mathbf{X}(\mathbf{w}^k)} = \frac{\left(\mathbf{Y}-\mathbf{Xq}^k\right)^T\mathbf{w}^k}{\left(\mathbf{Xw}^k\right)^T\mathbf{Xw}^k} = \frac{\left(\mathbf{Y}-\mathbf{Xq}^k\right)^T\mathbf{w}^k}{\left\|\mathbf{Xw}^k\right\|^2} \tag{4.143}$$

4.9.1.2 Polak–Ribiere CGM

The Polak–Ribiere (Polak and Ribière 1969) algorithm offers a slight change in the classic Fletcher–Reeves algorithm. Instead of using the inner product of the function gradient from the current (k) iteration with itself in the numerator for γ^k, the *change* in the gradient from the last iteration to the current iteration is used in the inner product with the current gradient.

$$\gamma^k = \frac{\left(\nabla_{\mathbf{q}}S^k - \nabla_{\mathbf{q}}S^{k-1}\right)^T \nabla_{\mathbf{q}}S^k}{\left(\nabla_{\mathbf{q}}S^{k-1}\right)\nabla_{\mathbf{q}}S^{k-1}} \tag{4.144}$$

The descent direction is computed using Eqs. (4.137) and (4.138). The step size is computed using Eq. (4.143).

Example 4.8 Solve Example 4.3 using SD and Fletcher–Reeves CGM. Use the piecewise constant assumption for the heat flux variation and perform 5 iterations using $\mathbf{q}^0 = \mathbf{1}$.

Solution
The dimensionless time step is $\Delta\tilde{t} = 0.5$, and the X22B-0T0 $\mathbf{X}$ matrix is

$$\mathbf{X} = \begin{bmatrix} 0.3348 & 0 & 0 & 0 \\ 0.4986 & 0.3348 & 0 & 0 \\ 0.5000 & 0.4986 & 0.3348 & 0 \\ 0.5000 & 0.5000 & 0.4986 & 0.3348 \end{bmatrix} \tag{4.145}$$

Table 4.5 shows the values obtained from the first five iterations of using the SD algorithm. The values of S^k drop dramatically over these first few iterations, but still is on the order of 100 on the last step. The $\mathbf{w}^k$ entries are equal to the ∇S for the SD algorithm. So, it is clear that neither S^k or ∇S^k are approaching zero after these five iterations. The rate of convergence is very slow, and, in fact, 267 iterations are required to reduce the RMS error of the gradient to 10^{-5}.

Table 4.6 shows the iteration results for the Fletcher–Reeves CGM. The algorithm finds the minimum of the quadratic function after four iterations, as it should. Both S^k and ∇S^k are zero after the $k=4$ iteration step.

Discussion

The exact heat fluxes, at times $t - \Delta t/2$ corresponding to the assumption of piecewise constant heat flux, are 25, 75, 125, and 175 W/m^2. The errors for the fully converged CG algorithm are 28.3, −3.8, 1.49, and −1.94%. These same relative errors result from the SD algorithm if the iterations are continued until convergence is achieved. These errors result from two

Table 4.5 Iteration results for steepest descent

k	S^k	w^k				ρ^k	q^k			
0	36 830	292.7	277.5	225.4	111.9	0.2931	1	1	1	1
1	3710	−29.4	−2.1	27.6	26.6	2.7616	86.8	82.3	67.1	33.8
2	486.8	30.1	25.2	20.7	13.8	0.2953	5.7	76.4	143.3	107.2
3	168.3	−1.6	−2.2	1.3	5.4	3.8598	14.6	83.8	149.4	111.3
4	94.5	8.8	4.9	2.7	3.9	0.3167	8.5	75.4	154.4	132.2
							11.3	77	155.3	133.4

Table 4.6 Iteration results for Fletcher–Reeves conjugate gradient.

k	S^k	w^k				ρ^k	q^k			
0	36 830	292.7	277.5	225.4	111.9	0.2931	1	1	1	1
1	3710	−26.3	0.7	29.9	27.7	3.0594	86.8	82.3	67.1	33.9
2	139.2	1.5	−2.0	−1.2	3.5	12.5623	6.2	84.5	158.6	118.6
3	15.0	−0.2	0.5	−0.6	0.5	35.1511	25.5	59.1	144.1	162.4
4	0.0	0.0	0.0	0.0	0.0	0.2898	17.9	77.9	123.1	178.4
							17.9	77.9	123.1	178.4

sources. Primarily, the data supplied are approximate – although these data are from the solution for a linear-in-time heat flux, the values are truncated to integer values and are therefore approximate. Secondly, the assumed piecewise constant heat flux for the direct solution model does not approximate the response with the relatively large $\Delta t = 0.5$ s. As the data time step is decreased, errors in the estimated heat fluxes decrease; however, the ill-posedness of the IHCP increases for smaller time steps. If highly accurate data are used (with 15 decimal places of accuracy) and the time step is reduced to $\Delta t = 0.0625$ s (with the number of time steps increased to 32), the average error in the main of the results is about 0.22%, but the errors for the first few and last few estimates increase significantly.

4.9.2 Adjoint Method (Nonlinear Problems)

Direct computation of the gradient ∇S is generally possible using the direct solver for the IHCP. For linear problems, as described in the previous section, the gradient computation is explicit. However, even in the case of temperature-dependent properties, a linearization around the properties based on the temperature solution at iteration k of the CGM allows update of the gradient ∇S from something like Eq. (4.133). The nonlinear governing equations will need to be solved at each iteration, and the matrix $\mathbf{X}$ will be updated at each iteration, but computation of the gradient is straightforward.

For other types of inverse problems, for example, determining the thermal properties as a function of temperature, the gradient of the objective function ∇S cannot be computed explicitly. The unknown function could be discretized as

$$k(T) = \sum_{i=1}^{n} k_i \zeta_i(T) \tag{4.146}$$

where ζ_i are known basis functions to interpolate the k_i values. One means of computing the needed gradient is to perturb the individual k_i values one at a time, solve the nonlinear governing equation each time, and approximate each of the n components of the gradient using finite differences. This procedure is computationally inefficient and significantly extends the time required to solve the CG problem.

A superior approach is the *adjoint method (AM)*, which uses concepts from optimization and functional calculus to solve for all the components of the gradient at once. The additional computational expense is approximately that of solving the governing equation one additional time (instead of n times needed for the finite difference approximation approach).

There are several important references for the AM. Ozisik and Orlande (2021) provide an excellent discourse, and Jarny and Orlande (2011) provide numerous illustrative examples. A series of unpublished lectures by Lamm (1990) directly addresses the IHCP in the context of the AM.

The AM uses the notion of Lagrange multipliers to augment the objective sum of squares function. The Lagrange multiplier has been called the *adjoint variable* by Jarny and others (Alifanov 1994; Alifanov et al. 1995; Jarny and Orlande 2011; Ozisik and Orlande 2021), and determination of the adjoint variable leads to computation of the gradient of the objective function, S.

4.9.2.1 Some Necessary Mathematics

The AM is described in functional calculus terms. As the name suggests, functional calculus is concerned with "functions of functions" rather than "functions of variables." A loose connection between familiar optimization and calculus concepts will be used to establish related ideas needed for functional calculus. This section reviews some definitions and concepts from variable calculus and optimization.

Consider a function f of several parameters, β_i:

$$f = f(\beta_1, \beta_2 ..., \beta_n) = f(\boldsymbol{\beta}) \tag{4.147}$$

The (partial) derivative of f with respect to β_i is defined formally as

$$\frac{\partial f}{\partial \beta_i} = \lim_{\Delta \beta_i \to 0} \frac{f(\beta_1, \beta_2, ..., \beta_i + \Delta \beta_i, ...\beta_n) - f(\boldsymbol{\beta})}{\Delta \beta_i} \tag{4.148}$$

Of course, in practice, familiar calculus rules for computing the derivative are used, when possible, to obtain the derivative without resorting to the formal definition.

The differential of the function f is

$$\begin{aligned} df &= \frac{\partial f}{\partial \beta_1} d\beta_1 + \frac{\partial f}{\partial \beta_2} d\beta_2 + \cdots + \frac{\partial f}{\partial \beta_n} d\beta_n \\ &= \left\lfloor \frac{\partial f}{\partial \beta_1} \ \frac{\partial f}{\partial \beta_2} \ \cdots \ \frac{\partial f}{\partial \beta_n} \right\rfloor \lfloor d\beta_1 \ \ d\beta_2 \cdots d\beta_n \rfloor^T \\ &= \langle \nabla f, d\boldsymbol{\beta} \rangle \end{aligned} \tag{4.149}$$

This last result expresses the differential of f, df, as the *inner product* of the gradient of f, ∇f, and the differential of the vector $\boldsymbol{\beta}$, $d\boldsymbol{\beta}$.

Another important needed concept comes from constrained optimization of a function using Lagrange multipliers. When a function is to be minimized or maximized subject to a constraint, suitable unconstrained optimization techniques (such as conjugate gradient) can be applied to an augmented objective function, which is called the *Lagrangian*. The Lagrangian contains a new term added to the objective function variable consisting of the product of the *Lagrange multiplier* and the constraint equation. If the constraint equation is algebraic, such as

$$g = g(\boldsymbol{\beta}) = 0 \tag{4.150}$$

and $f(\boldsymbol{\beta})$ is the function to be minimized, then the Lagrangian will be written as

$$L(\boldsymbol{\beta}, \psi) = f(\boldsymbol{\beta}) + \psi g(\boldsymbol{\beta}) \tag{4.151}$$

where ψ is the Lagrange multiplier. The usual necessary condition for extremum is applied to the Lagrangian, but considering ψ as a new unknown. The condition for the extremum allows evaluation of the Lagrange multiplier.

$$dL = \frac{\partial f}{\partial \boldsymbol{\beta}} d\boldsymbol{\beta} + \psi \frac{\partial g}{\partial \boldsymbol{\beta}} d\boldsymbol{\beta} = \left(\frac{\partial f}{\partial \boldsymbol{\beta}} + \psi \frac{\partial g}{\partial \boldsymbol{\beta}} \right) d\boldsymbol{\beta} \tag{4.152}$$

The differential $dL = 0$ is required for an extremum, so

$$\psi = \frac{-\left(\dfrac{\partial g}{\partial \boldsymbol{\beta}}\right)^T \dfrac{\partial f}{\partial \boldsymbol{\beta}}}{\left(\dfrac{\partial g}{\partial \boldsymbol{\beta}}\right)^T \left(\dfrac{\partial g}{\partial \boldsymbol{\beta}}\right)} \tag{4.153}$$

4.9.2.2 The Continuous Form of IHCP

In this section, the measured temperatures are considered a continuous function of time, $y(t)$, the unknown heat flux a continuous function of time, $q(t)$, and the model equation for temperature a continuous function of time and space, $T(x,t)$. The sum of squares objective function becomes a continuous integral from $t = 0$ until the final time $t = t_f$: (Note this is a function (S) of a function ($q(t)$), and is called a *functional*.)

$$S = \int_{t=0}^{t=t_f} \left(y_{x_m}(t) - T(x_m, t; q(t)) \right)^2 dt \tag{4.154a}$$

If $q(t)$ is approximated using known basis functions, as suggested in Eq. (4.146), or as pointwise approximations of the function value at each measurement time, then the parameterized functional can be considered dependent on the vector $\mathbf{q}$:

$$S = \int_{t=0}^{t=t_f} \left(y_{x_m}(t) - T(x_m, t; \mathbf{q}) \right)^2 dt \tag{4.154b}$$

The model equation for $T(x,t)$ is the solution to the partial differential equation

$$\frac{\partial}{\partial x} \left(k \frac{\partial T}{\partial x} \right) = C \frac{\partial T}{\partial t} \tag{4.155a}$$

with the following boundary and initial conditions:

$$-k \frac{\partial T}{\partial x} \bigg|_{x=0} = q(t) \tag{4.155b}$$

$$\frac{\partial T}{\partial x} \bigg|_{x=L} = 0 \tag{4.155c}$$

$$T(x, 0) = 0 \tag{4.155d}$$

4.9.2.3 The Sensitivity Problem

The "sensitivity problem" results from taking the *variation* of the governing equation and boundary conditions in Eqs. (4.155a)–(4.155d). The variation in functional calculus is analogous to the differential in variable calculus. The "sensitivity problem" (or, rather, its solution), relates changes in the dependent function $T(x,t;q(t))$ to changes in the unknown function $q(t)$. The sensitivity problems can be derived in a formal manner by application of its definition, which is similar to the definition of the derivative (Eq. 4.148) in variable calculus. Let $q(t)$ be perturbed by an amount Δq, which will result in a perturbation ΔT. By substituting $q + \Delta q$ and $T + \Delta T$ into Eqs. (4.155a)–(4.155d), and subtracting the original equation (Eqs. (4.155a)–(4.155d)) and discarding any higher order terms (Δq^2, $\Delta q \Delta T$, etc.), the variation of the problem in Eqs. (4.155a)–(4.155d) is obtained. However, the same result can be found more directly by taking the variation of the terms in Eqs. (4.155a)–(4.155d). The resulting sensitivity problem for T is

$$\frac{\partial}{\partial x} \left(k \frac{\partial (\Delta T)}{\partial x} \right) = C \frac{\partial (\Delta T)}{\partial t} \tag{4.156a}$$

with the following boundary and initial conditions:

$$-k \frac{\partial (\Delta T)}{\partial x} \bigg|_{x=0} = \Delta q(t) \tag{4.156b}$$

$$\frac{\partial (\Delta T)}{\partial x} \bigg|_{x=L} = 0 \tag{4.156c}$$

$$(\Delta T)(x, 0) = 0 \tag{4.156d}$$

This problem is driven by the variation in q, Δq. The solution of this problem is necessary in the iterative AM solution (see Section 4.9.2.6). However, the conditions imposed in the boundary and initial conditions are used in the derivation below to tease out the desired gradient from the Lagrangian.

4.9.2.4 The Lagrangian and the Adjoint Problem

The minimum of the function S in Eq. (4.154a) is sought subject to the constraint that the temperature field satisfies Eqs. (4.155a)–(4.155d). The Lagrangian functional, L, is constructed using the Lagrange multiplier $\psi(x,t)$.

$$L = S(T(x_m,\ t;\mathbf{q})) + \int\limits_{0}^{t_f} \int\limits_{x=0}^{x=L} \psi(x,\ t)\left(\frac{\partial}{\partial x}\left(k\frac{\partial T}{\partial x}\right) - C\frac{\partial T}{\partial t}\right)dxdt \tag{4.157}$$

It is important to recognize that when the computed temperature field $T(x,t)$ satisfies the constraint (i.e. is a solution of Eqs. (4.155a)–(4.155d), which is always the case), then the Lagrangian is identical to the objective (sum of squares) functional. This fact will be used later to identify the gradient of the objective function, ∇S.

As in variable calculus, the necessary condition for extremum is that the differential (here, the variation) must be zero. The Lagrange multiplier (also called the adjoint variable) is fixed and not dependent on $\mathbf{q}$ or T, therefore its variation is zero. Using familiarity with calculus differential rules, the variation of L can be written as

$$\Delta L = \Delta S(T(x_m,\ t;\mathbf{q})) + \int\limits_{0}^{t_f} \int\limits_{x=0}^{x=L} \psi(x,\ t)\left(\frac{\partial}{\partial x}\left(k\frac{\partial \Delta T}{\partial x}\right) - C\frac{\partial \Delta T}{\partial t}\right)dxdt \tag{4.158}$$

The first term in Eq. (4.158) can be evaluated easily, again using familiarity with calculus differentials:

$$\Delta S = -2\int\limits_{t=0}^{t_f} \left(y_{x_m}(t) - T(x_m,\ t;\mathbf{q})\right)(\Delta T(x_m,\ t))dt \tag{4.159}$$

Here, $\Delta T(x_m,t)$ is the *directional derivative of T in the direction of* Δq – the solution of the sensitivity problem (Eqs. (4.156a)–(4.156d) – this derivative is also called the *Frechet derivative*.

The remaining term is Eq. (4.158) is transformed using several integrations by parts in order to bring in the boundary and initial conditions from the sensitivity problem. The goal of these manipulations is to transfer derivatives onto the adjoint variable and allow identification of conditions $\psi(x,t)$ must satisfy. Integrating by parts twice on x in the first term, and by parts once on t in the last term provides the following result:

$$\int\limits_{0}^{t_f} \int\limits_{x=0}^{x=L} \psi(x,\ t)\left(\frac{\partial}{\partial x}\left(k\frac{\partial \Delta T}{\partial x}\right) - C\frac{\partial \Delta T}{\partial t}\right)dxdt =$$

$$\int\limits_{0}^{t_f} \int\limits_{x=0}^{x=L} \Delta T\left(\frac{\partial}{\partial x}\left(k\frac{\partial \psi}{\partial x}\right) + C\frac{\partial \psi}{\partial t}\right)dxdt + \int\limits_{0}^{t_f}\left(\left.\psi k\frac{\partial \Delta T}{\partial x}\right]_0^L - \left.\left(k\frac{\partial \psi}{\partial x}\Delta T\right]_0^L\right)dt \tag{4.160}$$

$$- \int\limits_{x=0}^{x=L} (\psi C\Delta T]_0^{t_f}\,dx$$

Some of the terms emanating from the integration by parts can be evaluated using the boundary and initial condition terms from the sensitivity problem (Eqs. (4.156b)–(4.156d)). Introducing these terms into (4.160) gives the result

$$\int\limits_{0}^{t_f} \int\limits_{x=0}^{x=L} \psi(x,\ t)\left(\frac{\partial}{\partial x}\left(k\frac{\partial \Delta T}{\partial x}\right) - C\frac{\partial \Delta T}{\partial t}\right)dxdt =$$

$$\int\limits_{0}^{t_f} \int\limits_{x=0}^{x=L} \Delta T\left(\frac{\partial}{\partial x}\left(k\frac{\partial \psi}{\partial x}\right) + C\frac{\partial \psi}{\partial t}\right)dxdt + \int\limits_{0}^{t_f}\psi(0,\ t)\Delta\mathbf{q} - \left.\left(k\frac{\partial \psi}{\partial x}\Delta T\right]_0^L\,dt \tag{4.161}$$

$$- C\int\limits_{x=0}^{x=L} \psi\left(x,\ t_f\right)\Delta T\left(x,\ t_f\right)dx$$

The variation of the Lagrangian in Eq. (4.158) can now be expressed using the results from Eqs. (4.159) and (4.161), with some rearrangement, as

$$\Delta L = \int\limits_{0}^{t_f} \int\limits_{x=0}^{x=L} \Delta T \left(\frac{\partial}{\partial x} \left(k \frac{\partial \psi}{\partial x} \right) + C \frac{\partial \psi}{\partial t} - 2 \big(y_{x_m}(t) - T(x,\ t; \mathbf{q}) \big) \delta(x - x_m) \right) dx dt +$$

$$\int\limits_{0}^{t_f} \psi(0, t) \Delta \mathbf{q} - \left(k \frac{\partial \psi}{\partial x} \Delta T \right]_{0}^{L} dt - C \int\limits_{x=0}^{x=L} \psi(x, t_f) \Delta T(x, t_f) dx \qquad (4.162)$$

where the relation $T(x_m, t; \mathbf{q}) = T(x, t; \mathbf{q}) \delta(x - x_m)$ has been introduced (δ is the Dirac delta function).

The variation must vanish as a necessary condition for an extremum. To force this variation to zero, a portion of it is defined as the *adjoint problem*, which is used to compute the adjoint variable (Lagrange multiplier) $\psi(x,t)$. Let

$$\frac{\partial}{\partial x} \left(k \frac{\partial \psi}{\partial x} \right) + C \frac{\partial \psi}{\partial t} - 2 \big(y_{x_m}(t) - T(x, t; \mathbf{q}) \big) \delta(x - x_m) = 0 \qquad (4.163a)$$

$$\left. \frac{\partial \psi}{\partial x} \right|_{x=0} = 0 \qquad (4.163b)$$

$$\left. \frac{\partial \psi}{\partial x} \right|_{x=L} = 0 \qquad (4.163c)$$

$$\psi(x, t_f) = 0 \qquad (4.163d)$$

Note that the solution of the Eqs. (4.163a)–(4.163d) is driven entirely by the residual error, $y_{x_m}(t) - T(x_m, t; \mathbf{q})$, so, when this residual is zero (when T matches y), then the adjoint variable vanishes. Notice also that Eqs. (4.163a)–(4.163d) are solved backward in time from the final condition (Eq. 4.163d) to zero. This can be accomplished using standard techniques by letting $\tau = t_f - t$ and integrating from $\tau = 0$ to $\tau = t_f$.

4.9.2.5 The Gradient Equation

One final piece of this mathematical puzzle reveals the desired result: calculation of the gradient of the sum of squares functional (Eqs. 4.154a and 4.154b) with respect to the vector $\mathbf{q}$.

When the conditions of Eqs. (4.163a)–(4.163d) are satisfied (when the adjoint problem is solved), then the variation of the Lagrangian (Eq. 4.162) reduces to

$$\begin{aligned} \Delta L &= \int\limits_{0}^{t_f} \psi(0, t) \Delta q(t)\ dt \\ &= \int\limits_{0}^{t_f} \psi(0, t) \Delta \mathbf{q}\ dt \\ &= \langle \psi(0,\ t), \Delta \mathbf{q} \rangle \end{aligned} \qquad (4.164)$$

Recall that, whenever the constraint equation (Eq. 4.155a) is satisfied (which is true when T satisfies the governing equations), then the Lagrangian is identical to the sum of squares functional (see Eq. (4.157)). With this recognition, rewrite Eq. (4.164) as

$$\Delta S = \langle \psi(0,\ t), \Delta \mathbf{q} \rangle \qquad (4.165)$$

which is the inner product of "something" and the variation of $q(t)$ (which has been parameterized as the vector $\mathbf{q}$). Analogous to the definition of the differential in variable calculus (Eq. 4.149)), the definition of the variation (directional derivative) of a functional is the inner product of the gradient of the function and a unit vector in a direction. The "direction" of $\Delta \mathbf{q}$ is the same as that of a unit vector in that direction, so the inner product in Eq. (4.164) can be identified as

$$\Delta S = \langle \nabla S, \Delta \mathbf{q} \rangle = \langle \psi(0,\ t), \Delta \mathbf{q} \rangle \qquad (4.166)$$

At last, the gradient of S can be identified by comparison of the two expressions in Eq. (4.166) as

$$\nabla S = \psi(0, t) \qquad (4.167)$$

4.9.2.6 Summary of IHCP solution by Adjoint Method
1) Establish basis functions to characterize $q(t)$ (viz. Eq. (4.155b); piecewise constant, etc.))
2) Make an initial guess for $\mathbf{q}^0$
3) Solve the forward problem using $\mathbf{q}^k$
4) Compute the residual vector
5) Solve the adjoint problem
6) Compute ∇S and the descent direction $\mathbf{w}^k$
7) Solve the sensitivity problem for ΔT
8) Compute optimal step size ρ^k
9) Update $\mathbf{q}^k$
10) Monitor convergence, return to step 3 as required.

4.9.2.7 Comments About Adjoint Method
The AM is powerful and robust and can be applied to many parameter or function identification problems.

The complex derivation leads to a method for evaluating the gradient of S without resorting to multiple finite difference computations.

The method never requires computation of the Jacobian (the $\mathbf{X}$ matrix from the linear problem, $X_{i,j} = \partial T_i/\partial q_j$).

The final condition for the adjoint variable (Eq. (4.163d)) is zero, and the adjoint variable at the active surface is the gradient of the S functional. This means the gradient of S remains zero at the final time, which in turn means there can be no change in the value of q_n at the final time (see Eqs. (4.130), (4.131), and (4.134)). Therefore, a judicious choice for the initial guess of the heat flux at the final time is needed, since this value will not change during iterations. This can be done, for example, if data are taken beyond the time of heating at the active surface, when the surface heat flux has stopped and has fallen to zero. Ozisik and Orlande (2021, p. 149) recommend extending the time frame for analysis beyond the time of interest by taking data longer than needed.

Example 4.9 Solve Example 4.3 using the AM. Use Crank–Nicolson finite difference formulation to solve the forward, adjoint, and sensitivity problems. Use a pointwise variation for the unknown heat flux and perform five iterations using $\mathbf{q}^0 = 200$.

Solution
The solution is obtained using the Crank–Nicolson algorithm with a coarse spatial mesh using 11 nodes ($\Delta x = 0.10$). The measurement location is $x_{\text{meas}} = 1.0$. For each of the five iterations of the adjoint/CGM method, the temperature at the measurement location, the solution for the adjoint variable at the active surface (which is the gradient ∇S; see Eq. (4.167)), and the sensitivity at the measurement location are obtained. These values are presented in Table 4.7 for each of the discrete measurement times ti = 0.5, 1.0, 1.5, and 2.0. As seen in Table 4.7, the largest change in any of these solution variables follows the first iteration, and there is little change in any these on subsequent iterations.

Table 4.8 shows the iteration results for the Fletcher–Reeves CGM used with the AM for this example. Consistent with the solution variables in Table 4.7, the first iteration produces the largest change in the q vector, with only small changes in subsequent iterations.

The values obtained for the q vector are 32.3, 105.5, 175.9, and 200 W/m2. These values are at the times $t_i = 0.5$, 1.0, 1.5, and 2.0s. The exact values of heat flux which were used to generate the data are 50, 100, 150, and 200 W/m2.

Table 4.7 Iteration results for adjoint method variables.

k	$T(x_{\text{meas}}, t_i)$				$\psi^k(0, t_i)$				$\Delta T(x_{\text{meas}}, t_i)$			
0	27.5	112.2	218.1	316.0	14.3	8.0	2.0	0.0	−1.97	−7.16	−11.2	−12.1
1	4.6	28.7	87.0	174.9	0.14	0.22	0.08	0.0	−0.02	−0.09	−0.18	−0.22
2	4.5	28.4	86.3	174.1	0.06	0.18	0.07	0.0	−0.02	−0.10	−0.21	−0.27
3	4.5	28.3	86.1	173.8	0.04	0.16	0.07	0.0	−0.02	−0.12	−0.28	−0.36
4	4.5	28.2	86.0	173.7	0.03	0.16	0.06	0.0	−0.02	−0.14	−0.34	−0.45

Table 4.8 Iteration results for adjoint method updates using Fletcher–Reeves conjugate gradient.

k	S^k	w^k				ρ^k	q^k			
0	44 697	−14.3	−8.0	−2.0	0.0	11.6648	200	200	200	200
1	80.7	−0.14	−0.22	−0.08	0.0	3.6940	33.1	106.7	176.4	200
2	79.5	−0.14	−0.30	−0.11	0.0	0.8443	32.6	105.9	176.1	200
3	79.4	−0.15	−0.41	−0.16	0.0	0.3846	32.4	105.6	176.0	200
4	79.3	−0.17	−0.52	−0.21	0.0	0.2354	32.4	105.5	175.9	200
							32.3	105.3	175.9	200

Discussion

Note in Table 4.7 that the value of the gradient $\nabla S = \psi^k(0, t_i)$ is zero at the final time, as was mentioned previously. This means that the value of q at the final time will not change with iterations, as is seen in Table 4.8. The "judicious choice" of the correct value of $q(t_f)$ as the initial guess provided the results shown. If, instead, the initial guess of $\mathbf{q} = 1$ was used, then the results q = {142.4, 97.7, 29.1, 1.0} are obtained! The "guess" for the final time step is critical.

4.9.3 General Comments about CGM

The CGMs provide an approach for direct solution of the ill-posed IHCP without any explicit regularization. The number of iterations for CGM solution is considered the regularizing parameter. The appropriate number of iterations is generally chosen based on the Discrepancy Principle (see Chapter 6).

For linear problems, when the gradient of the sum of squares objective function can be found explicitly, the CGM can be applied directly.

When mathematical access to the gradient ∇S is difficult, or in solution of nonlinear problems, the AM can be used to obtain the required gradient for CGM iterations. For the IHCP, the AM contains the peculiarity that the component of the gradient at the final time is zero. This poses challenges in resolution of the unknown heating function at the end of the data time interval.

4.10 Truncated Singular Value Decomposition Method

SVD is a numerical procedure for factoring a matrix into three parts to solve a system of linear algebraic equations. SVD can be applied to overdetermined systems (more data than equations) or under-determined systems (more equations than data), but the determined case (n equations and n unknowns) is considered here. Many texts on numerical methods can provide information on SVD procedures (see, for example, Press et al. 2007), and SVD subroutines or functions are commonly available in scientific computing packages, including MATLAB and Mathematica.

4.10.1 SVD Concepts

For a matrix $\mathbf{X}$, the SVD factorization can be written as

$$\mathbf{X} = \mathbf{U}\mathbf{W}\mathbf{V}^{\mathbf{T}} \tag{4.168}$$

In Eq. (4.168), $\mathbf{U}$ and $\mathbf{V}$ are column orthogonal matrix (such that $\mathbf{U}^{\mathbf{T}} \times \mathbf{U} = \mathbf{I}$ and $\mathbf{V}^{\mathbf{T}} \times \mathbf{V} = \mathbf{I}$). $\mathbf{W}$ is a diagonal matrix. The advantage of SVD factorization is that the inverse of $\mathbf{X}$ can be easily computed as:

$$\mathbf{X}^{-1} = \mathbf{V}\mathbf{W}^{-1}\mathbf{U}^{T} \tag{4.169}$$

Since **W** is a diagonal matrix, its inverse can be computed simply by taking the (scalar) inverse of each of the diagonal elements:

$$\mathbf{W}^{-1} = \begin{bmatrix} 1/w_{11} & 0 & \cdots & 0 \\ 0 & 1/w_{22} & \cdots & 0 \\ 0 & \cdots & \ddots & 0 \\ 0 & 0 & \cdots & 1/w_{nn} \end{bmatrix} \tag{4.170}$$

Some authors and practitioners prefer to work with a modified form of the discrete Eq. (4.16) with pre-multiplication by the matrix $\mathbf{X}^T$:

$$\mathbf{X}^T\mathbf{T} = \mathbf{X}^T\mathbf{X}\mathbf{q} \tag{4.171}$$

In this case, the $\mathbf{X}^T\mathbf{X}$ becomes the matrix which requires SVD and inversion. The present discussion will consider the basic form of Eq. (4.16) and inverse of $\mathbf{X}^{-1}$ in Eq. (4.169).

4.10.2 TSVD in the IHCP

Like the CGM, the SVD procedure can be used in the IHCP to invert Eq. (4.128) by directly matching the measured temperatures (**Y**) to the model-predicted values (**Xq**):

$$\hat{\mathbf{q}} = \mathbf{X}^{-1}\mathbf{Y} \tag{4.172}$$

The ill-posedness of the IHCP is asserted in the structure of the matrix **W**. The elements of **W** are in descending order, and those near the end that are zero, or nearly zero, are called *singular values*. By excluding the smallest of the singular values, an approximation to the inverse of the ill-conditioned matrix **X** can be obtained.

TSVD is the name given to this approach to solving the IHCP. Note that shrinking the size of the matrix $\mathbf{W}^{-1}$, the dimensions of **V** and $\mathbf{U}^T$ must be reduced accordingly. That is, if $\mathbf{W}^{-1}$ becomes an $(n - r) \times (n - r)$ matrix, then the rightmost r columns of **V** and the last r rows of $\mathbf{U}^T$ are excluded. This effect is accomplished by leaving zero entries for the last r rows of the $n \times n$ matrix $\mathbf{W}^{-1}$.

An excellent source for additional information on SVD methods for inverse problems is Hansen (1987, 1998).

Example 4.10 Solve Example 4.3 using SVD. Use piecewise constant assumption for the unknown heat flux.

Solution

The X matrix for this solution is the same as that in Example 4.8 and is given in Eq. (4.145). The condition number for this matrix is 11.4, which indicates that the matrix is not very ill-conditioned. Using MATLAB's "svd" function, the SVD component matrices are

$$\mathbf{U} = \begin{bmatrix} -0.1733 & -0.4818 & 0.6601 & -0.5496 \\ -0.4071 & -0.6503 & -0.0491 & 0.6394 \\ -0.5845 & -0.0810 & -0.6315 & -0.5030 \\ -0.6801 & 0.5817 & 0.4039 & 0.1896 \end{bmatrix} \tag{4.173}$$

$$\mathbf{W} = \begin{bmatrix} 1.3135 & 0 & 0 & 0 \\ 0 & 0.4042 & 0 & 0 \\ 0 & 0 & 0.2048 & 0 \\ 0 & 0 & 0 & 0.1155 \end{bmatrix} \tag{4.174}$$

$$\mathbf{V} = \begin{bmatrix} -0.6801 & -0.5817 & 0.4039 & -0.1896 \\ -0.5845 & 0.0810 & -0.6315 & 0.5030 \\ -0.4071 & 0.6503 & -0.0491 & -0.6394 \\ -0.1733 & 0.4818 & 0.6601 & 0.5496 \end{bmatrix} \tag{4.175}$$

The inverse of the matrix $\mathbf{X}$ can be found using the SVD inversion

$$\mathbf{X}^{-1} = \mathbf{V}\mathbf{W}^{-1}\mathbf{U}^T = \begin{bmatrix} 2.9869 & 0 & 0 & 0 \\ -4.4480 & 2.9869 & 0 & 0 \\ 2.1629 & -4.4480 & 2.9869 & 0 \\ -1.0390 & 2.1629 & -4.4480 & 2.9869 \end{bmatrix} \tag{4.176}$$

The heat fluxes are then found using the data $\mathbf{Y}$:

$$\hat{\mathbf{q}} = \mathbf{X}^{-1}\mathbf{Y} = \lfloor 18.2 \quad 78.3 \quad 123.4 \quad 175.8 \rfloor^T \tag{4.177}$$

The exact values are 25, 75, 125, and 175 W/m^2, with relative errors of 27.2, −4.3, 1.3, and −0.4%.

Discussion

The condition number of the $\mathbf{X}$ matrix is relatively small owing to the large data interval $\Delta t = 0.5$ s, and the corresponding dimensionless time interval of $\Delta \tilde{t} = 0.5$. The condition number is seen in the $\mathbf{W}$ matrix, as the ratio of the largest eigenvalue (1.3035) to the smallest eigenvalue (0.1155) of 11.37. The matrix can be inverted by any means, and there is no need for truncating the values.

If the time step is reduced to $\Delta t = 0.125$ s, the $\mathbf{X}$ matrix becomes noticeably ill-conditioned. The 16×16 matrix has a condition number of 1.38E+10, and the smallest eigenvalue in $\mathbf{W}$ is 8.696E−11. If direct inversion is attempted without removing any singular values, the resulting estimates for the heat fluxes are unrecognizable. By removing just the one smallest singular value (and all retained w_{ii} values are 0.0139 or larger), the resulting modified matrix has a condition number of 86.8, and a solution for the heat fluxes is readily obtained.

Table 4.9 shows the results obtained when the time step is 0.125 and one singular value is deleted. The resulting heat flux estimates are very good over the main time interval, but are noticeably erroneous at the end of the time span. IHCP solutions typically have difficulty at the end of the time span owing to the parabolic nature of the heat conduction process and the lack of "future" data.

4.10.3 General Remarks About TSVD

TSVD is a method to solve the IHCP using exact matching to measured data. Regularization is introduced into the solution through removal of the smallest singular values of the coefficient matrix ($\mathbf{X}$). The result is an approximation to the inverse of $\mathbf{X}$ that can be used to directly solve the system of equations.

4.11 Kalman Filter

In Chapter 5, the idea of an IHCP filter is introduced. This concept is separate from the other notions of the term, but within the context of IHCP problems often elicits association with the Kalman filter.

Kalman filtering was introduced in the early 1960s (Kalman 1960) and initially used primarily to estimate the state (location and velocity) of orbital and atmospheric objects using ground-based tracking measurements. The concept has grown in

Table 4.9 TSVD results when $\Delta t = 0.125$ deleting one singular value.

t_i	0.125	0.250	0.375	0.500	0.625	0.750	0.875	1.00
$\hat{q}_i$	4.65	19.48	31.00	43.83	56.23	68.76	81.25	93.74
q_{exact}	6.25	18.75	31.25	43.75	56.25	68.75	81.25	93.75
e_{rel} (%)	25.6	−3.9	0.8	−0.2	0.0	0.0	0.0	0.0
t_i	1.125	1.250	1.375	1.500	1.625	1.75	1.875	2.00
$\hat{q}_i$	106.27	118.67	131.53	142.77	159.62	157.08	221.64	54.01
q_{exact}	106.25	118.75	131.25	143.75	156.25	168.75	181.25	193.75
e_{rel} (%)	0.0	0.1	−0.2	0.7	−2.2	6.9	−22.3	72.1

application over the decades and is applied to data smoothing and filtering as well as state estimation. Though initially applied to linear problems, the basic method has evolved and can be applied to nonlinear problems.

References on KF are too numerous to give a reasonable list. Two suggested sources are Candy (1986) and Crassidis and Junkins (2004).

This section will provide a summary of KF as applied to linear estimation problems. A short discussion about two approaches to application of KF to solution of the IHCP will be given, followed by a more detailed exposition of the approach used by Scarpa and Milano (1995).

4.11.1 Discrete Kalman Filter

Kalman filter can be considered a sequential predictor–corrector integration method to advance the state of a system from time step $k - 1$ to time step k. The state of the system classically is the position and velocity of a moving object, but in the context of heat conduction may be considered temperatures.

A model of the system and measurements must be available in order to apply the method. Only discrete models are considered here. The classic notion of a *state transition matrix*, $\mathbf{\Phi}$, is used to advance the model states from one time step to the next, and an *observation matrix*, $\mathbf{H}$, is used to extract the model-predicted measurements from the state vector. Both of these relations are assumed to be contaminated with some zero-mean Gaussian noise.

The state of the system can be advanced from one time $k - 1$ to the next time step k using

$$\mathbf{x}_k = \mathbf{\Phi}_{k-1}\mathbf{x}_{k-1} + \Gamma_{k-1}\mathbf{u}_{k-1} + \mathbf{w}_{k-1} \tag{4.178}$$

$$\mathbf{y}_k = \mathbf{H}_k\mathbf{x}_k + \mathbf{v}_k \tag{4.179}$$

Here, $\mathbf{x}$ are the state variables which, in a conduction problem, might be the discrete values of temperature at computational points across a slab as might naturally arise from a numerical computation. The measurements $\mathbf{y}$ are linear combinations of the states, and typically in a heat conduction problem the matrix $\mathbf{H}$ is diagonal with only a few entries (or perhaps just one) of "1" and the remainder 0's. The vector $\mathbf{u}_{k-1}$ is a vector of known system inputs at time $k - 1$.

The measurement noise is $\mathbf{v}$ and the *process noise* is $\mathbf{w}$, and each are assumed to be Gaussian with zero mean and covariance of

$$E(\mathbf{v}\mathbf{v}^T) = E(v_i v_j) = \begin{cases} 0 & i \neq j \\ R^2 & i = j \end{cases} \tag{4.180a}$$

$$E(\mathbf{w}\mathbf{w}^T) = E(w_i w_j) = \begin{cases} 0 & i \neq j \\ Q^2 & i = j \end{cases} \tag{4.180b}$$

The current estimate of the state is $\hat{\mathbf{x}}_{k-1}$, and it is desired to update the estimate to $\hat{\mathbf{x}}_k$ when new measurements $\mathbf{y}_k$ become available. This will be done in two stages: a "predictor" step, based only on information from the previous time step, followed by a "corrector" step, which introduces measurement data from the current time step k.

Using Eq. (4.178) and the known information from step $k - 1$:

$$\hat{\mathbf{x}}_{k|k-1} = \mathbf{\Phi}_{k-1}\hat{\mathbf{x}}_{k-1|k-1} + \Gamma_{k-1}\mathbf{u}_{k-1} \tag{4.181a}$$

Here, $\hat{\mathbf{x}}_{k|k-1}$ is the prediction of the system's state at the new time step based only on information at hand from the previous time step. From this first-step prediction, a corresponding estimate for the measurement $\hat{\mathbf{y}}_{k|k-1}$ can be found using the measurement model Eq. (4.179). When a new measurement $\mathbf{y}_k$ becomes available, the *innovation* is defined as the discrepancy between the new measurement and the observation from the predicted state:

$$\mathbf{e}_k = \mathbf{y}_k - \hat{\mathbf{y}}_{k|k-1} = \mathbf{y}_k - \mathbf{H}_{k-1}\hat{\mathbf{x}}_{k|k-1} \tag{4.181b}$$

This "innovation" is used to drive the "corrector" step using a feedback matrix $\mathbf{K}$, which will be defined later:

$$\hat{\mathbf{x}}_{k|k} = \hat{\mathbf{x}}_{k|k-1} + \mathbf{K}_k\mathbf{e}_k \tag{4.181c}$$

So,

$$\hat{\mathbf{x}}_{k|k} = (\mathbf{I} - \mathbf{H}_{k-1})\hat{\mathbf{x}}_{k|k-1} + \mathbf{K}_k\mathbf{y}_k \tag{4.181d}$$

When combined with Eq. (4.181a), Eq. (4.181d) becomes

$$\hat{\mathbf{x}}_{k|k} = (\mathbf{I} - \mathbf{K}_k \mathbf{H}_{k-1})[\boldsymbol{\Phi}_{k-1}\hat{\mathbf{x}}_{k-1} + \boldsymbol{\Gamma}_{k-1}\mathbf{u}_{k-1}] + \mathbf{K}_k \mathbf{y}_k \tag{4.181e}$$

The covariance matrix for the state vector, $\mathbf{x}$, is

$$\mathbf{P} = \mathrm{cov}(\mathbf{x}, \mathbf{x}) = E\left((\mathbf{x} - \bar{\mathbf{x}})(\mathbf{x} - \bar{\mathbf{x}})^T\right) \tag{4.182a}$$

Or

$$\mathbf{P} = \begin{bmatrix} \sigma^2_{x_1} & \sigma^2_{x_1 x_2} & \cdots & \sigma^2_{x_1 x_{n-1}} & \sigma^2_{x_1 x_n} \\ \sigma^2_{x_2 x_1} & \sigma^2_{x_2} & \sigma^2_{x_2 x_3} & \cdots & \sigma^2_{x_2 x_n} \\ \vdots & \vdots & \ddots & & \vdots \\ \sigma^2_{x_{n-1} x_1} & \sigma^2_{x_{n-1} x_2} & \cdots & \sigma^2_{x_{n-1}} & \sigma^2_{x_{n-1} x_n} \\ \sigma^2_{x_n x_1} & \sigma^2_{x_n x_2} & \cdots & \sigma^2_{x_n x_{n-1}} & \sigma^2_{x_n} \end{bmatrix} \tag{4.182b}$$

$$\sigma^2_{x_i x_j} = E\left((x_i - \bar{x}_i)(x_j - \bar{x}_j)\right) \tag{4.182c}$$

The diagonal elements of $\mathbf{P}$ are the variances of the state estimates, and the off-diagonal elements are the covariances of the states. Ideally, $\mathbf{P}$ is strongly diagonal for uncorrelated states.

If the covariance matrix is known at step $k-1$, a "prediction" for the covariance matrix at k can be formed using the state transition matrix $\boldsymbol{\Phi}$ by ignoring the input at step k (see Eq. (4.178)):

$$\mathbf{P}_{k|k-1} = \boldsymbol{\Phi}_{k-1}\mathbf{P}_{k-1|k-1}\boldsymbol{\Phi}^T_{k-1} + \mathbf{Q} \tag{4.183a}$$

where $\mathbf{Q}$ is the diagonal matrix of the variance of the noise $\mathbf{w}$ in Eq. (4.178).

The *measurement* covariance is obtained from the state covariance using the observation matrix $\mathbf{H}$ (see Eq. (4.179) relating measurements to states), and recognizing the variance of the measurement noise $\mathbf{R}$:

$$\mathbf{S}_k = \mathbf{H}_k \mathbf{P}_{k|k-1} \mathbf{H}^T_k + \mathbf{R} \tag{4.183b}$$

The gain of the Kalman filter, $\mathbf{K}_k$, is found by minimizing the trace (diagonal elements) of the covariance matrix $\mathbf{P}$. The result is:

$$\mathbf{K}_k = \mathbf{P}_{k|k-1}\mathbf{H}^T_k + \mathbf{S}^{-1}_k \tag{4.184}$$

Then, the updated covariance matrix is

$$\mathbf{P}_{k|k} = (\mathbf{I} - \mathbf{K}_k \mathbf{H}_k)\mathbf{P}_{k|k-1} \tag{4.185}$$

The Kalman filter procedure is summarized as follows:

1) Use information from the previous time step $k-1$ to estimate the state $\mathbf{x}_{k|k-1}$ at the next time step k using Eq. (4.181a).
2) Use information from the previous time step $k-1$ to estimate the covariance matrix $\mathbf{P}_{k|k-1}$ using Eq. (4.183b).
3) Calculate the Kalman gain matrix $\mathbf{K}_k$ using Eqs. (4.183b) and (4.184).
4) On receipt of a new measurement $\mathbf{y}_k$, compute the error in the estimate $\mathbf{e}_k$ using Eq. (4.181b) and correct the state estimate using Eq. (4.181c) (or, equivalently, using Eq. (4.181e)).
5) Update the covariance matrix using Eq. (4.185).

4.11.2 Two Concepts for Applying Kalman Filter to IHCP

The most direct approach to solving IHCP using a Kalman filter is to consider unknown boundary condition information as additional components of the state vector. In this case, the unknown surface heat flux component at the next time step, q_k, will be included in the $\mathbf{x}_k$ vector along with the temperatures $\mathbf{T}_k$. This approach was perhaps first applied by Matsevity and Moultanovsky (1978) and later amplified by Scarpa and Milano (1995). This methodology is presented in Section 4.11.3.

A second approach for applying a Kalman filter for solution to the IHCP considers the unknown heat flux, q_k, as an input to the system through the vector $\mathbf{u}$ in Eq. (4.178). Since the vector $\mathbf{u}$ is not the object of the Kalman analysis (the state vector $\mathbf{x}$ is the desired output, and $\mathbf{u}$ is regarded as a known input), modification of the procedure is required. A recursive least-squares procedure for estimating q_k is coupled with the Kalman filter in this approach. Ji et al. (1997) and Tuan et al. (1997a,b) introduced this idea which has been applied by other researchers such as Noh et al. (2015). This methodology will not be presented in this text.

4.11.3 Scarpa and Milano Approach

This presentation follows the approach to Kalman Filter solution of the IHCP proposed by Scarpa and Milano (1995). The model for the state transition follows from a finite difference discretization of the computational domain. For example, for a fully implicit scheme

$$[\mathbf{A}_c]\{\mathbf{T}^k\} = [\mathbf{B}]\{\mathbf{T}^{k-1}\} + [\mathbf{C}]q^k \tag{4.186a}$$

So, the state transition model in the form of Eq. (4.178) is

$$\{\mathbf{T}^k\} = \underbrace{[\mathbf{A}_c]^{-1}[\mathbf{B}]}_{\Phi_{k-1}}\{\mathbf{T}^{k-1}\} + [\mathbf{A}_c]^{-1}\{\mathbf{C}\}q^k \tag{4.186b}$$

Notice that for this fully implicit formulation there is no dependence on the heat flux from the $k-1$ time step; this will be different if a Crank–Nicolson type scheme is employed. However, for either case, q^k is interpreted as the average heat flux over the time interval Δt. The $\mathbf{C}$ vector is all zeros except for $C(1)$: the heat flux directly impacts only the boundary node.

In the Scarpa and Milano approach, the unknown heat flux component at each time step is folded into the state vector, $\mathbf{x}$:

$$\mathbf{x}_k = \lfloor q_k \quad T_{1,k} \quad T_{2,k} \quad \cdots \quad T_{n-1,k} \quad T_{n,k} \rfloor^T \tag{4.187a}$$

where T_1, T_2, ..., T_n are the temperatures at discrete nodes in the one-dimensional domain.

The system dynamics in Eq. (4.178) are rewritten as

$$\mathbf{x}_k = \mathbf{A}_{k-1}\mathbf{x}_{k-1} \tag{4.187b}$$

where the matrix $\mathbf{A}$ is partitioned as

$$\mathbf{A} = \begin{bmatrix} 1 & 0 \\ \hline \mathbf{b} & \Phi \end{bmatrix} \tag{4.187c}$$

In Eq. (4.187c), the $\mathbf{b}$ vector is $n \times 1$ and reflects the influence of the surface heat flux q_k on the energy balance at each node. In terms of Eq. (4.186b), $\mathbf{b} = [\mathbf{A}]^{-1}\{\mathbf{C}\}$.

4.11.3.1 Kalman Filter

The predictor steps are analogous to Eqs. (4.181a) and (4.183a):

$$\hat{\mathbf{x}}_{k|k-1} = \mathbf{A}\hat{\mathbf{x}}_{k-1|k-1} \tag{4.188a}$$

$$\mathbf{P}_{k|k-1} = \mathbf{A}\mathbf{P}_{k-1|k-1}\mathbf{A}^T + \mathbf{Q}_w \tag{4.188b}$$

Here, the process noise matrix $\mathbf{Q}_w$ is assigned structure as

$$\mathbf{Q}_w = \begin{bmatrix} Q_q^2 & 0 & \cdots & 0 \\ 0 & 0 & & 0 \\ \vdots & & \ddots & \vdots \\ 0 & 0 & \cdots & 0 \end{bmatrix} \tag{4.188c}$$

The correction steps use the Kalman gain matrix (combining Eqs. (4.184) and (4.183b)), and the new measurement $\mathbf{y}_k$:

$$\mathbf{K}_k = \mathbf{P}_{k|k-1}\mathbf{H}_k^T + \left[\mathbf{H}_k\mathbf{P}_{k|k-1}\mathbf{H}_k^T + \mathbf{R}_v\right]^{-1} \tag{4.189a}$$

$$\hat{\mathbf{x}}_{k|k} = \hat{\mathbf{x}}_{k|k-1} + \mathbf{K}_k\left(\mathbf{y}_k - \mathbf{H}_k\hat{\mathbf{x}}_{k|k-1}\right) \tag{4.189b}$$

$$\mathbf{P}_{k|k} = [\mathbf{I} - \mathbf{K}_k\mathbf{H}_k]\mathbf{P}_{k|k-1} \tag{4.189c}$$

Figure 4.8 Scarpa and Milano results for varying process noise parameter. No smoothing is applied ($r = 0$). $\sigma_{noise} = 0.0025$.

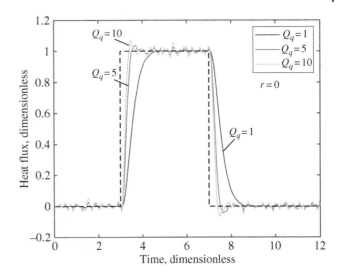

$\mathbf{R}_v$ is the measurement noise matrix and is diagonal with

$$\mathbf{R}_v = \sigma_v^2 \mathbf{I} \tag{4.189d}$$

Regularization in the Scarpa and Milano approach can be controlled through the process noise parameter, Q_q, in Eq. (4.188c). Figure 4.8 shows the results from a test problem solved with this technique. The exact heat flux (shown in Figure 4.8) is a step heat flux beginning at dimensionless time $t = 3$ and continuing until time $t = 7$. A time step in the data of 0.05 is used and random noise with $\sigma_Y = \sigma_v = 0.0025$ is added to the "data" from the exact heat flux. Three curves are shown in Figure 4.8 corresponding to $Q_q = 1$, 5, and 10. As seen in the figure, smaller values of Q_q result in smoother curves biased toward later times. Higher values of Q_q reduce the bias, but introduce more noise in the estimated values. This is the classic trade between bias and noise found in all IHCP solutions. An arbitrary assumption must be made for the initial covariance matrix (here $\mathbf{P}_{0|0} = \mathbf{I}$ has been used) and several steps of the Kalman procedure are needed to achieve reasonable estimates for $\mathbf{P}_{k|k}$.

4.11.3.2 Smoother

The Kalman filter calculations for the surface heat flux necessarily lag behind the actual heating action owing to the damping and lagging of the temperature measurements at subsurface locations. For a real-time application, nothing can be done to improve this, but if post-processing of the results is possible, application of a smoother procedure will improve the accuracy of the estimates.

Scarpa and Milano propose adoption of the maximum likelihood estimation scheme for smoothing the Kalman results (Rauch et al. 1965). This application introduces future time information by using the Kalman results from the next r times to modify the state estimates. Consider step k as the "current" time step, and assume that Kalman estimates have been obtained for step k, $k + 1$, ..., $k + r$ for $\hat{\mathbf{x}}_{i|i-1}$, $\hat{\mathbf{x}}_{i|i}$, $\mathbf{P}_{i|i-1}$, and $\mathbf{P}_{i|i}$, for $i = k, k + 1, ..., k + r = m$. Then, the smoothed estimate of the state at step k, considering all data up to and including m, is found recursively as

$$\tilde{\mathbf{x}}_{m-i|m} = \left[\hat{\mathbf{x}}_{m-i|m-i} + \mathbf{J}_{m-i}\left(\tilde{\mathbf{x}}_{m-i+1|m} - \hat{\mathbf{x}}_{m-i+1|m-i}\right)\right], \quad i = 1,2,...,r \tag{4.190a}$$

where $\tilde{\mathbf{x}}_{k|m}$ is the smoothed estimate of the state at time k taking account of r future time steps. The gain matrix in Eq. (4.190a) is

$$\mathbf{J}_{m-i} = \mathbf{P}_{m-i|m-i}\mathbf{A}^T\mathbf{P}_{m-i+1|m-i}^{-1} \tag{4.190b}$$

Regarding the smoother operation, Scarpa and Milano (1995) comment:

> "In this particular implementation, the Kalman smoother can be perceived as a zeroth-order sequential regularization method. With respect to the classic Tikhonov regularization procedure, the functional structure is extended to take into account all the stochastic terms eventually present in the physical model. As a consequence, the smoother provides estimates of the whole state vector (flux and temperature distribution)."

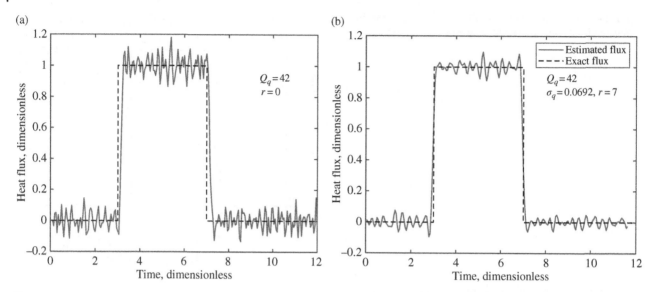

Figure 4.9 Effect of post-Kalman estimation smoothing on estimated heat flux. (a) Optimal value of R_q = 42, r = 0 gives minimum RMS error in heat flux. (b) Smoothing with r = 7 reduces RMS error in heat flux to new minimum. σ_{noise} = 0.0025.

Figure 4.9 illustrates the effect of the smoothing on the Kalman estimates. In Figure 4.9a, the regularization parameter Q_q is varied until the minimum in the RMS error between the exact (step input) and estimated heat flux is found. This corresponds to the value Q_q = 42. In Figure 4.9b, trial and error selection of r = 7 is found to further reduce the RMS error in the predicted heat flux. Comparison of the two figures shows that the smoother not only reduces fluctuations in the estimated values (reduces random error) but also reduces bias in the estimated history by effectively shifting the curve to the left. This is the result of adding future time information through the smoothing.

4.11.4 General Remarks About Kalman Filtering

KF can be applied to solution of IHCP in at least two ways. In the Scarpa and Milano approach, outlined above, the unknown heat flux component(s) at the k-th time step are folded into the state vector and estimated using the Kalman process. In another approach, the unknown heat flux components are included in the Kalman model as inputs and are estimated in parallel to the Kalman procedure using recursive least squares. This latter approach is not described further in this text.

Regularization in the Scarpa and Milano approach can be controlled through the process noise parameter, Q_q, in Eq. (4.188c). Small values of the Q_q parameter result in higher degrees of regularization, in the sense that more bias is introduced. Conversely, large values of Q_q reduce bias but introduce more noise in the estimates, which is consistent with less regularization.

Scarpa and Milano have suggested the post-estimation smoothing procedure which will help reduce the effects of noise brought on by decreasing regularization. This smoother introduces future time information for r time steps that allows the smoothing operation to reduce bias as well as noise. This smoothing operation effectively introduces r as a second regularization parameter.

4.12 Chapter Summary

The IHCP belongs to the class of ill-posed problems because the solution does not depend continuously on the data. Damping and lagging of temperature measurements from subsurface sensors contribute to this situation. To counter ill-posedness, some form of regularization must be introduced. Regularization modifies the original, ill-posed problem and transforms it into a "nearby" well-posed one.

IHCP solution techniques can be classified as sequential or whole-domain. Sequential methods rely on a temporally local subset of data and are applied repetitively over successive time steps. Whole-domain methods must have data supplied over the full time span of interest and these data are processed simultaneously.

The Stolz method and the FSM are sequential techniques. The Stolz method does not include any regularization and becomes unstable when time steps in the data are small. FSM adds regularization by considering a few future time intervals of data.

Discretization of the unknown surface disturbance in an IHCP solution can be accomplished considering the function as piecewise constant (stair-steps) or piecewise linear (connected line segments). The disturbance may be considered as either a temperature or heat flux variation.

TR is a family of whole-domain techniques. Regularization is added to the IHCP by including a weighted penalty term involving the unknown function to the sum of squares objective function. A sequential TR procedure includes features of sequential FSM and whole domain TR.

The CGM provides an iterative whole domain framework for solution of the IHCP through minimization of the sum-of-squares objective function. Regularization is provided by the iterative process.

The AM is an elegant mathematical procedure for developing the gradient of the objective function needed for the CGM. Thus, AM is an extension of the CGM which is useful for nonlinear problems or other inverse problems for which the required gradient cannot be computed easily.

The TSVD method is a whole-domain technique for directly solving the IHCP without a sum of squares objective function. Regularization is introduced by modifying the inverse of the sensitivity matrix by removing the smaller eigenvalues from the SVD procedure.

KF is used to solve the IHCP in at least two ways. In one scheme, attributed to Scarpa and Milano, the unknown heat flux is included in the state vector and estimated at each step using the Kalman procedure. In a second approach, the unknown heat flux is considered an input to the state model equation and is estimated using a parallel recursive least-squares procedure. In the first approach, regularization of the IHCP can be ascribed to the magnitude of the process noise parameter. Smaller process noise parameters result in more regularization (more bias) while larger process noise parameters reduce bias but increase sensitivity to noise. A post-estimation smoothing scheme can be used to mitigate the effect of noise introduced through regularization.

Problems

4.1 The exact solution for the X32B10T0 heat conduction problem, having a constant heat transfer coefficient at $x = 0$ and heated by a constant environment temperature T_∞, is

$$\tilde{T}(\tilde{x}, \tilde{t}) = \frac{T(x, t)}{T_\infty} = 1 - 2B \sum_{m=1}^{\infty} \frac{\left(\beta_m^2 + B^2\right) \cos\left(\beta_m(1 - \tilde{x})\right) \cos\left(\beta_m\right)}{\beta_m^2 \left(\beta_m^2 + B^2 + b\right)} e^{-\beta_m^2 \tilde{t}}$$

where $B = hL/k$, $\tilde{x} = x/L$, $\tilde{t} = \alpha t/L^2$, and β_m are the eigenvalues resulting from solution of the transcendental equation $\tan(\beta_m) = B/\beta_m$. (i) Derive expressions for the sensitivity of temperature to the Biot number, B. (ii) Compute numerical values from your solution at $x = L$ for $B = 1$, 10, and 100 and compare them to Figure 4.5.

4.2 For exact matching of the temperatures given below at $x = 0.25L$ in a steel slab, find the heat flux components $\hat{q}_1, \hat{q}_2$, and $\hat{q}_3$ in W/m^2.

i	1	2	3
t_i, s	1	2	3
Y_i, C	26.6	28	29

The initial temperature is 25 °C; the slab is 1 cm thick, insulated at $x = L$ and heated at $x = 0$; and the thermal properties are $k = 40$ W/m-C and $\alpha = 10^{-5}$ m^2/s.

4.3 Develop the gain coefficients to determine the surface temperatures using the $T =$ linear time variation assumption. Use these results and $r = 2$ to solve the same problem as Example 4.4.

4.4 An alternative integral to Duhamel's theorem is obtained by using Green's functions,

$$T(x, t) = T_0 + \frac{\alpha}{k} \int_0^t q(\lambda) G(x, t - \lambda) d\lambda$$

where $G(x, t)$ is the Green's function, $\alpha =$ thermal diffusivity, and $k =$ thermal conductivity. This can be approximated numerically using

$$T_M = T_0 + \sum_{n=1}^M q_n H_{M-n+1} \Delta t$$

where $T_M = T(x, t_M)$ and $q_M = q(t_{M-1/2})$ and

$$H_i = G\left(x, t_{i-1/2}\right) \frac{\alpha}{k}$$

Derive the following function specification algorithm for the constant q temporary assumption.

$$\hat{q}_M = \frac{\sum_{i=1}^r \left(Y_{M+i-1} - \hat{T}_{M+i-t}|_{q_M = \cdots = 0}\right) \sum_{j=1}^i G_j'}{\sum_{i=1}^r \left(\sum_{j=1}^i G_j'\right)^2}$$

$$G_j' = \frac{\alpha G_j \Delta t}{k}, \quad G_j = G\left(x, \frac{t_j - \Delta t}{2}\right)$$

What expression is used for $\hat{T}_{M+i=1}|_{q_M + \cdots = 0}$?

4.5 Calculate the gain coefficients, K_i, for $\Delta \tilde{t} = 0.05, 0.2,$ and 0.5 for the algorithm given in Problem 4.3. The Green's function for $x = L$ in a plate heated at $x = 0$ and insulated at $x = L$ is

$$G(L, t) = \frac{1}{L}\left[1 + 2 \sum_{m=1}^\infty e^{-m^2 \pi^2 \alpha t/L^2} (-1)^m\right]$$

Let α and k be unity.
Compare the values with those in Table 4.1.
a) Calculate for $r = 1, 2,$ and 3.
b) Calculate for $r = 4$.
c) Calculate for $r = 5$.

4.6 Derive a function specification algorithm for estimating $g(t)$ from two interior temperature measurement histories where $g(t)$ is the time-variable volume-energy generation term in a solid cylinder. The differential equation is

$$k \frac{\partial}{\partial r}\left(r \frac{\partial T}{\partial r}\right) + g(t) = \rho c \frac{\partial T}{\partial t} \tag{4.191}$$

Use the temporary assumption of $g(t)$ equal a constant for r future time steps. The solution of (4.191) is

$$T(r, t) = T_o + \int_0^t g(\lambda) \frac{\partial \theta(r, t - \lambda)}{\partial t} d\lambda$$

where $\theta(r, t)$ is the temperature rise for a unit step increase in $g(t)$ at time $t = 0$.

4.7 Give the matrix elements for Tikhonov Regularization, Eq. (4.108), if all three orders of regularization are included using $\alpha_0 = \alpha_1 = \alpha_2 = \alpha$.

4.8 The backward heat problem is the estimation of the initial temperature distribution in a body knowing one or more internal temperature histories and the boundary conditions. For the case of a flat plate with $T = T_0$ at $x = 0$ and $x = L$

and the initial temperature distribution $T(x, 0) = F(x)$, derive a zeroth-order whole domain regularization algorithm for estimating n components of $F(x)$,

$$F_i \equiv F\left(i\Delta x - \frac{\Delta x}{2}\right), \quad \Delta x = \frac{L}{n}$$

Consider the case of three interior sensors and m equally spaced time steps. The describing integral equation is

$$T(x, t) = T_0 + \int_0^L F(x')G(x, t, x')dx'$$

where $G(x, t, x')$ is a Green's function. Modify the notation of Problem 4.4 to permit multiple sensors.

4.9 Determine the **X** matrix for estimation of the surface temperature history from measured subsurface temperatures (the X12 formation) using piecewise constant surface temperature. Using $N = 8$ time steps and $\Delta t = 0.25$ s, with $k = 1$ W/m-C and $\alpha = 1$ m^2/s, evaluate this matrix and compute its condition number for a sensor located at $x/L = 1.0$. How does this compare with the condition number for the X22 case with the same parameters? What does this suggest about the ill-posedness of the X12 formulation compared to the X22 formulation?

4.10 Compare the piece-wise constant $q(t)$ assumption to the piece-wise linear assumption for Tikhonov Regularization. Use $\alpha_0 = 1E-6$, $1E-4$, and $1E-2$, and estimate surface heat fluxes for the data below. Use $k = 1$ W/m-C and $\alpha = 1$ m^2/s. The exact solution is $q(t) = 100t$. Compare the root mean square error for each of the three solutions. Plot the estimated heat fluxes along with the exact heat flux. Comment on the estimated values toward the end of the time interval.

Time, s	0.125	0.250	0.375	0.500	0.625	0.750	0.875	1.000
T, °C	1	1	3	6	12	17	26	36
Time, s	1.125	1.250	1.375	1.500	1.625	1.750	1.875	2.000
T, °C	47	59	74	89	107	126	147	169

4.11 Repeat Problem 4.10 using piecewise-constant approximation for $q(t)$ and the TSVD method. Compute for removing 1, 2, and 8 singular values. Compare the RMS error in heat flux for the three cases. How do the estimates behave toward the end of the time interval?

4.12 Repeat Problem 4.10 using piecewise-constant approximation for $q(t)$ and the Fletcher–Reeves CGM. Plot the RMS error in the heat flux against iteration number. Does there appear to be an optimal number of iterations for the best estimate of the heat flux history?

4.13 A fully implicit finite difference computation for a one-dimensional domain (X22B-0T0) with N nodes has the following equations:

$$T_1^k(1 + 2F_\Delta) - T_2^k 2F_\Delta = T_1^{k-1} + 2F_\Delta q^k \qquad i = 1$$
$$-T_{i-1}^k F_\Delta + T_i^k(2F_\Delta + 1) - T_{i+1}^k F_\Delta = T_i^{k-1} \quad 2 \le i \le N-1$$
$$T_N^k(1 + 2F_\Delta) - T_{N-1}^k 2F_\Delta = T_N^{k-1} \qquad i = N$$

where $F_\Delta = \Delta \tilde{t}/\Delta \tilde{x}^2$ is the cell Fourier number. For a grid with five node in $0 \le \tilde{x} \le 1$ and $\Delta \tilde{t} = 0.5$:
a) Determine the matrices **A**, **B**, and **C** for use in Eq. (4.186a).
b) Determine Φ and **b** for Eq. (4.187c).

4.14 Use results from Problem 4.13 to estimate the heat flux in Problem 4.10 using Kalman Filtering. Use $Q_q = 10$ and $r = 0$.

References

Alifanov, O. M. (1974) Regularization schemes for approximate solution of nonlinear reverse problem of heat conduction, *Journal of Engineering Physics*, 26(1), p. 90. https://doi.org/10.1007/BF00827295

Alifanov, O. M. (1975) Inverse boundary-value problems of heat conduction, *Journal of Engineering Physics*, 29(1), p. 821. https://doi.org/10.1007/BF00860617

Alifanov, O. M. (1983) Methods of solving ill-posed inverse problems, *Journal of Engineering Physics*, 45(5), p. 742. https://doi.org/10.1007/BF01254725

Alifanov, O. M. (1994) *Inverse Heat Transfer Problems*. International Series in Heat and Mass Transfer. Berlin, Heidelberg: Springer Berlin Heidelberg. https://doi.org/10.1007/978-3-642-76436-3.

Alifanov, O. M. and Artyukhin, E. A. (1975) Regularized numerical solution of nonlinear inverse heat-conduction problem, *Journal of Engineering Physics and Thermophysics*, 29(1), pp. 934–938. https://doi.org/10.1007/BF00860643

Alifanov, O. M., Artyukhin, E. A. and Rumyantsev, S. V. (1995) *Extreme Methods for Solving Ill-Posed Problems with Applications to Inverse Heat Transfer Problems*. New York: Begell House Inc.

Beck, J. V. (1961) *Calculation of Transient Thermocouple Temperature Measurements in Heat-Conducting Solids, Part II, The Calculation of Transient Heat Fluxes Using the Inverse Convolution*. Wilmington, MA: AVCO Corp. Research and Advanced Development Division.

Beck, J. V. (1962) Calculation of surface heat flux from an internal temperature history, in *ASME Paper No. 62-HT-46*. ASME.

Beck, J. V. (1968) Surface heat flux determination using an integral method, *Nuclear Engineering and Design*, 7(2), pp. 170–178. https://doi.org/10.1016/0029-5493(68)90058-7.

Beck, J. V. (1970) Nonlinear estimation applied to the nonlinear inverse heat conduction problem, *International Journal of Heat and Mass Transfer*, 13(4), pp. 703–716. https://doi.org/10.1016/0017-9310(70)90044-X.

Beck, J. V. and Arnold, K. J. (1977) *Parameter Estimation in Engineering and Science*. New York: John Wiley and Sons.

Beck, J. V. and Murio, D. A. (1986) Combined function specification-regularization procedure for solution of inverse heat conduction problem, *AIAA Journal*, 24(1), pp. 180–185. https://doi.org/10.2514/3.9240.

Bell, J. B. and Wardlaw, A. B. (1981a) Numerical solution of an ill-posed problem arising in wind tunnel heat transfer data reduction, *NSWC TR 82-32*.

Bell, J. B. and Wardlaw, A. B. (1981b) The noncharacteristic Cauchy problem for class of equations with time dependence. I. Problems in one space dimensions, *SIAM Journal on Mathematical Analysis*, 12, pp. 759–777.

Blackwell, B. F. (1981) An efficient technique for the numerical solution of the one-dimensional inverse problem of heat conduction, *Numerical Heat Transfer*, 4, pp. 229–239. https://doi.org/10.1080/01495728108961789

Brooks, C. D., Lamm, P. K. and Luo, X. (2010) Local regularization of nonlinear Volterra equations of Hammerstein type, *The Journal of Integral Equations and Applications*, 22(3), pp. 393–425. https://doi.org/10.1216/JIE-2010-22-3-393

Burggraf, O. R. (1964) An exact solution of the inverse problem in heat conduction theory and applications, *Journal of Heat Transfer*, 86(3), pp. 373–380. https://doi.org/10.1115/1.3688700

Candy, J. V. (1986) *Signal Processing: the Model-based Approach*. New York: McGraw-Hill.

Crassidis, J. L. and Junkins, J. L. (2004) *Optimal Estimation of Dynamic Systems*. 2nd edn. Taylor & Francis. https://doi.org/10.1201/b11154.

Dai, Z. and Lamm, P. K. (2008) Local regularization for the nonlinear inverse autoconvolution problem, *SIAM Journal on Numerical Analysis*, 46(2), pp. 832–868. https://doi.org/10.1137/070679247.

de Monte, F., Beck, J. V. and Amos, D. E. (2008) Diffusion of thermal disturbances in two-dimensional Cartesian transient heat conduction, *International Journal of Heat and Mass Transfer*, 51(25), pp. 5931–5941. http://10.0.3.248/j.ijheatmasstransfer.2008.05.015.

Fletcher, R. and Reeves, C. M. (1964) Function minimization by conjugate gradients, *Computer Journal*, 7(2), pp. 149–154.

Hadamard, J. (1923) *Lectures on Cauchy's Problem in Linear Partial Differential Equations*. Yale University. Mrs. Hepsa Ely Silliman memorial lectures. New Haven, CT: Yale University Press.

Hansen, P. C. (1987) The truncated SVD as a method for regularization, *BIT Numerical Mathematics*, 27(4), pp. 534–553. https://doi.org/10.1007/BF01937276

Hansen, P. C. (1998) *Rank-Deficient and Discrete Ill-Posed Problems*. Philadelphia, PA: SIAM.

Hensel, E. (1990) *Inverse Theory and Applications Engineering*. Prentice Hall.

Hoerl, A. E. and Kennard, R. W. (1970a) Ridge regression: applications to nonorthogonal problems, *Technometrics*, 12(1), pp. 69–82. https://doi.org/10.1080/00401706.1970.10488635.

Hoerl, A. E. and Kennard, R. W. (1970b) Ridge regression: biased estimation for nonorthogonal problems, *Technometrics*, 12(1), pp. 55–67. https://doi.org/10.2307/1267351.

Hoerl, A. E. and Kennard, R. W. (1976) Ridge regression iterative estimation of the biasing parameter, *Communications in Statistics: Theory & Methods*, 5(1), pp. 77–88. https://doi.org/10.1080/03610927608827333

Jarny, Y. and Orlande, H. R. B. (2011) Adjoint methods, in Orlande, H. R. B. *et al.* (eds) *Thermal Measurements and Inverse Techniques*. CRC Press, pp. 407–436.

Ji, C. C., Tuan, P. C. and Jang, H. Y. (1997) A recursive least-squares algorithm for on-line 1-D inverse heat conduction estimation, *International Journal of Heat and Mass Transfer*, 40(9), pp. 2081–2096. https://doi.org/10.1016/S0017-9310(96)00289-X.

Kalman, R. E. (1960) A new approach to linear filtering and prediction problems, *Journal of Basic Engineering*, 82(1), pp. 35–45. https://doi.org/10.1115/1.3662552

Lamm, P. K. (1990) Regularization and the "adjoint method" of solving inverse problems, in *Inverse Problems Symposium*, Michigan State University, East Lansing, MI.

Lamm, P. K. (2005) Full convergence of sequential local regularization methods for Volterra inverse problems, *Inverse Problems*, 21 (3), p. 785–803. https://doi.org/10.1088/0266-5611/21/3/001.

Lamm, P. K. and Dai, Z. (2005) On local regularization methods for linear Volterra equations and nonlinear equations of Hammerstein type, *Inverse Problems*. Tautenhahn, U. (ed), 21(5), pp. 1773–1790. https://doi.org/10.1088/0266-5611/21/5/016

Lamm, P. K. and Scofield, T. L. (2001) Local regularization methods for the stabilization of linear ill-posed equations of Volterra type, *Numerical Functional Analysis & Optimization*, 22(7/8), p. 913. http://10.0.4.57/NFA-100108315.

Lamm, P. K. and Elden, L. (1997) Numerical solution of first-kind Volterra equations by sequential Tikhonov regularization, *SIAM Journal on Numerical Analysis*, 34(4), p. 1432. https://doi.org/10.1137/S003614299528081X

Lawson, C. L. and Hanson, R. J. (1974) *Solving Least Squares Problems*. Prentice-Hall series in automatic computation. Prentice-Hall.

Lichtblau, D. and Weisstein, E. W. (2002) Condition number, *From MathWorld–A Wolfram Web Resource*. https://mathworld.wolfram.com/ConditionNumber.html

Marquardt, D. W. (1963) An algorithm for least-squares estimation of nonlinear parameters, *Journal of the Society for Industrial and Applied Mathematics*. Atiqullah, M. (ed), 11, p. 431.

Marquardt, D. W. (1970) Generalized inverses, ridge regression, biased linear estimation, and nonlinear estimation, *Technometrics*, 12(3), pp. 591–612. https://doi.org/10.1080/00401706.1970.10488699.

Matsevityi, Y. M. and Multanovskii, A. V. (1978) An iterative filter for solution of the inverse heat-conduction problem, *Journal of Engineering Physics*, 35(5), pp. 1373–1378. https://doi.org/10.1007/BF00859694.

Matz, A. W. (1964) Automating damped least squares to solve the equations determining refractive index of crystals, *Journal of the Royal Statistical Society: Series C: Applied Statistics*, 13(2), pp. 118–127. https://doi.org/10.2307/2985704.

Noh, J. H. *et al.* (2015) Inverse heat transfer analysis of multi-layered tube using thermal resistance network and Kalman filter, *International Journal of Heat and Mass Transfer*, 89, pp. 1016–1023. https://doi.org/10.1016/j.ijheatmasstransfer.2015.06.009.

Ozisik, M. N. and Orlande, H. R. B. (2021) *Inverse Heat Transfer: Fundamentals and Applications*. Boca Raton, FL: CRC Press.

Polak, E. and Ribière, G. (1969) *Note sur la Convergence de Méthodes de Directions Conjuguées*, Laurent, P. J. (ed), 3(16), p. 35.

Press, W. H. *et al.* (2007) *Numerical Recipes in C: The Art of Scientific Computing*. 3rd edn. Cambridge University Press.

Rauch, H. E., Tung, F. and Striebel, C. T. (1965) Maximum likelihood estimates of linear dynamic systems, *AIAA Journal*, 3(8), pp. 1445–1450. https://doi.org/10.2514/3.3166.

Samadi, F., Woodbury, K. A. and Beck, J. V. (2018) Evaluation of generalized polynomial function specification methods, *International Journal of Heat and Mass Transfer*, 122, pp. 1116–1127. https://doi.org/10.1016/j.ijheatmasstransfer.2018.02.018.

Scarpa, F. and Milano, G. (1995) Kalman smoothing technique applied to the inverse heat conduction problem, *Numerical Heat Transfer Part B: Fundamentals*, 28(1), pp. 79–96. https://doi.org/10.1080/10407799508928822.

Stolz, G. (1960) Numerical solutions to an inverse problem of heat conduction for simple shapes, *Journal of Heat Transfer*, 82(1), pp. 20–25. https://doi.org/10.1115/1.3679871

Tikhonov, A. N. (1943) The stability of inverse problems, *Doklady Akademii Nauk SSSR*, 39(5), 195–198.

Tikhonov, A. N. (1963) Regularization of ill-posed problems, *Doklady Akademii Nauk SSSR*, 153(1), pp. 49–52.

Tikhonov, A. N. and Arsenin, V. Y. (1977) *Solutions of Ill-Posed Problems*. Washington, DC: V.H. Winston & Sons.

Tuan, P. C., Fong, L. W. and Huang, W. T. (1997a) Application of Kalman filtering with input estimation technique to on-line cylindrical inverse heat conduction problems', *JSME International Journal Series B: Fluids and Thermal Engineering*, 40(1), pp. 126–133. https://doi.org/10.1299/jsmeb.40.126.

Tuan, P.-C., Lee, S.-C. and Hou, W.-T. (1997b) An efficient on-line thermal input estimation method using Kalman filter and recursive least square algorithm, *Inverse Problems in Engineering*, 5(4), pp. 309–333.

Weber, C. F. (1981) Analysis and solution of the ill-posed inverse heat conduction problem, *International Journal of Heat and Mass Transfer*, 24(11), pp. 1783–1792. https://doi.org/10.1016/0017-9310(81)90144-7.

Widder, D. V. (1975) *The Heat Equation*. Pure and Applied Mathematics, 67. Academic Press.

Woodbury, K. A., Najafi, H. and Beck, J. V. (2017) Exact analytical solution for 2-D transient heat conduction in a rectangle with partial heating on one edge, *International Journal of Thermal Sciences*, 112, pp. 252–262. https://doi.org/10.1016/j.ijthermalsci.2016.10.014.

5

Filter Form of IHCP Solution

5.1 Introduction

The solution techniques used for solving IHCPs in Chapter 4 can be re-formulated as digital filters. The filter coefficient formulation is not a new method, but rather a new perspective through which several of the existing methods can be investigated. The digital filter approach is important because it is much more computationally efficient than the basic methods. Due to its efficiency, it can be readily implemented in an on-line method of analysis. Heat flux measuring devices can incorporate the digital filter concept and visual digital output can be provided with a short delay (near real-time) (Woodbury and Beck 2013).

The digital filter approach can be used to address linear (Beck et al. 1985) and non-liner (Beck 2008) IHCPs. Non-linear IHCP refer to problems for which material properties of the domain are temperature dependent. Digital filters can also be used for solving complex IHCPs with moving boundary (Uyanna and Najafi 2022), multi-layer (Najafi et al. 2015a) and multi-dimensional domains (Najafi et al. 2015b). In this chapter, first the application of digital filter approach for linear IHCP is discussed and then the modified approach for solving the non-linear IHCP is explained.

5.2 Temperature Perturbation Approach

In Chapter 4, various techniques were explored for developing solutions to IHCPs. In this chapter, it will be shown that all of these solutions can be presented in the form of Eq. (5.1):

$$\hat{\mathbf{q}} = \mathbf{FY}, \tag{5.1}$$

where $\hat{\mathbf{q}}$ is the estimated heat flux vector, $\mathbf{Y}$ is the vector consisting of temperature measurement history at a particular location within the domain, and $\mathbf{F}$ is the filter matrix which will be further discussed in this chapter. The components of the filter matrix are known as *filter coefficients*. The filter coefficients can be calculated from the filter matrix by setting all the $\mathbf{Y}$ components equal to zero except the Mth component, Y_M, which should be set as one:

$$\mathbf{Y} = \lfloor 0 \ \cdots \ 0 \ 1 \ 0 \ \cdots \ 0 \rfloor^T \tag{5.2}$$

where $Y_M = 1$ and all other $Y_i = 0$, results in a vector of coefficients

$$f_{M,i} = \lfloor f_1 \ \ f_2 \ \ \cdots \ \ f_M \ \ f_{M+1} \ \ \cdots \ \ f_n \rfloor^T \tag{5.3}$$

These values can be interpreted as the contribution to each value in the heat flux vector when $Y_M = 1$ and all other $Y_i = 0$. They are the *sensitivities* of each heat flux component with respect to Y_M:

$$f_{M,i} = \left\lfloor \frac{\delta q_1}{\delta Y_M} \ \ \frac{\delta q_2}{\delta Y_M} \ \ \cdots \ \ \frac{\delta q_M}{\delta Y_M} \ \ \cdots \ \ \frac{\delta q_{n-1}}{\delta Y_M} \ \ \frac{\delta q_n}{\delta Y_M} \right\rfloor^T \tag{5.4}$$

If the process is repeated for $M = 1,2,..., n$, the sensitivities of all the q_is to all the measurements are obtained.

Then the heat fluxes required to produce an observed history of temperatures can be computed using superposition of these:

$$\hat{q}_1 = Y_1 f_{1,1} + Y_2 f_{1,2} + Y_3 f_{1,3} + \cdots + Y_M f_{1,M} + Y_{M+1} f_{1,M+1} + \cdots + Y_n f_{1,n} \tag{5.5a}$$

Inverse Heat Conduction: Ill-Posed Problems, Second Edition. Keith A. Woodbury, Hamidreza Najafi, Filippo de Monte, and James V. Beck.
© 2023 John Wiley & Sons, Inc. Published 2023 by John Wiley & Sons, Inc.

$$\hat{q}_2 = Y_1 f_{2,1} + Y_2 f_{2,2} + Y_3 f_{2,3} + \cdots + Y_M f_{2,M} + Y_{M+1} f_{2,M+1} + \cdots + Y_n f_{2,n} \tag{5.5b}$$

$$\hat{q}_M = Y_1 f_{M,1} + Y_2 f_{M,2} + Y_3 f_{M,3} + \cdots + Y_M f_{M,M} + Y_{M+1} f_{M,M+1} + \cdots + Y_n f_{M,n} \tag{5.5c}$$

Consequently, the sensitivities of the heat flux components with respect to $Y_1, Y_2,..., Y_n$ values can be defined as:

$$f_{1,i} = \left[\frac{\delta q_1}{\delta Y_1} \quad \frac{\delta q_2}{\delta Y_1} \quad \cdots \quad \frac{\delta q_M}{\delta Y_1} \quad \cdots \quad \frac{\delta q_{n-1}}{\delta Y_1} \quad \frac{\delta q_n}{\delta Y_1} \right]^T \tag{5.6a}$$

$$f_{2,i} = \left[\frac{\delta q_1}{\delta Y_2} \quad \frac{\delta q_2}{\delta Y_2} \quad \cdots \quad \frac{\delta q_M}{\delta Y_2} \quad \cdots \quad \frac{\delta q_{n-1}}{\delta Y_2} \quad \frac{\delta q_n}{\delta Y_2} \right]^T \tag{5.6b}$$

$$f_{M,i} = \left[\frac{\delta q_1}{\delta Y_M} \quad \frac{\delta q_2}{\delta Y_M} \quad \cdots \quad \frac{\delta q_M}{\delta Y_M} \quad \cdots \quad \frac{\delta q_{n-1}}{\delta Y_M} \quad \frac{\delta q_n}{\delta Y_M} \right]^T \tag{5.6c}$$

$$f_{n,i} = \left[\frac{\delta q_1}{\delta Y_n} \quad \frac{\delta q_2}{\delta Y_n} \quad \cdots \quad \frac{\delta q_M}{\delta Y_n} \quad \cdots \quad \frac{\delta q_{n-1}}{\delta Y_n} \quad \frac{\delta q_n}{\delta Y_n} \right]^T \tag{5.6d}$$

The filter matrix, $\mathbf{F}$, can be then formed by these entries, with columns corresponding to temperature impulses, and rows corresponding to heat flux components:

$$F = \begin{bmatrix} \frac{\delta q_1}{\delta Y_1} & \frac{\delta q_1}{\delta Y_2} & \cdots & \frac{\delta q_1}{\delta Y_M} & \frac{\delta q_1}{\delta Y_{M+1}} & \cdots & \frac{\delta q_1}{\delta Y_n} \\ \frac{\delta q_2}{\delta Y_1} & \frac{\delta q_2}{\delta Y_2} & \cdots & \frac{\delta q_2}{\delta Y_M} & \frac{\delta q_2}{\delta Y_{M+1}} & \cdots & \frac{\delta q_2}{\delta Y_n} \\ \vdots & \vdots & \ddots & \vdots & \vdots & \ddots & \vdots \\ \frac{\delta q_M}{\delta Y_1} & \frac{\delta q_M}{\delta Y_2} & \cdots & \frac{\delta q_M}{\delta Y_M} & \frac{\delta q_M}{\delta Y_{M+1}} & \cdots & \frac{\delta q_M}{\delta Y_n} \\ \vdots & \vdots & \ddots & \vdots & \vdots & \ddots & \vdots \\ \frac{\delta q_{n-1}}{\delta Y_1} & \frac{\delta q_{n-1}}{\delta Y_2} & \cdots & \frac{\delta q_{n-1}}{\delta Y_M} & \frac{\delta q_{n-1}}{\delta Y_{M+1}} & \cdots & \frac{\delta q_{n-1}}{\delta Y_n} \\ \frac{\delta q_n}{\delta Y_1} & \frac{\delta q_n}{\delta Y_2} & & \frac{\delta q_n}{\delta Y_M} & \frac{\delta q_n}{\delta Y_{M+1}} & & \frac{\delta q_n}{\delta Y_n} \end{bmatrix} \tag{5.7}$$

Therefore, the heat flux at time step M, can be found as:

$$\hat{q}_M = Y_1 \frac{\delta q_M}{\delta Y_1} + Y_2 \frac{\delta q_M}{\delta Y_2} + Y_3 \frac{\delta q_M}{\delta Y_3} + \cdots + Y_M \frac{\delta q_M}{\delta Y_M} + Y_{M+1} \frac{\delta q_M}{\delta Y_{M+1}} + \cdots + Y_n \frac{\delta q_M}{\delta Y_n} \tag{5.8}$$

Note the rows of $\mathbf{F}$ above correspond to the affected heat flux components, and the columns correspond to the perturbed temperature:

$$F_{ij} = \text{effect of temperature } Y_j \text{ on heat flux } q_i$$

So, the sum above can be written in terms of matrix components as

$$\hat{q}_M = Y_1 F_{M,1} + Y_2 F_{M,2} + Y_3 F_{M,3} + \cdots + Y_M F_{M,M} + Y_{M+1} F_{M,M+1} + \cdots + Y_n F_{M,n} \tag{5.9}$$

or, $\hat{\mathbf{q}} = \mathbf{FY}$, which was given in Eq. (5.1).

For many IHCP solution methods, the matrix $\mathbf{F}$ has a definitive structure:

$$F = \begin{bmatrix} f_0 & f_{-1} & \cdots & f_{1-M} & \cdots & \cdots & f_{1-n} \\ f_1 & f_0 & f_{-1} & f_{-2} & \cdots & & f_{2-n} \\ \vdots & \vdots & \ddots & \ddots & \vdots & \ddots & \vdots \\ f_{M-1} & f_{M-2} & \cdots & f_0 & f_{-1} & \cdots & f_{M-n} \\ \vdots & \vdots & \ddots & \vdots & \ddots & \ddots & \vdots \\ f_{n-2} & f_{n-3} & \cdots & f_2 & f_1 & f_0 & f_{-1} \\ f_{n-1} & f_{n-2} & \cdots & f_{n-M} & f_{M+1-n} & \cdots & f_0 \end{bmatrix} = f_{(row-col)} = f_{(flux-temp)} \tag{5.10}$$

for any time value t_M, the heat flux estimate is

$$\hat{q}_M = \sum_{i=1}^{n} f_{M-i} Y_i \tag{5.11}$$

if $\mathbf{F}$ has all $f_{M,j} \approx 0$ except for $M - m_p < j < M + m_f$

$$\hat{q}_M = \sum_{i=1}^{m_p + m_f} f_{m_p - i + 1} Y_{M - m_p + i - 1} \tag{5.12}$$

uses the left-to-right negative and positive numbering for the f's as used in Eq. (5.10).

In the next Section 5.3, the structure of the filter matrix for various IHCP solution methods is assessed. The close inspection of the filter matrix entries will lead us to use selected filter-based techniques for solving linear and non-linear IHCPs in a near real-time fashion.

5.3 Filter Matrix Perspective

The close assessment of the characteristics of the filter matrix resulting from different IHCP techniques provides invaluable insights into these solution approaches that are discussed in the present chapter and further elaborated in Chapter 7. It is particularly interesting that for some IHCP solution methods, such as Tikhonov regularization, that is traditionally a whole-domain method, the filter formulation facilitates a sequential form (Woodbury and Beck 2013). This allows the use of filter coefficient method for near real-time calculation of heat fluxes from a continuing stream of temperature data, which can be used for continuous control and monitoring purposes of a process. In this section, the use of filter matrix for various IHCP solution techniques are explored.

5.3.1 Function Specification Method

In the function specification method, when the heat flux is held constant for r future steps; the heat flux $\hat{q}_M$ at time t_M can be found using Eqs. (4.70a) and (4.70b).

The filter coefficients can be obtained by setting all the $\mathbf{Y}$ components equal to zero except the Mth value which is set equal to unity. The first value obtained from Eq. (4.70a) is given for $M = 1$:

$$\hat{q}_1 = \frac{\phi_r}{\sum_{i=1}^{r} \phi_i^2}. \tag{5.13}$$

The filter matrix for the function specification method has a persymmetric structure. In other words, the matrix is symmetric about the northeast-southwest diagonal:

$$F_{FS} = \begin{bmatrix} f_{-r+1} & 0 & \cdots & 0 & \cdots & 0 & 0 \\ f_{-r+2} & f_{-r+1} & \cdots & 0 & \cdots & 0 & 0 \\ \vdots & \vdots & \ddots & \vdots & \cdots & \vdots & \vdots \\ f_{-r+1} & f_{-1} & \cdots & f_{-r+1} & \cdots & 0 & 0 \\ \vdots & \vdots & \cdots & \vdots & \ddots & \vdots & \vdots \\ f_{-r+1} & f_{N-2} & \cdots & f_{-1} & \cdots & f_{-r+1} & 0 \\ f_{-r+1} & f_{N-1} & \cdots & f_0 & \cdots & f_{-r+2} & f_{-r+1} \end{bmatrix}. \tag{5.14}$$

It should be noted that $\hat{q}_1$, which can be found using Eq. (5.13), is actually the diagonal component of the filter coefficients.

Figure 5.1 shows columns of the $\mathbf{F}_{FS}$ matrix for the X22B10T0 case with $\tilde{x} = 0.5$, $\Delta \tilde{t} = 0.05$, and $r = 7$. Notice

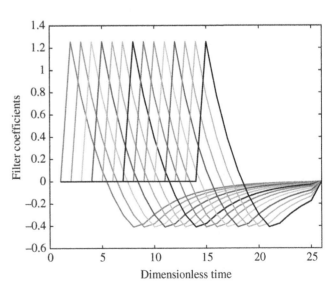

Figure 5.1 Columns 7–15 of the filter matrix for function specification method ($\tilde{x} = 0.5$, $\Delta \tilde{t} = 0.05$, and $r = 7$).

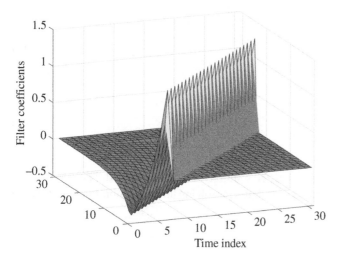

Figure 5.2 Surface plot of the filter coefficients for the function specification method ($\tilde{x}$ = 0.5, $\Delta\tilde{t}$ = 0.05, and r = 7).

that the same curve shape is repeated for successive columns. This is better illustrated in Figure 5.2, where the columns of the filter matrix are presented in a surface plot. Table 5.1 contains numerical values for the first column of the $\mathbf{F}_{FS}$ matrix for various values of r (r = 2, 3, 4, 6, 8, 10). Inspection of these values shows that as time increases the values approach zero. Also, the sums of the filter coefficients approaches to zero (last row of Table 5.1). Note that the latter is not a general property of the filter, but is related to the geometry and boundary conditions of the problem. For the X22B10T0 problem, the sum of the filter coefficients must equal zero to satisfy conservation of energy. For a temperature history ($\mathbf{Y}$) that is not changing with time and insulated boundary at $\tilde{x}$ = 1, the only way that the heat flux can be determined as zero using Eq. (5.1) is if the filter coefficients sums up to zero.

Table 5.1 Filter coefficients for the function specification method X22B10T0 case ($\tilde{x}$ = 0.5, $\Delta\tilde{t}$ = 0.05).

$\tilde{t}$	r = 2	r = 3	r = 4	r = 6	r = 8	r = 10
0.05	15.7998	6.9892	3.9008	1.7138	0.9576	0.6103
0.10	−11.5385	−0.2926	1.0334	0.9229	0.6353	0.4486
0.15	−4.2881	−2.9603	−0.6135	0.3438	0.3765	0.3121
0.20	−0.0030	−2.2086	−1.4390	−0.0794	0.1687	0.1968
0.25	0.0254	−0.9051	−1.2129	−0.3843	0.0019	0.0996
0.30	0.0038	−0.3692	−0.7027	−0.5729	−0.1318	0.0175
0.35	0.0004	−0.1503	−0.4068	−0.5213	−0.2372	−0.0518
0.40	0.0000	−0.0612	−0.2355	−0.3815	−0.3079	−0.1101
0.45	0.0000	−0.0249	−0.1363	−0.2792	−0.2883	−0.1584
0.50	0.0000	−0.0101	−0.0789	−0.2044	−0.2315	−0.1923
0.55	0.0000	−0.0041	−0.0457	−0.1496	−0.1859	−0.1827
0.60	0.0000	−0.0017	−0.0264	−0.1095	−0.1492	−0.1543
0.65	0.0000	−0.0007	−0.0153	−0.0801	−0.1198	−0.1302
0.70	0.0000	−0.0003	−0.0089	−0.0586	−0.0962	−0.1099
0.75	0.0000	−0.0001	−0.0051	−0.0429	−0.0773	−0.0928
0.80	0.0000	0.0000	−0.0030	−0.0314	−0.0620	−0.0783
0.85	0.0000	0.0000	−0.0017	−0.0230	−0.0498	−0.0661
0.90	0.0000	0.0000	−0.0010	−0.0168	−0.0400	−0.0558
0.95	0.0000	0.0000	−0.0006	−0.0123	−0.0321	−0.0471
1.00	0.0000	0.0000	−0.0003	−0.0090	−0.0258	−0.0398
1.05	0.0000	0.0000	−0.0002	−0.0066	−0.0207	−0.0336
1.10	0.0000	0.0000	−0.0001	−0.0048	−0.0166	−0.0283
1.15	0.0000	0.0000	−0.0001	−0.0035	−0.0133	0.0000
1.20	0.0000	0.0000	0.0000	−0.0026	−0.0107	0.0000

Table 5.1 (Continued)

$\tilde{t}$	$r = 2$	$r = 3$	$r = 4$	$r = 6$	$r = 8$	$r = 10$
1.25	0.0000	0.0000	0.0000	−0.0019	0.0000	0.0000
1.30	0.0000	0.0000	0.0000	−0.0014	0.0000	0.0000
1.35	0.0000	0.0000	0.0000	0.0000	0.0000	0.0000
1.40	0.0000	0.0000	0.0000	0.0000	0.0000	0.0000
1.45	0.0000	0.0000	0.0000	0.0000	0.0000	0.0000
1.50	0.0000	0.0000	0.0000	0.0000	0.0000	0.0000
1.55	0.0000	0.0000	0.0000	0.0000	0.0000	0.0000
Sum f	7.40E-14	2.64E-10	5.78E-06	3.78E-03	4.37E-02	1.53E-01

5.3.2 Tikhonov Regularization

As discussed in Chapter 4, the ith order Tikhonov IHCP solution method allows estimating the unknown heat fluxes as:

$$\hat{\mathbf{q}}_T = \left(\mathbf{X}^T\mathbf{X} + \alpha_{T,i}\mathbf{H}_i^T\mathbf{H}_i\right)^{-1}\mathbf{X}^T\mathbf{Y} = \mathbf{F}_{T,i}\mathbf{Y}$$
$$\mathbf{F}_{T,i} = \left(\mathbf{X}^T\mathbf{X} + \alpha_{T,i}\mathbf{H}_i^T\mathbf{H}_i\right)^{-1}\mathbf{X}^T$$

(5.15)

The 0th, 1st, and 2nd order regularization matrices, $\mathbf{H}_i$, are given in Eqs. (4.110a)–(4.110c).

In Eq. (5.15), $\alpha_{T,i}$ is the ith order Tikhonov regularization coefficient and $\mathbf{F}_{T,i}$ represents the ith order Tikhonov regularization filter matrix (T in the subscript refers to the Tikhonov regularization). The structure of the filter matrix for Tikhonov regularization method, $\mathbf{F}_T$, given in Eq. (5.10). Notice that the matrix's elements on each diagonal are the same. This is known as Toeplitz structure.

To assess these characteristics the values of filter coefficients are calculated for X22B10T0 case with $N = 51$ time steps, zeroth order regularization, $\tilde{x} = 0.5$, $\Delta\tilde{t} = 0.05$, and $\alpha_t = 0.001$. Several rows of the filter matrix ($\mathbf{F}_T$) are plotted in Figure 5.3. The entries of the 26th row (middle row) are specified with markers to better demonstrate the shape of filter coefficients. These values are also listed in Table 5.2.

Notice that the sum of the f-vector components given in Table 5.2 is 9.594E-13, which is very close to zero, as expected, given the boundary conditions of this case (as previously described in Section 5.2). Figure 5.3 shows selected rows of the filter matrix, and except for the first few and last few rows, all the rows have the same entries but are shifted in time.

The shape of the filter coefficients at the end of the time domain gives insight to the poor heat flux estimates attainable at the end of the data time window.

Figure 5.3 also illustrates that the filter coefficients approaches zero for some relatively small time window surrounding any current time. This characteristic of the Tikhonov filter coefficients will be used later to use it as a sequential approach for near real-time calculation of heat flux.

Figure 5.4 shows the surface plot of the Tikhonov filter coefficients. The two main characteristics of these coefficients are further illustrated in this figure: values of f approach zero from both ends and all rows of the filter matrix (with the exception of first few and last few rows)

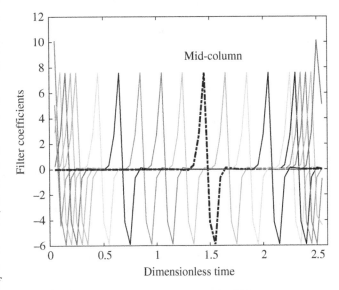

Figure 5.3 Filter coefficients for zeroth order Tikhonov regularization ($\tilde{x} = 0.5$, $\Delta\tilde{t} = 0.05$ and $\alpha_t = 0.001$).

Table 5.2 Values of filter coefficients (middle-column) for Tikhonov regularization method ($\bar{x} = 0.5$, $\Delta\bar{t} = 0.05$, and $\alpha_t = 0.001$).

$\bar{t}$	f	$\bar{t}$	f	$\bar{t}$	f
0.05	5.862E-14	0.85	8.131E-05	1.65	−7.322E-04
0.10	−5.329E-14	0.90	3.217E-04	1.70	−1.769E-04
0.15	3.197E-14	0.95	1.033E-04	1.75	−9.430E-06
0.20	3.446E-13	1.00	−5.623E-03	1.80	6.703E-06
0.25	1.119E-12	1.05	−3.099E-02	1.85	2.300E-06
0.30	−2.093E-12	1.10	−5.222E-02	1.90	2.696E-07
0.35	−3.203E-11	1.15	3.236E-01	1.95	−4.749E-08
0.40	−1.234E-10	1.20	2.669E+00	2.00	−2.689E-08
0.45	−2.576E-11	1.25	7.577E+00	2.05	−4.766E-09
0.50	2.231E-09	1.30	−4.134E+00	2.10	1.203E-10
0.55	1.200E-08	1.35	−5.890E+00	2.15	2.815E-10
0.60	1.921E-08	1.40	−6.658E-01	2.20	6.955E-11
0.65	−1.305E-07	1.45	1.283E-01	2.25	4.030E-12
0.70	−1.041E-06	1.50	6.936E-02	2.30	−2.526E-12
0.75	−2.872E-06	1.55	1.199E-02	$\sum_{i=1}^{n} f_i = 9.594\text{E-}13$	
0.80	5.108E-06	1.60	−3.912E-04		

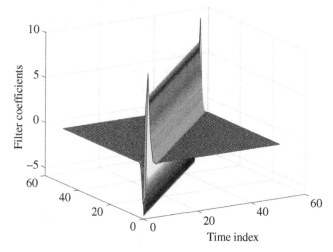

Figure 5.4 Surface plot of the rows of the filter matrix for Tikhonov regularization method ($\bar{x} = 0.5$, $\Delta\bar{t} = 0.05$, and $\alpha_t = 0.001$).

are identical. Using the filter form of TR for solving various IHCPs will be further discussed in this chapter.

5.3.3 Singular Value Decomposition

From the concepts in Section 4.10, the filter matrix for the singular value decomposition (SVD) method can be identified as:

$$\mathbf{F}_{SVD} = \mathbf{V}\mathbf{W}^{-1}\mathbf{U}^T \qquad (5.16)$$

where $\mathbf{U}$, $\mathbf{V}$, and $\mathbf{W}$ are the SVD components of the matrix $\mathbf{X}$.

The SVD method produces oscillatory results for f in time and space throughout the entire time domain. Figure 5.5 shows the main diagonal (lower curve) values for the filter matrix $\mathbf{F}_{SVD}$ as a function of index along the diagonal for the X22B10T0 problem with $\tilde{x} = 0.5$ for 31 time steps and 15 eigenvalues.

Figure 5.5 shows there are significant oscillations in the diagonal elements of $\mathbf{F}_{SVD}$, which is in contrast to the Toeplitz structure of Tikhonov and FS filter matrices. Also, the magnitude of these oscillations increases towards the beginning and end. A total of 15 peaks are seen which are associated with the 15 eigenvalues. The curves are symmetric about the middle index.

Figure 5.6 shows the filter coefficients on the mid-column and the two adjacent columns (mid-column+2 and mid-column−2) of the filter matrix associated with SVD method versus the dimensionless time. Due to persymmetry, these column values are the same as corresponding rows. Notice that, although filter coefficients become smaller in magnitude as they get closer to the two ends of the time domain, they do not really approach to zero (unlike the filter coefficients derived for the Tikhonov regularization method). This characteristic of the SVD filter matrix is further illustrated in Figure 5.7, which makes clear that the filter coefficients do not damp to zero towards the two ends of the time domain but instead continue to fluctuate with decreasing amplitude.

5.3.4 Conjugate Gradient

Filter coefficients for conjugate gradient method (CGM) can be constructed using the temperature perturbation approach by solving the inverse problem using a vector of data which is all zero except for a "1" at the current time location. By applying CGM for each successive time, the full filter matrix for CGM can be computed.

The number of iterations needed for CGM should be determined using some optimality criterion. This concept is described in Chapter 6. Optimal **F** can be determined by generating the full matrix for each iteration and using the resulting matrix to compute $\hat{\mathbf{q}}$ and check the optimality criterion.

To illustrate the characteristics of the filter matrix for the CGM, the values of filter coefficients are evaluated for X22B10T0 case with six iterations and $N = 31$, time steps, $\tilde{x} = 0.5$ and $\Delta\tilde{t} = 0.05$. Figure 5.8 shows the filter coefficients computed for CGM using the step heat flux input. The coefficients from the middle, middle minus one, and middle plus one rows are shown. As seen, the three curves are very similar, but not identical. The peak values are different, and the shapes of the

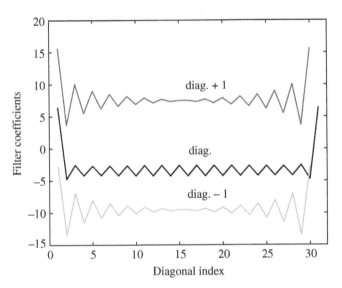

Figure 5.5 Values of the filter coefficients along the main diagonal, diagonal +1 and diagonal −1 for singular value decomposition ($\tilde{x} = 0.5$ for 31 time steps and 15 eigenvalues).

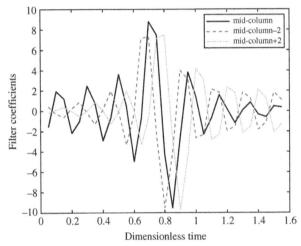

Figure 5.6 Filter coefficients for the mid column and plus and minus 2 adjacent ones for the SVD method ($\tilde{x} = 0.5$ for 31 time steps and 15 eigenvalues).

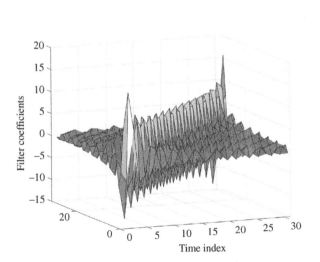

Figure 5.7 Surface plot of the rows of the filter matrix for SVD method ($\tilde{x} = 0.5$ for 31 time steps and 15 eigenvalues).

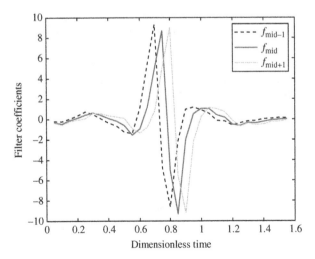

Figure 5.8 Filter coefficients for the mid-column and plus and minus 1 adjacent columns for the CGM using Fletcher–Reeves algorithm ($\tilde{x} = 0.5$, $\Delta\tilde{t} = 0.05$).

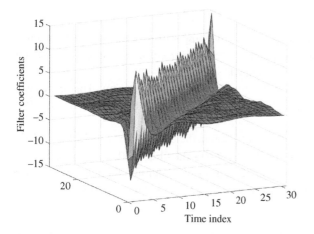

Figure 5.9 Surface plot of the columns of the CGM filter matrix ($\bar{x}$ = 0.5, $\Delta\bar{t}$ = 0.05 for 31 time steps).

oscillations at early time values are different. It is also clear that the f coefficients do not approach zero quickly which suggests that data from the far past and future are needed to compute the middle value of $\mathbf{q}_M$, unlike the filter coefficients for TR and FS. This is further illustrated in the surface plot shown in Figure 5.9 which presents all the columns of the filter matrix versus time index. Therefore, the CGM does not result in universal filter coefficients that can be used for continuous "near real-time" application, unlike the TR or FS. Numerical values of the middle column of the **F** matrix are listed in Table 5.3.

5.4 Sequential Filter Form

The filter form of some of the IHCP solution techniques can be used in sequential form, notably the function specification method and the Tikhonov regularization method. The filter approach is specially very valuable with the Tikhonov regularization technique. This is because the function specification method is already known as a sequential solution approach for IHCP, however, the Tikhonov regularization approach is well-known as a traditionally whole-time domain approach. In other words, the traditional way of using Tikhonov regularization method required the use of temperature data from the entire time domain (for the whole duration of the experiment) to calculate the surface heat flux all at once. However, the filter approach provides the possibility of using TR as a sequential technique.

The filter form of TR can be used for near real-time calculation of unknown heat flux. This means that, in order to calculate the heat flux on the surface of a domain at any point of time, only temperature measurement values from several previous and future time steps are needed, and not for the entire time domain. This filter characteristic is clear from inspection of Figures 5.3 and 5.4, where the filter coefficients approach zero from both ends of the time domain and there are only limited number of non-zero (i.e. significant) filter coefficients around the current time step.

The need for future time information in the TR filter is not surprising as this is characteristic of all IHCP solutions. However, only relatively few datapoints from future time intervals are needed which allows calculation of the heat flux with a relatively small time delay ("near real-time"). Note that the delay may vary from problem to problem, based on the geometry, sensor location, material properties, and time step. This will be further discussed under this section.

Table 5.3 Values of filter coefficients (middle-column) for CGM ($\bar{x}$ = 0.5, $\Delta\bar{t}$ = 0.05).

$\bar{t}$	f	$\bar{t}$	f	$\bar{t}$	f
0.05	−0.33050	0.55	−1.55373	1.05	1.03215
0.10	−0.46675	0.60	−0.94255	1.10	0.41684
0.15	−0.12094	0.65	1.19753	1.15	0.18414
0.20	0.11863	0.70	5.03379	1.20	−0.50348
0.25	0.41476	0.75	8.69408	1.25	−0.62215
0.30	0.62298	0.80	−5.03016	1.30	−0.32560
0.35	0.40601	0.85	−9.29195	1.35	−0.37544
0.40	0.08151	0.90	−1.98165	1.40	−0.22565
0.45	−0.17130	0.95	0.52396	1.45	−0.08327
0.50	−0.62024	1.00	1.06168	1.50	−0.00901
				1.55	0.03490

Since the rows of the TR filter matrix are all identical (Figures 5.3 and 5.4), the components of one middle row (filter coefficients) can be used for calculating the heat flux using Eq. (5.12). As seen in Eq. (5.12), surface heat flux calculation at any given time step (M) only requires temperature measurement data from limited number of time steps ($t_M - m_p$ to $t_M + m_f$) and the non-zero components of one row of the filter matrix, i.e. filter coefficients. Therefore, Eq. (5.12) can be used for calculating the surface heat flux in a near real-time fashion.

This will provide a very simple and computationally efficient approach for solving IHCPs. For any given geometry and boundary condition, the filter coefficients can be pre-calculated and used with an appropriate temperature sensor to evaluate the heat flux in a near real-time fashion. Note that m_p and m_f are the number of non-zero (i.e. significant) terms in the filter matrix previous and after the current time step, respectively. To illustrate use of the filter based TR approach Example 5.1 is provided.

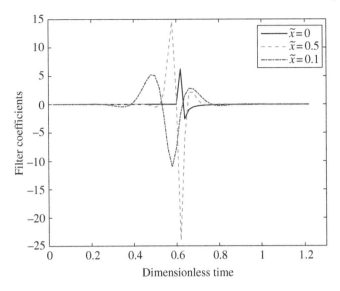

Figure 5.10 Mid-column of the TR-filter matrix for various dimensionless locations.

Example 5.1 Consider a one-dimensional slab which is exposed to a surface heat flux in form of constant heating and insulated from the backside (X22B10T0). Assume a thermocouple is used to measure the temperature and a random error in the amount of 0.25% of the temperature measured at each time step is present in the data with a dimensionless time step of 0.02 and for 61 time steps.

(A) Calculate the filter coefficients using filter-based TR if sensor is located at $\tilde{x} = 0, \tilde{x} = 0.5$ and $\tilde{x} = 1$. Discuss the values of m_p and m_f and how they should be selected. (B) Using the temperature measurement from $\tilde{x} = 0.5$, calculate the surface heat flux.

Solution

A) The filter matrix can be calculated using Eq. (5.15). The mid-column of the filter matrix is plotted for $\tilde{x} = 0, \tilde{x} = 0.5$ and $\tilde{x} = 1$ to provide an understanding regarding how they compare (Figure 5.10).

The number of significant (larger than zero) filter coefficients decreases as the sensor gets closer to the surface, indicating a reduction in the delay between the time that the heat flux is being applied until the time step at which the impact of the surface heat flux is being felt at a particular sensor location. The number of significant filter coefficients before and after the current time step are $m_p = 1$ and $m_f = 7$, $m_p = 8$ and $m_f = 8$, and $m_p = 18$ and $m_f = 12$ for the sensor located at $\tilde{x} = 0, \tilde{x} = 0.5$, and $\tilde{x} = 1$, respectively. As seen, as the sensor gets more distance from the surface, the values of m_p and m_f becomes larger, indicating that more values from previous and future time steps are needed to estimate surface heat flux at the current time step, which is expected (why?). Note that a larger m_f indicates a larger delay time in reporting the surface heat flux at a given time step.

B) the filter coefficients for the mid-plane sensor location ($\tilde{x} = 0.5$) are obtained as below (mid-column of the filter matrix that can also be determined using temperature perturbation technique, by setting all $\mathbf{Y}$ components as zero with the exception of $\mathbf{Y}(31) = 1$). Notice that the significant values (>0.01) are emphasized with bold font.

$$f = \begin{bmatrix} 0.00000 & 0.00000 & 0.00000 & 0.00000 & 0.00000 & 0.00000 & 0.00000 & 0.00000 & 0.00000 \\ 0.00000 & 0.00000 & 0.00000 & 0.00001 & 0.00003 & 0.00002 & -0.00011 & -0.00047 & -0.00084 \\ 0.00012 & \mathbf{0.00533} & \mathbf{0.01626} & \mathbf{0.02019} & \mathbf{-0.03183} & \mathbf{-0.22031} & \mathbf{-0.51205} & \mathbf{-0.32644} & \mathbf{1.98523} \\ \mathbf{8.23863} & \mathbf{14.59050} & \mathbf{1.26012} & \mathbf{-24.01075} & \mathbf{-5.89700} & \mathbf{2.09083} & \mathbf{2.11706} & \mathbf{0.78503} & \mathbf{0.06386} \\ \mathbf{-0.09864} & \mathbf{-0.06391} & \mathbf{-0.01769} & 0.00131 & 0.00370 & 0.00177 & 0.00033 & -0.00012 & -0.00012 \\ -0.00004 & 0.0000 & 0.00001 & 0.00000 & 0.00000 & 0.00000 & 0.00000 & 0.00000 & 0.00000 \\ 0.00000 & 0.00000 & 0.00000 & 0.00000 & 0.00000 & 0.00000 & 0.00000 & & \end{bmatrix}$$

The temperature response associated with the X22B10T0 case at $\tilde{x} = 0.5$ is presented in Figure 5.11.

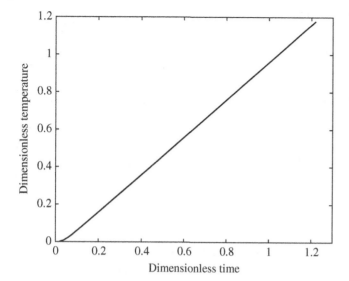

Figure 5.11 Temperature response for X22B10T0 case at mid-plane ($\tilde{x}$ = 0.5 and $\Delta\tilde{t}$ = 0.02).

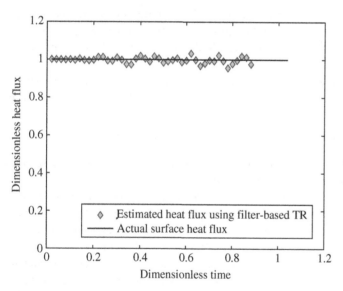

Figure 5.12 Calculated surface heat flux using filter-based TR versus exact values ($\tilde{x}$ = 0.5, $\Delta\tilde{t}$ = 0.02 and α = 0.0001).

The surface heat flux can be then estimated using noisy temperature data through Eq. (5.12). The estimated surface heat flux is presented in Figure 5.12. A good agreement can be observed between the exact heat flux values versus the estimated values that are produced using the TR-based filter coefficient method.

5.5 Using Second Temperature Sensor as Boundary Condition

In developing solutions for IHCPs, remote boundaries (i.e. $\tilde{x}$ = 1) are typically assumed to be insulated or cooled with a known heat transfer coefficient while temperature measurement is known at one interior point (i.e. $0 \leq \tilde{x}_1 \leq 1$) that is being used to calculate the surface heat flux. However, in many practical applications, it is not possible to insulate the remote boundary or assume it is perfectly insulated without introducing major error to the solution.

An alternative is to use an additional thermocouple for a secondary temperature measurement at $\tilde{x}_2 = L$ (location L may, or may not, coincide with the location of a physical boundary). In this section, a method is developed to incorporate the temperature measurement history from a second subsurface sensor as a remote boundary condition in an IHCP solution. The solution is then written in the form of digital filters using TR method, allowing near real-time calculation of surface heat flux.

5.5.1 Exact Solution for the Direct Problem

Consider the schematic given in Figure 5.13. The temperature at x_1 is a function of the surface heat flux $q(t)$ and the temperature at $x = L$ which can be calculated using Green's functions (Cole et al. 2011):

$$T(x_1, t) = \frac{\alpha}{k} \int_{\tau=0}^{t} q(\tau)G_{X21}(x_1, 0, t-\tau)d\tau + \alpha \int_{\tau=0}^{t} y(\tau)\left(-\frac{\partial G_{X21}}{\partial x'}(x_1, L, t-\tau)\right)d\tau \qquad (5.17)$$

The Green's function in this equation is for a boundary condition of the second kind at $x = 0$ and the first kind at $x = L$. The notation X21 denotes a Cartesian geometry with a boundary condition of the second kind at $x = 0$ and the first kind at $x = L$. The heat flux components, q_i and q_{i+1}, and also between adjacent temperature components, y_i and y_{i+1}, are specified as being constant between time points (piecewise constant function); but they could also be linear, quadratic, and cubic.

A solution for the X21B10T0 problem with a constant heat flux at $x = 0$ and zero temperature at $x = L$ is given in Eq. (2.35). As described in Section 3.4.1, this solution can be used as a building block to construct approximate solutions to arbitrary surface heat flux functions using a piecewise constant heat flux variation assumption. Here, the ϕ_i values (similar to the dimensional definition in Eq. (3.45a) are termed the *response function* and represent the *dimensionless* temperature rise at the measurement location due to a unit disturbance (step change) in the surface condition.

Likewise, in consideration of a step change in temperature at the boundary $x = L$, the solution for a constant temperature, T_0, at $x = L$ (i.e. X21B01T0 problem) can be obtained from the X12B10T0 solution in Section 2.3.6 by replacing x with $L-x$:

$$\frac{T(x, t)}{T_0} = 1 - 2 \sum_{m=1}^{\infty} e^{-\beta_m^2 \frac{\alpha t}{L^2}} \frac{\sin\left(\beta_m\left(1 - \frac{x}{L}\right)\right)}{\beta_m}$$

(5.18)

and the eigenvalues are $\beta_m = (m - 1/2)\pi$ seconds. The (dimensionless) response function at x_1 for a constant temperature disturbance T_0 at $x = L$, denoted as η_i, is

$$\eta_i = \frac{T(x_1, i\Delta t)}{T_0} = 1 - 2 \sum_{m=1}^{\infty} e^{-\beta_m^2 \frac{\alpha i \Delta t}{L^2}} \frac{\sin\left(\beta_m\left(1 - \frac{x_1}{L}\right)\right)}{\beta_m}$$

(5.19)

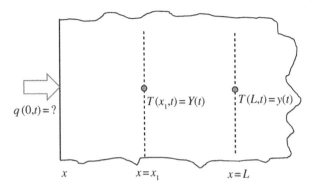

Figure 5.13 Schematic of the physical domain with second temperature sensor as boundary condition.

The temperature at x_1 can be found for any transient heat flux at $x = 0$ and transient temperature at $x = L$ by superposition of the effects from the two boundaries using thee response functions ϕ_i and η_i.

Due to linearity, the $q(t)$ and $y(t)$ inputs in Eq. (5.17) can be found separately. First, consider the temperature response at x_1 due to a piecewise constant heat flux at $x = 0$ with an initial temperature equal to zero, the temperatures at t_i, $i = 1, 2, ..., M$ are found through superposition as:

$$T_{q,M} = T_q(x_1, M\Delta t) = \frac{L}{k} \sum_{i=1}^{M} q_i \Delta\phi_{M-i}$$

(5.20)

where $\Delta\phi_i = \phi_{i+1} - \phi_i$ and $\phi_0 = 0$. The temperature at time step M can be also written as:

$$T_{q,M} = q_1 X_M + q_2 X_{M-1} + \cdots + q_{M-1} X_2 + q_M X_1 = \sum_{i=1}^{M} q_i X_{M-i+1}$$

(5.21)

The q subscript in Eqs. (5.20) and (5.21) on T refers to the fact that the contribution to the temperature at x_1 is caused by the unknown heat flux. The contribution to the temperature at x_1 caused by the temperature history at $x = L$ can be given as:

$$T_{y,M} = \sum_{i=1}^{M} y_i \Delta\eta_{M-i}$$

(5.22)

or

$$T_{y,M} = \sum_{i=1}^{M} y_i Z_{M-i+1}$$

(5.23)

Finally, the temperature at x_1 caused by the heat flux at $x = 0$ and the temperature at $x = L$ is the sum of the two contributions (superposition):

$$T_M = T_{q,M} + T_{y,M} = \frac{L}{k} \sum_{i=1}^{M} q_i \Delta\phi_{M-i} + \sum_{i=1}^{M} y_i \Delta\eta_{M-i} = \sum_{i=1}^{M} q_i X_{M-i+1} + \sum_{i=1}^{M} y_i Z_{M-i+1}$$

(5.24)

The step basis function representations for temperature can be given by the matrix equation as

$$\mathbf{T} = \mathbf{Xq} + \mathbf{Zy}$$

(5.25)

where

$$\mathbf{T} = \begin{bmatrix} T_1 \\ T_2 \\ \vdots \\ T_n \end{bmatrix}, \quad \mathbf{q} = \begin{bmatrix} q_1 \\ q_2 \\ \vdots \\ q_n \end{bmatrix}, \quad \mathbf{X} = \begin{bmatrix} X_1 & 0 & 0 & \cdots & 0 & 0 \\ X_2 & X_1 & 0 & \cdots & 0 & 0 \\ X_3 & X_2 & X_1 & \cdots & 0 & 0 \\ \vdots & \vdots & \vdots & & \vdots & \vdots \\ X_n & X_{n-1} & X_{n-2} & \cdots & X_2 & X_1 \end{bmatrix}$$

(5.26)

$$
\mathbf{y} = \begin{bmatrix} y_1 \\ y_2 \\ \vdots \\ y_n \end{bmatrix}, \quad \mathbf{Z} = \begin{bmatrix} Z_1 & 0 & 0 & \cdots & 0 & 0 \\ Z_2 & Z_1 & 0 & \cdots & 0 & 0 \\ Z_3 & Z_2 & Z_1 & \cdots & 0 & 0 \\ \vdots & \vdots & \vdots & & \vdots & \vdots \\ Z_n & Z_{n-1} & Z_{n-2} & \cdots & Z_2 & Z_1 \end{bmatrix}
\tag{5.27}
$$

Notice that $\mathbf{Z}$ is dimensionless, and $\mathbf{X}$ has a unit of temperature per unit heat flux. The components of the X and Z matrices are defined as:

$$
X_i = \frac{L}{k}\Delta\phi_{i-1}, \; Z_i = \Delta\eta_{i-1}.
\tag{5.28}
$$

5.5.2 Tikhonov Regularization Method as IHCP Solution

The Tikhonov regularization method is used to solve the IHCP. The sum of squares of the errors between the measured temperatures and estimated temperatures can be found as:

$$
S = (\mathbf{Y} - \mathbf{Xq} - \mathbf{Zy})^T (\mathbf{Y} - \mathbf{Xq} - \mathbf{Zy}) + \alpha_T \mathbf{q}^T \mathbf{H}^T \mathbf{Hq}
\tag{5.29}
$$

The IHCP solution can be achieved by minimizing S with respect to $\mathbf{q}$. Note that $\mathbf{Y}$ and $\mathbf{y}$ represent the temperature measurement vector at x_1 and the measured boundary temperature at $x = L$, respectively. An initial temperature of zero is considered. The α_T is the Tikhonov regularization parameter and $\mathbf{H}$ is the regularization matrix (first-order matrix is used here). The estimated heat flux vector, $\hat{\mathbf{q}}$, can be found by minimizing S with respect to $\mathbf{q}$ and is given by:

$$
\begin{aligned}
\hat{\mathbf{q}} &= \left[\mathbf{X}^T \mathbf{X} + \alpha_T \mathbf{H}^T \mathbf{H} \right]^{-1} \mathbf{X}^T (\mathbf{Y} - \mathbf{Zy}) \\
&= \mathbf{F}(\mathbf{Y} - \mathbf{Zy}) = \mathbf{FY} + \mathbf{Gy}
\end{aligned}
\tag{5.30}
$$

In the above equation, $\mathbf{F}$ is the filter matrix. Note that this filter matrix is defined with the $\mathbf{X}$ matrix from the X21B10T0 case. The $\mathbf{F}$ matrix for the X21B10T0 case ($\mathbf{F}_{X21}$) has many similar properties to the $\mathbf{F}$ matrix that was previously introduced for the X22 case (Section 5.3.2). Particularly, except for the first few and last few rows, the entries on each row are identical but shifted in time by one time step. The units of $\mathbf{F}$ are heat flux per unit temperature (W/m^2–K). It is noteworthy that in Eq. (5.30), the term $\mathbf{Zy}$ can be explained as a correction factor to the measured data $\mathbf{Y}$ to account for the contribution of the non-homogeneous boundary condition at $x = L$ to the measured Y_i values.

5.5.3 Filter Form of IHCP Solution

As previously discussed, the IHCP solution that is developed using TR can be written in the filter form to be used as a sequential algorithm. The part of the heat flux due to the measured temperature at x_1 for the time step, t_M, can be found as:

$$
\begin{aligned}
\hat{q}_{M,Y} &= f_1 Y_{M+m_f} + f_2 Y_{M+m_f-1} + \cdots + f_{m_f} Y_M + \cdots \\
&\qquad + f_{M+m_p-1} Y_{M-m_p+1} + f_{M+m_p} Y_{M-m_p} \\
&= \sum_{i=1}^{m_p+m_f} f_i Y_{M+m_f-i}
\end{aligned}
\tag{5.31}
$$

where the f's are elements of a *filter vector*, and m_f is the number of future times steps, and m_p refers to the number of past time steps for which the filter coefficients are significant (i.e. non-zero). The entries of the middle column of the $\mathbf{F}_{X21}$ matrix are plotted in Figure 5.14 for different sensor locations. The width or range of the filter, determined by the interval with non-zero values, increases as the sensor depth increases. This indicates that the number of time steps of data required for resolving the surface heat flux increases with distance of the sensor from the active surface, which is intuitive. Sensor locations closer to the surface will have higher sensitivity to the unknown heat flux and will result in a digital filter with the smallest width.

In addition to Eq. (5.31), another term is needed to account for the changing temperature at $x = L$, as shown in Eq. (5.30). This term, $\mathbf{Gy} = -\mathbf{FZy}$, can also be written in a filter form as

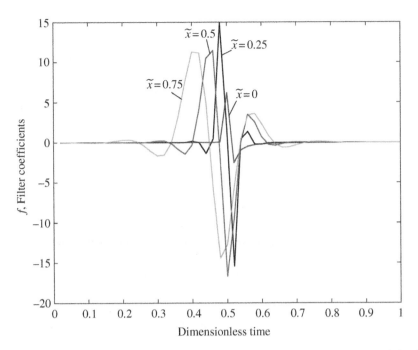

Figure 5.14 Dimensionless filter coefficients *f* for the surface heat flux at *x* = 0 for different sensor locations. First-order Tikhonov regularization α_T = 1e-4 and $\Delta\tilde{t}$ = 0.02.

$$
\begin{aligned}
\hat{q}_{M,y} &= g_1 y_{M+m_f} + g_2 y_{M+m_f-1} + \cdots + g_{m_f} y_M + \cdots \\
&\quad + g_{M+m_p-1} y_{M-m_p+1} + g_{M+m_p} y_{M-m_p} \\
&= \sum_{i=1}^{m_p+m_f} g_i y_{M+m_f-i}
\end{aligned}
\tag{5.32}
$$

where $\hat{q}_{M,y}$ represents the estimate for the part of the heat flux at $x = 0$ caused by the measured temperature at $x = L$. The properties of the filter matrix **G** are very comparable to the properties of the matrix **F**. In other words, the rows of the matrix, except for the first few and last few, have identical values, and each row is shifted by one time step. Also, the values on each row and column approaches to zero for small and large times.

Figure 5.15 illustrates the middle column of the **G** matrix for different sensor depths. The values plotted are the entries of the **g** vector. The width or range of the **g** filter is determined by the number of non-zero entries. As seen, the width of the filter increases with smaller sensor depths. This is because the measured temperature is at the remote boundary ($x = L$), and sensor locations further from this active boundary are at smaller values of x.

The filter form solution of the IHCP can be given as the sum of Eqs. (5.31) and (5.32):

$$
\begin{aligned}
\hat{q}_M &= \sum_{i=1}^{m_f+m_p} \left(f_i Y_{M+m_f-i} + g_i y_{M+m_f-i} \right) \\
&= f_1 Y_{M+m_f-1} + f_2 Y_{M+m_f-2} + \cdots + f_{m_f-1} Y_{M+1} + f_{m_f} Y_M \\
&\quad + f_{m_f+1} Y_{M-1} + \cdots + f_{m_p+m_f} Y_{M-m_p} \\
&\quad + g_1 y_{M+m_f-1} + g_2 y_{M+m_f-2} + \cdots + g_{m_f} y_M \\
&\quad + g_{m_f+1} y_{M-1} + \cdots + g_{m_p+m_f} y_{M-m_p}
\end{aligned}
\tag{5.33}
$$

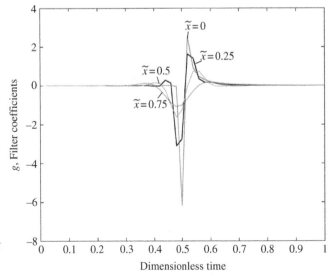

Figure 5.15 Dimensionless filter coefficients *g* for the surface heat flux at *x* = 0 for different sensor locations. First-order Tikhonov regularization α_T = 1e−4 and $\Delta\tilde{t}$ = 0.02.

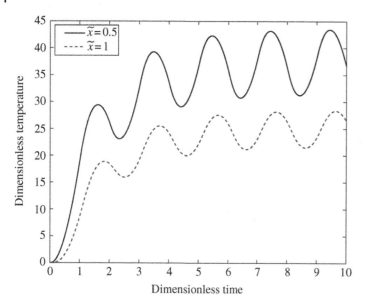

Figure 5.16 Temperatures for triangular heat flux profile ($\Delta \tilde{t} = 0.02$).

Note that all the f-filter coefficients can be found at one time by setting all the **Y** and **y** components equal to zero except the m_f component, Y_{m_f}, in Eq. (5.33) is set equal to one. To get the g-filter coefficients (those for the **y** vector), the same procedure is followed with now all the components of **Y** and **y** equal to zero except $y_{m_f} = 1$. For the TR method, it is also possible to obtain the coefficients as a column of the filter matrix (calculated using the temperature responses computed from the exact solution, as explained in Section 5.3.2). To demonstrate the application of this approach an example is presented.

Example 5.2 A finite domain of thickness 2L is considered which is exposed to a triangular heat flux on the surface ($x = 0$) and zero temperature at $x = 2L$. The dimensionless temperatures at $\tilde{x} = x/L = 0.5$, Y, and $\tilde{x} = 1$, y, are calculated and shown in Figure 5.16. In practice, temperature measurement errors are always present in the data. Therefore, normally distributed errors with a standard deviation of $\sigma = 0.5$ were added to the exact temperatures. Use the filter-based TR method Eq. (5.33) to calculate the surface heat flux using the noisy temperature data as input.

Solution

The values of m_p and m_f are selected as 50 and therefore computations start at the 51st time step and end 50 steps before the final time. The surface heat flux are calculated using both the whole time domain TR method (5.30) and the filter method Eq. (5.33) and plotted in Figure 5.17: using $\alpha_T = 0.01$. Excellent agreement between the results from whole-time domain TR and filter-based TR is observed. Also, even in the presence of the noise in the temperature data, the filter coefficient approach proved to be accurate, resulting in a RMS error of 2.54, which is about 10% of the average heat flux over the entire time domain.

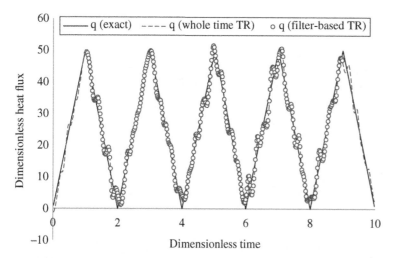

Figure 5.17 Comparison of estimated surface heat flux values by the whole-time domain TR and filter-based TR versus exact heat flux values ($\Delta \tilde{t} = 0.02$, $\alpha_T = 0.01$).

5.6 Filter Coefficients for Multi-Layer Domain

Conduction through multi-layer mediums has been discussed in several references. However, unlike direct problems, the solution of IHCPs for a multi-layer domain is only discussed in limited studies (Al-Najem and Özişik 1992; Taktak et al. 1993; Al-Najem 1997; Ruan et al. 2007). In this section, a filter-based solution approach for solving the IHCP in a two-layer medium is developed and discussed based on the work presented by Najafi et al. (Najafi 2015; Najafi et al. 2015a).

A schematic of a two-layer medium is shown in Figure 5.18 with two temperature sensors in the second (inner) material. The temperature measurement is performed using two sensors that are located at $x = x_1$ and $x = x_2$. The unknown parameter is the heat flux at the surface of the outer layer (layer 1, $x = 0$). Location x_2 may, or may not, be on the physical boundary of the inner layer.

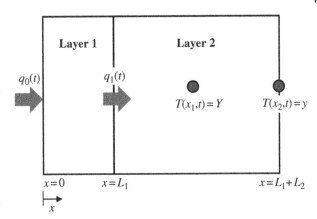

Figure 5.18 Schematic of the IHCP in a two-layer medium.

5.6.1 Solution Strategy for IHCP in Multi-Layer Domain

The solution is presented in two steps. First, a solution must be established for the IHCP in layer 2 (the inner layer). In other words, using the temperature measurements at x_1 and $x_2 = L_1 + L_2$, the heat flux, q_1, at $x = L_1$ must be determined. The solution to this IHCP can be obtained using the approach discussed in Section 5.5. As the result of solving this IHCP, the heat flux and temperature are both known at the interface (q_1 and T_1). These values are used as boundary conditions to solve the second IHCP associated with layer 1, where q at $x = 0$ (i.e. $q_0(t)$) is unknown.

5.6.1.1 Inner Layer

Two temperature measurements histories are assumed to be available at $x = x_1$ and $x = L_1 + L_2$ ($L_1 \leq x_1 \leq L_1 + L_2$) on layer 2.

$$T(x_1, t) = Y(t) \tag{5.34}$$

$$T(L_1 + L_2, t) = y(t) \tag{5.35}$$

The initial temperature is considered zero,

$$T(x, 0) = 0 \tag{5.36}$$

The temperature at x_1 ($L_1 \leq x_1 \leq L_1 + L_2$) is a function of the surface heat flux $q_1(t)$ and the temperature at $x = L_1 + L_2$.

$$\mathbf{T} = \mathbf{X}_2\mathbf{q}_1 + \mathbf{Z}\mathbf{y} \tag{5.37}$$

where

$$
T = \begin{bmatrix} T_1 \\ T_2 \\ \vdots \\ T_n \end{bmatrix}, \quad
q_1 = \begin{bmatrix} q_{1,1} \\ q_{1,2} \\ \vdots \\ q_{1,n} \end{bmatrix}, \quad
X_2 = \begin{bmatrix}
X_{2,1} & 0 & 0 & \cdots & 0 & 0 \\
X_{2,2} & X_{2,1} & 0 & \cdots & 0 & 0 \\
X_{2,3} & X_{2,2} & X_{2,1} & \cdots & 0 & 0 \\
\vdots & \vdots & \vdots & & \vdots & \vdots \\
X_{2,n} & X_{2,n-1} & X_{2,n-2} & \cdots & X_{2,2} & X_{2,1}
\end{bmatrix} \tag{5.38}
$$

$$
y = \begin{bmatrix} y_1 \\ y_2 \\ \vdots \\ y_n \end{bmatrix}, \quad
\mathbf{Z} = \begin{bmatrix}
Z_1 & 0 & 0 & \cdots & 0 & 0 \\
Z_2 & Z_1 & 0 & \cdots & 0 & 0 \\
Z_3 & Z_2 & Z_1 & \cdots & 0 & 0 \\
\vdots & \vdots & \vdots & & \vdots & \vdots \\
Z_n & Z_{n-1} & Z_{n-2} & \cdots & Z_2 & Z_1
\end{bmatrix} \tag{5.39}
$$

The estimated value heat flux vector at the interface, $\hat{\mathbf{q}}_1$, is found by minimizing the sum of the squares between the computed and measured temperature values at $x = L_1 + L_2$:

$$\hat{\mathbf{q}}_1 = \left[\mathbf{X}_2{}^T\mathbf{X}_2 + \alpha_T\mathbf{H}^T\mathbf{H}\right]^{-1}\mathbf{X}_2{}^T(\mathbf{Y} - \mathbf{Zy})$$

$$= \mathbf{F}_2(\mathbf{Y} - \mathbf{Zy}), \qquad \mathbf{F}_2 = \left[\mathbf{X}_2{}^T\mathbf{X}_2 + \alpha_T\mathbf{H}^T\mathbf{H}\right]^{-1}\mathbf{X}_2{}^T \tag{5.40}$$

Here the subscript 2 for $\mathbf{X}$ and $\mathbf{F}$ refers to layer 2. The value of $\hat{\mathbf{q}}_1$ will be used as the first known boundary condition for the IHCP associated with the layer 1 and the second boundary condition is the temperature at the interface which can be calculated as follows:

$$\hat{\mathbf{T}}_1 = \mathbf{X}_2\hat{\mathbf{q}}_1 + \mathbf{Zy} \tag{5.41}$$

where subscript 2 refers to correspondence to layer 2. By substituting $\hat{\mathbf{q}}_1$ from Eq. (5.40) in Eq. (5.41) $\hat{\mathbf{T}}_1$ can be found as:

$$\hat{\mathbf{T}}_1 = \mathbf{X}_2\mathbf{F}_2\mathbf{Y} + (\mathbf{Z} - \mathbf{X}_2\mathbf{F}_2\mathbf{Z})\mathbf{y} \tag{5.42}$$

The vector of heat fluxes at the remote surface ($x = 0$) is the unknown parameter in the second IHCP.

5.6.1.2 Outer Layer

The IHCP associated with the outer layer involves the use of $\mathbf{T}_1$ and $\mathbf{q}_1$ to calculate $\mathbf{q}_0$ (i.e. surface heat flux). The temperature at any point within layer 1 can be found as:

$$T(L_1, t) = \frac{\alpha_1}{k_1}\int_{\tau=0}^{t} q_0(\tau)G_{X22}(x_1, 0, t-\tau)d\tau + \frac{\alpha_1}{k_1}\int_{\tau=0}^{t} q_1(\tau)G_{X22}(x_1, L_1, t-\tau)d\tau \tag{5.43}$$

The above equation is the Green's function for a boundary condition of the second kind at both $x = 0$ and at $x = L_1$ (X22B–T0). A solution for the X22 case with a constant heat flux at $x = 0$ and zero heat flux at $x = L_1$ is given in Eq. (2.35). The same solution applies for a zero heat flux at $x = 0$ and a constant heat flux at $x = L_1$ through a simple change of variables, that is, $\xi = L_1 - x$. The temperature response at x_1 ($0 \leq x_1 \leq L_1$) can be found due to the heat flux histories $q_0(t)$ and $q_1(t)$. For assumed piecewise constant variation in these heat fluxes, the temperature can be computed from

$$T_M = T_{q_0,M} + T_{q_1,M} = \sum_{i=1}^{M} q_{0i}\Delta\phi_{M-i} + \sum_{i=1}^{M} q_{1i}\Delta\phi'_{M-i} \tag{5.44}$$

where

$$\phi(x_1, t) = \frac{\partial T_{X22}(x_1, t)}{\partial q_0}; \quad \phi'(x_1, t) = -\frac{\partial T_{X22}(L_1 - x_1, t)}{\partial q_1} \tag{5.45}$$

Equation (5.44) can be written in the matrix form as:

$$\mathbf{T} = \mathbf{X}_0\mathbf{q}_0 + \mathbf{X}_L\mathbf{q}_1 \tag{5.46}$$

where

$$\mathbf{T} = \begin{bmatrix} T_1 \\ T_2 \\ \vdots \\ T_n \end{bmatrix}, \quad \mathbf{q}_0 = \begin{bmatrix} q_{0,1} \\ q_{0,2} \\ \vdots \\ q_{0,n} \end{bmatrix}, \quad \mathbf{q}_1 = \begin{bmatrix} q_{1,1} \\ q_{1,2} \\ \vdots \\ q_{1,n} \end{bmatrix}, \tag{5.47}$$

$$\mathbf{X}_0 = \begin{bmatrix} X_{0,1} & 0 & 0 & \cdots & 0 & 0 \\ X_{0,2} & X_{0,1} & 0 & \cdots & 0 & 0 \\ X_{0,3} & X_{0,2} & X_{0,1} & \cdots & 0 & 0 \\ \vdots & \vdots & \vdots & & \vdots & \vdots \\ X_{0,n} & X_{0,n-1} & X_{0,n-2} & \cdots & X_{0,2} & X_{0,1} \end{bmatrix}$$

$$\mathbf{X}_L = \begin{bmatrix} X_{L,1} & 0 & 0 & \cdots & 0 & 0 \\ X_{L,2} & X_{L,1} & 0 & \cdots & 0 & 0 \\ X_{L,3} & X_{L,2} & X_{L,1} & \cdots & 0 & 0 \\ \vdots & \vdots & \vdots & & \vdots & \vdots \\ X_{L,n} & X_{L,n-1} & X_{L,n-2} & \cdots & X_{L,2} & X_{L,1} \end{bmatrix} \tag{5.48}$$

The components of the $\mathbf{X_0}$ and $\mathbf{X}_L$ matrices are related to the response basis functions as:

$$X_{0,i} = \Delta\phi_{i-1}, \quad X_{L,i} = \Delta\phi'_{i-1} \tag{5.49}$$

The whole domain Tikhonov regularization method is used to solve the IHCP. Later, the filter coefficients are found from the solution. The IHCP solution for Layer 1 starts with a matrix form for the sum of squares with an added regularization term given by

$$S = \left(\hat{\mathbf{T}}_{\mathbf{1}} - \mathbf{X}_0\mathbf{q}_0 - \mathbf{X}_{L_1}\hat{\mathbf{q}}_1\right)^T \left(\hat{\mathbf{T}}_{\mathbf{1}} - \mathbf{X}_0\mathbf{q}_0 - \mathbf{X}_{L_1}\hat{\mathbf{q}}_1\right) + \alpha_T \mathbf{q}_0^T \mathbf{H}^T \mathbf{H}\mathbf{q}_0 \tag{5.50}$$

This is minimized with respect to the parameter vector $\mathbf{q}_0$. The symbol $\hat{\mathbf{T}}_{\mathbf{1}}$ is the estimated temperature vector at L_1 and $\hat{\mathbf{q}}_{\mathbf{1}}$ is the estimated heat flux (from the solution of the IHCP in Layer 2) at $x = L_1$. The estimated value of the heat flux vector, denoted $\hat{\mathbf{q}}_0$, is then given by

$$\begin{aligned} \hat{\mathbf{q}}_0 &= \left[\mathbf{X}_0^T\mathbf{X}_0 + \alpha_T\mathbf{H}^T\mathbf{H}\right]^{-1}\mathbf{X}_0^T\left(\hat{\mathbf{T}}_{\mathbf{1}} - \mathbf{X}_{L_1}\hat{\mathbf{q}}_1\right) \\ &= \mathbf{F}_1\left(\hat{\mathbf{T}}_{\mathbf{1}} - \mathbf{X}_{L_1}\hat{\mathbf{q}}_1\right), \quad \mathbf{F}_1 = \left[\mathbf{X}_0^T\mathbf{X}_0 + \alpha_T\mathbf{H}^T\mathbf{H}\right]^{-1}\mathbf{X}_0^T \end{aligned} \tag{5.51}$$

5.6.1.3 Combined Solution

To achieve a single expression for a two-layer IHCP, Eq. (5.40) is substituted in Eq. (5.51). An expression valid for perfect contact between the layers is:

$$\hat{\mathbf{q}}_0 = \mathbf{F}_1\hat{\mathbf{T}}_1 - \mathbf{F}_1\mathbf{X}_L(\mathbf{F}_2\mathbf{Y} - \mathbf{F}_2\mathbf{Z}\mathbf{y}) \tag{5.52a}$$

where $\hat{\mathbf{T}}_1$ is the temperature at the interface and can be found from Eq. (5.42). Substituting Eq. (5.42) in Eq. (5.52a) gives the expression:

$$\begin{aligned} \hat{\mathbf{q}}_0 &= \mathbf{F}_1\hat{\mathbf{T}}_1 - \mathbf{F}_1\mathbf{X}_L(\mathbf{F}_2\mathbf{Y} - \mathbf{F}_2\mathbf{Z}\mathbf{y}) \\ &= \mathbf{F}_1(\mathbf{X_2} + \mathbf{X}_L)\mathbf{F}_2\mathbf{Y} + \mathbf{F}_1(\mathbf{Z} - \mathbf{X_2}\mathbf{F}_2\mathbf{Z} - \mathbf{X}_L\mathbf{F}_2\mathbf{Z})\mathbf{y} \end{aligned} \tag{5.52b}$$

The relation between dimensional quantities and dimensionless quantities assuming each region is non-dimensionalized using its own thickness and thermophysical properties can be given as:

$$\tilde{F}_2 = \frac{L_2}{k_2}F_2, \quad \tilde{F}_1 = \frac{L_1}{k_1}F_1, \quad \tilde{X}_2 = \frac{k_2}{L_2}X_2, \quad \tilde{X}_0 = \frac{k_1}{L_1}X_0, \quad \tilde{X}_L = \frac{k_1}{L_1}X_L$$
$$\tilde{Y} = \frac{1}{q_{ref}L_1/k_1}Y, \quad \tilde{y} = \frac{1}{q_{ref}L_1/k_1}y, \quad \tilde{R}_{int} = \frac{k_1}{L_1}R_{int} \tag{5.53}$$

5.6.2 Filter Form of the Solution

Equation (5.52b) can be written in filter form as:

$$\hat{q}_M = \sum_{j=1}^{m_p + m_f} \left(f_j Y_{M+m_f-j} + g_j y_{M+m_f-j}\right) \tag{5.54}$$

$$\begin{aligned} f &= \lfloor f_1 \quad f_2 \quad \cdots \quad f_{m_p+m_f-1} \quad f_{m_p+m_f} \rfloor = column[\mathbf{F}_1(\mathbf{X_2}\mathbf{F}_2 + \mathbf{X}_L\mathbf{F}_2)] \\ g &= \lfloor g_1 \quad g_2 \quad \cdots \quad g_{m_p+m_f-1} \quad g_{m_p+m_f} \rfloor = column[\mathbf{F}_1(\mathbf{Z} - \mathbf{X_2}\mathbf{F}_2\mathbf{Z} - \mathbf{X}_L\mathbf{F}_2\mathbf{Z})] \end{aligned} \tag{5.55}$$

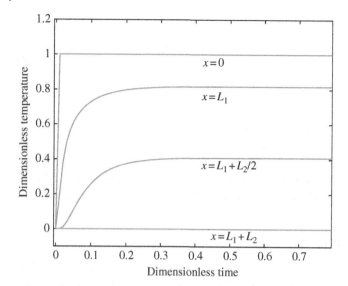

Figure 5.19 Dimensionless temperature for X1C11B10T0 case (t_d = 0.01).

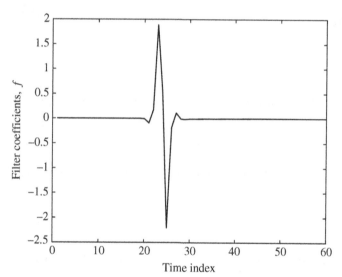

Figure 5.20 f filter coefficients for the two-layer domain (X1C11B10T0, t_d = 0.01, α_t = 0.005).

where f and g have the same characteristics as filter coefficients. Alternatively, all the f-filter coefficients can be found at one time by setting all the $\mathbf{Y}$ and $\mathbf{y}$ components equal to zero except the $m_p + 1$ component of $\mathbf{Y}$ is set equal to one ($Y_{m_p + 1} = 1$). The solution of the IHCP with this data gives the f coefficients. Similarly, to calculate the g-filter coefficients, the same procedure is followed with now all the components of $\mathbf{Y}$ and $\mathbf{y}$ equal to zero except $y_{m_p + 1} = 1$.

Example 5.3 Consider a two-layer domain with the following properties: layer 1 is 0.1 m with a thermal conductivity of 54 W/m−K and volumetric heat capacity of 3642.3 kJ/m^{3}K. Layer 2 is 0.2 m thick with thermal conductivity of 14.2 W/m−K and volumetric heat capacity of 3829.4 kJ/m^{3}K. The surface of the domain here is subject to a step change in temperature at the surface (X1C11B10T0). Here the notation "1C11" indicates a perfect contact between two regions ("C1") with type 1 conditions on either side of the composite medium. The solution to the direct problem is generated using the exact method for the two-layer one-dimensional conduction problem (Cole 2018). The temperature data are evaluated and used as inputs for the IHCP solution with a dimensionless time step of 0.01. The temperature values are plotted in Figure 5.19. Note that for this example, it is assumed that the temperature sensors are located at $x = L_1 + L_2/2$ and $x = L_1 + L_2$. A uniform random ±1% error in temperature measurement is added to the temperature data. The thermal contact resistance is neglected. Calculate the surface heat flux using filter-based TR method.

Solution

First-order regularization is used with α_t = 0.005. The filter coefficients f and g are obtained using Eq. (5.55) and plotted in Figures 5.20 and 5.21, respectively. The values of m_p and m_f are determined as 25 through assessment of the filter coefficients. Using Eqs. (5.52b) and (5.54) the surface heat flux is evaluated and compared with the exact solution, as plotted in Figure 5.22. The RMS between the exact solution and the estimated heat flux by using whole time domain and the filter method is evaluated as 5.4499 and 5.4591 W/m^2, respectively. A good agreement between the exact data and estimated values is achieved which is an indication for the successful performance of the filter-based TR approach.

5.7 Filter Coefficients for Non-Linear IHCP: Application for Heat Flux Measurement Using Directional Flame Thermometer

In the previous section, the possibility of developing filter-based TR for solving IHCP in multi-layer domains was explored and it was shown that the method can be effectively used for near real-time surface heat flux estimation. In this section, an example of industrial application of IHCP in a multi-layer domain with temperature dependent material properties is discussed: application of directional flame thermometer (DFT) for heat flux measurement.

Figure 5.21 g filter coefficients for the two-layer domain (X1C11B10T0, t_d = 0.01, α_t = 0.005).

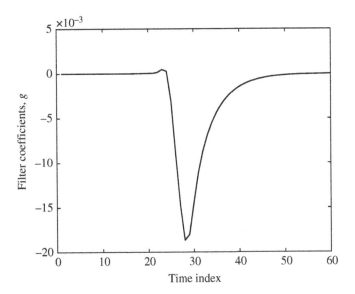

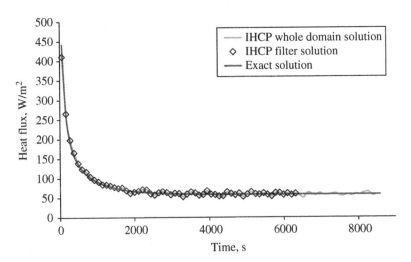

Figure 5.22 Calculated surface heat flux for (X1C11B10T0, t_d = 0.01, α_t = 0.005).

A schematic of a DFT is shown in Figure 5.23. Sandia National Laboratories advanced DFT design to use for both transient and quasi steady state heat flux measurement in large pool fires and other tests in extreme conditions (Keltner et al. 2010). Application of DFT has been also reported for heat flux measurement in wildland-urban interface fires (Manzello et al. 2010), fire resistance tests (Sultan 2010), and other applications (Lam et al. 2009).

Typically, evaluating the surface heat flux using temperature data obtained by a DFT was performed offline at the conclusion of the experiment through a whole-time domain approach. However, Najafi et al. (2015) developed a filter-based approach for solving the IHCP associated with a DFT, facilitating near real-time surface heat flux estimation. It should be noted that considering the wide range of temperature variation for DFT application, the material properties of the layers greatly vary during the experiment which causes non-linearity to the problem. Therefore, an effective IHCP solution for such application must be capable of

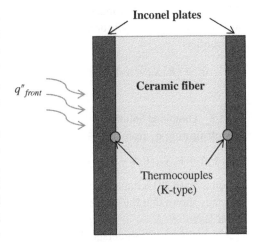

Figure 5.23 Schematic of a directional flame thermometer (DFT).

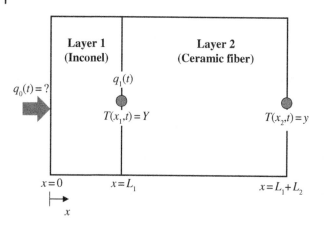

Figure 5.24 The two-layer IHCP associated with DFT application.

addressing the temperature-dependent material properties. The major advantage of the filter-based solution is its ability to account for nonlinearity of the problem through a simple and effective approach while providing near real-time results which allows active monitoring of heat flux that can be used for optimal control purposes. Availability of a near real-time algorithm for accurate reduction of data will allow for continual monitoring of the furnace during operation.

5.7.1 Solution for the IHCP

The solution procedure for solving the IHCP in DFT is similar to the procedure previously explained in Section 5.6 for IHCP in multi-layer domains. The front side of the DFT is exposed to heat flux which is to be determined and the temperature measurement data from two thermocouples at the interfaces between the layers are available. A schematic of the two-layer problem is given in Figure 5.24. Two IHCPs are solved separately for heat flux estimation on the front plate surface and the middle layer and eventually a coupled solution is derived for the two-layer problem associated with DFT. This is similar to the schematic that was previously given in Figure 5.18, but the front sensor is placed exactly at the interface between the two layers (Layers 1 and 2).

The solution is started from Layer 2 (Ceramic Fiber) and the unknown heat flux ($\mathbf{q_1}$) is evaluated using two temperature measurements at two boundaries. After solving the first IHCP, the heat flux and temperature are both known at the interface ($\mathbf{q_1}$ and $\mathbf{Y}$). These values will be used as boundary conditions to solve the second IHCP associated with the Inconel plate to calculate $\mathbf{q_0}$.

This problem is a special form of the multi-layer problem that was previously discussed. In the formulation that was explained in Section 5.6, the front sensor could be located anywhere within the second layer. For a DFT, the front sensor is simply located at the boundary.

5.7.1.1 Back Layer (Insulation)

As before, Tikhonov regularization method can be used to evaluate the heat flux at the interface, $\hat{\mathbf{q}}_1$, as:

$$
\begin{aligned}
\hat{\mathbf{q}}_1 &= \left[\mathbf{X}_2{}^T\mathbf{X}_2 + \alpha_T\mathbf{H}^T\mathbf{H}\right]^{-1}\mathbf{X}_2{}^T(\mathbf{Y} - \mathbf{Zy}) \\
&= \mathbf{F}_2(\mathbf{Y} - \mathbf{Zy})
\end{aligned}
\tag{5.56}
$$

where $\mathbf{Y}$ and $\mathbf{y}$ are the vectors of measured temperatures at $x = L_1$ and $x = L_1+L_2$, respectively. The subscript 2 for $\mathbf{X}$ and $\mathbf{F}$ refers to Layer 2 and $\mathbf{H}$ is the first order regularization matrix. Note that this equation is dimensional.

5.7.1.2 Front Layer (Inconel plate)

As before, using the whole domain Tikhonov regularization method, the estimated surface heat flux at the outer surface can be found as:

$$
\hat{\mathbf{q}}_0 = \left[\mathbf{X}_0^T\mathbf{X}_0 + \alpha_T\mathbf{H}^T\mathbf{H}\right]^{-1}\mathbf{X}_0^T(\mathbf{Y} - \mathbf{X}_{L_1}\mathbf{q}_1) = \mathbf{F}_1(\mathbf{Y} - \mathbf{X}_{L_1}\mathbf{q}_1)
\tag{5.57}
$$

5.7.1.3 Combined Solution

By substituting $\hat{\mathbf{q}}_1$ from Eq. (5.56) into Eq. (5.57), the following equation can be obtained:

$$
\hat{\mathbf{q}}_0 = \mathbf{F}_1\mathbf{Y} + \mathbf{F}_1\mathbf{X}_{L_1}(\mathbf{F}_2\mathbf{Y} - \mathbf{F}_2\mathbf{Zy}) = \mathbf{F}_1(\mathbf{I} + \mathbf{X}_{L_1}\mathbf{F}_2)\mathbf{Y} - \mathbf{F}_1\mathbf{X}_{L_1}\mathbf{F}_2\mathbf{Zy}
\tag{5.58}
$$

Equation (5.58) is dimensional: $\mathbf{F}$ has units of $\text{W/m}^2–\text{K}$, $\mathbf{X}$ has units of $\text{K}–\text{m}^2/\text{W}$, and $\mathbf{Z}$ is dimensionless. Equation (5.58) basically shows that the temperature measurements from the inner layer can be used to evaluate the surface heat flux of the outer layer.

It is useful to also understand the dimensionless form of Eq. (5.58). This is particularly useful since the two layers may have different properties so non-dimensionalization must be done carefully. The relation between dimensional quantities and dimensionless quantities by different thermal properties in the two regions are defined as below. The dimensionless quantities, denoted with a tilde "~", are:

$$\tilde{\mathbf{F}}_2 = \frac{L_2}{k_2}\mathbf{F}_2, \quad \tilde{\mathbf{F}}_1 = \frac{L_1}{k_1}\mathbf{F}_1, \quad \tilde{\mathbf{X}}_2 = \frac{k_2}{L_2}\mathbf{X}_2, \quad \tilde{\mathbf{X}}_0 = \frac{k_1}{L_1}\mathbf{X}_0,$$

$$\tilde{\mathbf{X}}_L = \frac{k_1}{L_1}\mathbf{X}_L, \quad \tilde{\mathbf{Y}} = \frac{1}{q_{ref}L_1/k_1}\mathbf{Y}, \quad \tilde{\mathbf{y}} = \frac{1}{q_{ref}L_1/k_1}\mathbf{y}, \quad \tilde{\mathbf{Z}} = \mathbf{Z} \tag{5.59}$$

The non-dimensional form of Eq. (5.58) can be then written as:

$$\hat{\mathbf{q}}_0 = \tilde{\mathbf{F}}_1\left(\mathbf{I} + \tilde{\mathbf{X}}_{L_1}\tilde{\mathbf{F}}_2\frac{\tilde{k}}{\tilde{L}}\right)\tilde{\mathbf{Y}} - \left(\tilde{\mathbf{F}}_1\tilde{\mathbf{X}}_{L_1}\tilde{\mathbf{F}}_2\tilde{\mathbf{Z}}\frac{\tilde{k}}{\tilde{L}}\right)\tilde{\mathbf{y}} \tag{5.60}$$

5.7.2 Filter form of the solution

Equation (5.58) can be written in filter form as follows:

$$\hat{\mathbf{q}}_0 = \mathbf{FY} + \mathbf{Gy} \tag{5.61}$$

where $\mathbf{F} = \mathbf{F}_1(\mathbf{I} + \mathbf{X}_{L_1}\mathbf{F}_2)$ and $\mathbf{G} = -\mathbf{F}_1\mathbf{X}_{L_1}\mathbf{F}_2\mathbf{Z}$ are the filter matrices and have the same characteristics as filter matrix which was discussed earlier in this chapter. The number of non-negligible coefficients before and after the current time step is denoted as m_p and m_f, respectively. Therefore, the heat flux at the current time step is only a function of temperature data associated to the time steps with non-negligible filter coefficients (several previous time step and a few future time steps) and can be found as:

$$\hat{q}_M = \sum_{j=1}^{m_p + m_f}\left(f_j Y_{M + m_f - j} + g_j y_{M + m_f - j}\right) \tag{5.62}$$

where the vectors of the filter coefficients are

$$\mathbf{f} = \lfloor f_1 \quad f_2 \quad \cdots \quad f_{m_p + m_f - 1} \quad f_{m_p + m_f} \rfloor = row(\mathbf{I} + \mathbf{X}_{L_1}\mathbf{F}_2) \tag{5.63a}$$

$$\mathbf{g} = \lfloor g_1 \quad g_2 \quad \cdots \quad g_{m_p + m_f - 1} \quad g_{m_p + m_f} \rfloor = row(-\mathbf{F}_1\mathbf{X}_{L_1}\mathbf{F}_2\mathbf{Z}) \tag{5.63b}$$

5.7.3 Accounting for Temperature-Dependent Material Properties

When DFT is used for heat flux measurement, the material experiences a very large temperature variation during the test, starting from ambient temperature and reaching to more than 1000 K. Similar scenarios are expected for many real-world applications where temperature variation is significant (laser therapy of tumors, re-entry vehicle, etc.). One of the strengths of the filter approach is it's capability to account for temperature dependent material properties through simple interpolation. In this approach, the filter coefficients for a given problem need to be generated for several different temperatures initially and before start of the experiment. When using the filter method for near real-time heat flux estimation, at each time step, the associated filter coefficients must be interpolated between the pre-calculated values for the neighboring temperatures to accurately account for the temperature dependent material properties. Here, this concept is further elaborated through the application of IHCP solution in a DFT.

Considering the material properties of an actual DFT given in Table 5.4, time step of 5 seconds and Tikhonov regularization parameter $\alpha_T = 0.001$, the "**f**" and "**g**" filter coefficients are determined for different temperatures. The calculated filter coefficients are plotted versus the time index in Figures 5.25 and 5.26 for different temperatures.

From Figures 5.25 and 5.26, it is clear that the filter coefficients at different temperatures follow a similar trend, but are different in values. As expected, there are limited number of non-zero filter coefficients. m_p is the number of non-negligible coefficients before the current time step and m_f is the number of non-negligible coefficients after the current time step. Any coefficient less than $\varepsilon = 5 \times 10^{-5}$ is assumed to be negligible.

The selection of m_p and m_f must be done carefully as they depend on the boundary conditions, material properties, sensor location, time step, and regularization parameter. To determine appropriate m_p and m_f, filter coefficients must be closely assessed. This is further illustrated in Figures 5.25 and 5.26. Clearly, the smaller is the number of required future time steps

Table 5.4 Characteristics of the layers of a directional flame thermometer (DFT).

Thickness, m (K)	Layer 1 (Inconel)	Layer 2 (Ceramic Fiber)
	0.0031	0.019
T = 300	k = 14.9 W/m−K, α = 3.9e-6 m^2/s	k = 0.046 W/m−K, α = 3.2e-7 m^2/s
T = 450	k = 17.4 W/m−K, α = 4.2e-6 m^2/s	k = 0.065 W/m−K, α = 3.9e-7 m^2/s
T = 800	k = 22.6 W/m−K, α = 4.9e-6 m^2/s	k = 0.13 W/m−K, α = 6.9e-7 m^2/s
T = 1000	k = 25.4 W/m−K, α = 5.3e-6 m^2/s	k = 0.19 W/m−K, α = 7.9e-7 m^2/s
T = 1050	k = 26.1 W/m−K, α = 5.31e-06 m^2/s	k = 0.21 W/m−K, α = 1.04e-06 m^2/s
T = 1200	k = 28.2 W/m−K, α = 5.57e-6 m^2/s	k = 0.27 W/m−K, α = 1.2e-6 m^2/s

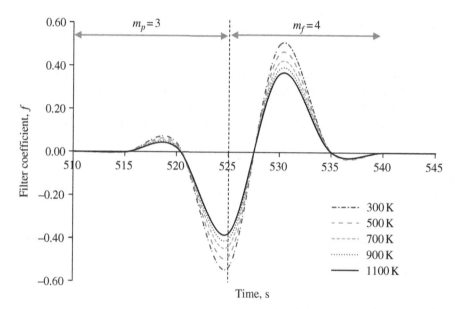

Figure 5.25 f filter coefficients for different temperatures.

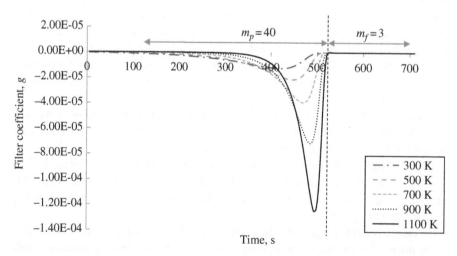

Figure 5.26 g filter coefficients for different temperatures.

(m_f) for calculating the heat flux at the current time step, the closer the algorithm can operate to the real time. Careful observation of Figures 5.25 and 5.26 demonstrate that m_f is 4 and 3 for **f** and **g** filter coefficients, respectively. Therefore, the maximum of these two values, which is 4, should be used to achieve accurate results. Similarly, m_p can be determined by close observation of the filter coefficients shown in Figures 5.25 and 5.26 which suggest that m_p is 3 and 40 for **f** and **g**, respectively. Therefore, the bigger value of 40 should be used for calculations. It should be noted that a bigger value of m_p does not affect how fast (closer to real-time) the algorithm can operate as it only decides how many data points from previous time steps must be used to determine the heat flux at the current time.

The regularization parameter,α_t, should also be selected carefully to obtain an accurate estimation of the heat flux at the surface. Optimal selection of the regularization parameter is discussed in Chapter 6. Several methods are available for selecting regularization parameter, including L-curve (Hansen and O'Leary 1993), generalized cross validation and discrepancy principle. Woodbury and Beck (2013) showed that the optimal selection of the regularization parameter by minimizing the heat flux error produces results comparable to the Morozov principle. It was shown that the appropriate selection of the order of magnitude of the regularizing parameter is important to minimize errors, but the precise selection of the regularization does not make a large impact for accurate estimation of heat flux. When using filter approach for real-time heat flux estimation, it is necessary to have a preliminary understanding of the problem and the nature of the heat flux profile (whether it is uniform or involves sudden or smooth changes, etc.) to pick the appropriate regularization parameter and corresponding m_p and m_f values.

As seen, for the given geometry, material properties and sampling time step of 5 seconds in DFT application, the temperature data from four future time steps ($m_f = 4$) is needed to do the calculations (including the current time step). This suggests that the method can report the heat flux with a 15 seconds delay. Also, it is determined that for accurate calculation of heat flux, the temperature data from 40 previous time steps ($m_p = 40$) is needed (equivalent to 200 seconds).

As seen in Figures 5.25 and 5.26, the filter coefficients vary considerably as the temperature changes. Beck (2008) showed that when the material properties are temperature dependent, one can interpolate filter coefficients at each time step based on the measured temperature to achieve accurate results and handle the non-linearity of the problem. Since the temperature greatly varies in DFT application, it is necessary to find the filter coefficients at each time step with a particular temperature and use them accordingly. This can be done by finding the filter coefficients for a set of temperatures and interpolate between those at each time step. Here, the filter coefficients for 300, 400, ..., 1300 K are calculated. At each time step, the associated filter coefficients for the temperature at the current time is calculated using linear interpolation. The obtained filter coefficients are used to estimate the heat flux. Later it is shown in the test cases that accounting for temperature dependent material properties can greatly improve the accuracy of the estimated heat flux values.

5.7.4 Examples

In order to demonstrate the use of filter method for solving non-linear IHCPs and particularly the DFT problem, two examples are presented with two different heat flux profile: triangular heat flux profile produced by numerical simulation in ANSYS and heat flux data obtained experimentally using DFT in field test.

Example 5.4 A one-dimensional two-layer medium is created in ANSYS and a known heat flux profile is applied on the front surface. The temperature values at the two interfaces are then calculated and obtained from ANSYS and then used as inputs to the filter-based solution Eq. (5.62). The characteristics of the two layers, including the temperature-dependent material properties, are listed in Table 5.4, the thermocouples' locations are considered the same as the DFT case (Figure 5.24). A triangular heat flux profile is applied to the front surface while the back surface of the second layer is maintained at 300 K. The temperature histories (time step of one second for a period of 185 seconds) are then obtained at the desired surfaces using ANSYS simulation. Figure 5.27 depicts the evaluated temperature values at $x = L_1$ and $x = L_1 + L_2$ (**Y** and **y,** respectively). Calculate the surface heat flux using filter-based TR approach.

Solution

The IHCP is solved first using a constant set of filter coefficients associated with one value of temperature including 300, 500, 800, and 1000 K and the resulting surface heat flux is calculated using Eq. (5.62) and plotted against the ANSYS input heat flux in Figure 5.28. The regularization parameter for this example is set as 0.01, the time step is one second and the number of previous and future time steps are 150 and 8, respectively. It can be observed that for cases with large temperature

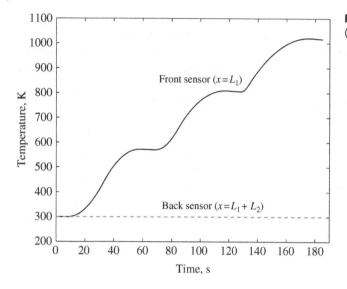

Figure 5.27 Temperature measurements on front (**Y**) and back (**y**) side of the second layer (Example 5.4).

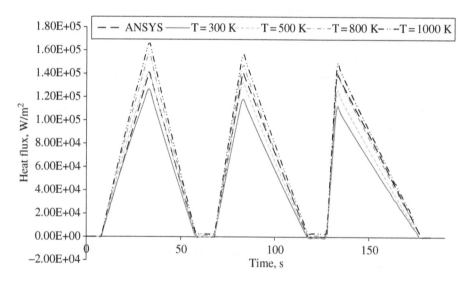

Figure 5.28 Surface heat flux evaluated for 5.2 with constant material properties at different temperatures (sampling time: 1s, $m_p = 40$, $m_f = 4$, $\alpha_t = 10^{-2}$).

variations, when temperature-dependent material properties are not accounted in the solution, significant discrepancy in the results will be inevitable.

To improve the results, the filter coefficients are linearly interpolated at each time step and used accordingly to account for temperature dependent material properties. The results are plotted in Figure 5.29 and compared against the ANSYS simulation inputs as well as constant filter coefficients that are generated at the overall average temperature of 691 K (average temperature for the entire duration of the experiment). It can be seen that the interpolation of filter coefficients successfully addresses the non-linearity of the problem and results in the lowest possible errors. The RMS error between the calculated heat flux values using filter method (linear with constant filter coefficients and non-linear with linear interpolation between pre-calculated filter coefficients) and the ANSYS data are determined and listed in Table 5.5. As anticipated, the RMS error associated with the case which used filter interpolation to address the non-linearity of the problem resulted in the lowest RMS error. This shows the importance of accounting for the variation of material properties with temperature, particularly

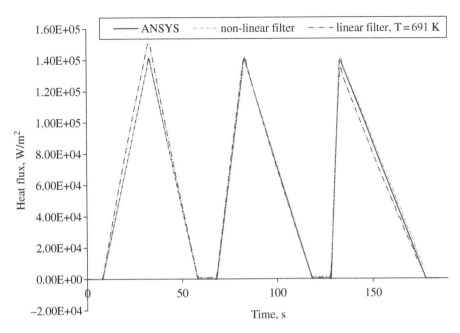

Figure 5.29 Calculated surface heat flux for Example 5.2 accounting for temperature dependent material properties (non-linear filter), ANSYS inputs and filter coefficients for constant material properties at average temperature of 691 K (sampling time: 1 seconds, m_p = 150, m_f = 8, α_t = 10^{-2}).

Table 5.5 Calculated RMS error between the estimated heat fluxes using the developed filter method and the exact values (Example 5.4).

		Linear					Non-Linear
Example 5.4	**Temperature (K)**	**300**	**500**	**800**	**691 (T_{Avg})**	**1000**	
	RMS Error (W/m²)	9,213.0	4,727.4	4,842.6	3,834.8	7,973	2,484.7

for applications that experience large temperature changes. The value of Tikhonov regularization parameter is set as 0.01 for all cases which yielded the minimum RMS value. The RMS error is calculated as:

$$E_{RMS} = \left(\frac{\sum_{i=1}^{n} \left(q_{exact,i} - q_{estimated,i} \right)^2}{n} \right)^{\frac{1}{2}} \tag{5.64}$$

where n is the number of time steps.

Example 5.5 For this example, actual field test data obtained from Sultan (2010). The temperature measurements from the front and backside thermocouples are plotted in Figure 5.30. The sampling time is 5 seconds, and the data is obtained for 3655 seconds. Calculate the surface heat flux using filter-based TR approach.

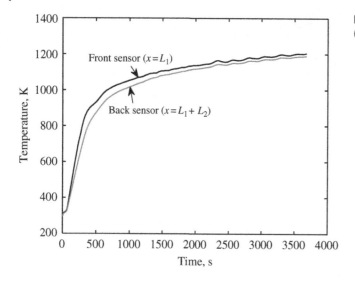

Figure 5.30 Temperature measurement data for DFT field test (Example 5.5)

Solution

The temperature data are substituted in Eq. (5.62) to calculate the surface heat flux. The filter coefficients are pre-calculated as presented in Figures 5.25 and 5.26. Similar to Example 5.2, first, the impact of not accounting for temperature dependent material properties is explored. The filter method is used with filter coefficients that are calculated using a constant temperature (300, 800, and 1200 K) to determine the surface heat flux, as shown in Figure 5.31.

Results are also presented based on analysis using the industry code "IHCP1D." This code was written by James Beck and is used by Sandia National Labs to analyze data from the DFT. The IHCP1D code solves the non-linear IHCP problem iteratively using finite difference numerical solution for the forward heat conduction problem.

All cases with no consideration of temperature dependent material properties deviate significantly from the IHCP1D results (which is expected). By accounting for temperature dependent material properties, the heat flux is estimated using the filter coefficient method and compared with the results from IHCP1D in Figure 5.32. A very good agreement between the results from the two methods is observed. These results show how the filter approach can be used to effectively account for temperature dependent material properties through simple interpolation. In this case, the Tikhonov regularization

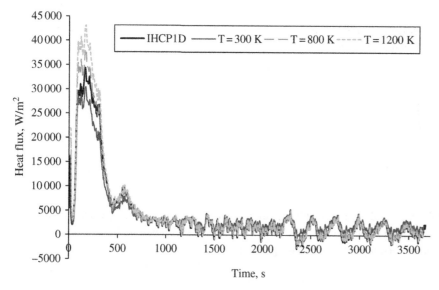

Figure 5.31 Calculated surface heat flux for DFT field data-constant material properties at different temperatures (sampling time: 5 seconds, m_p = 40, m_f = 4, α_t = 10^{-7}).

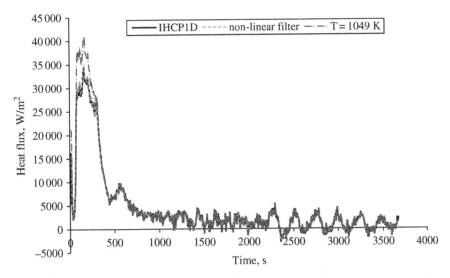

Figure 5.32 Comparison of calculated surface heat flux for DFT field data by accounting for temperature dependent material properties (non-linear filter), constant material properties at the average temperature of 1049 K and results from IHCP1D whole time analysis (sampling time: 5 seconds, $m_p = 40$, $m_f = 4$, $\alpha_t = 10^{-7}$).

Table 5.6 Calculated RMS error between the estimated heat fluxes using the developed filter method and the results from IHCP1D (Example 5.3).

		Linear					Non-linear
Example 5.5	**Temperature (K)**	**300**	**500**	**800**	**1049 (T_{Avg})**	**1200**	
	RMS Error (W/m^2)	1476.2	965.5	903.7	1412.9	1820.8	151.0

parameter is selected as 10^{-7}, $m_f = 4$ and $m_p = 40$. The value of m_f specifies the delay for heat flux estimation, which is 15 seconds for this simulation, as discussed earlier. The RMS error between the results from filter method (linear and non-linear) and results from IHCP1D is listed in Table 5.6. It is observed that the non-linear solution leads to a much lower RMS error which results in significantly higher accuracy.

5.8 Chapter Summary

Chapter 5 is focused on developing digital filter form of IHCP solution techniques. In estimating the surface heat flux using various IHCP solution techniques, it is possible to write the solution in the form of a filter matrix multiplied by the vector of temperature measurement data. The characteristics of the filter matrix are unique for each solution method. Chapter 8 started by exploring the filter form solution for FSM, TR, SVD, and CGM and the characteristics of the filter matrix for each of these methods is discussed. It is shown that for certain methods (FSM and TR) the characteristics of filter coefficients facilitate the use of digital filters for near real-time surface heat flux estimation which is fast and computationally efficient and therefore appropriate for online monitoring and control purposes. The filter-based TR method is then further developed for using with problems that have temperature measurement on their remote boundary and IHCPs in multi-layer domains. Additionally, application of filter-based TR for solving non-linear IHCPs (i.e. domains with temperature dependent material properties) through linear interpolation is explained in detail. Overall, the filter form solution of IHCP proved to be a very effective and computationally efficient technique for solving both linear and non-linear IHCPs.

Problems

5.1 Consider $\mathbf{R}$ matrix which is defined as $\mathbf{R} = \mathbf{XF}$. Calculate the $\mathbf{R}$ matrix for X22B10T0 case with $N = 51$ time steps, zeroth order regularization, $\tilde{x} = 0.5$, and $\Delta \tilde{t} = 0.05$ when using Tikhonov regularization method assuming $\alpha_t = 0.001$. Plot several rows of the $\mathbf{R}$ matrix and discuss your observations. What do you expect for the summation of the r coefficients (components of one row of the $\mathbf{R}$ matrix) which is defined as:

$$\sum_{j=1}^{N} r_{ij} = ? \text{ (for any row } i \text{ other than the first few and last few rows)}$$

5.2 Consider $\mathbf{P}$ matrix which is defined as $\mathbf{P} = \mathbf{FX}$ for the conditions described in Problem 5.1. Plot several rows of the $\mathbf{P}$ matrix and discuss your observations. What do you expect for the summation of the p coefficients (components of one row of the $\mathbf{P}$ matrix) which is defined as:

$$\sum_{j=1}^{N} p_{ij} = ? \text{ (for any row } i \text{ other than the first few and last few rows)}$$

5.3 For the conditions described in Problem 5.1, calculate the mid-column of the filter matrix for $\tilde{x} = 0.25$, $\tilde{x} = 0.5$, and $\tilde{x} = 0.75$, assuming $\alpha_t = 0.001$. Describe your observations by comparing the resulting filter coefficients. Calculate the sum of filter coefficients for each sensor location and explore how does changing the regularization parameter between 0.000001 and 0.1 affects the filter coefficients and summation of filter coefficients at each sensor location.

5.4 It was described in Section 5.3.1 that the sum of filter coefficients (components of a single row of the filter matrix) for X22B10T0 problem is 0. Such concept can be used for intrinsic verification. Explore what would be the sum of the filter coefficients for the X21B10T0 case.

5.5 The filter form of the CGM explained in Section 5.3.4 and the filter coefficients when using Fletcher–Reeves algorithm were plotted in Figure 5.8. Use alternative gradient methods including steepest decent and Polak–Ribiere, calculate the filter coefficients and plot the mid column as well as plus and minus 1 adjacent columns and describe your observations. Can filter form of CGM be used as a sequential method?

5.6 Repeat Example 5.3 assuming the temperature sensors are now located at $x = L_1 + L_2$ and $x = L_1$. Calculate the filter coefficients and the surface heat flux using the filter-based TR method and discuss how your filter coefficients compared with the ones presented for Example 5.3. The numerical values for dimensionless temperature are available in the companion website.

5.7 Consider a one-dimensional slab (2 cm thick) is subject to heat flux on it's surface. Temperature measurements are obtained at $x = L/2$ and $x = L$. Calculate the filter coefficients at 300, 450, 600, 750, and 950 K for this problem assuming the slab is made of Inconel with material properties listed in Table 5.4.

5.8 Repeat Example 5.4 and first calculate the filter coefficients for 400, 600, and 900 °C. Evaluate the surface heat flux using each set of these filter coefficients (without interpolation, similar to Figure 5.28) and calculate the associated RMS between estimated and exact values of surface heat flux. Second, use the interpolation of filter coefficients to account for nonlinearity of the IHCP and compare the resulting surface heat flux and RMS value against what you previously calculated. The numerical values for temperature and heat flux are available in the companion website.

References

Al-Najem, N. M. (1997) Whole time domain solution of inverse heat conduction problem in multi-layer media, *Heat and Mass Transfer* 33(3), pp. 233–240. https://doi.org/10.1007/S002310050183.

Al-Najem, N. M. and Özişik, M. N. (1992) 'Estimating unknown surface condition in composite media', *International Communications in Heat and Mass Transfer*, 19(1), pp. 69–77. https://doi.org/10.1016/0735-1933(92)90065-P.

Beck, J. V. (2008) Filter solutions for the nonlinear inverse heat conduction problem, *Inverse Problems in Science & Engineering*, 16 (1), pp. 3–20. https://doi.org/10.1080/17415970701198332.

Beck, J. V., Blackwell, B. and St Clair Jr, C. R. (1985) *Inverse Heat Conduction: Ill-Posed Problems*. New York: A Wiley-Interscience.

Cole, K. D. (2018) *ExACT Analytical Conduction Toolbox* (December 1, 2021). Available at: http://exact.unl.edu.

Cole, K. D. *et al.* (2011) *Heat Conduction using Green's Functions.*, 2nd edn. Boca Raton, FL: CRC Press.

Hansen, P. C. and O'Leary, D. P. (1993) The use of the L-curve in the regularization of discrete Ill-posed problems, *SIAM Journal on Scientific Computing*, 14(6), pp. 1487–1503. https://doi.org/10.1137/0914086.

Keltner, N. R., Beck, J. V. and Nakos, J. T. (2010) Using directional flame thermometers for measuring thermal exposure, *Journal of ASTM International*, 7(2). https://doi.org/10.1520/JAI102280.

Lam, C. S. *et al.* (2009) 'Steady-state heat flux measurements in radiative and mixed radiative–convective environments', *Fire and Materials*, 33(7), pp. 303–321. https://doi.org/10.1002/FAM.992.

Manzello, S. L., Park, S.-H. and Cleary, T. G. (2010) Development of rapidly deployable instrumentation packages for data acquisition in wildland-urban interface (WUI) fires, *Fire Safety Journal*, 45, pp. 327–336. https://doi.org/10.1016/j.firesaf.2010.06.005.

Najafi, H. (2015) *Real-Time Heat Flux Estimation Using Filter Based Solutions for Inverse Heat Conduction Problems*. University of Alabama.

Najafi, H. *et al.* (2015) Real-time heat flux measurement using directional flame thermometer, *Applied Thermal Engineering*, 86, pp. 229–237. https://doi.org/10.1016/j.applthermaleng.2015.04.053.

Najafi, H., Woodbury, K. A. and Beck, J. V. (2015a) A filter based solution for inverse heat conduction problems in multi-layer mediums, *International Journal of Heat and Mass Transfer*, 83, pp. 710–720. https://doi.org/10.1016/j.ijheatmasstransfer.2014.12.055.

Najafi, H., Woodbury, K. A. and Beck, J. V. (2015b) Real time solution for inverse heat conduction problems in a two-dimensional plate with multiple heat fluxes at the surface, *International Journal of Heat and Mass Transfer*, 91, pp. 1148–1156. https://doi.org/10.1016/j.ijheatmasstransfer.2015.08.020.

Ruan, Y., Liu, J. C. and Richmond, O. (2007) Determining the unknown cooling condition and contact heat transfer coefficient during solidification of alloys, 1(1), pp. 45–69. https://doi.org/10.1080/174159794088027572.

Sultan, M. A. (2010) Performance of different temperature sensors in standard fire resistance test furnaces, *Fire Technology*, 46(4), pp. 853–881. https://doi.org/10.1007/S10694-010-0166-9.

Taktak, R., Beck, J. V. and Scott, E. P. (1993) Optimal experimental design for estimating thermal properties of composite materials, *International Journal of Heat and Mass Transfer*, 36(12), pp. 2977–2986. https://doi.org/10.1016/0017-9310(93)90027-4.

Uyanna, O. and Najafi, H. (2022) A novel solution for inverse heat conduction problem in one-dimensional medium with moving boundary and temperature-dependent material properties, *International Journal of Heat and Mass Transfer*, 182, p. 122023. https://doi.org/10.1016/j.ijheatmasstransfer.2021.122023.

Woodbury, K. A. and Beck, J. V. (2013) Estimation metrics and optimal regularization in a Tikhonov digital filter for the inverse heat conduction problem, *International Journal of Heat and Mass Transfer*, 62, pp. 31–39. https://doi.org/10.1016/j.ijheatmasstransfer.2013.02.052.

6

Optimal Regularization

Chapter 4 described and demonstrated the need for some form of regularization to stabilize solution of inverse heat conduction problems (IHCPs). Regularization introduces stability, but at the expense of introducing bias in the estimation.

The first part of this chapter formalizes the trade-off between minimum bias and minimum variance. The optimal amount of regularization required for solving the IHCP strikes a balance between these two.

The remainder of the chapter will outline several methods which can be used to determine the optimal degree of regularization. Two main approaches require minimization of the expected value of the residual in heat flux or temperature estimates. Two additional approaches are the graphical L-curve method and the Generalized Cross Validation (GCV) method.

6.1 Preliminaries

6.1.1 Some Mathematics

Discussions in the following sections pertain to the *expected value* of a random variable, and the *variance* of a random variable. Also, the distinction between a random variable, a parameter, and the estimate of a parameter needs to be drawn. These concepts are quickly reviewed in this section.

Consider an experiment or measurement of an outcome, and let that outcome be denoted by the variable y. If the procedure is repeated N times, then outcomes y_i, for $i = 1,2,...,N$ will be obtained. Presumably, there will be some differences in the outcome measures y_i obtained; the outcome y can be considered a *random variable*.

The *expected value* of y is the one most likely to be observed in any sequence of repeated trials. For the N observations, it is the mean, or averaged, value:

$$E(y) = \frac{1}{N}\sum_{i=1}^{N} y_i = \bar{y} \tag{6.1}$$

The actual individual observations, y_i, will form a pattern or distribution around $\bar{y}$. The amount of scatter or variability of the pattern is quantified by the *variance* of y, denoted $V(y)$. This measure is the average of the square of the distance away from $\bar{y}$ that any observation will fall. For a population of N values, the variance for a population is

$$V(y) = E\{(y - E(y))^2\}$$
$$= \frac{1}{N}\sum_{i=1}^{N} (y_i - \bar{y})^2 = \sigma_y^2 \tag{6.2}$$

which is the same as the square of the standard deviation, σ.

A mathematical model for the outcome of a process will involve constants, or *parameters*, that will be used to predict the outcome. For example, a linear model $y = \beta_1 x + \beta_2$ contains two parameters, β_1 and β_2, and models the outcome y in terms of an independent variable x. When some procedure is used to estimate the values of the parameters, the " ^ " decoration is used to distinguish the estimate, $\hat{\beta}$ from the true, generally unknown, value, β.

Inverse Heat Conduction: Ill-Posed Problems, Second Edition. Keith A. Woodbury, Hamidreza Najafi, Filippo de Monte, and James V. Beck.
© 2023 John Wiley & Sons, Inc. Published 2023 by John Wiley & Sons, Inc.

As described in Chapters 3 and 4, in the IHCP, the unknown function $q(t)$ is often parameterized by assuming either a piecewise-constant (stairsteps) or piecewise-linear (line segments) approximation. In either case, the unknown function is represented as a vector of parameters.

Another concept that is used in the following sections is *bias*. Bias refers to the idea that an estimated value for a parameter differs from its expected value. Consider a parameter, β, and an estimate of this parameter, denoted $\hat{\beta}$. The bias of this estimate might be expressed mathematically as

$$\text{bias} = \left(\beta - E(\hat{\beta})\right) \tag{6.3}$$

Finally, a matrix identity will be used later. For any quadratic form,

$$Q = \mathbf{Y}^T \mathbf{\Psi} \mathbf{Y} = tr\left[\mathbf{\Psi} \mathbf{Y} \mathbf{Y}^T\right] \tag{6.4}$$

where tr denotes the trace which is the sum of the diagonal elements. (See the quadratic form relation e.g. page 224 (Beck and Arnold 1977)).

6.1.2 Design vs. Experimental Setting

The methods for optimization in this chapter can generally be used in a *design setting*, where numerical experiments are conducted prior to any physical experiments to choose the IHCP method and regularization parameter(s). This can be done by assuming one or more heat flux variations that are similar to those anticipated for the planned experiment and using accurate solutions to the forward model to generate ideal temperature responses. These ideal responses can be combined with assumptions for the noise in the measurements (typically Gaussian with an assumed known constant standard deviation) to generate artificial data. Knowledge of the exact heat flux and characterization of noise in the data allows the analyst to test and design the IHCP algorithm for application to data from upcoming experiments.

Some of the optimization methods in this chapter are suitable for application in an *experimental setting*, where measurement data are available. Generally, characterization of the level of noise in the measurements is also required and this is most often taken as the standard deviation of the measurement error. Of the methods presented in this chapter, only the GCV method does not require knowledge of the standard deviation of the measurement errors.

6.2 Two Conflicting Objectives

6.2.1 Minimum Deterministic Bias

One of the characteristics of good estimators is that of minimum bias; in fact, in parameter estimation problems, an unbiased estimator is usually sought. Mathematically, it is desirable for the expected value of the estimator, $\hat{\beta}$, of a parameter, β, to be the expected value of β,

$$E(\hat{\beta}) = E(\beta) = \beta \tag{6.5}$$

In other words, the expected value of the estimate is the true value. If Eq. (6.5) is true, the estimator $\hat{\beta}$ is said to be unbiased and "on the average" the estimate given by $\hat{\beta}$ would be near β. If $\hat{\beta}$ were a biased estimator of β, $\hat{\beta}$ would, on the average, tend to be either higher or lower than the true value of β.

In ill-posed problems, it is preferable for the estimators for the heat flux components, $q_1,..., q_n$, to have low bias. It is *not* required that the bias be zero, however. In ill-posed problems, the bias and variability (variance) of the estimators for the unknown parameters are interrelated. Consequently, the bias is only one aspect of these estimators that must be considered.

The bias of an estimator can be investigated when the random measurement errors are set equal to zero. This bias or error in the estimator can be called the *deterministic bias*. It is advantageous to make it as small as possible and yet not make the variance unacceptably large.

6.2.2 Minimum Sensitivity to Random Errors

Another characteristic of a good estimator, in addition to minimum bias, is that of minimum variance. If $\hat{\beta}$ is the estimator of the parameter β, then this means that the variance of $\hat{\beta}$,

$$V(\hat{\beta}) = E\left\{\left[\hat{\beta} - E(\hat{\beta})\right]^2\right\} \tag{6.6}$$

should be a minimum.

6.2.3 Balancing Bias and Variance

For linear parameter estimation problems with the first, second, seventh, and eighth standard assumptions satisfied (see Chapter 1), the estimator that is a linear function of the measurements and has unbiased and minimum variance is called the Gauss–Markov estimator (Beck and Arnold 1977, p. 232). For ill-posed problems, better estimates are obtained if the requirement of unbiased estimators is not imposed. An alternative objective metric is needed for the IHCP and this is discussed in Section 6.3.

For ill-posed problems, the common requirements of zero bias and minimum variance do not yield satisfactory estimators. For zero bias, the estimators are very sensitive to measurement errors; that is, the variance of $\hat{\beta}$ is large. There are ways to reduce the variance for the IHCP such as requiring that *all* the components of q be equal, or

$$\hat{q}_1 = \hat{q}_2 = \cdots = \hat{q}_n = \text{constant} \tag{6.7}$$

If a requirement such as Eq. (6.7) is introduced, the variance of q_i is relatively small but it usually has a large bias. Hence, the deterministic bias and variance of the estimator are interrelated and an optimal strategy should consider both aspects.

6.3 Mean Squared Error

A function that considers both bias and variability is the mean squared error of the estimated heat flux. Let $\hat{q}_M$ be a component of the estimated $\hat{\mathbf{q}}$ vector and let q_M be the true value of the component. The mean squared error of $\hat{q}_M$ is

$$\mathscr{R}_M^2 = E\left[(\hat{q}_M - q_M)^2\right] \tag{6.8}$$

Estimators that minimize $\mathscr{R}_M^2$ for all values of M, $M = 1, 2,..., N$, are sought. In such case, the mean squared error of the vector $\hat{\mathbf{q}}$ is of interest:

$$R_q^2 = \frac{1}{N}\sum_{i=1}^{N}(\hat{q}_i - q_i)^2 = \frac{1}{N}(\hat{\mathbf{q}} - \mathbf{q})^T(\hat{\mathbf{q}} - \mathbf{q}) \tag{6.9}$$

However, insights can be drawn by considering the mean squared error of a single q_M value, indicated in Eq. (6.8).

It is important to observe that Eq. (6.8) does not give the variance of q_M; that is, $\mathscr{R}_M^2 \neq V(\hat{q}_M)$ except for the special case when the expected value of $\hat{q}_M$ is the true value, q_M; this only occurs if $\hat{q}_M$ is an unbiased estimator of q_M. Since for ill-posed problems the estimator is usually biased, Eq. (6.8) is not, in general, the variance of $\hat{q}_M$.

Eq. (6.8) includes two parts: deterministic and stochastic, as will be shown next. Adding and subtracting the expected value of $\hat{q}_M$ inside the right side of Eq. (6.8) gives

$$\mathscr{R}_M^2 = E\left(\left\{[\hat{q}_M - E(\hat{q}_M)] - [q_M - E(\hat{q}_M)]\right\}^2\right) \tag{6.10a}$$

$$= E\left\{[\hat{q}_M - E(\hat{q}_M)]^2\right\} - 2E\left\{[\hat{q}_M - E(\hat{q}_M)][q_M - E(\hat{q}_M)]\right\} + E\left\{[q_M - E(\hat{q}_M)]^2\right\} \tag{6.10b}$$

The first term on the right side of Eq. (6.10b) is the variance of the estimator, $\hat{q}_M$,

$$V(\hat{q}_M) = E\left\{[\hat{q}_M - E(\hat{q}_M)]^2\right\} \tag{6.11}$$

The second term on the right side of Eq. (6.10b) can be shown to be zero because $q_M - E(\hat{q}_M)$ is not a random variable and consequently

$$E[\hat{q}_M - E(\hat{q}_M)] = E(\hat{q}_M) - E(\hat{q}_M) = 0 \tag{6.12}$$

The last term in Eq. (6.10b) is the square of a bias and the outer expected value symbol can be dropped; the resulting expression is given the symbol $\mathscr{D}_M^2$,

$$\mathscr{D}_M^2 = [q_M - E(\hat{q}_M)]^2 \tag{6.13}$$

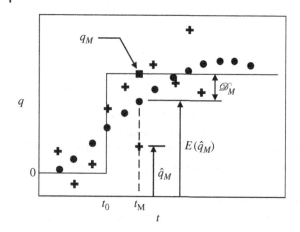

Figure 6.1 The true heat flux q_M at time t_M, the estimated value q_M, and the mean of the estimated value, $E(\hat{q}_M)$.

Hence the mean squared error of $\hat{q}_M$ is composed of two parts: a variance and the square of a deterministic error, and the relationship between them is

$$\mathscr{R}_M^2 = V(\hat{q}_M) + \mathscr{D}_M^2 \tag{6.14}$$

The significance of the components of the mean squared error given by Eq. (6.14) can be illustrated using Figure 6.1. The continuous solid line is a representation of a "true" heat flux that starts at time zero and at time t_0 jumps to a constant value. The true heat flux at time t_M is shown by a solid square in Figure 6.1 and is denoted q_M. The dots illustrate estimated values of the heat flux curve for errorless measurements or, equivalently, the expected (i.e. theoretical average) value of the estimated heat flux, $E(\hat{q}_i)$. Note that the $E(\hat{q}_i)$ values are biased since at most times the values are either greater or less than the true values. For example, at time t_M, the mean estimated value is low. The difference between q_M and $E(\hat{q}_M)$ is the deterministic bias, $\mathscr{D}_M$. The crosses in Figure 6.1 show simulated and calculated heat fluxes with measurement errors; the value at time t_M is denoted $\hat{q}_M$. Data from other sets of measurements would yield a set of estimated heat fluxes at t_M that would center about $E(\hat{q}_M)$, rather than the true value q_M. A measure of the variability of the $\hat{q}_M$ values with respect to $E(\hat{q}_M)$ is called the standard deviation of $\hat{q}_M$ and the square of the standard deviation is the variance of $\hat{q}_M$ which is denoted $V(\hat{q}_M)$. The mean squared error given by Eq. (6.14) contains both the effect of measurement errors which is described by the variance, $V(\hat{q}_M)$, and the effect of the deterministic bias, $\mathcal{D}_M$, which is due to a biased IHCP algorithm. In the IHCP, some bias is accepted in order to reduce the extreme sensitivity to measurement errors, particularly as the time steps are made small.

6.4 Minimize Mean Squared Error in Heat Flux R_q^2

The previous sections discussed the two competing components of the error in an estimated heat flux component: deterministic error (bias) and random error (variance). The amount of regularization introduced into the IHCP solution will dictate the magnitudes of each. Generally, when there is no regularization, the bias will be minimum, but the variance will be very large. At the other extreme, when the amount of regularization is significant, the bias will be maximal, but the variance will be very small.

This section focuses on using the sum of the squares of the differences between the estimated heat fluxes ($\hat{\mathbf{q}}$) and the true values ($\mathbf{q}$) as a measure of goodness for the IHCP solution. This measure is defined in Eq. (6.9).

6.4.1 Definition of R_q^2

Because the desired outcome of the IHCP solution is accurate estimation of the unknown surface heat flux, the mean sum of squares given by Eq. (6.9) is a logical choice for a measure of goodness for the estimate. Of course, application of Eq. (6.9) requires knowledge of the exact heat flux components, which are not known in any particular experiment. However, the R_q^2 measure can be used in a numerical or design setting to determine the optimal amount of regularization using an exact solution with a known heat flux input.

As discussed in Chapter 5, IHCP solution algorithms can generally be represented with a digital filter form, for which

$$\hat{\mathbf{q}} = \mathbf{FY} \tag{6.15}$$

Here, $\mathbf{F}$ is the filter matrix and $\mathbf{Y}$ is the vector of measured temperatures. Recall that the measurements $\mathbf{Y}$ differ from the exact temperatures $\mathbf{T}$ due to errors $\boldsymbol{\varepsilon}$:

$$\mathbf{Y} = \mathbf{T} + \boldsymbol{\varepsilon} \tag{6.16}$$

Using Eq. (6.15), the mean squared error in Eq. (6.9) becomes

$$R_q^2 = \frac{1}{N}(\mathbf{FY}-\mathbf{q})^T(\mathbf{FY}-\mathbf{q}) \tag{6.17}$$

To find the optimal amount of regularization, a Monte Carlo approach can be used to find the minimum of Eq. (6.17) as the degree of regularization is varied. Using an exact solution with a known $q(t)$, exact temperature responses at a specified location can be computed. These temperatures can be perturbed with random Gaussian noise with zero mean and specified standard deviation to create the data $\mathbf{Y}$. This process can be repeated many times, using different random noise sets, and the IHCP solved each time, and the actual value of R_q^2 found for each level of regularization. The average or expected value of R_q^2 over a large number of random trials can be used as a measure of goodness.

The foregoing procedure is tedious and time intensive. A simpler approach is to consider the expected value of R_q^2, $E(R_q^2)$, which is considered in the next section.

6.4.2 Expected Value of R_q^2

The expected value of the mean squared error follows from expansion of Eq. (6.17):

$$E\left(R_q^2\right) = \frac{1}{N}E\left[(\mathbf{FY})^T\mathbf{FY}\right] - [\mathbf{F}E(\mathbf{Y})]^T\mathbf{q} - \mathbf{q}^T\mathbf{F}E(\mathbf{Y}) + \mathbf{q}^T\mathbf{q}] \tag{6.18}$$

The simplification of Eq. (6.18) is given in the appendix of Woodbury and Beck (2013), but an outline of the steps is given below.

Using the quadratic relation in Eq. (6.4) and making the standard statistical assumptions regarding the measurement errors (see Chapter 1), the first term in Eq. (6.18) is

$$E\left[(\mathbf{FY})^T\mathbf{FY}\right] = tr\left[\mathbf{F}E(\mathbf{YY}^T)\mathbf{F}^T\right] = tr\left[\mathbf{F}(\mathbf{Xqq}^T\mathbf{X}^T + \sigma_Y^2\mathbf{I})\mathbf{F}^T\right] \tag{6.19a}$$

The quadratic form relation can now be applied to the first group of terms in Eq. (6.19a) to remove the *tr*:

$$E\left[(\mathbf{FY})^T\mathbf{FY}\right] = (\mathbf{FXq})^T(\mathbf{FXq}) + \sigma_Y^2 tr(\mathbf{F}^T\mathbf{F}) \tag{6.19b}$$

Because $E(\mathbf{Y}) = \mathbf{T} = \mathbf{Xq}$, the last three terms in Eq. (6.18) can be written as

$$-[\mathbf{F}E(\mathbf{Y})]^T\mathbf{q} - \mathbf{q}^T\mathbf{F}E(\mathbf{Y}) + \mathbf{q}^T\mathbf{q} = -(\mathbf{FXq})^T\mathbf{q} - \mathbf{q}^T\mathbf{FXq} + \mathbf{q}^T\mathbf{q} \tag{6.20}$$

Combining Eqs. (6.19b) and (6.20) into Eq. (6.18) gives, after some manipulation,

$$E\left(R_q^2\right) = \frac{1}{N}\left([(\mathbf{FX}-\mathbf{I})\mathbf{q}]^T[(\mathbf{FX}-\mathbf{I})\mathbf{q}] + \sigma_Y^2 tr(\mathbf{F}^T\mathbf{F})\right) \tag{6.21}$$
$$= E_{q,bias}^2 + E_{q,rand}^2$$

where

$$E_{q,bias}^2 = \frac{1}{N}[(\mathbf{FX}-\mathbf{I})\mathbf{q}]^T[(\mathbf{FX}-\mathbf{I})\mathbf{q}] \tag{6.22a}$$

$$E_{q,rand}^2 = \frac{\sigma_Y^2}{N} tr(\mathbf{F}^T\mathbf{F}) \tag{6.22b}$$

Equation (6.21) consists of two distinct terms. The first, given in Eq. (6.22a), is the bias component. This term depends on the actual heat flux history which must be known or specified in order to evaluate $E_{q,bias}^2$. The second term, Eq. (6.22b), is due to random errors in the temperature data $\mathbf{Y}$ and is proportional to the variance of these errors. Notice that $\mathbf{FX}$ is a dimensionless quantity, and that $\mathbf{F}$ has units of W/m²-°C.

This result provides a powerful way to examine the components of the errors in the estimated heat fluxes. Woodbury and Beck (Woodbury and Beck 2013) demonstrate that, in the case of no regularization, the $E_{q,bias}^2$ term goes to zero. Also, in the case of "infinite" regularization, $E_{q,rand}^2$ tends to zero. Their results are specific to Tikhonov regularization (TR) but can be interpreted as generalizations to any IHCP solution that can be represented in a filter form.

Equations (6.21) and (6.22a,b) are given in matrix form. As described in Chapter 5, an IHCP filter can be applied sequentially (locally in time) considering the m_p past time steps and m_f future time steps. For such application, the bias and random error components of the expected value of the mean squared error of heat flux can be computed as

$$E^2_{q,bias} = \frac{1}{N - m_f - m_p} \sum_{M = m_p + 1}^{N - m_f} \left(\sum_{i = -m_p}^{+m_f} [(p_i - \delta_{0i})q_{M-i}]^2 \right) \tag{6.23a}$$

$$E^2_{q,rand} = \sigma_Y^2 \sum_{-m_p}^{+m_f} f_i^2 \tag{6.23b}$$

In Eq. (6.23a), δ_{0i} is the Kronecker delta which is equal to 1.0 when $i = 0$ and is equal to 0.0 for all other values of i. Also, p is a vector equal to the product of the row of $\mathbf{F}$ and the corresponding column number of $\mathbf{X}$ used to define the vector.

Equations (6.22), or equivalently (6.23), provide means for estimating the bias and variance of the estimated heat flux from an IHCP solution.

The average expected error in the heat flux calculation is the square root of $E(R_q^2)$ and is sometimes called the standard error or the expected value of the root-mean-squared (RMS) error:

$$s_q = E_{q,RMS} = \sqrt{E\left(R_q^2\right)} = \sqrt{E^2_{q,bias} + E^2_{q,rand}} \tag{6.24}$$

Example 6.1 Estimate the bias and variance of the zeroth-order Tikhonov IHCP solution with $\alpha_0 = 0.0001$ for application to a linear in time heat flux over the interval $0 \leq t \leq 2$ seconds:

$$q(t) = 100t \ \mathrm{W/m^2}, t \text{ in seconds} \tag{6.25}$$

The heat flux is imposed on a finite plate insulated at the remote boundary with $\alpha = 1 \ \mathrm{m^2/s}$, $k = 1 \ \mathrm{W/m\text{-}°C}$, and $L = 1$ m. This is the same X22B20T0 case considered in several examples in Chapter 4. Consider measurements are made at $x = L$ every 0.25 seconds and the standard deviation of the measurements is $\sigma = 0.5 \ °C$. Use the piecewise constant approximation for the heat flux history.

Solution

For this case, assuming piecewise constant heat fluxes, the $\mathbf{X}$ matrix is

$$X = \begin{bmatrix} 0.1005 & 0 & 0 & 0 & \cdots & \cdots & \cdots & 0 \\ 0.2343 & 0.1005 & \ddots & \ddots & 0 & \cdots & \cdots & 0 \\ 0.2487 & 0.2343 & \ddots & \ddots & 0 & \cdots & \cdots & \vdots \\ 0.2499 & 0.2487 & \ddots & \ddots & 0 & \ddots & \cdots & \vdots \\ 0.2500 & 0.2499 & \ddots & \ddots & 0.1005 & \ddots & \ddots & \vdots \\ \vdots & 0.2500 & \ddots & \ddots & 0.2343 & \ddots & \ddots & \vdots \\ \vdots & \vdots & \ddots & \ddots & 0.2487 & \ddots & \ddots & 0 \\ 0.2500 & 0.2500 & \cdots & \cdots & 0.2499 & 0.2487 & 0.2343 & 0.1005 \end{bmatrix} \tag{6.26}$$

With $\alpha_0 = 1\text{E}{-}4$, the filter matrix is

$$\mathbf{F} = \left(\mathbf{X}^T\mathbf{X} + \alpha_0\mathbf{I}\right)^{-1}\mathbf{X}^T \tag{6.27}$$

$$F = \begin{bmatrix} 6.498 & 2.651 & -1.961 & 1.421 & -0.996 & 0.652 & -0.362 & 0.113 \\ -12.493 & 1.600 & 6.250 & -4.535 & 3.179 & -2.082 & 1.154 & -0.362 \\ 11.082 & -8.608 & -1.196 & 8.169 & -5.734 & 3.754 & -2.082 & 0.652 \\ -9.656 & 8.291 & -6.683 & -2.399 & 8.747 & -5.734 & 3.179 & -0.996 \\ 8.991 & -7.742 & 7.089 & -6.105 & -2.399 & 8.169 & -4.535 & 1.421 \\ -9.056 & 7.799 & -7.164 & 7.089 & -6.683 & -1.196 & 6.250 & -1.961 \\ 9.857 & -8.489 & 7.799 & -7.742 & 8.291 & -8.608 & 1.600 & 2.651 \\ -11.445 & 9.857 & -9.056 & 8.991 & -9.656 & 11.082 & -12.493 & 6.498 \end{bmatrix} \tag{6.28}$$

Eq. (6.22a) will be used to compute the bias. For this equation, the heat flux components for the true (exact) solution are needed. For the piecewise constant assumption, these are evaluated at times $t - \Delta t/2$. Therefore,

$$\mathbf{q} = \lfloor 12.5 \ \ 37.5 \ \ 62.5 \ \ 87.5 \ \ 112.5 \ \ 137.5 \ \ 162.5 \ \ 187.5 \rfloor^T \tag{6.29}$$

and

$$(\mathbf{FX} - \mathbf{I})\mathbf{q} = [1.311 \ \ -4.001 \ \ 7.145 \ \ -10.87 \ \ 15.48 \ \ -21.36 \ \ 28.96 \ \ -38.66]^T \tag{6.30}$$

so the expected value of the bias error is

$$E_{q,bias}^2 = \frac{1}{N} \left[[\mathbf{FX} - \mathbf{I}]\mathbf{q} \right]^T \left[[\mathbf{FX} - \mathbf{I}]\mathbf{q} \right] = \frac{3{,}216}{8} = 402.0 \, \text{W}^2/\text{m}^4 \tag{6.31}$$

Eq. (6.22b) will be used to compute the variance of the estimate in the presence of noise. The diagonal elements of $\mathbf{F}^T\mathbf{F}$ are

$$\text{diag}(\mathbf{F}^T\mathbf{F}) = \lfloor 805.3 \ \ 442.4 \ \ 333.4 \ \ 323.4 \ \ 332.9 \ \ 316.8 \ \ 234.2 \ \ 56.67 \rfloor^T \tag{6.32}$$

The sum of these is the trace of $\mathbf{F}^T\mathbf{F}$. The expected value of the variance due to random measurement errors is

$$E_{q,rand}^2 = \frac{\sigma_Y^2}{N} tr(\mathbf{F}^T\mathbf{F}) = \frac{(0.5^{\circ}\text{C})^2}{8} \left[2{,}845.0 \frac{\text{W}^2}{\text{m}^4 - {^{\circ}}\text{C}^2} \right] = 88.9 \ \text{W}^2/\text{m}^4 \tag{6.33}$$

Discussion

All calculations are performed in MATLAB to machine precision and have been rounded here for compact presentation. The average expected error (RMS error) in this heat flux calculation is:

$$s_q = E_{q,RMS} = \sqrt{454.0 + 88.9} = 23.3 \, \text{W/m}^2 \tag{6.34}$$

The expected error values are affected by the choice of time step and the number of data points. Table 6.1 summarizes the expected errors for this application as the data time step is decreased (note that the number of data points increases also in order to cover the same $0 \le t \le 2$ seconds time span). As the time step decreases, the magnitude of the entries in the **X** matrix decreases. As seen in Table 6.1, this affects the magnitude of the expected values of $E_{q,rand}^2$ and $E_{q,bias}^2$. However, the expected value of the standard error becomes relatively constant.

6.4.3 Optimal Regularization Using $E(R_q^2)$

Equations (6.21) and (6.24) provide powerful tools for optimizing the degree of regularization for an IHCP solution that can be represented in a filter. The optimal regularization is the amount that balances the bias and the sensitivity to random errors by minimizing the expected value of the RMS error in the heat flux estimate given in Eq. (6.24).

Table 6.1 Expected error values for Example 6.1 for $\alpha_0 = 1E-4$ for various data time intervals.

Δt	N	$E^2_{q,bias}$	$E^2_{q,rand}$	$E_{q,RMS}$
0.50	4	0.0539	6.515	2.563
0.25	8	454.0	88.91	23.30
0.125	16	1430.0	185.7	40.20
0.0625	32	1376.2	232.3	40.10
0.03125	64	1511.9	136.7	40.60

This approach does require knowledge of the exact heat flux history, **q**, and is therefore limited to a design setting. It does not require solution of the IHCP for the assumed **q**, but the IHCP solution may be required to determine the filter matrix **F** (or, alternatively, the filter vector **f**; see Chapter 5).

For TR, any numerical algorithm for minimizing a scalar function of a single variable can be used to determine the optimal regularization coefficient α. For other methods, such as function specification (FS) or conjugate gradient method (CGM), a directed search, considering one value after another of r or the number of iterations, will be needed to determine the optimum.

Consider the case of Example 6.1 with $\Delta t = 0.25$ seconds. Figure 6.2 shows the result of varying the TR parameter and computing the expected values of the bias Eq. (6.22a) and random Eq. (6.22b) components of $E(R^2_q)$. In Figure 6.2, when very little regularization is present, the random component of the expected error dominates. At very large values, the bias is dominant. In the middle of the range at about $\alpha_0 = 5.4E-5$, the minimum in the $E(R^2_q)$ curve corresponds to the optimal amount of regularization.

Example 6.2 Determine the optimal value of the regularization parameter α_0 for zeroth-order TR using piecewise constant heat flux with $\Delta t = 0.25$ seconds. The heat flux is imposed on a finite plate insulated at the remote boundary with $\alpha = 1$ m^2/s, $k = 1$ W/m-°C, and $L = 1$ m. Measurements will be made at the remote boundary. Consider $\sigma_Y = 0.5$ °C and three different heat flux histories (time t is in seconds; q in W/m^2):

a) $q(t) = 100t$ b) $q(t) = \begin{cases} 0 & t < 0.5 \\ 100 & t \geq 0.5 \end{cases}$ c) $q(t) = \begin{cases} 0 & t < 0.5 \\ 400(t - 0.5) & 0.5 \leq t < 1.0 \\ 200(1 - 2(t - 1)) & 1.0 \leq t < 1.5 \\ 0 & t \geq 1.5 \end{cases}$

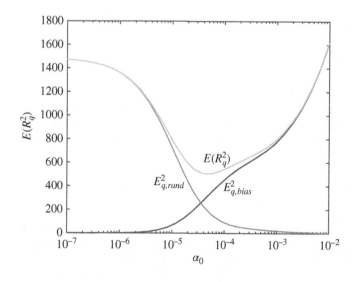

Figure 6.2 Expected value of the average sum-squared error in heat flux with random and bias components for base case of Example 6.2.

Table 6.2 Optimal α_0 and expected error values for three different heat fluxes for Example 6.2.

$q(t)$		α_0	$E_{q,bias}^2$	$E_{q,rand}^2$	$E_{q,RMS}$
case a)	ramp	5.363E−5	300.0	165.6	21.6
case b)	step	1.944E−4	90.2	52.6	11.9
case c)	triangle	1.175E−3	7.5	17.6	5.0

Solution

The MATLAB function fminbnd is used to determine the value of α_0 that minimizes $E_{q,RMS}$ (or equivalently $E(R_q^2)$). Table 6.2 summarizes the results.

Discussion

Case a) has the smoothest variation of $q(t)$ of the three with a constant slope and no discontinuities. The optimal regularization parameter is the smallest of the three for this case, but the standard error is the highest. Case b) has the most abrupt discontinuity in $q(t)$, with a step change in function value at $t = 0.5$ seconds. Case c) has three slope discontinuities in the triangular pulse and the resulting optimal α_0 is the largest of the three cases. Cases b) and c) have lower $E_{q,bias}^2$ and $E_{q,RMS}$ because the actual heat flux is zero over 25% (case b)) and 50% (case c)) of the time duration. These zero components in $\mathbf{q}$ make no contribution to the bias computation in Eq. (6.22a).

Example 6.3 Find the optimal number of future times for FS using the minimum expected error in the heat flux criterion for the same heat fluxes and parameters in Example 6.2.

Solution

A directed search solution is needed to find the minimum of $E_{q,RMS}$. Here, a whole domain version of the FS method is considered by computing the filter matrix $\mathbf{F}$ using the temperature perturbation ideas from Chapter 5. The IHCP is solved N times, and each time the "data" for the problem consists of a vector of zeros with a single value of "1" in the column. (These data are successive columns of the $N \times N$ identity matrix.) As an example, when $r = 3$, for $q = $ const, the $\mathbf{F}$ matrix is:

$$F = \begin{bmatrix} 0.217 & 0.724 & 1.261 & 0.000 & 0.000 & 0.000 & 0.000 & 0.000 \\ -0.119 & -0.178 & 0.350 & 1.261 & 0.000 & 0.000 & 0.000 & 0.000 \\ -0.055 & -0.301 & -0.496 & 0.350 & 1.261 & 0.000 & 0.000 & 0.000 \\ -0.024 & -0.136 & -0.443 & -0.496 & 0.350 & 1.261 & 0.000 & 0.000 \\ -0.108 & -0.061 & -0.199 & -0.443 & -0.496 & 0350 & 1.261 & 0.000 \\ -0.005 & -0.027 & -0.088 & -0.199 & -0.443 & -0.496 & 0.350 & 1.261 \\ 0.000 & 0.000 & 0.000 & 0.000 & 0.000 & 0.000 & 0.000 & 0.000 \\ 0.000 & 0.000 & 0.000 & 0.000 & 0.000 & 0.000 & 0.000 & 0.000 \end{bmatrix} \tag{6.35}$$

Notice that the last $r − 1$ rows are all zeros as the last $r{-}1$ components of $\mathbf{q}$ cannot be estimated due to insufficient future time data. Accordingly, the last $r{-}1$ rows of the $(\mathbf{FX}-\mathbf{I})\mathbf{q}$ vector must be explicitly zeroed in computing $E_{q,bias}^2$ in Eq. (6.22a). By successively considering $r = 1,2,...$ and evaluating $E_{q,RMS}^2$, the value of r that minimizes $E_{q,RMS}^2$ can be found.

Table 6.3 presents the results of the optimization.

Discussion

For the dimensionless time step $\Delta\tilde{t} = 0.25$ seconds, the optimal number of future time steps required for each of the heat flux histories is $r = 2$: only one additional time step of data beyond exact matching (Stolz method) is required to stabilize the solution. Notice that the magnitudes of the bias terms in Table 6.3 for FS are generally significantly larger than the bias terms in Table 6.2 for TR, so the optimal FS solution brings more bias than the optimal TR solution.

Regularization for FS can only be applied in discrete steps, whereas TR has a continuous variable to apply regularization. This means that TR can be optimized to a greater degree.

Table 6.3 Optimal r and expected error values for three different heat fluxes for Example 6.3.

$q(t)$		r_{opt}	$E_{q,bias}^2$	$E_{q,rand}^2$	$E_{q,RMS}$
Case a)	Ramp	2	9.6	3.7	3.7
Case b)	Step	2	270.0	3.7	16.5
Case c)	Triangle	2	411.8	3.7	20.4

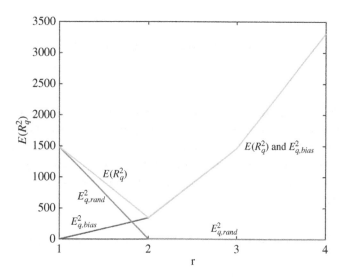

Figure 6.3 Expected value of the average sum-squared error in heat flux with random and bias components for triangular heat flux case of Example 6.3.

Figure 6.3 shows the components of the average expected value of the sum-squared error for the triangular heat flux in this example. The immediate trade-off between random and bias error as r increases from 1 to 2 is apparent. When $r = 1$ (exact matching), the bias error is zero, but the random error is about 1500 W^2/m^4. When r increases to $r = 2$, the random error is driven to nearly zero, but at the expense of the bias error component rising to about 400 W^2/m^4. For and r values greater than $r = 2$, the bias error dominates and increases $E(R_q^2)$ dramatically.

6.5 Mean Squared Error in Temperature R_T^2

Another metric that can be used to characterize the goodness of an IHCP solution is the error in the estimated temperature measurement, $\hat{\mathbf{T}}$. This metric is more indirect because the desired (estimated) function $\hat{\mathbf{q}}$ is not evaluated. However, this measure has the advantage that it uses experimental data as reference, and hence can be used in an experimental setting.

6.5.1 Definition of R_T^2

The mean squared error of the temperature residuals in discrete form is given by

$$R_T^2 = \left(\mathbf{Y} - \hat{\mathbf{Y}}\right)^T \left(\mathbf{Y} - \hat{\mathbf{Y}}\right)/N \tag{6.36}$$

where $\mathbf{Y}$ is measured data defined in Eq. (6.16) and $\hat{\mathbf{Y}}$ are the estimated values of the measurements. $\hat{\mathbf{Y}}$ is related to the estimated heat flux as

$$\hat{\mathbf{Y}} = \mathbf{X}\hat{\mathbf{q}} \tag{6.37}$$

For an IHCP solution that can be represented as a filter, Eq. (6.37) can be written as

$$\hat{\mathbf{Y}} = \mathbf{XFY} \tag{6.38}$$

With Eq. (6.38), Eq. (6.36) can be written as

$$\begin{aligned} R_T^2 &= \frac{1}{N}(\mathbf{Y} - \mathbf{XFY})^T(\mathbf{Y} - \mathbf{XFY}) \\ &= \frac{1}{N}((\mathbf{I} - \mathbf{XF})\mathbf{Y})^T((\mathbf{I} - \mathbf{XF})\mathbf{Y}) \\ &= \frac{1}{N}\mathbf{Y}^T\left((\mathbf{I} - \mathbf{XF})^T(\mathbf{I} - \mathbf{XF})\right)\mathbf{Y} \end{aligned} \tag{6.39}$$

6.5.2 Expected Value of R_T^2

For a particular set of measured data, $\mathbf{Y}$, Eq. (6.36) can be used to compare the "estimated" measurements to the actual ones. However, when data conform to the standard statistical assumptions of additive uncorrelated errors with zero mean and constant variance, the expected value of R_T^2 provides a useful measure.

The expected value $E(R_T^2)$ is derived from Eq. (6.39) by inserting the measurement relation Eq. (6.16) and making use of the quadratic form relation in Eq. (6.4). The derivation can be found in the appendix of Woodbury and Beck (2013). The result is

$$
\begin{aligned}
E(R_T^2) &= \frac{1}{N}\mathbf{T}^T(\mathbf{I}-\mathbf{XF})^T(\mathbf{I}-\mathbf{XF})\mathbf{T} + \frac{1}{N}\sigma_Y^2 tr\left((\mathbf{I}-\mathbf{XF})^T(\mathbf{I}-\mathbf{XF})\right) \\
&= \frac{1}{N}\mathbf{q}^T((\mathbf{I}-\mathbf{XF})\mathbf{X})^T((\mathbf{I}-\mathbf{XF})\mathbf{X})\mathbf{q} + \frac{1}{N}\sigma_Y^2 tr\left((\mathbf{I}-\mathbf{XF})^T(\mathbf{I}-\mathbf{XF})\right)
\end{aligned}
\tag{6.40a, b}
$$

The two terms in either Eq. (6.40a) or Eq. (6.40b) have similar roles to those in Eq. (6.21). The first term of Eq. (6.40a,b) is the bias in the estimate due to regularization and the second term is due to the effect of random measurement errors. The bias term depends both on the structure of the IHCP (regularization, time step size, and assumption about variation of the heat flux) and on the true surface heat flux history $\mathbf{q}$ or, equivalently, the true temperature history $\mathbf{T}$. The second term is directly proportional to the variance of the measurement errors and is independent of the actual process (true heat flux or true temperature history).

6.5.3 Morozov Discrepancy Principle

The Morozov discrepancy principle (Morozov 1966, 1967, 1984; Tikhonov and Arsenin 1977) states that the correct amount of regularization is that which makes the value of Eq. (6.36) consistent with the level of error, δ, in the measured temperatures. Often this error measure δ is taken as the variance of the measured temperatures or

$$
R_T^2 = \frac{1}{N}\left(\mathbf{Y}-\hat{\mathbf{Y}}\right)^T\left(\mathbf{Y}-\hat{\mathbf{Y}}\right) = \delta \approx \sigma_Y^2
\tag{6.41}
$$

Equation (6.41) can be used for a particular set of data, $\mathbf{Y}$. However, for design purposes, when the exact temperature or heat flux history is available, a better approach is to consider the expected value of R_T^2:

$$
\sqrt{E(R_T^2)} = \sigma_Y
\tag{6.42}
$$

Utilizing Eq. (6.40a,b) in Eq. (6.42) gives

$$
\sqrt{\frac{1}{N}\left[\mathbf{T}^T(\mathbf{I}-\mathbf{XF})^T(\mathbf{I}-\mathbf{XF})\mathbf{T} + \sigma_Y^2 tr\left((\mathbf{I}-\mathbf{XF})^T(\mathbf{I}-\mathbf{XF})\right)\right]} = \sigma_Y
\tag{6.43a}
$$

or in another form

$$
\left\{\mathbf{T}^T(\mathbf{I}-\mathbf{XF})^T(\mathbf{I}-\mathbf{XF})\mathbf{T} + \sigma_Y^2\left[tr\left((\mathbf{I}-\mathbf{XF})^T(\mathbf{I}-\mathbf{XF})\right) - N\right]\right\}^2 = 0
\tag{6.43b}
$$

which is forced to zero by finding the appropriate degree of regularization. Due to the quadratic character of Eq. (6.43b), the solution can be found easily using a minimization algorithm such as MATLAB's fminbnd().

Example 6.4 Use Eq. (6.41) and the Morozov discrepancy principle to find the optimal value of the regularization parameter α_0 for the TR using piecewise constant heat flux with $\Delta t = 0.25$ seconds. Measurements are made at the remote boundary, and $\alpha = 1\ m^2/s$, $k = 1\ W/m\text{-}°C$, and $L = 1\ m$. Consider $\sigma_Y = 0.5\ °C$ with the data shown in Table 6.4.

Table 6.4 Data for Example 6.4.

Time (s)	0.25	0.50	0.75	1.00	1.25	1.50	1.75	2.00
Y (°C)	0.7	6.1	16.5	35.7	58.6	89.0	125.6	168.0

Solution

Rewrite Eq. (6.41) as

$$\alpha_{0,opt} = \arg\min \left(\frac{1}{N} \left(\mathbf{Y} - \hat{\mathbf{Y}}\right)^T \left(\mathbf{Y} - \hat{\mathbf{Y}}\right) - \sigma_Y^2 \right)^2 \tag{6.44}$$

Now, use MATLAB's fminbnd() function (or similar) to minimize this function by varying the α_0 parameter. The optimal value of α_0 is found as $\alpha_{0,opt,Mz} = 4.80\text{E}{-}4$.

Discussion

The exact heat flux used to generate data for this example is "case a)" in Example 6.2, and the set of data used was noised by a set of random Gaussian numbers with zero mean and $\sigma = 0.5\,°\text{C}$. Using the expected value of the standard error in temperature in the discrepancy principle as in either of Eq. (6.43a, b) to determine the optimal regularization parameter results in $\alpha_{0,opt,ET} = 6.95\text{E}{-}4$.

Figure 6.4 shows the location of these optimization points in the context of the average expected value of the residual sum of squares of the heat flux function $q(t)$. Also shown in the figure is the optimal value $\alpha_{0,opt,Eq} = 4.72\text{E}{-}5$.

Some observations can be made. One is that the discrepancy principle is relatively simple to apply in an experimental setting based on any set of measured data and knowledge of the noise in the data, σ_Y. However, for any particular set of measured data, the optimal value of the regularization parameter will generally differ from the average value that would be obtained considering many sets of data (this corresponds to using $E\left(R_T^2\right)$ in the discrepancy principle). Figure 6.5 shows

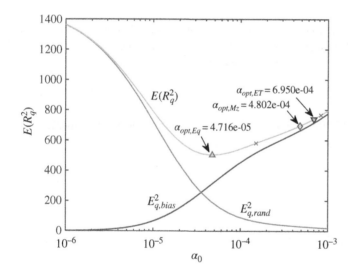

Figure 6.4 Comparison of results from three regularization techniques applied to Example 6.4.

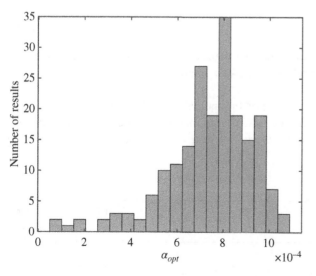

Figure 6.5 Histogram of results from 200 trials using the Discrepancy Principle on different sets of random Gaussian error in Example 6.4 (ET_IHCP_Tik.m, fig4).

the histogram of results obtained from Monte Carlo simulation using 200 different sets of data with Gaussian noise. This distribution is skewed toward the high end of the range for $\alpha_{0,opt}$ but shows that values as small as 1/7 or as great as 1.5 times the expected value ($\alpha_{0,opt,ET} = 6.95E-4$) might be obtained for any single set of data.

Finally, the success of the Morozov method in producing the optimal for a given set of data depends also on accurate knowledge of the value of the standard deviation of the measurements. In experimental settings, this can be difficult to determine. Figure 6.5 displays two small "x" symbols that indicate the location for the "optimal" regularization if the standard deviation used is in error by ±25%. If the assumed standard deviation is low by –25%, the "optimal" value of α_0 determined is 1.51E–4 (difference of 61%), and if the standard deviation used is high by +25%, the "optimal" value of α_0 determined is 8.32E–4 (difference of 73%).

A final observation in regard to Figure 6.4 is that for this heat flux (X22B20) and data, the Discrepancy Principle tends to over-regularize in regard to achieving the minimum $E(R_q^2)$. This means more bias is introduced which pushes the solution beyond the minimum in Figure 6.4.

6.6 The L-Curve

The "L-curve" is a graphical method for visualizing the compromise between smoothing (bias) and sensitivity to noise (random errors) as regularization is increased. Excellent references on the subject are due to Hansen (Hansen and O'Leary 1993; Hansen 1998) and Vogel (2002).

6.6.1 Definition of L-Curve

To construct the L-curve, plot the squared norm of the regularized residual on the vertical axis against the squared norm of the regularized solution on the horizontal axis over a range of regularization. Log-Log coordinates or scaling should be used, and the resulting graph has a classic "L" shape; however, with small sets of data, the curve may degenerate significantly and bear only slight resemblance to an "L."

More specifically, define the L-curve using the following coordinates:

$$
\begin{aligned}
x_{Lcurve}(\alpha) &= \log_{10}\left(\|\mathbf{X}\hat{\mathbf{q}}_\alpha - \mathbf{Y}\|_2^2\right) \\
y_{Lcurve}(\alpha) &= \log_{10}\left(\|\hat{\mathbf{q}}_\alpha\|_2^2\right)
\end{aligned}
\tag{6.45}
$$

Here, the subscript "α" is reminiscent of TR but is intended to generically represent any form of regularization. The $\|\cdot\|_2^2$ represents the L_2 norm squared, which is the sum of the squared components of the vector.

Figure 6.6 shows an L-curve generated for an IHCP solution using zeroth-order TR. The curve displaying discrete points is generated from one sample of artificial data generated from the exact solution using simulated measurement errors. The data used were generated from "case a)" in Example 6.2 using 32 data points over the interval $0 \leq t \leq 2.0$ with artificial Gaussian noise (zero mean, $\sigma = 0.5\,°C$) added. One hundred values of the regularization parameter α_0 were used to generate the L-curve data points using Eqs. (6.45). Each point in the plot corresponds to a particular value of α_0, and the plot points are in monotonically increasing values of α_0 from left to right. With small values of α_0, random errors are large, resulting in large values of y_{Lcurve}. Conversely, for large values of α_0, bias errors dominate, resulting in large values of $x_{L\text{-}curve}$.

(Some researchers choose to represent the L-curve with axes switched. See, for example, Kindermann and Raik(2020). The curve still presents as an "L" shape.)

Of course, different sets of noisy data for the same problem and solution will result in different L-curves. In a design setting, the expected values of the L-curve coordinates can be used to generate the curve.

6.6.2 Using Expected Value to Define L-Curve

For discrete solution of the IHCP using $\mathbf{T} = \mathbf{Xq}$, the L-curve coordinates can be expressed as:

$$
\begin{aligned}
x_{Lcurve}(\alpha) &= \log_{10}\left((\mathbf{X}\hat{\mathbf{q}}_\alpha - \mathbf{Y})^T(\mathbf{X}\hat{\mathbf{q}}_\alpha - \mathbf{Y})\right) \\
y_{Lcurve}(\alpha) &= \log_{10}\left(\hat{\mathbf{q}}_\alpha^T\hat{\mathbf{q}}_\alpha\right)
\end{aligned}
\tag{6.46}
$$

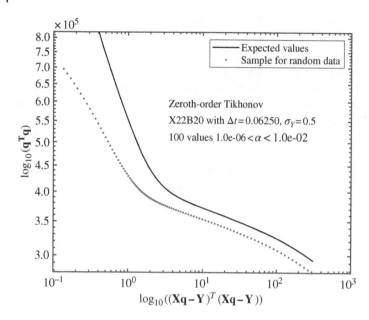

Figure 6.6 Typical L-curve for 1-D IHCP.

When the IHCP solution can be represented as a filter ($\hat{\mathbf{q}} = \mathbf{FY}$), and measurements are modeled as $\mathbf{Y} = \mathbf{T} + \boldsymbol{\varepsilon}$, Eqs. (6.46) becomes

$$x_{Lcurve}(\alpha) = \log_{10}\left(\mathbf{Y}^T(\mathbf{XF}_\alpha - \mathbf{I})^T(\mathbf{XF}_\alpha - \mathbf{I})\mathbf{Y}\right)$$
$$y_{Lcurve}(\alpha) = \log_{10}\left((\mathbf{F}_\alpha[\mathbf{Xq} + \boldsymbol{\varepsilon}])^T(\mathbf{F}_\alpha[\mathbf{Xq} + \boldsymbol{\varepsilon}])\right) \tag{6.47}$$

In Eq. (6.47), $\mathbf{q}$ is the true heat flux, and the subscript "α" is inserted to indicate quantities that depend on the degree of regularization.

Using the standard statistical assumptions for the measurement errors $\boldsymbol{\varepsilon}$ (uncorrelated errors with zero mean and constant variance σ_Y^2), the expected value of the L-curve coordinates can be written:

$$
\begin{aligned}
E(x_{Lcurve}(\alpha)) &= \log_{10}\left[E\left(\mathbf{Y}^T(\mathbf{XF}_\alpha - \mathbf{I})^T(\mathbf{XF}_\alpha - \mathbf{I})\mathbf{Y}\right)\right] \\
&= \log_{10}\left[\mathbf{q}^T\left((\mathbf{I} - \mathbf{XF}_\alpha)\mathbf{X}\right)^T((\mathbf{I} - \mathbf{XF}_\alpha)\mathbf{X})\mathbf{q} + \sigma_Y^2\,tr\left((\mathbf{I} - \mathbf{XF}_\alpha)^T(\mathbf{I} - \mathbf{XF}_\alpha)\right)\right] \\
&= \log_{10}\left[\mathbf{q}^T(\mathbf{B}_\alpha\mathbf{X})^T(\mathbf{B}_\alpha\mathbf{X})\mathbf{q} + \sigma_Y^2\,tr(\mathbf{B}_\alpha^T\mathbf{B}_\alpha)\right] \qquad \mathbf{B}_\alpha = (\mathbf{I} - \mathbf{XF}_\alpha) \\
E(y_{Lcurve}(\alpha)) &= \log_{10}\left[E\left((\mathbf{F}_\alpha\mathbf{Xq} + \mathbf{F}_\alpha\boldsymbol{\varepsilon})^T(\mathbf{F}_\alpha\mathbf{Xq} + \mathbf{F}_\alpha\boldsymbol{\varepsilon})\right)\right] \\
&= \log_{10}\left[\mathbf{q}^T\mathbf{X}^T\mathbf{F}_\alpha^T\mathbf{F}_\alpha\mathbf{Xq} + \sigma_Y^2\,tr(\mathbf{F}_\alpha^T\mathbf{F}_\alpha)\right]
\end{aligned} \tag{6.48}
$$

Eq. (6.48) allows determination of the L-curve in a design setting for a specified and known heat flux history $q(t)$ and a specified level of noise in the measurement data σ_Y. Notice that the vertical axis coordinate is exactly the $N \times E(R_T^2)$ (cf. Eq. (6.40a,b)).

The solid line in Figure 6.6 is the expected value of the L-curve for application to $q(t) = 100t$ heat flux using zeroth-order TR with $\sigma_Y = 0.5\,°C$. This is the curve that will result from averaging a large number of Monte Carlo simulations with different simulated noise sets (only one such sample result is depicted in Figure 6.6 as the dotted line). For designing an IHCP solution, the expected value of the L-curve provides an efficient means for generating the average behavior of the solution.

6.6.3 Optimal Regularization Using L-Curve

The optimal amount of regularization corresponds to the amount that produces the point in the elbow of the curve. The "elbow" in the L-curve in Figure 6.6 is apparent, but the precise location is both difficult to identify automatically and can be expensive to compute.

Figure 6.7 Various "optimal" regularizations presented on the L-curves of Figure 6.6.

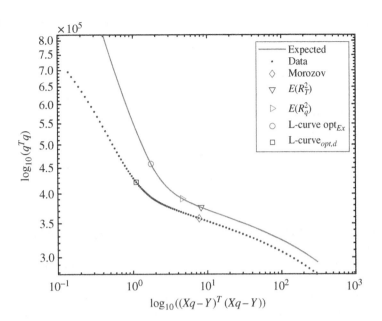

Hansen and O'Leary(1993) recommend selecting as "optimal" the amount of regularization that produces the maximum curvature in the L-curve. A naïve approach is to use the discrete points for the curve and compute the second derivative using finite differences. It is easy to search the set of results for the maximum, and identify the corresponding value of α (or degree of regularization). This approach will produce a value in the neighborhood of the elbow, but perhaps not close (depending on the sharpness of the L-curve).

A library of MATLAB routines authored by Hansen is archived on GitHUB (Hansen n.d.). This library includes the function "corner.m" which uses a progressive algorithm (Hansen and O'Leary 1993) to identify and refine the location of the "corner" of the L-curve. This routine was used to identify the optimal value of α for the cases in Figure 6.6.

Figure 6.7 displays the L-curves from Figure 6.6 with several markers indicating the "optimal" amount of regularization. On the discrete (lower) curve, for a single set of data in an experimental setting, the Hansen procedure for finding the corner point results in the point denoted by the square. Importantly, the L-curve procedure to find this optimal requires only a set of experimental data and the corresponding results from the IHCP analysis. If the additional information of the amount of noise in the experimental data, characterized by σ_Y, is known, then the Morozov discrepancy principle can be used with the experimental data to identify the optimal degree of regularization. This point is identified on the curve as the diamond. Although both are "in the neighborhood" of the visual optimal in the curve, neither method approximates the optimal well. The Hansen procedure produces an α before (above) the elbow, which will be under-regularized, while the Morozov procedure produces a value below the elbow, which will be over-smooth.

In a design setting, using an assumed $q(t)$ and σ_Y, the expected value of the L-curve can be generated. This is shown as the solid curve in Figure 6.7. The Hansen procedure result for optimal using this curve is identified with the open circle, and that found using Morozov's discrepancy principle, coupled with the $E(R_T^2)$, is identified with the inverted triangle. Also shown as a right-pointing triangle on this curve is the optimal regularization found by minimizing the average sum-squared error in the heat flux Eq. (6.21) without any reference to the L-curve. Although still not at the "elbow" of the curve, this point is much closer to the elbow than the others.

6.7 Generalized Cross Validation

GCV is an ingenious scheme to utilize errors in the data to select parameters in the estimation process. The idea was originally developed for application to parameter estimation using least squares (Allen 1971, 1974), including use in ridge regression (Golub et al. 1979). GCV does not require an estimate of the standard deviation of the measurement errors,

σ^2. GCV has been advocated by a number of inverse problems practitioners, including Trujillo and Busby (1997) and Coles and Murio (2000, 2005).

6.7.1 The GCV Function

The optimal degree of regularization for solution of an ill-posed problem is found by minimizing the GCV function. For an inverse procedure that can be represented as a filter matrix, with a discrete model equation $\mathbf{T} = \mathbf{Xq}$, and estimator $\hat{\mathbf{q}} = \mathbf{FY}$, the GCV function (Golub et al. 1979) can be expressed as

$$V_\alpha = \frac{\frac{1}{N}(\mathbf{Y} - \mathbf{XF}_\alpha\mathbf{Y})^T(\mathbf{Y} - \mathbf{XF}_\alpha\mathbf{Y})}{\left[\frac{1}{N}tr(\mathbf{I} - \mathbf{XF}_\alpha)\right]^2} = \frac{N(\mathbf{I} - \mathbf{XF}_\alpha)^T\mathbf{Y}^T\mathbf{Y}(\mathbf{I} - \mathbf{XF}_\alpha)}{[N - tr(\mathbf{XF}_\alpha)]^2} \tag{6.49}$$

Here, α is suggestive of TR, but is used to denote any form of regularization. The subscripts denote dependence of the quantity on the degree of regularization.

Equation (6.49) is derived through consideration of the estimation of parameters from a set of data using a least-squares objective function. Trujillo and Busby (1997) give a good summary of this derivation. The basic idea is to remove one data point from the set and estimate the parameters, and then use the omitted data point to quantify the accuracy of the estimate. By repeating this process N times, selectively eliminating one data point each time, then considering the ensemble of these trials, the performance measure in Eq. (6.49) is obtained.

The matrix $\mathbf{XF}$ relates the value of the measurement computed through the estimated heat fluxes directly to the measurements, that is, $\hat{\mathbf{Y}} = \mathbf{XFY}$. In GCV literature, the matrix $\mathbf{XF}$ is called the *influence* matrix.

The numerator of Eq. (6.49) is the mean squared error of the measurement residuals, R_T^2 (viz., Eq. (6.36)). When regularization is absent or mild, random errors are large and so is this numerator. As regularization is increased, the residuals become smaller, until the effects of bias in the estimates appear, then this numerator term increases. The denominator of Eq. (6.49) is the square of the deviation of the diagonal elements of the influence matrix $\mathbf{XF}$ from unity. As regularization increases, the diagonal elements of $\mathbf{F}$ generally increase. The optimal degree of regularization results when the ratio of the residuals to the increasing size of $\mathbf{F}$ is minimum. Examples 6.5 and 6.6 illustrate GCV for optimizing regularization.

Example 6.5 Use the information and data from Example 6.4 to determine the optimal α_0 for TR using GCV.

Solution

The $\mathbf{X}$ matrix is the same as in Example 6.1, Eq. (6.26). The GCV function in Eq. (6.49) can be written as a MATLAB function of the scalar variable α_0 using zeroth-order TR. The MATLAB function fminbnd() is used to find the minimum of the function. Some caution in application of fminbnd() is needed, however, because the default convergence tolerance on the solution variable is too coarse in consideration of the magnitude of the Tikhonov parameter, so the "TolX" precision is set to 1E-15.

The resulting optimal value of α_0 is $\alpha_{0,opt,GCV} = 3.96\text{E-}4$.

Discussion

This optimal value of α_0 is of the same order of magnitude as that found using the Morozov criteria in Example 6.4, and although not shown in Figure 6.4, the point $\alpha_{0,opt,GCV}$ will lie just to the left of the point indicated for $\alpha_{0,opt,MZ}$. Importantly, the GCV result is obtained directly from the data and no knowledge of the standard deviation of the noise in the data is required. Figure 6.4 indicates that the optimal $\alpha_{0,opt,GCV}$ over-regularizes in relation to minimizing the expected error in the mean residual of the heat flux.

Example 6.6 Use the information and data from Example 6.4 with GCV to optimize the solution using a) TSVD and b) Fletcher–Reeves CGM.

Solution

a) Truncated Singular Value Decomposition

For this analysis, the model equation used for the SVD solution is premultiplied by the matrix $\mathbf{X}^T$ (as indicated in Eq. (4.171)):

$$\mathbf{X}^T\mathbf{T} = \left(\mathbf{X}^T\mathbf{X}\right)\mathbf{q} \tag{6.50}$$

So, in the exact matching of the data $\mathbf{Y}$, as part of the TSVD method, the estimates for heat flux are

$$\hat{\mathbf{q}} = \left(\mathbf{X}^T\mathbf{X}\right)^{-1}\mathbf{X}^T\mathbf{Y} \tag{6.51}$$

where the indicated inverse is computed by eliminating n_s singular values. So, the $\mathbf{F}$ matrix for this application is

$$\mathbf{F} = \left(\mathbf{V}\mathbf{W}_{n_s}^{-1}\mathbf{U}^T\right)\mathbf{X}^T \tag{6.52}$$

where $\mathbf{U}$, $\mathbf{W}$, and $\mathbf{V}$ are the SVD components of the decomposition of the matrix $\mathbf{X}^T\mathbf{X}$, and the subscript n_s indicates that a number of the diagonal components of the $\mathbf{W}^{-1}$ are zeroed. For reference, the $\mathbf{F}$ matrix with $n_s = 2$ is

$$F = \begin{bmatrix} 4.214 & 4.027 & -2.179 & 0.511 & 0.667 & -1.169 & 0.996 & -0.377 \\ -5.796 & -2.406 & 6.957 & -2.090 & -1.381 & 2.941 & -2.601 & 0.996 \\ 0.904 & -2.476 & -2.563 & 5.136 & 0.240 & -2.923 & 2.941 & -1.169 \\ 2.265 & 0.812 & -4.292 & -0.236 & 3.594 & 0.240 & -1.381 & 0.667 \\ -2.922 & 0.373 & 3.140 & -5.834 & -0.236 & 5.136 & -2.090 & 0.511 \\ 1.723 & -0.536 & -1.171 & 3.140 & -4.292 & -2.563 & 6.957 & -2.179 \\ 0.262 & 0.155 & -0.536 & 0.373 & 0.812 & -2.476 & -2.406 & 4.027 \\ -1.935 & 0.262 & 1.723 & -2.922 & 2.265 & 0.904 & -5.796 & 4.214 \end{bmatrix} \tag{6.53}$$

Regularization for the TSVD method can only be applied in discrete steps (number of singular values eliminated, n_s). Unconstrained optimization routines based on continuous real-valued variables, such as fminbnd, will not produce reliable results, and the significant overhead of a genetic algorithm search is unnecessary. A "brute force" approach is used for this example, whereby the GCV function is computed for every value of n_s and the results stored. Afterward, a simple search to find the minimum value of the resulting vector of values is used to find the optimal solution.

Figure 6.8 shows a plot of the GCV function values computed for $1 \leq n_s \leq 5$ for this example. The diamond marker indicates the minimum of these values as identified using the MATLAB min() function at $n_{s,opt,GCV} = 2$. For comparison, also shown as an open circle marker is the optimal $n_{s,opt,Eq} = 1$ found by minimizing the expected value of the mean residual

Figure 6.8 GCV function values for SVD regularization in Example 6.6.

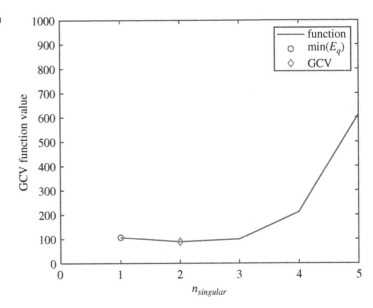

error in the heat flux. (Determination of $n_{s,opt,Eq}$ requires knowledge of the exact heat flux, which in this case is the linear ramp heating $q(t) = 100t$. Of course, in a purely experimental setting, this exact heat flux cannot be known.) Consistent with the findings from the Tikhonov regularization, GCV tends to over-regularize with respect to minimizing the mean residual error in the heat flux.

b) Fletcher–Reeves Conjugate Gradient Method

This optimization is confounded in two regards: first, this CGM does not naturally convey the **F** matrix. However, using the temperature perturbation techniques of Chapter 5, each column of the full matrix can be computed. Second, as with the SVD method, regularization only occurs in discrete steps, as each iteration adds additional regularization.

Like the SVD optimization with GCV, the approach applied here is to compute the **F** matrix at each step of the conjugate gradient iteration and evaluate the value of the GCV function. These GCV function values are stored and can be searched to find the minimum after completing some number of iterations.

Figure 6.9 shows the plot of the GCV function for iterations 1 through 8 (the GCV function value at iteration 8 is beyond the limit of the vertical axis and thus does not appear). The diamond marker denotes the optimal number of iterations found using GCV, $n_{iter,opt,GCV} = 7$. For comparison, an open circle is also in the figure denoting the optimal number of iterations found by minimizing the mean residual of the heat flux (relative to the true heat flux). In this case, the two methods identified the same number of iterations as optimal.

For reference, the optimal **F** matrix is

$$F = \begin{bmatrix} 6.185 & 3.049 & -2.431 & 1.889 & -1.406 & 0.964 & -0.552 & 0.176 \\ -11.370 & 0.374 & 7.635 & -5.934 & 4.416 & -3.030 & 1.736 & -0.553 \\ 8.774 & -6.451 & -3.399 & 10.378 & -7.727 & 5.307 & -3.042 & 0.970 \\ -5.879 & 5.018 & -3.672 & -5.246 & 11.303 & -7.736 & 4.414 & -1.403 \\ 3.494 & -3.167 & 3.200 & -2.719 & -5.343 & 10.526 & -6.039 & 1.926 \\ -4.469 & 3.726 & -3.319 & 3.394 & -3.241 & -4.118 & 8.227 & -2.652 \\ 0.385 & -0.887 & 2.006 & -3.797 & 6.095 & -7.743 & 1.427 & 2.649 \\ 1.445 & -0.237 & -1.585 & 3.853 & -6.457 & 9.367 & -11.773 & 6.324 \end{bmatrix}$$

(6.54)

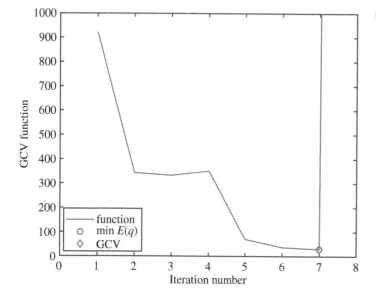

Figure 6.9 GCV function values for CGM regularization in Example 6.6.

6.8 Chapter Summary

Optimization of regularization in IHCPs strives to strike a balance between sensitivity to noise (random errors) and introduction of bias (deterministic errors) on the solution procedure. Optimization can be done in a *design setting*, using assumed heat fluxes, accurate direct solutions, and assumed measurement errors, or in an *experimental setting*, using measured temperature data and (generally) estimated standard deviation of the measurement errors.

The mean squared error function provides a vehicle for striking the balance between random and bias errors. In a design or analysis setting, when a true value of heat flux is known, the expected value of the mean squared error in the heat flux, R_q^2, provides the best measure of the accuracy of an IHCP solution. When measurement errors are normally distributed with zero mean and constant variance, the expected value of this mean squared error, $E(R_q^2)$, has two distinct parts: the bias error Eq. (6.22a) and the random error Eq. (6.22b). With no or little regularization, the random portion dominates. As regularization is increased, the random errors decrease while the bias error grows. The resulting sum, $E(R_q^2)$, has a minimum value, and the optimal amount of regularization needed for the IHCP produces this minimum.

In experimental settings, where the heat flux history is unknown, the mean squared error in the measured temperature (R_T^2, Eq. (6.36)) provides a mechanism for assessing the accuracy of the IHCP solution. Under standard statistical assumptions regarding the measurement errors (normally distributed, zero mean, constant variance), the expected value of R_T^2 has two distinct parts: random and bias.

The Morozov discrepancy principle is widely used to optimize regularization in IHCP methods by selecting the degree of regularization that matches R_T^2 to the variance of the measurement errors, σ_Y^2. This can be done in one of two ways. For a particular set of data (and assumed value of σ_Y), the degree of regularization is chosen to satisfy Eq. (6.41). Alternatively, the expected value of $E(R_T^2)$ can be used for the assumed value of σ_Y to determine the best amount of regularization for any experimental measurements taken from a normally distributed set with that standard deviation. However, this last option requires knowledge of the exact heat flux and so is limited to a design setting.

The L-curve method for optimization of regularization provides a graphical depiction of the trade-off between random error and bias. The classic L-curve plots the norm of the regularized residual versus norm of the regularized solution as regularization is varied. The amount of regularization that produces the point at the elbow of this curve is the optimum. Automation of the selection of the optimal degree of regularization using the L-curve is challenging and can be computationally time-consuming. The L-curve optimization is developed for use in experimental settings, but the expected values of the axes can be used to optimize for a family of normally distributed errors characterized by σ_Y.

GCV is a statistical scheme to produce the optimal amount of regularization. It is used in experimental settings and, importantly, does not require characterization of the magnitude of noise σ_Y in the data. Rather, GCV uses information in the data directly to evaluate the optimal. For examples considered in this section, GCV produced optima very similar to that from minimizing $E(R_q^2)$.

All of the techniques in this chapter can be applied to discrete IHCP solutions which are characterized by direct model $\mathbf{T} = \mathbf{Xq}$ and estimator $\hat{\mathbf{q}} = \mathbf{FY}$.

Problems

6.1 A quadratic heat flux pulse can be created by superposition of a linear and quadratic heat flux ($p = 1$ and $p = 2$ in Eq. (2.3)). A MATLAB function that computes the heat flux input called q_2pulse(t, tstart, tstop) is

```
function [ flux ] = q_2pulse(t, tstart, tstop)
num = max(size(t));
flux = zeros(size(t));
ta = (tstart+tstop)/2 - tstart;
trel = t-tstart;
trel_a = trel/ta;
for i = 1:num
    if( t(i) <= tstart )
        flux(i) = 0;
```

```
        elseif( t(i) <= 2*ta + tstart )
            flux(i) = 2*trel_a(i) - trel_a(i)^2;
        else
            flux(i) = 0;
        end
    end
end
```

Estimate the bias and variance of the zeroth-order Tikhonov IHCP solution with $\alpha_0 = 0.1$ for application to this heat flux using piecewise constant assumption with $\Delta t = 0.25$ seconds over the interval $0 \leq t \leq 10$ seconds, with tstart = 2 seconds and tstop = 8 seconds. Consider dimensionless variables with $\alpha = 1$ m^2/s, $k = 1$ W/m-°C, $L = 1$ m, $q_0 = 1$ W/m^2, and $\sigma_Y = 0.05$ °C.

6.2 Expand on Problem 6.1 by plotting the bias error, the variance error, and the expected value of the RMS error in the heat flux estimate for varying α_0. What value of α_0 minimizes the expected RMS error?

6.3 Repeat Problem 6.2 for piecewise linear heat flux approximation.

6.4 Calculate the expected value of the mean squared error in the estimated surface temperature for Problem 6.1 using Eq. (6.40a,b).

6.5 Measurements are made at the insulated surface in a one-dimensional domain for an unknown heat flux. Properties are $\alpha = 1$ m^2/s, $k = 1$ W/m-°C, $L = 1$ m, and $q_0 = 1$ W/m^2. The following data are obtained:

t	-0.24	-0.18	-0.12	-0.06	0.00	0.06	0.12	0.18
Y	0.0093	-0.0010	-0.0014	-0.0034	0.0049	-0.0065	0.0010	0.0047
t	0.24	0.30	0.36	0.42	0.48	0.54	0.60	0.66
Y	0.0074	0.0019	0.0011	0.0001	0.0116	0.0344	0.0788	0.1161
t	0.72	0.78	0.84	0.90	0.96	1.02	1.08	1.14
Y	0.1645	0.2174	0.2621	0.3169	0.3603	0.3669	0.3827	0.3988
t	1.20	1.26	1.32	1.38	1.44	1.50		
Y	0.3996	0.3980	0.4011	0.4001	0.4061	0.4130		

Use $q = const$ assumption with zeroth-order Tikhonov and the Morozov discrepancy principle to find the optimal value of α_0 if a) $\sigma_Y = 0.002$ C and b) $\sigma_Y = 0.005$ C. Discuss your results in light of the fact that "a)" is the true value used to generate these data.

6.6 Use GCV and the data from Problem 6.5 to find the optimal α_0 for zeroth-order TR using piecewise constant heat flux assumption.

References

Allen, D. M. (1971) Mean square error of prediction as a criterion for selecting variables, *Technometrics*, 13(3), p. 475. https://doi.org/10.2307/1267161.

Allen, D. M. (1974) The relationship between variable selection and data agumentation and a method for prediction', *Technometrics*, 16(1), p. 127. https://doi.org/10.2307/1267500.

Beck, J. V. and Arnold, K. J. (1977) *Parameter Estimation in Engineering and Science*. New York: John Wiley and Sons.

Coles, C. and Murio, D. A. (2000) Identification of parameters in the 2-D IHCP, *Computers and Mathematics with Applications*, 40 (8), pp. 939–956. https://doi.org/10.1016/S0898-1221(00)85005-1.

Coles, C. and Murio, D. (2005) Numerical solutions of inverse spatial Lotka–Volterra systems, *Mathematical and Computer Modelling*, 42(13), pp. 1411–1420. https://doi.org/10.1016/J.MCM.2005.02.007.

Golub, G. H., Heath, M. and Wahba, G. (1979) Generalized cross-validation as a method for choosing a good ridge parameter', *Technometrics*, 21(2), pp. 215–223. https://doi.org/10.2307/1268518.

Hansen, P. C. (1998) *Rank-Deficient and Discrete Ill-Posed Problems*. Philadelphia, PA: SIAM.

Hansen, P. C. (n.d.) *GitHub – hadiTab/regu: A MATLAB package for analysis and solution of discrete ill-posed problems, developed by Prof. Per Christian Hansen, DTU Compute, Technical University of Denmark*. Available at: https://github.com/hadiTab/regu (accessed 5 January 2022).

Hansen, P. C. and O'Leary, D. P. (1993) The use of the L-curve in the regularization of discrete Ill-posed problems', *SIAM Journal on Scientific Computing*, 14(6), pp. 1487–1503. https://doi.org/10.1137/0914086.

Kindermann, S. and Raik, K. (2020) A simplified L-curve method as error estimator, *Electronic Transactions on Numerical Analysis*, 53, pp. 217–238. https://doi.org/10.1553/ETNA_VOL53S217.

Morozov, V. A. (1966) Regularization of ill-posed problems and the choice of regularization parameter, *Journal of Mathematics and Mathematical Physics*, 6(1), pp. 242–251. https://doi.org/10.1016/0041-5553(66)90046-2.

Morozov, V. A. (1967) The choice of parameter in solving functional equations by the regularization method, *USSR Doklady Akad. Nauk SSSR*, 175(6), pp. 1000–1003.

Morozov, V. A. (1984) *Methods for Solving Incorrectly Posed Problems*. Edited by Z. Nashed. New York, NY, USA: Springer-Verlag.

Tikhonov, A. N. and Arsenin, V. Y. (1977) *Solutions of Ill-Posed Problems*. Washington, DC: V.H. Winston & Sons.

Trujillo, D. M. and Busby, H. R. (1997) *Practical Inverse Analysis in Engineering*. CRC Press. https://doi.org/10.1201/9780203710951.

Vogel, C. R. (2002) *Computational Methods for Inverse Problems*. Philadelphia, PA: SIAM.

Woodbury, K. A. and Beck, J. V (2013) Estimation metrics and optimal regularization in a Tikhonov digital filter for the inverse heat conduction problem, *International Journal of Heat and Mass Transfer*, 62, pp. 31–39. https://doi.org/10.1016/j.ijheatmasstransfer.2013.02.052.

7

Evaluation of IHCP Solution Procedures

7.1 Introduction

Previous chapters presented general procedures for treating the inverse heat conduction problem (IHCP) and for mathematically modeling the physical problem. The IHCP can be viewed as the estimation of the surface disturbance (heat flux or temperature) from transient temperature measurements inside a heat-conducting solid. It is an ill-posed problem which is characterized by extreme sensitivity of the surface heat flux (or temperature) to small variations in the interior temperatures. Several methods for solving the IHCP were given in Chapter 4, and each method affords some mechanism for regularizing the solution. Some of these methods are used in this chapter.

Of the various ways of modeling transient heat conduction in solid bodies, the one used in this chapter employs a convolution integral equation based on superposition principles. This method requires the problem to be linear; that is, the thermal properties (k, ρ, and c) are not functions of temperature but can be functions of position.

Advantages of the superposition approach are that the body can have an arbitrary shape and the thermal properties can be functions of position (Figure 7.1). The temperature distribution can be one-, two-, or three-dimensional. The only requirement is that the *influence function* $\phi(\mathbf{x}, t)$ be known. [$\phi(\mathbf{x}, t)$ is the temperature rise at x due to a unit step disturbance at the surface at time zero.] In Figure 7.1a and b, the temperature distributions are functions of only one spatial independent variable; in Figure 7.1a, the generic coordinate $\mathbf{x}$ becomes x and in Figure 7.1b $\mathbf{x}$ becomes r, the radial coordinate for a cylinder or for a sphere.

The interfaces between dissimilar materials can have perfect or imperfect contact characterized by h_c (Figure 7.1c).

The boundary conditions that are permitted include insulation, a constant surface temperature that is equal to the initial temperature, T_i, or a convective boundary condition provided the ambient temperature, T_∞, is equal to the initial body temperature, T_i (see Figure 7.1). The conditions of a single prescribed surface heat flux, which is a function of time only, and a uniform initial temperature ensures that there is a single convolution integral in the superposition. The only condition that causes the temperature to change is the surface disturbance, typically the heat flux, $q(t)$.

The heat flux $q(t)$ is considered uniform over the surfaces where it is applied, and it can be applied to surfaces that are not single planes as shown by Figure 7.1c and over surfaces that are not completely covered as shown by Figure 7.1d. These two cases are examples of two-dimensional heat flow. Even though the temperature distribution is not one-dimensional, superposition can be used because it is valid for multidimensional linear cases. Multidimensional heat conduction and the IHCP solution are addressed in Chapter 8.

As previously stated, the influence function $\phi(\mathbf{x}, t)$ must be known. The $\phi(\mathbf{x}, t)$ solution depends on the geometry, and the interface and boundary conditions; it is the temperature rise for a disturbance (heat flux or temperature) of unity at the heated surface and may be solved in many ways. For example, $\phi(\mathbf{x}, t)$ can be determined by some exact method of solution such as application of Green's function or separation of variables. It can also be found numerically using finite differences or finite elements. The surface element method can be used (Keltner and Beck 1981; Beck et al. 1985; Litkouhi and Beck 1985). The $\phi(\mathbf{x}, t)$ history is needed only at the sensor locations.

There are many ways to numerically approximate the surface disturbance in the convolution integral, some of which were discussed in Chapter 4. Both piecewise constant and piecewise linear approximations are utilized in this chapter.

Even with the restrictions indicated, many algorithms can be applied to the IHCP. Most of the methods described in Chapter 4 are explored in this chapter, specifically Function Specification (FS), Tikhonov Regularization, Conjugate Gradient, Truncated Singular Value Decomposition (TSVD), and Kalman Filter methods. The potential for each of these methods for application sequentially using the digital filter concept will be explored.

Inverse Heat Conduction: Ill-Posed Problems, Second Edition. Keith A. Woodbury, Hamidreza Najafi, Filippo de Monte, and James V. Beck.
© 2023 John Wiley & Sons, Inc. Published 2023 by John Wiley & Sons, Inc.

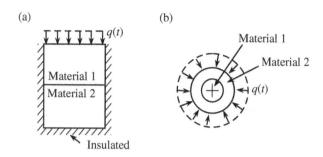

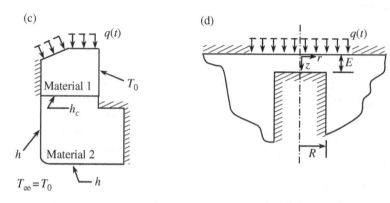

Figure 7.1 Various geometries, boundary conditions, and interface conditions that can be treated using superposition and convolution. Thermal properties are independent of temperature, (a), composite plate; (b) composite cylinder or sphere; (c) composite body of irregular shape; (d) axisymmetric semi-infinite body with cylindrical void and surface partially heated.

The plan for this chapter is to first define a suite of test problems for evaluation of IHCP algorithms. In subsequent sections, the data from the test cases are used to estimate the corresponding heat fluxes for five IHCP methods: FS, Tikhonov regularization (TR), conjugate gradient method (CGM), TSVD method, and KF method. For each method except KF, the degree of regularization in the IHCP solution will be determined by minimizing the expected value of the mean squared error of the heat flux estimate. (As pointed out in Chapter 6, this requires access to the exact (true) heat flux history, which is possible in these analyses). The chapter closes with a summary comparing the results of these methods.

7.2 Test Cases

7.2.1 Introduction

In this section, several test cases are used to evaluate the performance of various IHCP solution techniques. These selected test cases are chosen from a broader suite of potential test cases which are available in MATLAB codes on the companion website for this book.

One test case is for a step increase in the surface heat flux. The second case is for a heat flux that varies in time in a triangular fashion. The third test case is for a quartic heat flux pulse (Woodbury and Beck 2013), which is similar to the triangular heat flux, but is smooth in the sense that it has continuous first derivatives. The last test case is for the temperature perturbation described in Chapter 5 using input temperatures equal to zero except the temperature at time t_M, and this test case is used to find the filter coefficients.

The geometry for each of these test cases is a flat plate which is heated at $x = 0$ and insulated at $x = L$: the X22B-0T0 case. The measurements are at $x = L$, which is the greatest possible distance from the heated surface and hence poses a greater challenge than any another sensor location for an IHCP algorithm.

For each test case (except temperature perturbation), exact analytical solutions are used to compute the response at a sensor location to a high degree of accuracy (generally 10^{-15}). For each of these cases, artificial noise in the form of normally distributed random numbers with constant standard deviation is added to the exact values to produce simulated measured data. The magnitude of the standard deviation of the noise is computed as 0.5% of the maximum temperature rise resulting from the heat flux input.

Results in this chapter are presented in terms of dimensionless variables (see Eqs. (2.4)–(2.6)). Dimensionless time steps of $\Delta \tilde{t} = 0.06$ are used, which are "small" for the IHCP. Using the penetration time (see Chapter 4) of $\Delta \tilde{t}_{pen} = 0.06$ implies a temperature rise of about $10^{-(0.1/0.06)} = 0.02$ dimensionless temperature units. In other words, at the end of one time step, the dimensionless temperature at the sensor location will have increased by only about 0.02. This also implies required measurement accuracy or resolution: if the measuring instrument cannot detect the small change in temperature, then no new information is gained. Some test cases in this chapter are used with $\Delta \tilde{t} = 0.03$, for which the implied measurement sensitivity is 0.0005, and $\Delta \tilde{t} = 0.015$, for which the temperature change is 2.2×10^{-7}. This latter case presents an extreme challenge for the IHCP due to these very small time step.

These test cases use a limited number of data points to facilitate easy comparison of results. For $\Delta \tilde{t} = 0.06$, the number of data points used is 30. When $\Delta \tilde{t} = 0.03$ is considered, the number of time steps used is 60, and for $\Delta \tilde{t} = 0.015$, the number of time steps is 120. In all cases, the "initial time" is considered time $\tilde{t} = -0.24$, and the final time is then $\tilde{t} = 1.56$. For each heat flux case, the heating begins at time $\tilde{t} = 0$, so there is no heat flux for a short period before heating begins. For the triangular and the quartic cases, the heat flux returns to zero at time $\tilde{t} = 1.2$, and for the step change, the heat flux remains at the value 1.0 over the entire time.

The important dimension for the IHCP is the distance from the disturbance to the measurement location. For a sensor at a position $x = E$, where $0 < E < L$, an IHCP estimation procedure for q_M gives rather similar results to the case for a sensor at $x = L$ if the dimensionless time step, $\Delta \tilde{t}_E = \alpha t / E^2$, for the $x = E$ location is equal to the dimensionless time step, $\Delta \tilde{t} = \alpha t / L^2$, for the $x = L$ sensor location. In other words, the test cases in this chapter with the sensor at $x = L$ give insight for cases with a sensor at other locations if the dimensionless time step is based on the distance of the sensor from the heated surface.

7.2.2 Step Change in Surface Heat Flux

A basic test case is for a step change at $t = 0$ in the surface heat flux, which is shown in Figure 7.2. This is sometimes called a "constant" heat flux because the heat flux is the constant value of q_c for $t > 0$. The use of the words "step change" makes clear, however, that the heat flux is zero for some time, in this case when $t < 0$. For linear problems, neither the sign (positive or negative) nor the magnitude of q_c is important because the estimated heat flux values are linearly proportional to q_c. Considering q_c as the reference value, the dimensionless heat flux for this case is $\tilde{q} = 1.0$. For greater generality, the examples in this chapter are solved in terms of dimensionless variables.

The dimensionless temperature response for this test case is the X22B10T0 solution. The computational analytical solution for this problem is outlined in Section 2.3.8.1.

7.2.3 Triangular Heat Flux

The second test case is for heat flux that varies in time in a triangular fashion (Figure 7.3). Before $\tilde{t} = 0$, the heat flux is zero. For $0 \leq \tilde{t} \leq 0.6$, the surface q increases linearly with time, and for $\tilde{t} > 0.6$, the flux decreases linearly to zero at $\tilde{t} = 1.2$ and remains zero thereafter.

The linear portion of the heat flux for $0 \leq \tilde{t} \leq 0.6$ is described by

$$\tilde{q} = \tilde{t} \tag{7.1}$$

where

$$\tilde{q} = \frac{q}{q_{ref}}, \tilde{t} = \frac{\alpha t}{L^2} \tag{7.2}$$

and q_{ref} is a nominal value of heat flux; namely, the value of $q(t)$ associated with $\tilde{t}$ equal to unity in Eq. (7.1). The dimensionless temperature is denoted by

$$\tilde{T} \equiv \frac{T - T_0}{\left(q_{ref} L / k \right)} \tag{7.3}$$

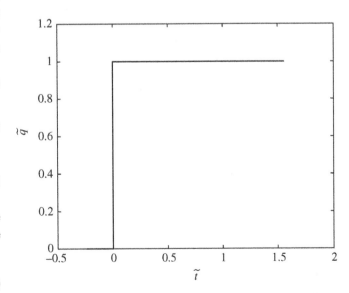

Figure 7.2 Heat flux test case for a step change in surface heat flux.

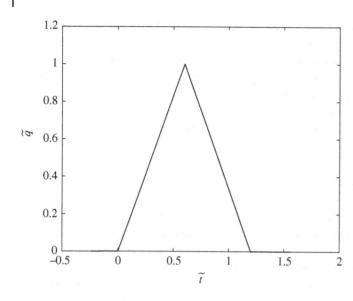

Figure 7.3 Triangular heat flux for test case. Finite insulated plate.

The temperatures at $\tilde{x} = 0$ and 1 for the linear heat flux given by Eq. (7.1) are[1]

$$\tilde{T}(0, \tilde{t}) = \tilde{\phi}(0, \tilde{t}), \quad \tilde{T}(1, \tilde{t}) = \tilde{\phi}(1, \tilde{t}) \tag{7.4}$$

$$\tilde{\phi}(0, \tilde{t}) = \frac{1}{2}(\tilde{t})^2 + \frac{1}{3}\tilde{t} - \frac{1}{45} + \frac{2}{\pi^4} \sum_{n=1}^{\infty} \frac{1}{n^4} \exp\left(-\pi^2 n^2 \tilde{t}\right) \tag{7.5}$$

$$\tilde{\phi}(1, \tilde{t}) = \frac{1}{2}(\tilde{t})^2 + \frac{1}{6}\tilde{t} + \frac{7}{360} + \frac{2}{\pi^4} \sum_{n=1}^{\infty} \frac{(-1)^n}{n^4} \exp\left(-\pi^2 n^2 \tilde{t}\right) \tag{7.6}$$

These expressions can be used for $0 \leq \tilde{t} \leq 0.6$. For $0.6 \leq \tilde{t} \leq 1.2$, the temperature is given by

$$\tilde{T}(\tilde{x}, \tilde{t}) = \tilde{\phi}(\tilde{x}, \tilde{t}) - 2\tilde{\phi}(\tilde{x}, \tilde{t} - 0.6) \tag{7.7}$$

and for $\tilde{t} > 1.2$,

$$\tilde{T}(\tilde{x}, \tilde{t}) = \tilde{\phi}(\tilde{x}, \tilde{t}) - 2\tilde{\phi}(\tilde{x}, \tilde{t} - 0.6) + \tilde{\phi}(\tilde{x}, \tilde{t} - 1.2) \tag{7.8}$$

These expressions are derived using simple superposition. The temperature histories for the heated and insulated surfaces are shown in Figure 7.4.

The surface temperature history shown in Figure 7.4 is quite different from that at the insulated surface. It responds immediately both to the onset of heating and to changes in the heating rate; the slope of $\tilde{T}(0, \tilde{t})$ changes at both $\tilde{t} = 0.6$ and 1.2. The $\tilde{x} = 0$ curve also has a maximum at an intermediate time, about $\tilde{t} = 0.9$. The insulated surface temperature is negligible until $\tilde{t} = 0.18$ (damped and lagged) and there are no abrupt changes in slope at $\tilde{t} = 0.6$ and 1.2. Both temperatures continue changing until sometime after heating ceases. Numerical values of $\tilde{T}(1, \tilde{t})$ for the triangular heat flux are given in Table 7.1 for $\Delta\tilde{t} = 0.06$.

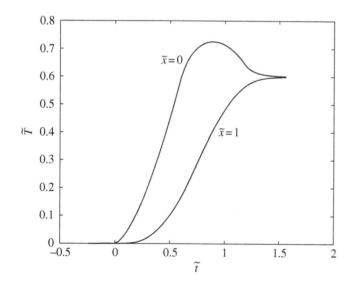

Figure 7.4 Heated and insulated surface temperatures for triangular heat flux test case.

---

1 These expressions are for times greater than the second deviation time. For small values of time, alternate expressions from the computational analytical solution (Chapter 2) can be used.

Table 7.1 Temperatures at an insulated surface of a finite plate heated with the triangular heat flux shown in Figure 7.4.

			$\Delta\tilde{t} = 0.06$			
$\tilde{t}$	$\bar{T}(1,\tilde{t})$	$\tilde{t}$	$\bar{T}(1,\tilde{t})$	$\tilde{t}$	$\bar{T}(1,\tilde{t})$
0.06	0.000007	0.66	0.127200	1.26	0.348823
0.12	0.000374	0.72	0.157880	1.32	0.353762
0.18	0.002171	0.78	0.189293	1.38	0.356545
0.24	0.006323	0.84	0.219593	1.44	0.358089
0.30	0.013381	0.90	0.247680	1.50	0.358941
0.36	0.023656	0.96	0.272931	1.56	0.359415
0.42	0.037319	1.02	0.295006		
0.48	0.054465	1.08	0.313714		
0.54	0.075145	1.14	0.328954		
0.60	0.099389	1.20	0.340666		

7.2.4 Quartic Heat Flux

The quartic heat flux, shown in Figure 7.5, is one of several "pulse" test cases constructed from exact solutions to polynomial heat fluxes using superposition (Woodbury and Beck 2013). The quartic pulse has the same characteristics (start time, peak time, and end time) as the triangular flux but has continuous first derivatives at the transition points.

The quartic heat flux has $q(t)$ equal to zero at time zero and has a maximum of q_{ref} at time $t = t_0$ and back to zero at time $t = 2t_0$. At both $t = 0$ and $2t_0$, the derivative of the heat flux is zero. The heat flux that achieves these criteria is

$$\tilde{q}(\tilde{t}) = \frac{q(t)}{q_{ref}} = \left(\frac{\tilde{t}}{\tilde{t}_0}\right)^2 \left(\frac{\tilde{t}}{\tilde{t}_0} - 2\right)^2 = \left(\frac{\tilde{t}}{\tilde{t}_0}\right)^4 - 4\left(\frac{\tilde{t}}{\tilde{t}_0}\right)^3 + 4\left(\frac{\tilde{t}}{\tilde{t}_0}\right)^2, \quad 0 < \tilde{t} < 2\tilde{t}_0 \tag{7.9}$$

In the application in this chapter, $\tilde{t}_0 = 0.6$ and $\tilde{q}(\tilde{t}_0) = 1$.

Figure 7.5 Quartic heat flux pulse for test case. Finite insulated plate.

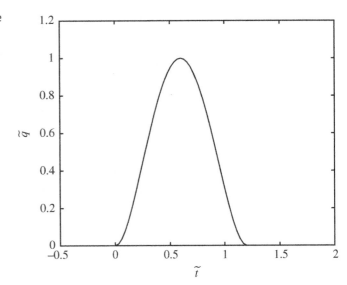

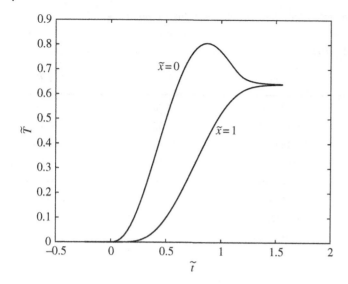

Figure 7.6 Heated and insulated surface temperatures for quartic heat flux pulse test case.

The temperature responses at the heated surface and at $x = L$ due to the quartic heat flux pulse are shown in Figure 7.6. These temperature responses for the heat flux history described in Eq. (7.9) can be constructed by adding the temperature component solutions from power heat flux input over finite time using the same coefficients as in Eq. (7.9). In words, the temperature response for the heat flux given in Eq. (7.9) is that for a fourth-order heat flux input minus four times that for a cubic heat flux input plus four times that for a quadratic heat flux input. For $t \leq 2t_0$, the continuous heating form of the solution is used, and for times greater than the heating duration, $t > 2t_0$, the finite-heating-time fundamental solutions are used:

$$
\tilde{T}_{\text{quartic}}(\tilde{x}, \tilde{t}) = \begin{cases} \tilde{T}_{\text{X22B340T0}} - 4\tilde{T}_{\text{X22B330T0}} + 4\tilde{T}_{\text{X22B320T0}} & \text{for } 0 < \tilde{t} \leq 2\tilde{t}_0 \\ \tilde{T}_{\text{X22B(34,5)0T0}}\big|_{\tilde{t}_1 = 2\tilde{t}_0} - 4\tilde{T}_{\text{X22B(33,5)0T0}}\big|_{\tilde{t}_1 = 2\tilde{t}_0} + 4\tilde{T}_{\text{X22B(32,5)0T0}}\big|_{\tilde{t}_1 = 2\tilde{t}_0} & \text{for } \tilde{t} > 2\tilde{t}_0 \end{cases} \tag{7.10}
$$

To compute the response in Eq. (7.10) when $\tilde{t} \leq 2\tilde{t}_0$, the following component solutions are needed. These are solutions to the X22B-0T0 problem with the active boundary condition of Eq. (2.3) with $p = 2, 3, 4$, respectively (only the "long time" solutions are presented here):

$$
\tilde{T}_{\text{X22B320T0}} = \frac{T^{(2)}(\tilde{x}, \tilde{t}, \tilde{t}_0)}{q_N L/k} = \frac{2}{\tilde{t}_0^2} \left[\begin{array}{c} \dfrac{\tilde{t}^3}{6} + \tilde{t}^2 S_{\text{X22}}^{(0)}(\tilde{x}) - 2\tilde{t} S_{\text{X22}}^{(1)}(\tilde{x}) + 2S_{\text{X22}}^{(2)}(\tilde{x}) \\ -2 \displaystyle\sum_{m=1}^{\infty} e^{-(m\pi)^2 \tilde{t}} \dfrac{\cos(\beta_m \tilde{x})}{\beta_m^6} \end{array} \right] \tag{7.11}
$$

$$
\tilde{T}_{\text{X22B330T0}} = \frac{T^{(3)}(\tilde{x}, \tilde{t}, \tilde{t}_0)}{q_N L/k} = \frac{2}{\tilde{t}_0^3} \left[\begin{array}{c} \dfrac{\tilde{t}^4}{8} + \tilde{t}^3 S_{\text{X22}}^{(0)}(\tilde{x}) - 3\tilde{t}^2 S_{\text{X22}}^{(1)}(\tilde{x}) + 6\tilde{t} S_{\text{X22}}^{(2)}(\tilde{x}) - 6S_{\text{X22}}^{(3)}(\tilde{x}) \\ +6 \displaystyle\sum_{m=1}^{\infty} e^{-(m\pi)^2 \tilde{t}} \dfrac{\cos(\beta_m \tilde{x})}{\beta_m^8} \end{array} \right] \tag{7.12}
$$

$$
\tilde{T}_{\text{X22B340T0}} = \frac{T^{(4)}(\tilde{x}, \tilde{t}, \tilde{t}_0)}{q_N L/k} = \frac{2}{\tilde{t}_0^4} \left[\begin{array}{c} \dfrac{\tilde{t}^5}{10} + \tilde{t}^4 S_{\text{X22}}^{(0)}(\tilde{x}) - 4\tilde{t}^3 S_{\text{X22}}^{(1)}(\tilde{x}) + 12\tilde{t}^2 S_{\text{X22}}^{(2)}(\tilde{x}) \\ -24\tilde{t} S_{\text{X22}}^{(3)}(\tilde{x}) + 24 S_{\text{X22}}^{(4)}(\tilde{x}) - 24 \displaystyle\sum_{m=1}^{\infty} e^{-(m\pi)^2 \tilde{t}} \dfrac{\cos(\beta_m \tilde{x})}{\beta_m^{10}} \end{array} \right] \tag{7.13}
$$

The $S_{X22}^{(i)}(\tilde{x})$ functions are infinite series of terms which can be expressed in closed form as

$$S_{X22}^{(0)}(\tilde{x}) = \frac{2 - 6\tilde{x} + 3\tilde{x}^2}{12} \tag{7.14}$$

$$S_{X22}^{(1)}(\tilde{x}) = \frac{8 - 60\tilde{x}^2 + 60\tilde{x}^3 - 15\tilde{x}^4}{720} \tag{7.15}$$

$$S_{X22}^{(2)}(\tilde{x}) = \frac{32 - 168\tilde{x}^2 + 210\tilde{x}^4 - 126\tilde{x}^5 + 21\tilde{x}^6}{30\ 240} \tag{7.16}$$

$$S_{X22}^{(3)}(\tilde{x}) = \frac{128 - 640\tilde{x}^2 + 560\tilde{x}^4 - 280\tilde{x}^6 + 120\tilde{x}^7 - 15\tilde{x}^8}{1\ 209\ 600} \tag{7.17}$$

$$S_{X22}^{(4)}(\tilde{x}) = \frac{2560 - 12672\tilde{x}^2 + 10560\tilde{x}^4 - 3696\tilde{x}^6 + 990\tilde{x}^8 - 330\tilde{x}^9 + 33\tilde{x}^{10}}{239\ 500\ 800}$$

$$\approx \frac{1}{\pi^{10}}\left[\frac{\cos(\pi\tilde{x})}{1} + \frac{\cos(2\pi\tilde{x})}{2^{10}} + \frac{\cos(3\pi\tilde{x})}{3^{10}}\right] \tag{7.18}$$

To compute the response in Eq. (7.10) when $\tilde{t} > 2\tilde{t}_0$, the following component solutions are needed. These are solutions to the X22B3p0T0 cases for finite heating until time t_1:

$$\tilde{T}_{X22B(32,\ 5)0T0} = \frac{1}{\tilde{t}_0^2}\left(\frac{\tilde{t}_1^3}{3} + 2\sum_{m=1}^{\infty}\cos(\beta_m\tilde{x})\frac{-2e^{-\beta_m^2\tilde{t}} + \left(2 - 2\beta_m^2\tilde{t}_1 + \beta_m^4\tilde{t}_1^2\right)e^{-\beta_m^2(\tilde{t}-\tilde{t}_1)}}{\beta_m^6}\right) \tag{7.19}$$

$$\tilde{T}_{X22B(33,\ 5)0T0} = \frac{\tilde{t}_1^4}{4\tilde{t}_0^3} + \frac{2}{\tilde{t}_0^3}\sum_{m=1}^{\infty}\cos(\beta_m\tilde{x})\frac{6e^{-\beta_m^2\tilde{t}} + \left(\begin{array}{c}-6 + 6\beta_m^2\tilde{t}_1 - \\ 3\beta_m^4\tilde{t}_1^2 + \beta_m^6\tilde{t}_1^3\end{array}\right)e^{-\beta_m^2(\tilde{t}-\tilde{t}_1)}}{\beta_m^8} \tag{7.20}$$

$$\tilde{T}_{X22B(34,\ 5)0T0} = \frac{\tilde{t}_1^5}{5\tilde{t}_0^4} + \frac{2}{\tilde{t}_0^4}\sum_{m=1}^{\infty}\cos(\beta_m\tilde{x})\frac{-24e^{-\beta_m^2\tilde{t}} + \left(\begin{array}{c}24 - 24\beta_m^2\tilde{t}_1 + 12\beta_m^4\tilde{t}_1^2 \\ -4\beta_m^6\tilde{t}_1^3 + \beta_m^8\tilde{t}_1^4\end{array}\right)e^{-\beta_m^2(\tilde{t}-\tilde{t}_1)}}{\beta_m^{10}} \tag{7.21}$$

Numerical values of $\tilde{T}(1, \tilde{t})$ for the quartic heat flux are given in Table 7.2 for $\Delta\tilde{t} = 0.06$.

Table 7.2 Temperatures at an insulated surface of a finite plate heated with the quartic heat flux pulse shown in Figure 7.6.

		$\Delta\tilde{t} = 0.06$			
$\tilde{t}$	$\tilde{T}(1, \tilde{t})$	$\tilde{t}$	$\tilde{T}(1, \tilde{t})$	$\tilde{t}$	$\tilde{T}(1, \tilde{t})$
0.06	0.005402	0.66	0.699656	1.26	0.654863
0.12	0.027891	0.72	0.750985	1.32	0.648194
0.18	0.069858	0.78	0.785674	1.38	0.644530
0.24	0.129773	0.84	0.802913	1.44	0.642505
0.30	0.204221	0.90	0.803118	1.50	0.641386
0.36	0.288823	0.96	0.788079	1.56	0.640766
0.42	0.378779	1.02	0.761104		
0.48	0.469241	1.08	0.727167		
0.54	0.555576	1.14	0.693047		
0.60	0.633579	1.20	0.667474		

7.2.5 Random Errors

To make the previous test cases more realistic, errors are added to the exact temperatures. The simulated temperatures are given by additive errors,

$$Y_i = T_i + \varepsilon_i \tag{7.22}$$

$$\varepsilon_i = \sigma_Y u_i \tag{7.23}$$

where u_i is a random number with zero mean and unity standard deviation, and σ_Y is the standard deviation of the measurement errors (ε_i). For examples in this chapter, a standard deviation of 0.5% of the maximum temperature rise during the heating process is used. This is consistent with a measurement error of 0.5 °C for temperatures measured in the range of 100 °C.

In most heat transfer problems, a reasonable simulation of the errors ε_i is one that has a constant variance with i. This means that the same magnitude of error occurs for low as well as high temperatures. This is usually much better than assuming multiplicative errors, for example,

$$Y_i = (T_i - T_0)(1 + \varepsilon_i) + T_0 \tag{7.24}$$

where again the variance of ε_i is a constant. The assumption in Eq. (7.24) is that the *relative* errors are constant; for example, Eq. (7.24) implies an error of 0.001 °C when $T_i - T_0 = 0.1$ °C is as likely as an error of 1 °C when $T_i - T_0 = 100$ °C. This is not reasonable when using the same sensor covering the temperature range of $T_i - T_0 = 0$–100 °C in a single transient experiment.

Likewise, in most measurement applications, the distribution of the errors about the mean value is approximately normal, or Gaussian. This means that smaller errors from the mean are observed more often than greater errors, although all errors contribute to the standard deviation σ_Y. The errors used in this chapter to simulate measured temperatures are generated using legacy MATLAB code:

```
state = 372;            % always use this seed
randn('state', state); % initialize the generator
noise = randn(nt,1)*sigma_noise;
```

For reference, the numbers used for 30 simulated measurements in the case of $\Delta \tilde{t} = 0.06$ are listed in Table 7.3. When more measurements are used (60 or 120), the same MATLAB code shown above is used with nt = 60 or 120, that is, the first 30 numbers are the same as Table 7.3 but additional numbers in the sequence are also obtained.

Table 7.3 Random numbers used to generate simulated measurements for nt = 30.

		$\Delta \tilde{t} = 0.06$			
$\tilde{t}$	$\tilde{\varepsilon}_i$	$\tilde{t}$	$\tilde{\varepsilon}_i$	$\tilde{t}$	$\tilde{\varepsilon}_i$
−0.18	1.866926	0.42	0.220125	1.02	1.990298
−0.12	−0.197011	0.48	0.445238	1.08	−1.047074
−0.06	−0.271328	0.54	0.253079	1.14	−0.374020
0.00	−0.682878	0.60	0.157191	1.20	1.469407
0.06	0.980267	0.66	1.876098	1.26	0.872546
0.12	−1.298371	0.72	0.502068	1.32	0.130606
0.18	0.191776	0.78	0.019142	1.38	0.511344
0.24	0.949967	0.84	0.727167	1.44	0.184551
0.30	1.470885	0.90	0.693047	1.50	1.309387
0.36	0.377479	0.96	0.667474	1.56	2.647314

Table 7.4 Table of materials, L, and q_0 for putting the dimensionless test case results into dimensional terms.

Material	$\alpha \times 10^5 (\text{m}^2/\text{s})$	k (W/m-°C)	L (cm) for $\Delta t = \Delta \tilde{t}$	$q_0 (\text{W/m}^2)$ for T(°C) = $\tilde{T}$
Brick	0.04	1	0.0632	1580
Steel	1.1	40	0.332	12 100
Copper	11	380	1.05	36 200

7.2.6 Temperature Perturbation

Another basic or fundamental case is the temperature perturbation test case, which can be used to generate the filter coefficients for the IHCP method (see Chapter 5). For this test case,

$$Y_i = \begin{cases} 1 & \text{for } i = r \\ 0 & \text{for } i \neq r \end{cases} \tag{7.25}$$

and r is chosen in the middle of the range (for example, $r = 15$ for 30 measurements).

This case was discussed in Section 5.2 in connection with sequential filter algorithms.

7.2.7 Test Cases with Units

The test cases in this chapter are given in dimensionless forms. This is most general since each case represents many possible plate materials, plate thicknesses, and surface heat fluxes. Nevertheless, it is helpful to be able to interpret the results in relation to physical cases. A way to relate the dimensionless results to those with units is presented.

To cover a wide range of heat conducting characteristics, three materials with dissimilar properties are selected; namely, brick, steel, and copper. See Table 7.4 in which the thermal diffusivities and thermal conductivities are listed.

One way to relate the dimensionless times,

$$\Delta \tilde{t} = \frac{\alpha \Delta t}{L^2} \quad \text{and} \quad \tilde{t} = \frac{\alpha t}{L^2} \tag{7.26}$$

to time steps and times with units of seconds is to establish the following equations:

$$\Delta \tilde{t} = \Delta t, s \quad \text{and} \quad \tilde{t} = t, s \tag{7.27}$$

These relations are valid if L is chosen such that

$$L = \alpha^{1/2} \tag{7.28}$$

The values of L in cm are given in Table 7.4 for the three materials listed. Consequently, for all the test cases for the finite plate, the dimensionless time steps (and times) can be interpreted as time steps (and times) in seconds for the cases listed in Table 7.4.

In the case of the triangular heat flux and quartic heat flux pulse examples, the nominal heat flux, q_{ref}, in Eqs. (7.3), (7.9) and (7.11)–(7.13), is treated in the same manner as q_0 for the step increase in the surface heat flux test case.

7.3 Function Specification Method

Regularization in the FS method is introduced through the number of future times r. The method is inherently sequential and cannot directly applied to minimize $E(R_q^2)$ in Eq. (6.21) to find the optimal degree of regularization. However, by employing the temperature perturbation method from Chapter 5, the equivalent filter matrix $\mathbf{F}$ for the FS method can be found for any value of r. This approach is used in this section to determine the optimal number of future times r_{opt} for each of the test cases.

7.3.1 Step Change in Surface Heat Flux

Figures 7.7 and 7.8 display results for the step heat flux test case with $\Delta \tilde{t} = 0.06$ and 0.03, respectively. Results for both piecewise constant and piecewise linear heat flux assumptions are presented in each figure. In each of the four cases, the optimal number of future time steps required is determined using the minimum expected value of the mean squared error in the heat flux, $\min(E(R_q^2)$; see Eq. (6.21)). The optimum r_{opt} is determined using the expected value, but the points in the figures are estimated for a particular set of noisy data, as described in Section 7.2.5. The magnitude of $E(R_q^2)$ is the square of the average distance of the estimated values from the true value for a large number of sets of noisy data.

Table 7.5 summarizes the results for time steps of $\Delta \tilde{t} = 0.06, 0.03$, and 0.015. In this table, and subsequent tables for other IHCP methods, $E(R_q)$ is used as the measure of goodness for each case considered. $E(R_q)$ is the square root of $E(R_q^2)$ and is sometimes called the root mean squared (RMS) error ($E_{q,RMS}$), or standard error (s_q).

The optimal number of future times is $r_{opt} = 4$ for both constant and linear approximations when $\Delta \tilde{t} = 0.06$, $r_{opt} = 7$ and 8 for $\Delta \tilde{t} = 0.03$ and increase to $r_{opt} = 11$ and 13 for $\Delta \tilde{t} = 0.015$. These results support the idea that the "look ahead time"

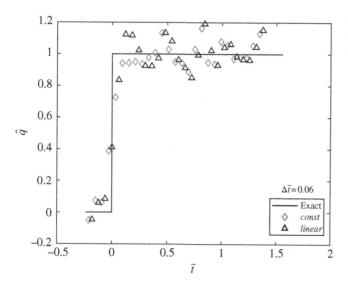

Figure 7.7 Estimated heat fluxes using function specification for step change test case with $\Delta \tilde{t} = 0.06$ using either piecewise constant or piecewise linear assumptions.

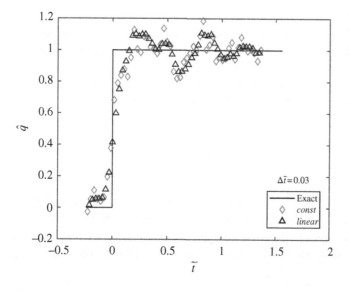

Figure 7.8 Estimated heat fluxes using function specification for step change test case with $\Delta \tilde{t} = 0.03$ using either piecewise constant or piecewise linear assumptions.

Table 7.5 Summary of results for step heat flux case using function specification for σ_{noise} = 0.0070.

Case	Assumption	$\Delta\tilde{t}$	r_{opt}	$E(R_q)$	$p = r_{opt}\Delta\tilde{t}$
step	$q = const$	0.060	4	0.16848	0.24
step	$q = const$	0.030	7	0.16269	0.21
step	$q = const$	0.015	13	0.15449	0.195
step	$q = linear$	0.060	4	0.13059	0.24
step	$q = linear$	0.030	8	0.11839	0.24
step	$q = linear$	0.015	14	0.11164	0.21

$p = r_{opt}\Delta t$ is approximately constant for different data frequencies (Woodbury and Thakur 1996). That is, when the time step is reduced by half, the number of future times needed is approximately doubled.

Table 7.5 indicates that the piecewise linear approximation for heat flux offers an advantage over the piecewise constant assumption in accurately estimating the step change in heat flux. For both time steps considered, the value of $E(R_q)$ is reduced when using the $q = linear$ assumption. Because the value of the heat flux step is 1.0 over most of the time interval, the magnitude of $E(R_q)$ suggests that the average error in the estimates is in the range of 11–17%.

7.3.2 Triangular Heat Flux

Calculated heat flux values for the triangular flux test case are displayed in Figure 7.9 for $\Delta\tilde{t} = 0.06$ and Figure 7.10 for $\Delta\tilde{t} = 0.03$. Both piecewise constant and piecewise linear heat flux assumptions are considered, and the minimum of the mean residual heat flux error is used to determine the optimal number of future time steps needed.

Table 7.6 summarizes the results from the four triangular heat flux test cases. Also shown in the table is another case with $\Delta\tilde{t} = 0.015$. Visual inspection of Figures 7.9 and 7.10 suggests that the linear heat flux assumption yields a better estimation, and the $E(R_q)$ values in Table 7.6 confirm that the piecewise linear assumption gives the best result for both time step cases. These results also support the idea that the "look-ahead" time, $p = r_{opt}\Delta t$, required is approximately constant.

7.3.3 Quartic Heat Flux

Heat flux estimates for the quartic heat flux case are shown in Figures 7.11 and 7.12. These are obtained using the optimal value of r determined by minimizing the mean squared error in the heat flux. The results in these figures are strikingly similar to those for the triangular case in Figures 7.9 and 7.10, including the tendency for the $q = $ linear case to produce

Figure 7.9 Estimated heat fluxes using function specification for triangular test case with $\Delta\tilde{t} = 0.06$ using either piecewise constant or piecewise linear assumptions.

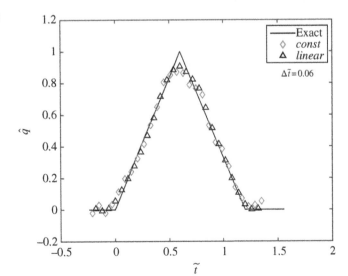

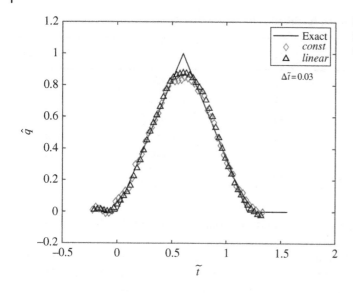

Figure 7.10 Estimated heat fluxes using function specification for triangular test case with $\Delta \tilde{t} = 0.03$ using either piecewise constant or piecewise linear assumptions.

Table 7.6 Summary of results for triangular heat flux case using function specification for $\sigma_{noise} = 0.0030$.

Case	Assumption	$\Delta \tilde{t}$	r_{opt}	$E(R_q)$	$p = r_{opt} \Delta \tilde{t}$
Triangle	$q = const$	0.060	4	0.03842	0.24
Triangle	$q = const$	0.030	8	0.03356	0.24
Triangle	$q = const$	0.015	15	0.02918	0.225
Triangle	$q = linear$	0.060	5	0.03343	0.30
Triangle	$q = linear$	0.030	9	0.02732	0.27
Triangle	$q = linear$	0.015	16	0.02323	0.24

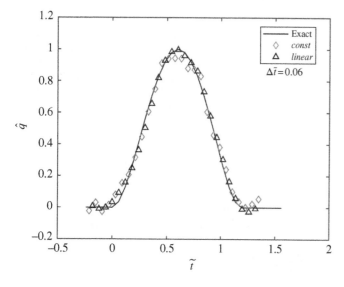

Figure 7.11 Estimated heat fluxes using function specification for quartic test case with $\Delta \tilde{t} = 0.06$ using either piecewise constant or piecewise linear assumptions.

Figure 7.12 Estimated heat fluxes using function specification for quartic test case with $\Delta \tilde{t} = 0.03$ using either piecewise constant or piecewise linear assumptions.

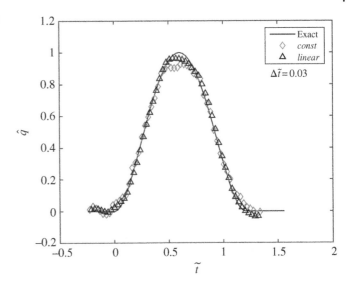

Table 7.7 Summary of results for quartic heat flux case using function specification for $\sigma_{noise} = 0.0032$.

Case	Assumption	$\Delta \tilde{t}$	r_{opt}	$E(R_q)$	$p = r\Delta \tilde{t}$
quartic	$q = const$	0.060	4	0.03901	0.24
quartic	$q = const$	0.030	8	0.03334	0.24
quartic	$q = const$	0.015	15	0.02834	0.225
quartic	$q = linear$	0.060	5	0.02836	0.30
quartic	$q = linear$	0.030	9	0.02237	0.27
quartic	$q = linear$	0.015	16	0.01831	0.24

higher and truer estimates than the $q = const$ case near the peak of the heat flux. In fact, the expected value of the RMS error $E(R_q)$ is smaller for the linear assumption.

Table 7.7 summarizes the results from the four quartic heat flux test cases and includes results for an additional small time step with $\Delta \tilde{t} = 0.015$. These results also strongly agree with those found for the triangular heat flux case. The "look-ahead" time p is relatively constant and covers the range $(0.225 \leq p \leq 0.30)$.

7.3.4 Temperature Perturbation

The temperature perturbation test case (data $Y_i = 0$ for all $i \neq M$, but $Y_M = 1$) can be used to generate the filter coefficients for any IHCP method. This idea was explained in Chapter 5 and was used to compute the filter matrix for the FS method in order to optimize the number of future times required. However, the signature of the filter coefficient vector for a single perturbation can give insights into the physics and stability of an IHCP solution technique. This section presents filter coefficient vectors for the FS method.

Figures 7.13 and 7.14 display the filter coefficients which are optimal for $\Delta \tilde{t} = 0.06$ and 0.03, respectively.

Figure 7.13 offers a direct comparison of the filter coefficients for the FS method using $q = const$ and $q = linear$ assumptions for $r = 4$. For these calculations, the time of the perturbation is $\tilde{t}_M = 0.66$, and precisely four time steps ahead of this time all of the components of the filter vector $f_i = 0$. Of course, this is consistent with the $r = 4$ future times used in the FS algorithm. This means that the calculation of the temperature response at t_M requires data in the future from t_{M+r-1} (the current time step M and three additional time steps). Similarly, the values of f_i before $\tilde{t}_M = 0.66$ are nonzero back to about $\tilde{t} = 0.24$, so about seven *past* time steps of data are required to compute the heat flux at time t_M. This relatively compact

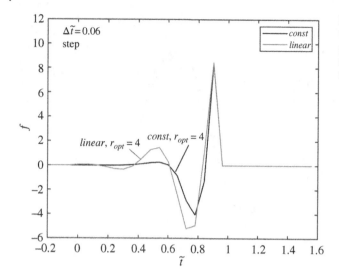

Figure 7.13 Filter coefficients for function specification obtained from temperature perturbation case with $\Delta\tilde{t}$ = 0.06 using either piecewise constant or piecewise linear assumptions. r_{opt} = 4 is shown which is optimal for step, triangular, and quartic cases using noisy data with σ_{noise} = 0.5% × T_{max}.

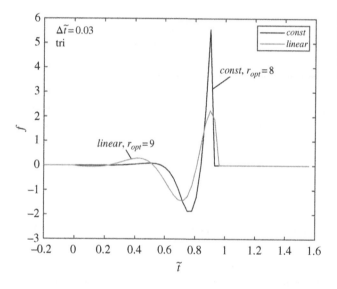

Figure 7.14 Filter coefficients for function specification obtained from temperature perturbation case with $\Delta\tilde{t}$ = 0.03 using either piecewise constant or piecewise linear assumptions. r_{opt} = 6 and r_{opt} = 8 are shown. Optimal for step, triangular, and quartic cases using noisy data with σ_{noise} = 0.5% × T_{max} is $7 \leq r_{opt} \leq 9$.

footprint for the IHCP filter coefficient vector mirrors physics and is desirable: the temperatures in the future and those in the distant past are not affected by the heat flux at the "current" time.

Figure 7.14 presents the filter vectors resulting from optimization for the triangular heat flux test case when $\Delta\tilde{t} = 0.03$. For the $q = const$ assumption, $r_{opt} = 8$, and for the $q = linear$ assumption, $r_{opt} = 9$. The time of the perturbation is still $\tilde{t}_M = 0.66$, and the time in the past at which the filter coefficients are nearly zero is $\tilde{t}_M = 0.27$, which is 13 time steps, although the $q = linear$ curve shows small variations until $\tilde{t}_M = 0.06$. The overall "width" of these filters in terms of the time span they cover is very similar to those from the $\Delta\tilde{t} = 0.06$ test cases.

As described in Chapter 5, provided the algorithm is stable, the sum of the f_i values is zero for the X22 IHCP. In a sense, this also verifies conservation of energy because the initial and final internal energies for this test case are zero.

The shape of the filter coefficient curves depend on the algorithm and the degree of regularization and not on the amount of noise in the data. (The optimal number of future times required does depend on the noise in the data, of course.) Two additional figures, Figures 7.15 and 7.16, are included to show features of the filter coefficients for FS method.

Figure 7.15 gives the filter coefficients for the $q = linear$ assumption using time step $\Delta\tilde{t} = 0.06$ for $r = 3$, 4, and 5. The erratic behavior of the curve for $r = 3$ indicates onset of instability. In fact, for $r = 2$ the filter coefficients are completely unstable. The cases for $r = 4$ and $r = 5$ are typical and show that, in general, as the width of the filter increases (due to

Figure 7.15 Filter coefficients for function specification obtained from temperature perturbation case with $\Delta\tilde{t} = 0.06$ using piecewise linear assumption. $r = 3, 4, 5$ are shown.

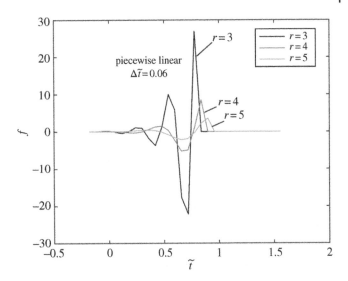

Figure 7.16 Filter coefficients for function specification obtained from temperature perturbation case with $\Delta\tilde{t} = 0.03$ using piecewise constant assumption. $r = 6, 7, 8, 9$ are shown.

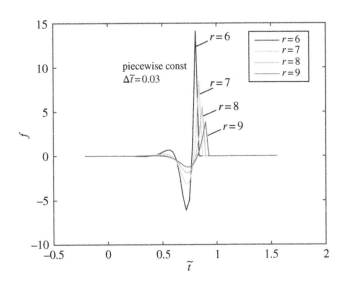

inclusion of more future time steps), the magnitude of the individual values decrease. This effect helps explain the trade of decreasing sensitivity to noise (lower filter coefficient magnitudes) for greater bias in the estimates (wider filter footprint due to more future data) as r is increased.

Figure 7.16 shows the coefficients for the $q = const$ assumption using time step $\Delta\tilde{t} = 0.03$ for $r = 6, 7, 8,$ and 9. The trend cited in the previous discussion regarding the magnitude of the coefficients as r increases is apparent.

7.3.5 Function Specification Test Case Summary

The amount of regularization for the FS method is controlled through the number of future time steps, r. The values of r can only be discrete, so the ability to finely tune the optimization of regularization using FS is limited by this fact.

The optimal number of future times needed for FS using different time steps (frequency of data) is roughly correlated with the dimensionless "look-ahead" time, $p = r\Delta t$. Over a range of relatively small dimensionless time steps for cases having measurement noise on the order of 0.5% of the maximum temperature rise, the value of p required for optimal estimation varies over the approximate range $0.20 < p < 0.30$. This can be used as a rule-of-thumb to estimate the optimal number of future times needed for different data frequencies producing $\Delta\tilde{t} \leq 0.06$. For larger time steps ($\Delta\tilde{t} > 0.06$), $r = 3$ or $r = 2$ is nearly optimal for many cases.

The $q = linear$ (piecewise linear) heat flux assumption does show some advantage over the $q = const$ (piecewise constant) assumption for the test problems and time steps considered. The average error in the heat flux estimate $E(R_q)$ is smaller for the $q = linear$ for every test case when using the same time step as in the $q = const$ case. The magnitude of $E(R_q)$ can be reduced with smaller time steps.

7.4 Tikhonov Regularization

TR is controlled by the magnitude of the regularization coefficient, α_i. The filter matrix **F** is easily calculated for TR, and this matrix is a continuous function of the regularization coefficient, hence the optimal degree of regularization needed to minimize $E(R_q^2)$ can be found using, for example, fminbnd() in MATLAB.

In this section, the performance of zeroth- and first-order TR methods are evaluated using both $q = const$ and $q = linear$ assumptions for heat flux variation.

7.4.1 Step Change in Surface Heat Flux

Figure 7.17 displays the results from the optimal heat flux estimation using noisy data and TR for the step change in heat flux test case using $\Delta\tilde{t} = 0.06$. The standard deviation of the noise is 0.5% of the maximum temperature rise resulting from the heating pulse. The curves with diamonds and triangles are obtained using zeroth-order TR, and the plus ("+") and cross ("×") symbol results are obtained using first-order TR. Results from $q = const$ (diamond and plus) and $q = linear$ (triangle and cross) assumptions for heat flux variation are indicated.

The results from the four estimations in Figure 7.17 are remarkably similar, except near the end of the time span. Inspection reveals that the $q = linear$ combination with zeroth-order regularization appears to have the largest envelope in the oscillations about the true value of $\tilde{q} = 1.0$ for $\tilde{t} > 0$. Inspection also suggests that the results, though optimized for minimum heat flux estimation error, are overly smooth at early time steps as the initial zero value of heat flux is not detected. This can be attributed to the relatively large ($\sigma = 0.070$) amount of additive noise in the data which becomes significant when added to the initially zero temperature data.

The behavior of the results at the end of the time interval is connected to the parabolic nature of heat conduction and the lack of data beyond the end of the time span. These results underscore the fact that *some future time data (information) is needed to estimate the heat flux at the current time.* This is a consequence of the time lagging of the effect of the heat flux on the response of a subsurface sensor. For sensors closer to the surface, this effect is diminished.

The difference in the behavior of the estimates at the end of the time interval for zeroth-order TR (open symbols) and first-order TR (crosses and plusses) is due to the specific bias underlying each approach. Zeroth-order TR penalizes the function

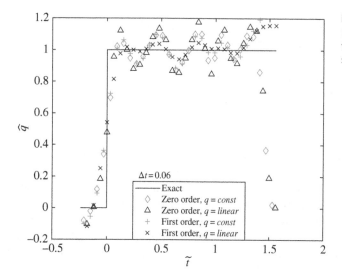

Figure 7.17 Optimal estimated heat fluxes using Tikhonov Regularization for step change test case with $\Delta\tilde{t} = 0.06$. Results for piecewise constant and piecewise linear assumptions are shown using both zeroth- and first-order regularization.

value and biases the results toward a zero value. First-order TR penalizes the function derivative and biases the results toward a constant value. These effects are evident in Figure 7.17. The duration of this time is similar to the "look-ahead" time discussed in the FS section and suggests that heat flux estimates for times within about 0.15 dimensionless units of the final time are significantly biased. This nominal value is larger (perhaps 0.18 dimensionless time units) for greater noise levels and is smaller (perhaps 0.12 dimensionless time units) for smaller data time intervals.

Table 7.8 summarizes the values of the optimal regularization coefficient for different data time intervals and combinations of heat flux variation assumption and order of regularization for the step test case. These results indicate that "better" estimates (those having smaller average error $E(R_q)$) for heat flux history are obtained with smaller data time intervals for all combinations. First-order regularization with $q = const$ assumption produced the largest average error, but produced the lowest $E(R_q)$ using the $q = linear$ assumption. Note that all zeroth-order regularization cases require smaller values of α as data time interval decreases, but all first-order regularization combinations display the opposite trend.

7.4.2 Triangular Heat Flux and Quartic Heat Flux

Results for estimated heat fluxes for the triangular test case and the quartic test case are shown in Figures 7.18 and 7.19, respectively, for data time interval $\Delta \tilde{t} = 0.06$. For both of these test cases, the optimized regularization results for various combinations of heat flux variation assumption and order of TR produce visually similar results.

Differences between zeroth-order and first-order regularization can be seen in Figures 7.18 and 7.19 at the end of the time interval because the zeroth order biases the result to $q = 0$, and the first order biases the heat flux toward a constant value, though not necessarily zero. This behavior is the same as observed for the step heat flux change but is not as obvious since the true heat flux is zero at the end of the data time interval. Where possible, this fact can be used to minimize erroneous estimates near the end of the data time period by taking data until the heat flux excitation is reduced to zero.

Table 7.9 summarizes the results of estimations from combinations of zeroth- and first-order TR with $q = const$ and $q = linear$ heat flux variation assumption. For each line in the table, the magnitude of the optimal regularization coefficient is very similar for both the triangular and quartic heat flux test cases. As with the step heat flux test case, the optimal regularization coefficient decreases with decreasing data time step for zeroth-order regularization, but increases with decreasing data time step for first-order regularization.

The need for increase in the regularization coefficient for the $q = linear$ assumption as the data time step decreases can be understood in light of the significant increase in the condition number of the filter matrix as $\Delta \tilde{t}$ decreases. In particular, the diagonal elements of the $\mathbf{X}$ matrix for the $q = linear$ assumption are much smaller than those for the $q = const$ assumption,

Table 7.8 Summary of results for step heat flux case using Tikhonov regularization.

Order	Assumption	$\Delta \tilde{t}$	Step σ_{noise} = 0.0070	
			α_{opt}	$E(R_q)$
0th	$q = const$	0.060	8.13E-04	0.14568
0th	$q = const$	0.030	5.39E-04	0.14152
0th	$q = const$	0.015	3.46E-04	0.13458
0th	$q = linear$	0.060	4.36E-04	0.27886
0th	$q = linear$	0.030	2.77E-04	0.26860
0th	$q = linear$	0.015	1.92E-04	0.25913
1st	$q = const$	0.060	8.50E-04	0.15014
1st	$q = const$	0.030	1.62E-03	0.14477
1st	$q = const$	0.015	3.51E-03	0.13682
1st	$q = linear$	0.060	2.76E-03	0.10943
1st	$q = linear$	0.030	5.43E-03	0.10854
1st	$q = linear$	0.015	7.44E-03	0.10295

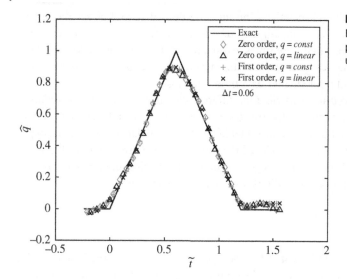

Figure 7.18 Optimal estimated heat fluxes using Tikhonov Regularization for triangular test case with $\Delta \tilde{t} = 0.06$. Results for piecewise constant and piecewise linear assumptions are shown using both zeroth- and first-order regularization.

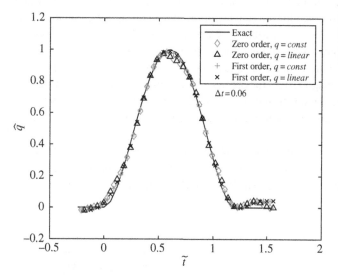

Figure 7.19 Optimal estimated heat fluxes using Tikhonov Regularization for quartic test case with $\Delta \tilde{t} = 0.06$. Results for piecewise constant and piecewise linear assumptions are shown using both zeroth- and first-order regularization.

and become smaller faster as $\Delta \tilde{t}$ decreases, contributing to a faster increase in the condition number for the linear case in comparison to the $q = const$ case.

The root mean squared error, $E(R_q)$, in Table 7.9 decreases with decreasing data time step. For the triangular heat flux test case, all of the combinations in the table produce very similar values of $E(R_q)$ for each $\Delta \tilde{t}$. However, for the quartic test case, the value of $E(R_q)$ for the first-order regularization produces noticeable improvements over the corresponding calculations using zeroth-order regularization, which suggests that the first order has some advantage for smoothly varying heat fluxes.

7.4.3 Temperature Perturbation

Figure 7.20 shows the filter coefficients for the optimal level of regularization for the triangular test case with $\Delta \tilde{t} = 0.06$ for all of the combinations of order of regularization and heat flux assumption listed in Table 7.9. These coefficients are obtained from the middle row of the filter matrix resulting from TR (see Chapter 5), but are the same as those obtained using the temperature perturbation test case.

Some observations can be made from Figure 7.20. The optimal filter coefficients for zeroth regularization are generally larger in magnitude than those for first-order regularization. Larger filter coefficients mean greater sensitivity to measured

Table 7.9 Summary of results for triangular and quartic heat flux cases using Tikhonov Regularization.

Order	Assumption	$\Delta \tilde{t}$	Triangular σ_{noise} = 0.0030		Quartic σ_{noise} = 0.0032	
			α_{opt}	$E(R_q)$	α_{opt}	$E(R_q)$
0th	q =const	0.060	1.61E-03	0.02822	1.67E-03	0.02670
0th	q =const	0.030	1.25E-03	0.02450	1.35E-03	0.02204
0th	q =const	0.015	9.13E-04	0.02106	1.04E-03	0.01806
0th	$q = linear$	0.060	1.56E-03	0.03013	1.61E-03	0.02641
0th	$q = linear$	0.030	1.24E-03	0.02489	1.32E-03	0.02192
0th	$q = linear$	0.015	9.14E-04	0.02111	1.04E-03	0.01803
1st	$q = const$	0.060	4.29E-03	0.02670	4.79E-03	0.02158
1st	$q = const$	0.030	7.81E-03	0.02372	1.31E-02	0.01762
1st	$q = const$	0.015	1.74E-02	0.02042	3.64E-02	0.01436
1st	$q = linear$	0.060	4.14E-03	0.02979	4.67E-03	0.02218
1st	$q = linear$	0.030	7.87E-03	0.02450	1.30E-02	0.01783
1st	$q = linear$	0.015	1.77E-02	0.02060	3.63E-02	0.01443

Figure 7.20 Filter coefficients for Tikhonov Regularization obtained from temperature perturbation case with $\Delta \tilde{t}$ = 0.06 using either piecewise constant or piecewise linear assumptions with either zeroth- or first-order regularization. Optimal values for regularization coefficient shown for triangular test case using noisy data with σ_{noise} = 0.5% × T_{max}.

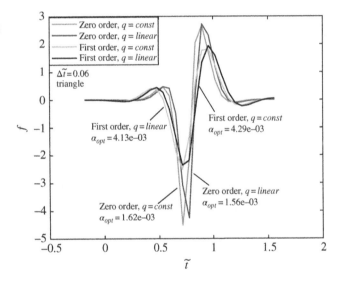

data, which contains both true information and measurement noise. But, in general, larger filter coefficients are better. Another observation is that the first-order regularization coefficients have a wider "footprint" on time axis. This means that first-order regularization requires more data in the future (to the right) and in the past (to the left) than zeroth-order regularization.

Figure 7.21 displays the filter coefficients for the $q = const$ assumption using zeroth-order regularization, and Figure 7.22 displays the filter coefficients for the $q = linear$ assumption using zeroth-order regularization. Both of these figures are generated using a data time increment of $\Delta \tilde{t} = 0.03$. Comparison of the figures confirms that, for the same value of α, the filter coefficients for the linear, first-order combination have larger magnitudes than the const, zeroth-order combination. This can be understood considering the differences in magnitudes of the $\mathbf{H}^T\mathbf{H}$ matrices which are used in the two regularization methods: $\mathbf{H_0}^T\mathbf{H_0}$ is populated with 1s on the diagonal, while $\mathbf{H_1}^T\mathbf{H_1}$ is a banded matrix of $-1, 2, -1$, so the same value of α adds more to the $\mathbf{X}^T\mathbf{X}$ matrix for first order than for zeroth order. This significant difference in the structure of the $\mathbf{H}^T\mathbf{H}$ matrix is also responsible for the wider footprint of the filter coefficients in Figure 7.22 over that in Figure 7.21 for every

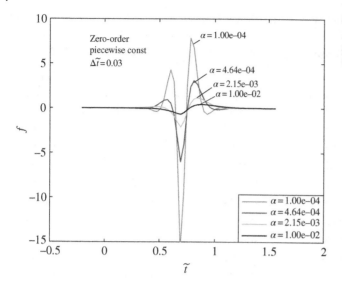

Figure 7.21 Filter coefficients for Tikhonov Regularization with $\Delta \tilde{t} = 0.03$ using piecewise constant heat flux assumption with zeroth-order regularization for a range of regularization coefficients.

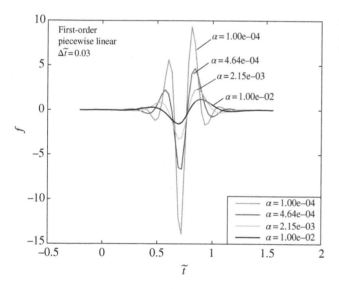

Figure 7.22 Filter coefficients for Tikhonov Regularization with $\Delta \tilde{t} = 0.03$ using piecewise linear heat flux assumption with first-order regularization for a range of regularization coefficients.

value of α: first-order regularization requires information from time steps ahead and behind the current time step, which pushes the envelope of the window out further than zeroth-order regularization.

7.4.4 Tikhonov Regularization Test Case Summary

TR is controlled through the regularization coefficient α. This is a continuous parameter that can be finely adjusted to optimize the heat flux estimation. Commercial routines such as MATLAB's fminbnd() can easily be used to determine the optimal value of α.

Heat flux estimates for TR are biased according to the order of regularization. Zeroth-order regularization biases the estimates toward a zero value of heat flux, and first-order regularization biases the estimates toward a constant value. These biases are especially evident at the end of the data time interval when the results from lack of future time information take effect. Heat flux estimates within a dimensionless time interval of about 0.12–0.18 at the end of the data record are generally affected by this biasing and should be considered unreliable.

The $q = const$ heat flux assumption coupled with the zeroth-order regularization shows good performance in accurate estimation, as indicated by the value of $E(R_q)$, of heat flux histories for a variety of test cases. However, the $q = linear$ heat flux assumption in combination with first-order regularization provides superior performance for two specific cases: the smoothly varying quartic heat flux and the discontinuous step change.

For the same values of α, the filter coefficients for the $q = linear$ assumption are larger than those for the $q = const$ assumption. However, the optimal values for the regularization coefficients for these two assumptions are different, and the resulting optimal filter coefficients for the $q = linear$ assumption are smaller than those for the $q = const$. The width of the filter on the time axis for the $q = linear$ assumption is wider than that for the $q = const$ case, which means the linear assumption requires more data in the future and in the past than the constant assumption.

7.5 Conjugate Gradient Method

Regularization in the CGM is controlled by the number of iterations used in the minimization procedure. Hansen (1998, pp. 145–154) discusses the regularization properties of CG iterations. Like the FS method, the regularization control can only be applied in discrete steps which limits the ability to refine the degree of optimization of regularization. Also like FSM, the optimization can only be investigated by considering a sequence of steps (iterations) and examining the history to determine the optimum *a posteriori*.

The filter matrix **F** for CGM does not appear explicitly in the development but can be computed based on superposition using the temperature perturbation technique as discussed in Chapter 5. The filter matrix is computed for each iteration in CGM and is used to find the optimal iteration number by minimizing mean squared error $E(R_q^2)$ or, equivalently, $E(R_q)$.

In this section, the performance of the CGM using both $q = const$ and $q = linear$ heat flux variation assumptions is investigated for three standard heat flux histories. Both Steepest Descent (SD) and Fletcher-Reeves (FR) (Fletcher and Reeves 1964) iterations are investigated. SD is not a true conjugate gradient algorithm but is included here and described as such (see Chapter 4). The filter coefficients for CGM are also discussed.

7.5.1 Step Change in Surface Heat Flux

Figure 7.23 shows the optimized estimates for the heat flux from the step heat flux test case for both the SD and FR iterative methods for both $q = const$ and $q = linear$ heat flux assumptions. All of the results are strongly similar, though the FR results (open symbols) display more variation for $\tilde{t} > 0.7$ or so. All of the cases bias toward zero at the end of the data time interval, which is again manifestation of the parabolic nature of the heat conduction problem and the dependence of the IHCP solution on future time data. From Figure 7.23, the last two data points for each of the four cases are significantly biased, corresponding to a dimensionless time interval of 0.12.

Figure 7.23 Optimal estimated heat fluxes using Conjugate Gradient Method for step change test case with $\Delta\tilde{t} = 0.06$. Results for piecewise constant and piecewise linear assumptions are shown using both Steepest Descent and Fletcher-Reeves methods.

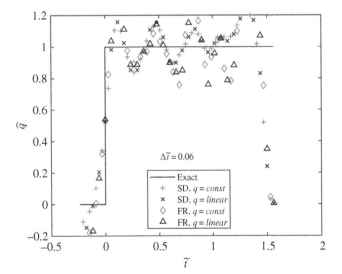

Table 7.10 Summary of results for step heat flux case using Conjugate Gradient methods.

Method[a]	Assumption	$\Delta \tilde{t}$	Step σ_{noise} = 0.0070	
			n_{opt}	$E(R_q)$
SD	$q = const$	0.060	291	0.14997
SD	$q = const$	0.030	401	0.14458
SD	$q = const$	0.015	581	0.13695
SD	$q = linear$	0.060	333	0.27128
SD	$q = linear$	0.030	623	0.25823
SD	$q = linear$	0.015	957	0.24898
FR	$q = const$	0.060	6	0.15727
FR	$q = const$	0.030	6	0.15050
FR	$q = const$	0.015	581	0.13695
FR	$q = linear$	0.060	6	0.27998
FR	$q = linear$	0.030	7	0.26689
FR	$q = linear$	0.015	957	0.24898

[a] SD, Steepest Descent; FR, Fletcher-Reeves

Table 7.10 summarizes the optimal number of iterations required for the estimates shown in Figure 7.23 and for the same heat flux test case using smaller data time intervals. In every case, the SD iteration method requires a significantly "large" number of iterations to produce the optimal result, and this is consistent with the known slow convergence of SD. However, for larger time steps, SD offers modest improvement in the $E(R_q)$ metric over corresponding cases using FR. Notice that, for the smallest time steps, the number of iterations needed for FR to achieve optimality is identical to that for SD, and the results for $E(R_q)$ are then identical.

Table 7.10 shows that the $q = linear$ assumption does not perform as well as the $q = const$ assumption. The $E(R_q)$ metric is roughly twice as large for corresponding cases. This is also apparent for the specific set of noisy data in Figure 7.23, as the $q = linear$ results (triangles and crosses) stray farther from the exact solution than the $q = const$ results (diamonds and plusses).

7.5.2 Triangular Heat Flux and Quartic Heat Flux

Results for the triangular heat flux test case using SD and FR with both $q = const$ and $q = linear$ heat flux are shown in Figure 7.24, and results for the quartic heat flux test case are in Figure 7.25. These results are optimized using the minimum

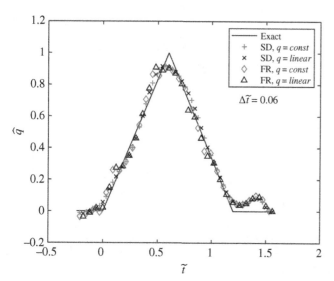

Figure 7.24 Optimal estimated heat fluxes using conjugate gradient method for triangular test case with $\Delta \tilde{t} = 0.06$. Results for piecewise constant and piecewise linear assumptions are shown using both Steepest Descent and Fletcher-Reeves methods.

Figure 7.25 Optimal estimated heat fluxes using conjugate gradient method for quartic test case with $\Delta \tilde{t} = 0.06$. Results for piecewise constant and piecewise linear assumptions are shown using both Steepest Descent and Fletcher-Reeves methods.

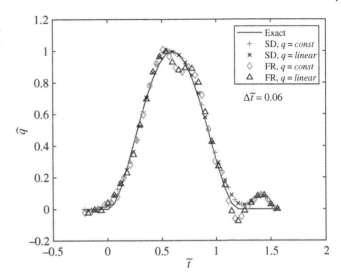

of the expected mean squared error, but the plotted points result from processing of a specific set of noisy data as described in Section 7.2.5.

For the triangular test case results in Figure 7.24, all of the results appear similar. However, for the quartic heat flux case in Figure 7.25, the FR results for both heat flux assumptions (triangles and diamonds) depart from the general trend in two places: just at the peak ($\tilde{t} = 0.6$) and at the last transition ($\tilde{t} = 1.2$).

Table 7.11 lists the quantitative results from the optimizations for Figures 7.24 and 7.25 as well as those obtained for smaller data time steps of $\Delta \tilde{t} = 0.030$ and $\Delta \tilde{t} = 0.015$. As with the step heat flux case, for the larger time steps, the FR method requires far fewer iterations for optimality, but for the smallest time step, the number of iterations needed for FR and SD is similar. Also similar to the step results, Table 7.11 shows that, for larger time steps, SD iteration produces a lower $E(R_q)$ than FR, while for the smallest time step, the $E(R_q)$ values are similar. For the triangular test case, the $q = linear$ assumption results for $E(R_q)$ are inferior to those for $q = const$, but for the quartic test case, the $E(R_q)$ results for corresponding time steps and methods for these two assumptions are similar.

Table 7.11 Summary of results for triangular and quartic heat flux cases using Conjugate Gradient methods.

Method[a]	Assumption	$\Delta \tilde{t}$	Triangular σ_{noise} = 0.0030		Quartic σ_{noise} = 0.0032	
			n_{opt}	$E(R_q)$	n_{opt}	$E(R_q)$
SD	$q = const$	0.060	135	0.03111	117	0.02956
SD	$q = const$	0.030	191	0.02834	153	0.02643
SD	$q = const$	0.015	273	0.02500	211	0.02326
SD	$q = linear$	0.060	141	0.03336	117	0.02922
SD	$q = linear$	0.030	193	0.02894	153	0.02636
SD	$q = linear$	0.015	273	0.02517	211	0.02328
FR	$q = const$	0.060	5	0.04442	6	0.04774
FR	$q = const$	0.030	6	0.03648	6	0.03936
FR	$q = const$	0.015	273	0.02500	211	0.02326
FR	$q = linear$	0.060	5	0.04406	6	0.04734
FR	$q = linear$	0.030	7	0.03669	7	0.03845
FR	$q = linear$	0.015	273	0.02517	211	0.02328

[a] SD, Steepest Descent; FR, Fletcher-Reeves.

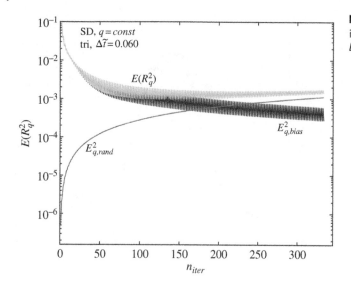

Figure 7.26 $E(R_q^2)$ versus iteration number for Steepest Descent in the triangular heat flux test case with $\Delta \tilde{t} = 0.060$. Optimal $E(R_q^2)$ indicated with a plus "+" marker.

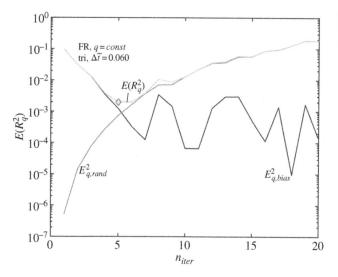

Figure 7.27 $E(R_q^2)$ versus iteration number for Fletcher-Reeves CGM in the triangular heat flux test case with $\Delta \tilde{t} = 0.060$. Optimal $E(R_q^2)$ indicated with a diamond marker.

Some insights into the CGM method can be seen in consideration of Figures 7.26 and 7.27. Each figure shows the bias and random error contributions to $E(R_q)$ along with the sum of these as iterations progress for the SD method (Figure 7.26) and the FR CGM (Figure 7.27). These results are generated using the exact heat flux for the triangular test case, but results for other heat fluxes are similar. A significant observation for these two figures is that as the degree of regularization increases (n_{iter} increases), the contribution of the random error *increases*, and the contribution of the bias error *decreases*. This is in stark contrast to the FS method or TR method for solving the IHCP, which have high random errors and low bias errors for under-regularized solutions, and random errors decrease while bias errors increase with increased regularization (cf. Figures 6.2 and 6.3). Regarding SD iterations in Figure 7.26, the random error grows smoothly with iteration number, but the bias error oscillates from one iteration to the next, resulting in the sawtooth curves for both $E_{q,bias}$ and $E(R_q)$.

7.5.3 Temperature Perturbation

Figure 7.28 shows the optimal filter coefficient for the quartic test case with dimensionless data time step $\Delta \tilde{t} = 0.060$. Both SD iteration and FR CGM are shown and each with both $q = const$ and $q = linear$ heat flux assumption. Since the optimal number of iterations is the same under both assumptions for each method ($n_{opt} = 117$ for SD and $n_{opt} = 6$ for FR), differences in these results are only due to the difference in heat flux variation assumption. The $q = linear$ assumption uses slightly more

Figure 7.28 Filter coefficients for conjugate gradient methods obtained from temperature perturbation case with $\Delta \tilde{t} = 0.06$ using either piecewise constant or piecewise linear assumptions using either Steepest Descent (SD) or Fletcher-Reeves (FR) iterations. Optimal number of iterations shown for quartic test case using noisy data with $\sigma_{noise} = 0.5\% \times T_{max}$.

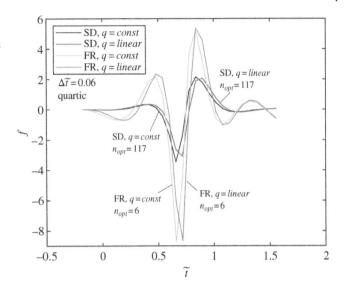

future information than $q = const$ as evidenced by the slight shift to the right for these coefficients. Also, the $q = linear$ coefficients have larger magnitude than the $q = const$ ones. As mentioned earlier, this provides greater sensitivity to measurements, including errors.

The results in Figure 7.28 are taken from the middle (15th) row of the filter matrix created using temperature perturbation for each data time (see Chapter 5). This row contains the sensitivity coefficients $\delta q_{15}/\delta Y_j$ needed to compute the 15th component of the heat flux vector which occurs at time $\tilde{t} = 0.66$ using all of the Y_j temperature measurements. The FR filter coefficients are nonzero basically from the beginning of the data time window until the very end. This indicates that virtually all the data are needed to compute the heat flux at the 15th time step; the width of the filter covers the entire data time domain. While not as apparent from Figure 7.28, the filter coefficients for SD iteration also have a very wide range as they tend slowly toward zero at the beginning and end of the data time window.

The filter coefficients for the middle row of the filter matrix for both SD and FR CGM using $\Delta \tilde{t} = 0.030$ are displayed in Figures 7.29 and 7.30, respectively. Only the $q = const$ assumption is considered in these two plots. Three curves are shown for each method at different stages of iterations illustrating the effect of increasing regularization on the coefficients. The curves for FR in Figure 7.30 change quickly as the number of iterations increases, consistent with the fast convergence of CG method, while the curves for SD require hundreds of iterations to achieve a similar change. This can be viewed as providing

Figure 7.29 Filter coefficients for Steepest Descent iterations with $\Delta \tilde{t} = 0.03$ using piecewise constant heat flux assumption for increasing number of iterations n_{iter}.

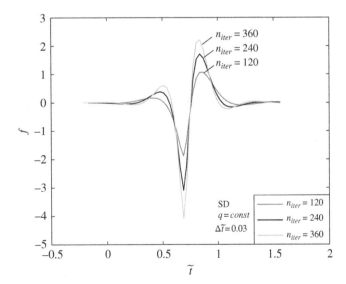

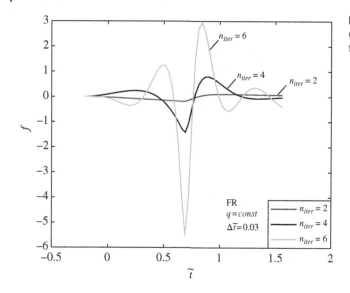

Figure 7.30 Filter coefficients for Fletcher-Reeves Conjugate Gradient iterations with $\Delta \tilde{t} = 0.03$ using piecewise constant heat flux assumption for increasing number of iterations n_{iter}.

the SD with higher resolution for optimization: small changes in regularization result for each iteration, but a large number of iterations are required to achieve the optimal.

Finally, as discussed in Chapter 5, filter coefficients for the CG method (including SD) are not suitable for implementation in a sequential algorithm for two reasons. The coefficients for each heat flux component are sufficiently different from others, so there is no universal set of coefficients that can be used for this purpose. Also, the width of the filter is effectively the width of the data time interval, which means that a particular heat flux component can only be determined through access of all the temperature measurements.

7.5.4 Conjugate Gradient Test Case Summary

Regularization is controlled in the CGM through the number of iteration steps. Unlike other IHCP solution methods, as regularization is increased, bias error in the estimated heat flux decreases, while random error in the estimation increases.

Although not truly a CGM, the SD algorithm is included in this section and is compared to the FR CG method. The SD method requires many more iterations to achieve optimal results but delivers better performance in terms of minimizing $E(R_q)$.

The $q = linear$ assumption does not appear to offer improved performance over that from the $q = const$ assumption for any of the test cases considered.

The filter coefficients for the CG methods (including SD) are not well suited for adaptation into sequential method.

The larger number of iterations required for optimization of the SD method over the FR CG method is disadvantageous, but these smaller sized iteration steps allow closer control over the degree of regularization in the solution.

7.6 Truncated Singular Value Decomposition

Regularization in the TSVD (or simply SVD) method is introduced by ignoring the smaller eigenvalues in the singular value decomposition of the coefficient matrix $\mathbf{X}$ for the discrete linear model $\mathbf{T} = \mathbf{Xq}$. Both Hansen (1998, Chapter 3) and Vogel (2002, section 2.2) discuss SVD regularization. TSVD shares the discrete regularization limitation of FS and CG methods, since eigenvalues can be removed only in finite integer numbers.

Like TR, the filter matrix $\mathbf{F}$ for TSVD appears explicitly in the solution of the IHCP. Once the matrix $\mathbf{X}$ is factored using SVD into components (as in Eq. (4.168)), the filter matrix is the inverse $\mathbf{X}^{-1}$ computed with the removed eigenvalues:

$$\mathbf{F} = \mathbf{X}_{n_{sing}}^{-1} = \mathbf{V W}_{n_{sing}}^{-1} \mathbf{U}^T \qquad (7.29)$$

Here n_{sing} denotes the quantity with the number n_{sing} eigenvalues removed.

In this section, the performance of the TSVD using both $q = const$ and $q = linear$ heat flux variation assumptions is investigated for three standard heat flux histories. The filter coefficients for TSVD are also discussed.

7.6.1 Step Change in Surface Heat Flux

Heat flux estimates obtained for the step heat flux case using the optimal number of singular values to remove are shown in Figure 7.31 for both the $q = const$ and the $q = linear$ heat flux assumptions. The optimal number of singular values to remove is determined by minimizing the expected value of the mean squared error between the exact and estimated heat flux. Results for the two heat flux assumptions using the standard test case data and noise are similar. The bias toward zero heat flux near the end of the data time interval is due to the lack of future time information. For TSVD, this time interval at the end of the data window is about 0.12 or 0.18 dimensionless time units.

Table 7.12 summarizes metrics for the step test case for TSVD under each heat flux variation assumption and with smaller data time intervals. The number of singular values to remove, n_{sing}, for the two assumptions is very similar for corresponding data time steps. Note that the number of singular values *to remove* is given, and this number is a majority of the total number of singular values for each case (viz., 22/30, 49/60, and 107/120 for the $q = const$ case). The $E(R_q)$ values indicate that a modest reduction in this metric is seen with smaller data time steps for the $q = const$ assumption, but $E(R_q)$ is nearly constant with time step for the $q = linear$ assumption. Importantly, the $q = linear$ assumption produces smaller $E(R_q)$ for the step test case for all time intervals listed.

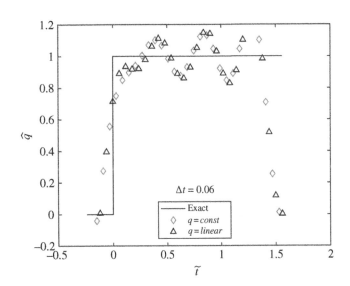

Figure 7.31 Optimal estimated heat fluxes using Truncated Singular Value Decomposition for step change test case with $\Delta \tilde{t} = 0.06$. Results for piecewise constant and piecewise linear assumptions are shown.

Table 7.12 Summary of results for step heat flux case using Truncated SVD method.

Method	Assumption	$\Delta \tilde{t}$	Step σ_{noise} = 0.0070	
			n_{opt}	$E(R_q)$
TSVD	$q = const$	0.060	22	0.15607
TSVD	$q = const$	0.030	49	0.14614
TSVD	$q = const$	0.015	107	0.13916
TSVD	$q = linear$	0.060	22	0.10523
TSVD	$q = linear$	0.030	53	0.10668
TSVD	$q = linear$	0.015	113	0.10974

7.6.2 Triangular and Quartic Heat Flux

Results for the triangular heat flux test case with both $q = const$ and $q = linear$ heat flux assumptions are shown in Figure 7.32, and the results for the quartic heat flux test case are in Figure 7.33. These results are optimized using the minimum of the expected mean squared error, but the plotted points result from processing of a specific set of noise added to exact data for each test case. In both figures, results for the $q = const$ and $q = linear$ heat flux assumptions are similar.

Table 7.13 summarizes the metrics for the triangular and quartic heat flux case for a range of data time steps. The number of singular values to remove, n_{sing}, for each time step is very similar for each time step and not much different from those for the step heat flux case (Table 7.12). The TSVD produces better results (lower $E(R_q)$) for the smoothly varying quartic heat flux than for the triangular heat flux. The $q = linear$ heat flux assumption degrades $E(R_q)$ for the triangular case, but offers modest improvement in $E(R_q)$ for the quartic heat flux case.

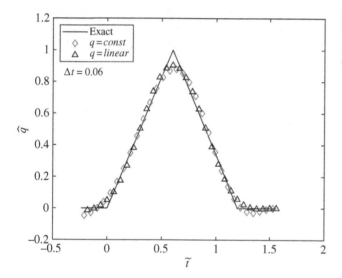

Figure 7.32 Optimal estimated heat fluxes using Truncated Singular Value Decomposition for triangular test case with $\Delta\tilde{t} = 0.06$. Results for piecewise constant and piecewise linear assumptions are shown.

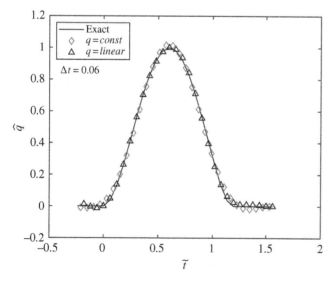

Figure 7.33 Optimal estimated heat fluxes using Truncated Singular Value Decomposition for quartic test case with $\Delta\tilde{t} = 0.06$. Results for piecewise constant and piecewise linear assumptions are shown.

Table 7.13 Summary of results for triangular and quartic heat flux cases using Truncated SVD method.

Method	Assumption	$\Delta \tilde{t}$	Triangular σ_{noise} = 0.0030		Quartic σ_{noise} = 0.0032	
			n_{opt}	$E(R_q)$	n_{opt}	$E(R_q)$
TSVD	$q = const$	0.060	26	0.02552	25	0.01918
TSVD	$q = const$	0.030	51	0.02259	53	0.01522
TSVD	$q = const$	0.015	111	0.01787	113	0.01218
TSVD	$q = linear$	0.060	25	0.02970	23	0.01859
TSVD	$q = linear$	0.030	51	0.02287	53	0.01394
TSVD	$q = linear$	0.015	111	0.01764	113	0.01132

7.6.3 Temperature Perturbation

Figure 7.34 displays the optimal filter coefficients for TSVD for the quartic test case with $\Delta \tilde{t} = 0.060$ for both the $q = const$ and $q = linear$ heat flux assumptions. These results are obtained using the temperature perturbation technique to find the sensitivity of q_{15} at $\tilde{t} = 0.66$ to each of the measurements Y_i (see Chapter 5).

Most striking about Figure 7.34 is that the values are non-zero just after the beginning until the very end. The value of the estimated heat flux component at time $\tilde{t} = 0.66$ is found by multiplying the measurement vector by these filter coefficients. This means that the heat flux component at the middle of the time span depends on measurements from the distant past until the far future, which is counter to intuition and physics.

This figure suggests a large change in the nature of the filter coefficients for incremental change in the number of singular values removed, n_{sing}. For the same value of n_{sing}, the curves for these two heat flux assumptions are very similar, with the $q = linear$ curve shifted slightly to the right (using a little more future information). As shown in Figure 7.34, removing two additional singular values for the $q = const$ assumption results in the lower magnitude of the filter coefficients and altered dependence on past values (note the "phase shift" of the $q = const$ curve relative to the $q = linear$ curve for times before $\tilde{t} = 0.66$).

Figure 7.35 better illustrates the sensitivity of the TSVD filter coefficients to the number of singular values removed. This figure shows the filter coefficients for the $q = const$ assumption for n_{sing} values around the optimal value for $\Delta \tilde{t} = 0.03$. As the number of singular values removed increases, the magnitude of the filter coefficients diminishes, and the frequency of the wave decreases.

Figure 7.34 Filter coefficients for Truncated SVD method with $\Delta \tilde{t} = 0.06$ using either piecewise constant or piecewise linear assumptions. Optimal number of singular values to remove shown for quartic test case using noisy data with $\sigma_{noise} = 0.5\% \times T_{max}$.

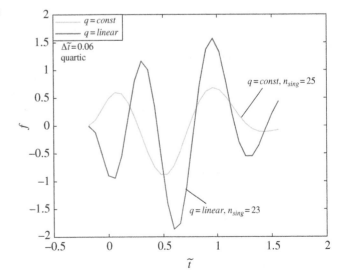

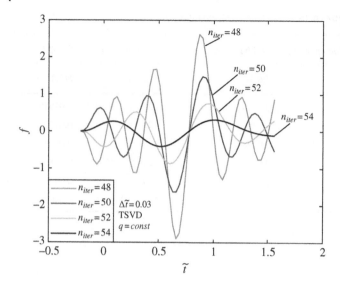

Figure 7.35 Filter coefficients for Truncated SVD with $\Delta \tilde{t} = 0.03$ using piecewise constant heat flux assumption for increasing number of removed eigenvalues n_{sing}.

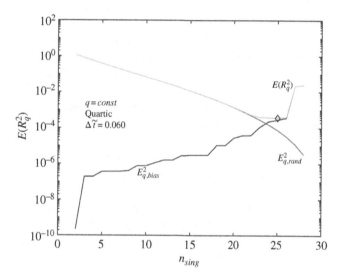

Figure 7.36 $E(R_q^2)$ and its random and bias error components as n_{sing} is increased for Truncated SVD with $\Delta \tilde{t} = 0.060$ using piecewise constant heat flux assumption with the quartic heat flux test case.

Figure 7.36 shows how the random and bias errors vary as the number of singular values removed increases in TSVD. These results are for the quartic heat flux and correspond to the same conditions of Figure 7.34. When very few singular values are removed, random errors are large and bias errors are negligible so the sum of these, the expected value of the root mean square error $E(R_q^2)$, is dominated by random error. As more singular values are removed, random errors decrease and bias errors grow. Only when the bias error becomes of similar magnitude as the random error does, the sum $E(R_q^2)$ curve flatten. The diamond symbol in Figure 7.36 identifies the optimal $n_{sing} = 25$ which corresponds to the $q = const$ filter coefficients in Figure 7.34. If more singular values are removed, the random errors become insignificant, and the bias error dominates $E(R_q^2)$.

7.6.4 TSVD Test Case Summary

Regularization in the SVD method is induced by ignoring the effects of the near singular eigenvalues from the decomposition in computing the inverse of the sensitivity matrix **X**. This is achieved by zeroing the smaller entries of the eigenvalue matrix.

For the test cases considered in this section, the number of singular values required to minimize the mean squared error is relatively large: between 70 and 90% of the eigenvalues should be ignored.

Test case results for the step change heat flux indicate the $q = linear$ assumption provides a smaller average expected error in the heat flux estimation, and that no reduction in $E(R_q)$ results from using higher frequency (smaller time step) data. On the contrary, for both the triangular and quartic test cases, the $q = linear$ assumption may provide modest improvement in minimizing the heat flux estimation error, and using smaller data time steps can reduce the average estimation error.

TSVD, like CGM, can only be regularized discretely by removing an integral number of singular values. Small changes in the number of values ignored result in significant change in the filter coefficient curve, altering the dependence of each particular component of heat flux on the data vector.

Filter coefficients for TSVD, like those for CGM, have a very wide footprint on the time axis and therefore requires data from the entire data time window to compute each heat flux component.

7.7 Kalman Filter

Regularization in the Kalman Filter solution to the IHCP is introduced primarily through the process noise parameter, Q_q Eqs. (4.188b) and (4.188c). Larger values of this parameter admit more noise in the estimation in exchange for lower bias. An optional post-processing smoothing step uses a number of future times, r, to improve the estimates, and this parameter reduces both bias and variance in the estimates.

Optimization for the Kalman Filter IHCP solution cannot be performed in the same manner as the preceding methods since the system model for KF is based on discrete numerical solution and not on superposition of exact solutions. Specifically, although the filter matrix, $\mathbf{F}$, for KF can be computed using the temperature perturbation technique, there is no equivalence to the sensitivity matrix, $\mathbf{X}$, for the KF solution. Therefore, the concept of minimizing the expected value of the residual sum of squares heat flux error introduced in Chapter 6 cannot be readily applied. An equivalent, but more computationally expensive method, using Monte Carlo simulation will be used to find the optimal parameters for KF.

Additionally, there are two regularizing parameters for KF with smoothing: Q_q and r. The approach used here is to first find the optimal value of Q_q with no smoothing (with $r = 0$), and then use this optimal Q_q and find the best value r to use to further reduce the residual sum of squares heat flux error.

In this section, the optimal regularizing parameters for KF in solution of the three test cases will be determined and the resulting heat flux histories displayed. The heat fluxes for KF are most similar to those for the $q = linear$ assumption since the values computed are coincident with each discrete time. There is no alternative "$q = const$" assumption. The "filter coefficients" from Chapter 5 are also computed for the optimal parameters. The effect of the regularizing parameters on these filter coefficients is also investigated.

7.7.1 Step Change in Surface Heat Flux

Figure 7.37 presents the results of the optimization of the Kalman Filter IHCP solution based on the known step increase in surface temperature using 30 "data points" with data time step $\Delta\tilde{t} = 0.060$. The optimal values of $Q_q = 29.2$ and $r = 2$ are obtained through 200 Monte Carlo trials, with a different set of random noise taken from a Gaussian distribution for each trial. However, the results presented in Figure 7.37 are computing using the "standard" noise (see Section 7.2.5) which is used for all test case problems presented in this chapter.

The smoother parameter, r, reduces the number of time steps that can be computed. Consequently, Figure 7.37 shows only $30 - 2 = 28$ estimated values (this behavior is analogous to the effect of the future times parameter r in FS).

Table 7.14 summarizes the optimal parameters for solution of the step heat flux case for varying time steps. As the size of the time step decreases, the magnitude of Q_q decreases, and the number of future times needed for smoothing, r, increases. There is little change in the RMS error in the heat flux estimate as the data time step decreases.

7.7.2 Triangular and Quartic Heat Flux

Figures 7.38 and 7.39 present the results computed using Kalman Filter with optimal parameters for the triangular and quartic heat flux test cases. The optimal parameters were determined through 200 Monte Carlo trials, and the specific results in the figures are based on exact "data" noised with the standard noise used for all test cases in this chapter.

For both of these test cases, the optimal smoothing parameter $r = 7$, which results in the truncated estimations in the two figures.

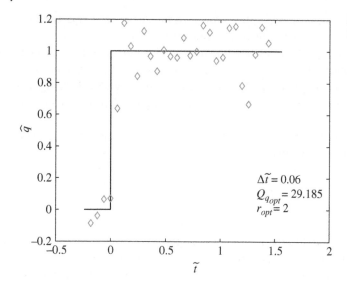

Figure 7.37 Optimal estimated heat fluxes using Kalman Filter for step change test case with $\Delta \tilde{t} = 0.06$.

Table 7.14 Summary of results for step heat flux case using Kalman Filter. RMS error in heat flux based on 200 Monte Carlo trials.

	Step σ_{noise} = 0.0070		
$\Delta \tilde{t}$	$Q_{q,opt}$	r_{opt}	$E_{q,RMS}$
0.060	28.5	2	0.2158
0.030	21.2	3	0.2256
0.015	18.3	5	0.2235

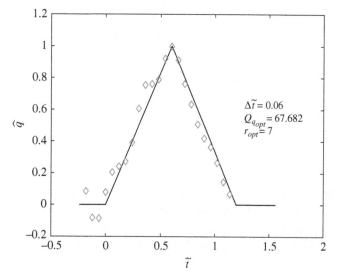

Figure 7.38 Optimal estimated heat fluxes using Kalman Filter for triangular test case with $\Delta \tilde{t} = 0.06$.

Figure 7.39 Optimal estimated heat fluxes using Kalman Filter for quartic test case with $\Delta \tilde{t} = 0.06$.

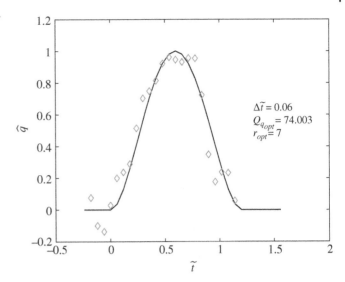

Table 7.15 Summary of results for triangular and quartic heat flux cases using Kalman Filtering. RMS error in heat flux based on 200 Monte Carlo trials.

$\Delta \tilde{t}$	Triangular σ_{noise} = 0.0030			Quartic σ_{noise} = 0.0032		
	$Q_{q,opt}$	r_{opt}	$E_{q,RMS}$	$Q_{q,opt}$	r_{opt}	$E_{q,RMS}$
0.060	68.9	7	0.06546	74.9	7	0.07240
0.030	44.4	13	0.03970	48.1	13	0.04382
0.015	36.6	25	0.02703	39.5	25	0.02797

Table 7.15 summarizes the optimal parameters identified through Monte Carlo simulation for each test case as the data time step is decreased. For these test cases, smaller time steps are effective in reducing the average error in the estimated heat flux. Notice also that the "look ahead" time $p = r\Delta \tilde{t}$ is approximately constant as the time step decreases.

7.7.3 Temperature Perturbation

The temperature perturbation technique can be used to compute the filter coefficients for the Kalman Filter solution to the IHCP. These filter coefficients can be used to process new streams of data sequentially, with or without smoothing applied, so that near real-time results can be obtained. Some insights can be gained by examining the filter coefficients for KF.

Figures 7.40 and 7.41 present the filter coefficients computed for the optimal parameters resulting from the step and triangular test cases for $\Delta \tilde{t} = 0.060$. Notice the character of these two curves are different and, as will be seen shortly, this is primarily due to the difference in magnitude of Q_q. The "current" time in these figures (time at which $Y = 1$, with all other $Y = 0$) is $\tilde{t} = 0.66$. Examination of the figures confirm that $r = 2$ time steps ahead of $\tilde{t} = 0.66$ are nonzero in Figure 7.40, and $r = 7$ time steps ahead of $\tilde{t} = 0.70$ are nonzero in Figure 7.41.

Figure 7.42 shows the effect of the regularizing parameter Q_q on the filter coefficients. The smoothing parameter $r = 0$ for these results. When $Q_q = 0$, the magnitude of the filter coefficients is very small, and any small value of Q_q will cause the curves to amplify. Successive linear increases in magnitude of this parameter, as seen in the figure, result in diminishing the increase in magnitude of these coefficients. In Figure 7.42, the "current" time is $\tilde{t} = 0.66$, and the most significant effect of increasing Q_q is to increase the weight (magnitude of the filter coefficient) on the data at the current and just-previous time. As Q_q is increased further, more weight is assigned to more data from the past, including both positive and negative components (compare the shape of $Q_q = 10$ curve to $Q_q = 30$ in Figure 7.42, especially at early times). Recall that as Q_q is

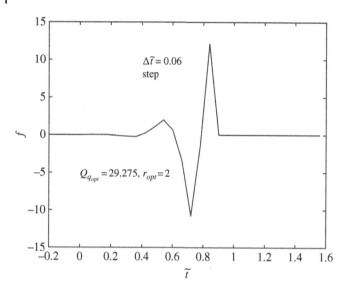

Figure 7.40 Filter coefficients for Kalman Filter solution with $\Delta \tilde{t} = 0.06$ optimized with the step heat flux test case.

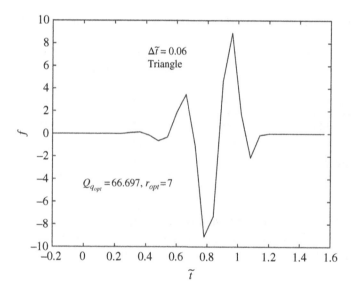

Figure 7.41 Filter coefficients for Kalman Filter solution with $\Delta \tilde{t} = 0.06$ optimized with the triangular heat flux test case.

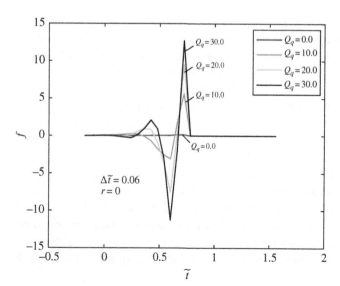

Figure 7.42 Filter coefficients for Kalman Filter solution with $\Delta \tilde{t} = 0.06$ and $r = 0$ (no smoothing) showing effect of the regularizing process noise parameter Q_q.

Figure 7.43 Filter coefficients for Kalman Filter solution with $\Delta \tilde{t} = 0.06$ and $Q_q = 68.3$ (optimal for triangular with $\Delta \tilde{t} = 0.060$) showing effect of the smoothing parameter r.

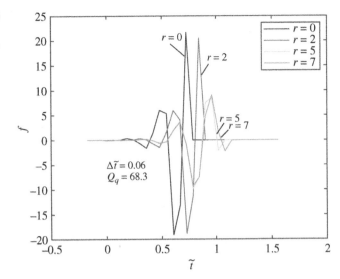

increased, bias in the estimated flux decreases, but noise in the estimates increases. So, increasing weight on past data in the estimation brings about these effects.

Figure 7.43 shows the effect of the smoothing parameter r on the filter coefficients. This figure is generated keeping the value of $Q_q = 68.3$, which was determined as an optimal value for the triangular heat flux in one of the Monte Carlo simulations. The "current" time is again $\tilde{t} = 0.66$ and corresponds to the peak in the curve for $r = 0$. As r increases, the peak in the curve shifts forward in time, indicating increased weights on data from future times, and magnitudes of all weights are diminished. The shift in the peak toward future times biases the estimation toward future times by anticipating changes in the surface heat flux – this is compensating for the lagging of information in the measured temperature to the surface heat flux. The decrease in magnitude of these coefficients results in reduction in sensitivity to noise in the measured data.

Example 7.1 The filter coefficients for Kalman Filter in Table 7.16 are typical of those resulting from Monte Carlo optimization for Q_q and r using the triangular heat flux test case with $\Delta \tilde{t} = 0.06$. The "current time" for these coefficients is the midpoint of the interval, $\tilde{t} = 15$:

Use these filter coefficients to estimate the heat flux for the temperature history in Table 7.17. For these data, the measurement noise is $\sigma = 0.001$.

Solution

There are a couple of approaches to applying the filter coefficients to the stream of data. One method, which relies on no heating activity before time t_1, is to create a padded set of data from the measurements by prepending n zeros and appending

Table 7.16 Filter coefficient for Example 7.1.

t_1–t_{10}	0.000	0.000	0.000	0.000	0.000	0.000	0.000	−0.041	0.069	0.212
t_{11}–t_{20}	−0.023	−0.714	−0.653	1.742	3.998	−0.435	−9.493	−8.251	4.689	9.760
t_{21}–t_{30}	1.579	−2.370	−0.046	0.000	0.000	0.000	0.000	0.000	0.000	0.000

Table 7.17 Measurement data for Example 7.1.

t_1–t_{10}	0.002	0.000	0.000	−0.001	0.001	0.001	0.007	0.014	0.025	0.037
t_{11}–t_{20}	0.050	0.062	0.072	0.080	0.086	0.085	0.082	0.076	0.064	0.053
t_{21}–t_{30}	0.042	0.026	0.017	0.012	0.007	0.003	0.002	0.001	0.002	0.003

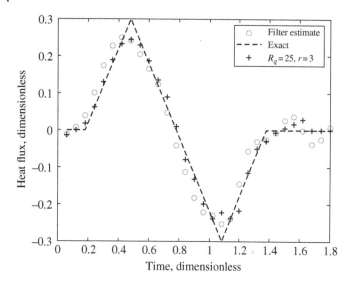

Figure 7.44 Estimated and exact heat fluxes for Example 7.1.

n zeros to the vector of data, where n is the number of data points (and the number of filter coefficients). This results in an extended set of data of length $3n$.

Next, process the data by multiplying the filter coefficients by overlapping segments of the extended data:

$$\text{for } i = 1 : 2 \times n; q_{ext}(i) = f \times Y_{ext}(i : i + n - 1); \text{end};$$

Here f is the $1 \times n$ vector of filter coefficients and Y_{ext} is the $3n \times 1$ vector of extended data. This sequence of operations results in a vector q_{ext} of length $2n$. However, the first $n/2 - 1 = 14$ entries are meaningless as the first estimated component corresponds to the "current" time in the filter, $n/2 = 15$. The estimated heat fluxes at the measurement times can be recovered using

$$\text{for } i = n/2 : n/2 + n; q_{hat}(i - n/2 + 1) = q_{ext}(i); \text{end};$$

Figure 7.44 shows the results of these calculations as the open circle symbols. The exact heat flux for this case is a sawtooth input also shown in the figure.

Discussion

The recovered heat fluxes in Figure 7.44 are acceptable, but not outstanding. The results are overly smooth, resulting from a value of r which is too large in construction of the filter ($r = 7$ was used to generate Table 7.16). Also shown in the figure as plus signs "+" are results obtained with direct processing of the measurements using the Kalman Filter algorithm (not with filter coefficients). These results are obtained with smaller values of $Q_q = 25$ and $r = 3$ than were used to generate the filter coefficients The filter coefficient results can be improved significantly by ensuring that a test case with similar magnitude of temperature responses and similar levels of noise in the data is used to optimize and compute the filter coefficients. Of course, the data time interval for the test case used to optimize must match the data time interval to be used in the application.

7.7.4 Kalman Filter Test Case Summary

Kalman Filter IHCP solutions to standard test problems cannot be optimized readily using the expected value of the residual heat flux error. However, Monte Carlo simulation can be used to perform a similar optimization. Because there are two regularizing parameters, the approach suggested here is to first vary Q_q with $r = 0$ to minimize the RMS error in the heat flux, and then use this optimal value of $Q_{q,opt}$ and vary r to further reduce the error metric.

The heat flux estimates from KF are best associated with the measurement times (similar to piecewise linear assumption).

Filter coefficients can be computed for Kalman Filter IHCP solutions using the temperature perturbation technique of Chapter 5. Examination of these filter coefficients reveals that the regularizing parameter Q_q amplifies the coefficients, and the smoothing parameter r shifts the filter coefficients into future times.

Proper selection of the regularizing parameters for the type of heat flux anticipated in the application is important to achieve good results. The expected magnitude of the measured response and level of noise in the measurements are especially important factors when selecting a test case for optimization.

7.8 Chapter Summary

This chapter presents examples of IHCP heat flux estimation using FS, TR, CG, SVD, and KF methods. For each IHCP method except KF, both piecewise constant ("$q = const$") and piecewise linear ("$q = linear$") assumption for the unknown heat flux are investigated.

Each method is used to estimate the heat flux from a suite of test problems including a step increase in surface temperature, a triangular heat flux history, and a "quartic" polynomial heat flux history.

For each IHCP method and heat flux history, knowledge of the exact heat flux history and the standard deviation of the measurement noise are used to optimize the regularization for the IHCP method. Additive measurement errors with zero mean and standard deviation of 0.5% of the maximum temperature rise for each test case are used. This optimization minimizes the expected value of the mean squared error between the exact heat flux and the estimated heat flux.

Using the determined optimal regularization parameter, each IHCP method is applied to a particular set of artificial data resulting from the exact solution for the prescribed heat flux test case and an assumed standard set of random numbers to simulate measurement errors.

Optimizations and IHCP calculations are performed for a nominal data time step with $\Delta \tilde{t} = 0.060$ and two smaller data time intervals of 0.030 and 0.015. For all methods evaluated, improved estimations (measured by smaller $E(R_q)$) are possible by decreasing the data time interval (higher data frequency).

Filter coefficients are examined for each method. The width and shape of the filter coefficient curve give insight to the dependence of a heat flux component to the data. In particular, these coefficients indicate how much future and past time data are needed to compute the "current" heat flux component. This information ultimately determines whether, or not, the IHCP method is suitable for implementation in a sequential algorithm. Of the five methods investigated in this chapter, FS, TR, and KF are suitable for sequential estimation, but CG and SVD methods are not.

All IHCP methods will produce poor estimates at the end of the data time window. This effect is easy to see in the estimated heat flux components for the step heat flux case as the exact (imposed) heat flux remained at a constant nonzero value, but the estimates generally biased toward zero. (The exception is for first-order TR which biases toward a constant value.) This effect is consistently observed over the last 0.12–0.18 dimensionless time units over a range of data time steps. For this purpose, the distance from the active surface to the measurement location should be used in the definition of the dimensionless time interval. This information can be used to discard poor estimates at the end of the data time interval. Alternatively, when possible, design experiments so the surface heat flux at the end of the data time window goes to zero.

The step change test case provides an idealization for heat fluxes that are discontinuous or have large rapid changes in heating. Table 7.18 summarizes the $E(R_q)$ metrics for the optimal conditions for each method using the nominal $\Delta \tilde{t} = 0.060$ time step. The $q = linear$ assumption provides the best results for this heat flux when using either first-order TR or SVD (indicated with bolded entries in Table 7.18). However, the $q = linear$ assumption provides the poorest results in the table for zeroth-order TR or either CG method.

The triangular and quartic heat flux test cases provide idealizations of continuously varying heat fluxes. These are very similar, but the quartic case has smooth derivatives while the triangular case has discontinuities in the derivative. Table 7.19 summarizes metrics from all the methods considered in this chapter for the triangular test case, and Table 7.20 contains the same information for the quartic heat flux case. The lowest few instances of $E(R_q)$ are highlighted in each table with bold entries. Common superior results from both tables are found for $q = const$ assumption using either first-order TR method or SVD. For the smoother quartic case, good results are also obtained with $q = linear$ assumption for first-order TR or SVD.

Table 7.18 Summary of results for step heat flux case for $\Delta \tilde{t} = 0.060$ and with $\sigma_{noise} = 0.5\% \times T_{max}$.

Method	Option	Assumption	Optimal	$E(R_q)$
Step test case				
FS		$q = const$	$r = 4$	0.16848
FS		$q = linear$	$r = 4$	0.13393
TR	0th order	$q = const$	$\alpha = 8.13\text{E-}04$	0.14568
TR	0th order	$q = linear$	$\alpha = 4.36\text{E-}04$	0.27886
TR	1st order	$q = const$	$\alpha = 8.50\text{E-}04$	0.15014
TR	1st order	$q = linear$	$\alpha = 2.76\text{E-}03$	**0.10943**
CG	SD	$q = const$	$n_{iter} = 291$	0.14997
CG	SD	$q = linear$	$n_{iter} = 333$	0.27128
CG	FR	$q = const$	$n_{iter} = 6$	0.15727
CG	FR	$q = linear$	$n_{iter} = 6$	0.27998
SVD		$q = const$	$n_{sing} = 22$	0.15607
SVD		$q = linear$	$n_{sing} = 22$	**0.10523**
KF			$Q_q = 28.5, r = 2$	0.2158[a]

[a] Based on 200 Monte Carlo trials, not expected value.

Table 7.19 Summary of results for triangular heat flux case for $\Delta \tilde{t} = 0.060$ and with $\sigma_{noise} = 0.5\% \times T_{max}$.

Method	Option	Assumption	Optimal	$E(R_q)$
Triangular test case				
FS		$q = const$	$r = 4$	0.03842
FS		$q = linear$	$r = 5$	0.03343
TR	0th order	$q = const$	$\alpha = 1.61\text{E-}03$	**0.02822**
TR	0th order	$q = linear$	$\alpha = 1.56\text{E-}03$	0.03013
TR	1st order	$q = const$	$\alpha = 4.29\text{E-}03$	**0.02670**
TR	1st order	$q = linear$	$\alpha = 4.14\text{E-}03$	0.02979
CG	SD	$q = const$	$n_{iter} = 135$	0.03111
CG	SD	$q = linear$	$n_{iter} = 141$	0.03336
CG	FR	$q = const$	$n_{iter} = 5$	0.04442
CG	FR	$q = linear$	$n_{iter} = 5$	0.04406
SVD		$q = const$	$n_{sing} = 26$	**0.02552**
SVD		$q = linear$	$n_{sing} = 25$	0.02970
KF			$Q_q = 68.9, r = 7$	0.06546[a]

[a] Based on 200 Monte Carlo trials, not expected value.

Table 7.20 Summary of results for quartic heat flux case for $\Delta\tilde{t} = 0.060$ and with $\sigma_{noise} = 0.5\% \times T_{max}$.

		Quartic test case		
Method	Option	Assumption	Optimal	$E(R_q)$
FS		$q = const$	$r = 4$	0.03901
FS		$q = linear$	$r = 5$	0.02836
TR	0th order	$q = const$	$\alpha = 1.67\text{E-}03$	0.02670
TR	0th order	$q = linear$	$\alpha = 1.61\text{E-}03$	0.02641
TR	1st order	$q = const$	$\alpha = 4.79\text{E-}03$	**0.02158**
TR	1st order	$q = linear$	$\alpha = 4.67\text{E-}03$	**0.02218**
CG	SD	$q = const$	$n_{iter} = 117$	0.02956
CG	SD	$q = linear$	$n_{iter} = 117$	0.02922
CG	FR	$q = const$	$n_{iter} = 6$	0.04774
CG	FR	$q = linear$	$n_{iter} = 6$	0.04734
SVD		$q = const$	$n_{sing} = 25$	**0.01918**
SVD		$q = linear$	$n_{sing} = 23$	**0.01859**
KF			$Q_q = 74.9, r = 7$	0.07240^a

^{a} Based on 200 Monte Carlo trials, not expected value.

Problems

7.1 A 5-mm-thick steel ($\alpha = 1.1\text{E-}5$ m^2/s, $k = 40$ W/m-K) plate is used as a calorimeter. The plate is initially at 25 °C. Data are taken every 5 seconds and the thermocouple attached to the back of the insulated steel plate produces the following readings:

t, s	5	10	15	20	25	30	35	40
Y, C	24.48	25.05	24.73	25.43	27.94	31.67	35.02	37.89
t, s	45	50	55	60	65	70	75	80
Y, C	41.30	43.96	47.36	50.13	53.31	55.57	58.54	60.68
t, s	85	90	95	100	105	110	115	120
Y, C	63.17	65.36	67.23	68.76	69.85	71.39	72.57	73.45
t, s	125	130	135	140	145	150		
Y, C	73.66	74.70	73.47	73.89	74.11	73.47		

a) Calculate the dimensionless time step for this data. Let $q_{ref} = 12\,100$ W/m^2. Does the magnitude of $\Delta\tilde{t}$ suggest that significant regularization will be needed?

b) Use the function specification method to process these data sequentially to estimate the heat flux. What is the minimum number of future times needed to get a stable/smooth solution?

7.2 Use zeroth-order Tikhonov regularization with $q = const$ assumption to process the data in Problem 7.1. Use a value of $\alpha_0 = 1\text{E-}4$.

7.3 Use $q = linear$ assumption and SVD to process the data in Problem 7.1. What is the magnitude of the smallest eigenvalue of the **X** matrix? What is the smallest number of singular values that must be removed to get a stable/smooth solution.

For the following problems, use the same Y data from Problem 7.1, but consider that these data are acquired every 1 second, instead of every 5 seconds.

7.4 Repeat Problem 7.1 considering that the data are obtained every 1 second.

7.5 Repeat Problem 7.2 considering that the data are obtained every 1 second. Use a value of $\alpha_0 = 0.1$.

7.6 Repeat Problem 7.3 considering that the data are obtained every 1 second.

7.7 Use the KF filter coefficients from Example 7.1 to process the data from Problem 6.5.

References

Beck, J. V., Keltner, N. R. and Schisler, I. P. (1985) Influence functions for the unsteady surface element method, *AIAA Journal*, 23(12), pp. 1978–1982. https://doi.org/10.2514/3.9205

Fletcher, R. and Reeves, C. M. (1964) Function minimization by conjugate gradients, *Computer Journal*, 7(2), pp. 149–154.

Hansen, P. C. (1998) *Rank-Deficient and Discrete Ill-Posed Problems*. Philadelphia, PA: SIAM.

Keltner, N. R. and Beck, J. V. (1981) Unsteady surface element method, *Journal of Heat Transfer*, 103(4), pp. 759–764. https://doi.org/10.1115/1.3244538

Litkouhi, B. and Beck, J. V. (1985) Intrinsic thermocouple analysis using multinode unsteady surface element method, *AIAA Journal*, 23(10), pp. 1609–1614. https://doi.org/10.2514/3.9131

Vogel, C. R. (2002) *Computational Methods for Inverse Problems*. Philadelphia, PA: SIAM.

Woodbury, K. A. and Beck, J. V. (2013) Heat conduction in a planar slab with power-law and polynomial heat flux input, *International Journal of Thermal Sciences*, 71, pp. 237–248. https://doi.org/10.1016/j.ijthermalsci.2013.04.009

Woodbury, K. A. and Thakur, S. K. (1996) Redundant data and future times in the inverse heat conduction problem, *Inverse Problems in Engineering*, 2(4), pp. 319–333.

8

Multiple Heat Flux Estimation

8.1 Introduction

In the previous chapters, the case of the single unknown surface heat flux history in a one-dimensional domain was considered. In this chapter, the inverse heat conduction problem (IHCP) in a two-dimensional domain with multiple unknown surface heat fluxes is treated.

A typical multiple heat flux IHCP is for the case of a body exposed to a heat flux that is both space and time variable. Thus, the temperature distribution is two-dimensional. Any coordinate system can be used, and the analysis is not restricted to two-dimensional cases. The body can be composed of several materials and can be irregular. The only requirement is that a method of solving the direct problem (*known* surface heat flux time and space variation) is available which could be through finite differences, finite elements, or Duhamel's theorem.

The solution of IHCPs in multidimensional medium has been explored in a few studies. An approximate method using Laplace transform is used by Imber (2012) for solving IHCP's in two-dimensional cylindrical geometries. Application of a sequential gradient method for solving multidimensional inverse problem is studied by Dowding and Beck (1999).

Osman et al. (1997) developed a solution through function specification method and regularization method for solving IHCP in a two-dimensional domain with arbitrary geometry. The solution of IHCP in a two-dimensional domain in the form of a rectangle is studied by Monde et al. (2003). Wang et al. (2011) developed a novel approach using decentralized fuzzy inference for solving steady state IHCPs in two-dimensional domains. García et al. (2009) investigated the solution of IHCP in two-dimensional domains with irregular shape and temperature-dependent thermal properties using sequential singular value decomposition (SVD).

The plan for this chapter is to discuss the IHCP with multiple unknown heat fluxes in a domain with two-dimensional temperature distribution. The solution approach that is presented here is based on the method that is developed by Najafi et al. (2015) using the filter form of Tikhonov regularization method which was previously introduced in Chapter 5.

8.2 The Forward and the Inverse Problems

The forward problem that allows calculation of temperature distribution caused by partial heating of one surface of a rectangular domain as a function of x, y, and t is used as the building block to evaluate the temperature responses that are needed to establish the matrices of sensitivity coefficients for the solution of the inverse problem. The forward problem and inverse problem are described in the following sections.

8.2.1 Forward Problem

The X22B(y1pt1)0Y22B00T0 problem describes a rectangular region partially heated on one edge (the $x = 0$ surface) with a constant heat flux. This problem is described mathematically in Eqs. (2.47a)–(2.47f). In this chapter, only the homogeneous initial condition case ($T_{in} = 0$) is considered. A schematic of the problem is given in Figure 2.10 but is reproduced in Figure 8.1.

Inverse Heat Conduction: Ill-Posed Problems, Second Edition. Keith A. Woodbury, Hamidreza Najafi, Filippo de Monte, and James V. Beck.
© 2023 John Wiley & Sons, Inc. Published 2023 by John Wiley & Sons, Inc.

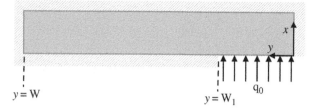

Figure 8.1 Schematic of a rectangular domain, partially heated from the bottom surface.

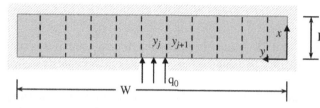

Figure 8.2 Schematic of partial heating of a rectangular plate from the bottom surface.

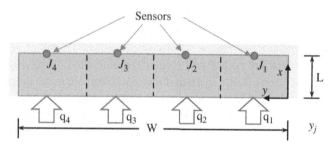

Figure 8.3 Schematic of an IHCP with four unknown heat fluxes and four temperature sensors.

The exact solution to this problem is documented in Appendix B, and the computational analytical solution is summarized in Section 2.4.3. In this chapter, the solution to this problem will be denoted as $T_{X22Y22}(x,y,t)$.

The solution to the above problem is the building block of the problem which is demonstrated in Figure 8.2. Superposition principles can be exploited to find the temperature response to the single heat pulse anywhere on the $x = 0$ surface. The solution to constant heating over the surface between y_j and y_{j+1} can be given as:

$$T_{y_j \to y_{j+1}}(x, y, t) = T_{0 \to y_{j+1}}(x, y, t) - T_{0 \to y_j}(x, y, t)$$

(8.1)

Similarly, the temperature response at any location (x,y) as the result of a series of p heat flux pulses on the surface can be found by summing the contributions from each individual pulse:

$$T(x, y, t) = \sum_{i=1}^{p} T_{y_{i-1} \to y_i}(x, y, t)$$

(8.2)

This can be used to generate temperature responses for the solution of IHCP which is discussed in Section 8.2.2.

8.2.2 Inverse Problem

For the inverse problem, multiple heat fluxes are assumed at the bottom surface of the plate and all other surfaces are perfectly insulated. The temperature sensors are located at the top surface of the plate. To achieve a unique solution, the number of temperature sensors (J) must not be smaller than the number of unknown heat fluxes (P). An example of an IHCP with four unknown heat fluxes and four temperature sensors is shown in Figure 8.3.

The unknown heat flux can be evaluated using measured temperature data. This can be mathematically described as follows:

$$-k\frac{\partial T}{\partial y}(0, y, t) = \mathbf{q} = \text{unknown} = \begin{cases} q_1(t), 0 < y < y_1 \\ q_2(t), y_1 < y < y_2 \\ \vdots \\ q_P(t), y_{P-1} < y < y_{P+1} \end{cases}$$

(8.3)

$$T\left(L, y_j, t\right) = \text{known} \quad j = 1,...,J$$

(8.4)

The solution to the discrete forward problem can be written in matrix form, as in Eq. (3.54):

$$\mathbf{T} = \mathbf{Xq}$$

Here $\mathbf{X}$ is the sensitivity matrix and $\mathbf{q}$ is the estimated heat flux. These matrices can be written as:

$$\mathbf{T} = \begin{bmatrix} \mathbf{T}(1) \\ \mathbf{T}(2) \\ \vdots \\ \mathbf{T}(N_t) \end{bmatrix}, \qquad \mathbf{T}(i) = \begin{bmatrix} T_1(i) \\ T_2(i) \\ \vdots \\ T_J(i) \end{bmatrix}$$

(8.5)

Here i refers to the time index and N_t is the total number of time steps. The components of vectors $\mathbf{T}(i)$ given in (8.5) are the temperatures at different sensor locations $(1, 2,...,J)$ at the ith time step.

The matrix of the heat fluxes can be given as:

$$\mathbf{q} = \begin{bmatrix} \mathbf{q}(1) \\ \mathbf{q}(2) \\ \vdots \\ \mathbf{q}(N_t) \end{bmatrix}, \qquad \mathbf{q}(i) = \begin{bmatrix} q_1(i) \\ q_2(i) \\ \vdots \\ q_P(i) \end{bmatrix} \qquad q_j(i) = q_{x_{j-1} \to x_j}(t_i) \tag{8.6}$$

The matrix of sensitivity coefficients, $\mathbf{X}$, can be found as:

$$\mathbf{X} = \begin{bmatrix} \mathbf{a}(1) & 0 & 0 & \cdots & 0 \\ \mathbf{a}(2) & \mathbf{a}(1) & 0 & \cdots & 0 \\ \mathbf{a}(3) & \mathbf{a}(2) & \mathbf{a}(1) & & \vdots \\ \vdots & \vdots & \vdots & \ddots & 0 \\ \mathbf{a}(N_t) & \mathbf{a}(N_t - 1) & \cdots & & \mathbf{a}(1) \end{bmatrix},$$

$$\mathbf{a}(i) = \begin{bmatrix} a_{11}(i) & a_{12}(i) & \cdots & a_{1P}(i) \\ a_{21}(i) & a_{22}(i) & \cdots & a_{2P}(i) \\ \vdots & \vdots & \cdots & \vdots \\ a_{J1}(i) & a_{J2} & \cdots & a_{JP}(i) \end{bmatrix}, \quad a_{jk}(i) = \frac{\partial T(x_j, t_i)}{\partial[q_k(1)]} \tag{8.7}$$

The sum of the squares of the errors between the estimated temperatures and known temperatures with the addition of the regularization term for time and space can be calculated as:

$$S = (\mathbf{Y} - \mathbf{T})^T(\mathbf{Y} - \mathbf{T}) + \alpha_T[\mathbf{Hq}]^T[\mathbf{Hq}] + \alpha_s[\mathbf{H_s q}]^T[\mathbf{H_s q}] \tag{8.8}$$

Here α_T and α_s are the regularization parameters used for time and space terms. $\mathbf{H}$ and $\mathbf{H_s}$ can be given as:

$$\mathbf{H} = \begin{bmatrix} -\mathbf{I} & \mathbf{I} & 0 & \cdots & 0 \\ 0 & -\mathbf{I} & \mathbf{I} & \cdots & 0 \\ \vdots & \vdots & \vdots & \cdots & \vdots \\ 0 & 0 & 0 & \cdots & 0 \end{bmatrix}, \quad \mathbf{I} = \begin{bmatrix} 1 & 0 & 0 & \cdots & 0 \\ 0 & 1 & 0 & \cdots & 0 \\ \vdots & \vdots & \vdots & \cdots & 0 \\ 0 & 0 & 0 & \cdots & 1 \end{bmatrix} \tag{8.9}$$

$$\mathbf{H_s} = \begin{bmatrix} \overline{\mathbf{H}}_\mathbf{s} & 0 & 0 & \cdots & 0 \\ 0 & \overline{\mathbf{H}}_\mathbf{s} & 0 & \cdots & 0 \\ \vdots & \vdots & \vdots & \cdots & \vdots \\ 0 & 0 & 0 & \cdots & \overline{\mathbf{H}}_\mathbf{s} \end{bmatrix}, \quad \overline{\mathbf{H}}_\mathbf{s} = \begin{bmatrix} -1 & 1 & 0 & \cdots & 0 & 0 \\ 0 & -1 & 1 & \cdots & 0 & 0 \\ \vdots & \vdots & \vdots & \cdots & 0 & 0 \\ 0 & 0 & 0 & \cdots & -1 & 1 \\ 0 & 0 & 0 & \cdots & 0 & 0 \end{bmatrix} \tag{8.10}$$

Minimizing the sum of the squares of the errors that was given in Eq. (8.8) with respect to $\mathbf{q}$ leads to:

$$\hat{\mathbf{q}} = [\mathbf{X}^T\mathbf{X} + \alpha_T\mathbf{H}^T\mathbf{H} + \alpha_s\mathbf{H_s}^T\mathbf{H_s}]^{-1}\mathbf{X}^T\mathbf{Y} \tag{8.11}$$

8.2.3 Filter Form of the Solution

The filter form solution of IHCP was previously discussed in Chapter 5 as a computationally efficient approach that allows near real-time heat flux estimation. The heat flux vector in Eq. (8.11) can be also written as follows:

$$\hat{\mathbf{q}} = \mathbf{FY}, \qquad \mathbf{F} = [\mathbf{X}^T\mathbf{X} + \alpha_T\mathbf{H}^T\mathbf{H} + \alpha_s\mathbf{H_s}^T\mathbf{H_s}]^{-1}\mathbf{X}^T \tag{8.12}$$

Here **F** is the filter matrix. Notice that $\hat{\mathbf{q}}$ has a dimension of $(N_t \times P) \times 1$. The dimension of **F** and **Y** matrices are $(N_t \times P) \times (N_t \times J)$ and $(N_t \times J) \times 1$, respectively. The structure of the filter matrix, **F**, can be shown as:

$$
\mathbf{F} = \begin{bmatrix}
\mathbf{f}_0 & \mathbf{f}_{-1} & \mathbf{f}_{-2} & \cdots & & \mathbf{f}_{2-N_t} & \mathbf{f}_{1-N_t} \\
\mathbf{f}_1 & \mathbf{f}_0 & \mathbf{f}_{-1} & & & \cdots & \mathbf{f}_{2-N_t} \\
\vdots & \ddots & \ddots & \ddots & & & \vdots \\
\vdots & \cdots & & \mathbf{f}_0 & & & \vdots \\
\vdots & \vdots & \vdots & & \ddots & \vdots & \vdots \\
\mathbf{f}_{N_t-2} & \mathbf{f}_{N_t-3} & \cdots & \vdots & \mathbf{f}_1 & \mathbf{f}_0 & \mathbf{f}_{-1} \\
\mathbf{f}_{N_t-1} & \mathbf{f}_{N_t-2} & \cdots & \cdots & \mathbf{f}_2 & \mathbf{f}_1 & \mathbf{f}_0
\end{bmatrix}
\tag{8.13}
$$

Each component of the **F** matrix (i.e. $\mathbf{f_n}$ where $n = (1 - N_t), (2 - N_t), \ldots, 0, 1, 2, \ldots, (N_t - 2), (N_t - 1)$) is a $P \times J$ block of entities, and therefore, each row of the **F** matrix is a block of P-rows and $N_t \times J$ columns. As seen in Eq. (8.13), the rows of filter matrix are identical but shifted in time.

To illustrate the characteristics of the filter matrix for a two-dimensional IHCP with multiple heat flux, an example is assessed in detail with four unknown heat fluxes ($P = 4$) and four temperature sensors ($J = 4$), as depicted in Figure 8.3. A dimensionless time step of 0.05 is considered with the dimensionless time interval between 0.05 and ends at 2 ($N_t = 40$). For this example, the dimension of the filter matrix (**F**) is a 160×160 matrix. Each block of four (P) rows of this matrix are being repeated, but shifted in time. To illustrate this, the 41st, 42nd, 43rd, and 44th rows (associated with 11th time step) as well as the 105th, 106th, 107th, and 108th rows (associated with 27th time step) of the filter matrix are plotted and shown in Figure 8.4.

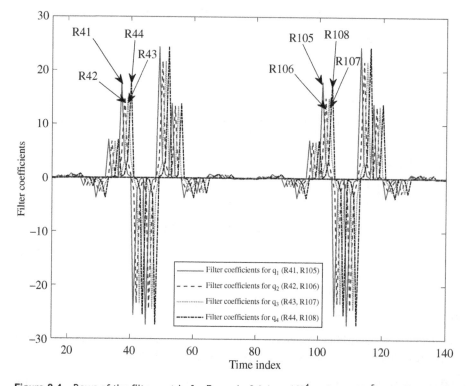

Figure 8.4 Rows of the filter matrix for Example 8.1 ($\alpha_t = 10^{-4}$ and $\alpha_s = 10^{-5}$, dimensionless time step: 0.05).

As seen, the rows of each of the blocks are identical to one another (e.g. values of filter coefficients on 41st row and 105th row are identical). Note that in **F** matrix 1st, 5th, 9th, ... $n + 4$th rows are associated with $\mathbf{q_1}$, 2nd, 6th, 10th, ... $n + 4$th are associated with $\mathbf{q_2}$; and the coefficients associated with $\mathbf{q_3}$ and $\mathbf{q_4}$ can be written similarly. Also, in each row of the **F** matrix, each column represents one of the temperature sensors. In other words, 1st, 5th, 9th, ... $n + 4$th columns are associated with first sensor, and 2nd, 6th, 10th, ... $n + 4$th columns are associated with the second sensor and so forth.

As seen, the filter coefficients associated with $\mathbf{q_1}$ and $\mathbf{q_3}$ are similar and filter coefficients associated with $\mathbf{q_2}$ and $\mathbf{q_4}$ are also similar. This is due to the fact that their locations are symmetric on the rectangular domain. Every four points along the y-axis in Figure 8.4 represent four sensor locations for only one time step. To provide a better understanding of this concept, a three-dimensional plot of filter coefficients is presented in Figure 8.5. Here, the 41st, 42nd, 43rd, and 44th rows of **F** matrix are presented. As seen, for $\mathbf{q_1}$, the largest filter coefficients are associated with the temperature at J_1 location, and as the sensor location gets more distance from the location of the heat source $\mathbf{q_1}$, the filter coefficients are getting smaller. Same observation can be made for the filter coefficients associated with $\mathbf{q_2}$, $\mathbf{q_3}$, and $\mathbf{q_4}$.

The values of the filter coefficients approach zero toward both ends. The number of non-negligible filter coefficients before and after the current time steps are denoted as m_p and m_f, respectively. Equation (8.12) can be written in filter form as:

$$\hat{q}_{p,M} = \hat{q}_p(t_M) = \sum_{i=1}^{(m_p + m_f) \times J} \left(f_{p,i} Y_{M + (m_f \times J) - i} \right) \tag{8.14}$$

Here $f_{p,i}$ refers to the filter coefficients in the ith column from one row of **F** associated with the pth unknown heat flux. Also, M denotes the current time step. Since all the rows of the filter matrix are identical but shifted in time, instead of using the whole matrix **F** to calculate the heat fluxes, the non-negligible terms from one row of this matrix can be used. Therefore, one can use the non-negligible terms from one row of the filter matrix $(f_1, f_2, \dots f_{(m_p + m_f) \times J})$ associated to the pth unknown heat flux, multiply those in the corresponding temperature values from temperature sensors, and use the summation of these terms to calculate each of the unknown heat fluxes ($\mathbf{q_1}$, $\mathbf{q_2}$,..., $\mathbf{q_P}$) at a specific time step (M).

As seen in Figure 8.5, the values of m_f and m_p are both equal to 9. The smaller the m_f is, the closer to real time the algorithm can operate. It should be noted that m_p and m_f are not necessarily equal for all different problems. Also, it is important to emphasize that a bigger value of m_p does not affect how fast (closer to real-time) the algorithm can operate as it only decides how many data points from previous time steps must be used to determine the heat flux at the current time.

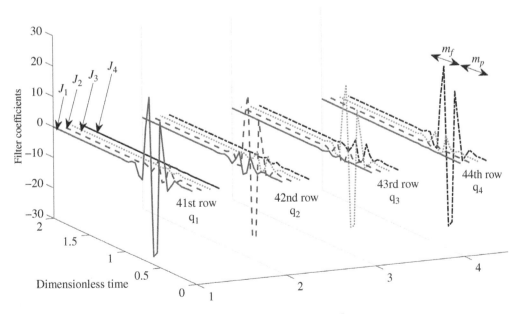

Figure 8.5 Filter coefficients for Example 8.1 ($\alpha_t = 10^{-4}$ and $\alpha_s = 10^{-5}$, dimensionless time step: 0.05).

8.3 Examples

Two examples are considered for which the forward problems are solved using ANSYS. For the forward problem, a rectangular domain is developed, and four heat flux components are applied to the bottom surface of the rectangle while all other surfaces are perfectly insulated (Figure 8.3). It is also assumed that the temperature can be measured by four temperature sensors installed at the top surface of the plate. Two different aspect ratios (L/W) are considered. For Examples 8.1 and 8.2, the aspect ratio is assumed to be 0.1 and 0.5, respectively.

Example 8.1 A dimensionless time interval between 0.05 and 2 is considered with a dimensionless time step of 0.05. To create a simple dimensionless test case, thermal conductivity and thermal diffusivity are assumed to be equal to 1. The filter coefficients for such problem are plotted and presented previously in Figure 8.4 and Figure 8.5. The heat fluxes that are applied on the surface as well as the corresponding temperature values that are measured by each temperature sensor are shown in Figures 8.6 and 8.7. These temperature data are used as inputs for the developed filter method Eq. (8.14). Calculate the surface heat flux using filter-based TR for the 2D IHCP.

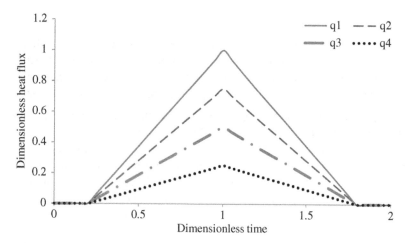

Figure 8.6 Input surface heat fluxes for Example 8.1

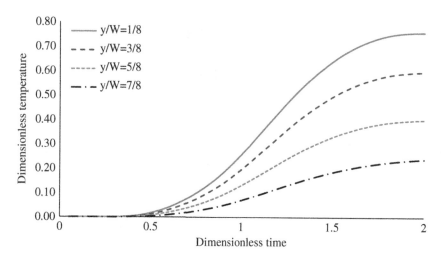

Figure 8.7 Temperature data at four locations for Example 8.1.

Solution

Inspection of Figure 8.4 and Figure 8.5 reveals that m_f and m_p are both equal to 9.

Using Eq. (8.14) and assuming $m_p = m_f = 9$ and regularization parameters are set as $\alpha_t = 10^{-4}$ and $\alpha_s = 10^{-5}$, the surface heat fluxes are evaluated using the temperature data. The calculated surface heat fluxes at different locations are shown in Figure 8.8.

As seen in Figure 8.8, the filter method is able to calculate all heat fluxes accurately.

The average root mean square (RMS) error calculated between exact and estimated values of heat fluxes are found as 0.0208, 0.0175, 0.0122, and 0.0061 for $\mathbf{q_1}$, $\mathbf{q_2}$, $\mathbf{q_3}$, and $\mathbf{q_4}$, respectively. The RMS error is calculated using the following equation:

$$E_{RMS} = \left(\frac{\sum_{i=1}^{n} \left(q_{exact,i} - q_{estimated,i} \right)^2}{n} \right)^{1/2} \tag{8.15}$$

where n is the number of time steps for which heat flux is estimated.

To assess the performance of the presented approach, a random error equivalent to 0.5% of the maximum temperature is added to the temperature data. The heat fluxes are then estimated again using the filter-based approach, and the calculated values are presented in Figure 8.9 as a three-dimensional plot. For this case, the regularization parameters are set as $\alpha_t = \alpha_s = 10^{-4}$ and the RMS error between the exact and calculated heat flux values are found as 0.0369, 0.0210, 0.0120, and 0.0091 for $\mathbf{q_1}$, $\mathbf{q_2}$, $\mathbf{q_3}$, and $\mathbf{q_4}$, respectively, demonstrating good agreement between the two estimated and exact heat flux inputs.

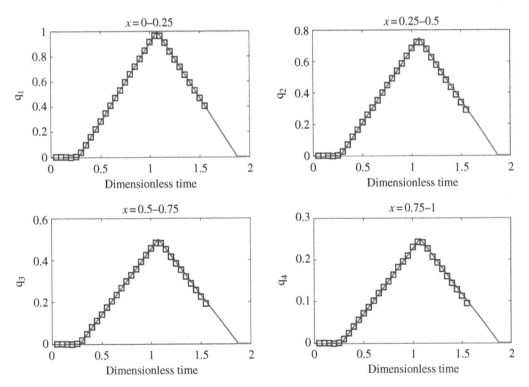

Figure 8.8 Estimated surface heat flux values for Example 8.1 (q_1, q_2, q_3, and q_4).

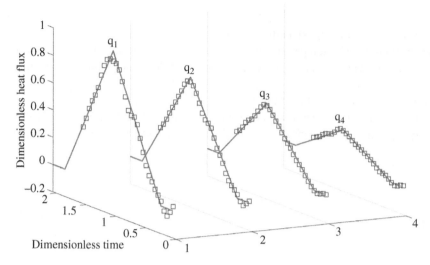

Figure 8.9 Estimated heat flux values for Example 8.1 in the presence of 0.5% error in temperature data.

Example 8.2 In the second example, different heat flux profile shapes are used to further assess the robustness of the presented approach in estimating different types of heat flux profiles. For q_1, q_2, q_3, and q_4, a rectangular, a parabolic, straight lines, and an arbitrary shape are considered, respectively. The aspect ratio for Example 8.2 is assumed as 0.5. This is purposely selected to be significantly larger than the value used for Example 8.1. Figures 8.10 and 8.11 show the heat fluxes that are applied on the bottom surface and the corresponding temperature values calculated on the top surface, respectively. The dimensionless time step is 0.05, and the experiment is assessed for $0.05 < \tilde{t} < 2$. Calculate the surface heat flux using the filter-based TR approach for the 2D IHCP.

Solution

The filter coefficients for Example 8.2, assuming regularization parameters as $\alpha_t = 10^{-4}$ and $\alpha_s = 10^{-5}$, are shown in Figure 8.12. A similar pattern to Example 8.1 can be observed. The largest filter coefficients associated with each heat flux corresponds to the temperature data from the sensor located at the closest distance to that that particular heat flux. The

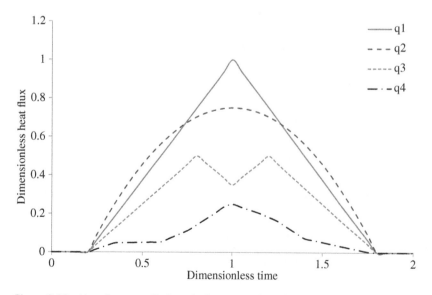

Figure 8.10 Heat fluxes applied on the bottom surface of the domain (Example 8.2).

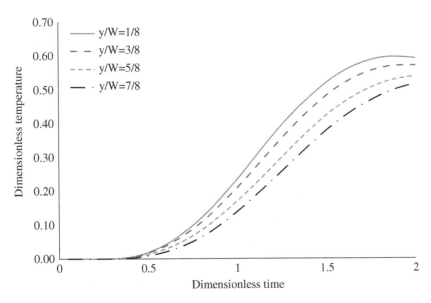

Figure 8.11 Calculated temperatures for the sensors at different locations (Example 8.2).

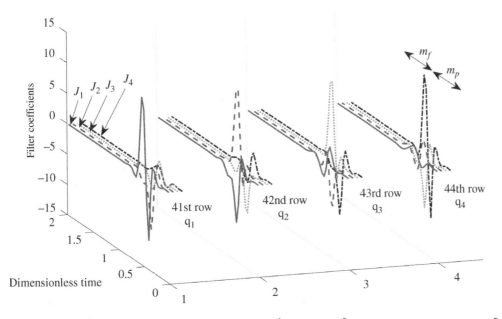

Figure 8.12 Filter coefficients for Example 8.2 ($\alpha_t = 10^{-4}$ and $\alpha_s = 10^{-5}$ and dimensionless time step $\Delta \tilde{t} = 0.05$).

number of non-negligible coefficients is found to be $m_p = m_f = 10$ through careful observation of the filter coefficients that are presented in Figure 8.12

By adding a random uniform error equivalent to 0.5% of the maximum temperature to the temperature data, the surface heat fluxes are estimated using Eq. (8.14) and presented in Figure 8.13. The RMS error between the exact and estimated heat fluxes are determined as 0.0364, 0.0274, 0.0175, and 0.0220 for $\mathbf{q_1}$, $\mathbf{q_2}$, $\mathbf{q_3}$, and $\mathbf{q_4}$, respectively. As seen, the proposed filter solution successfully calculated the unknown heat fluxes. It should be noted that Example 8.2, similar to Example 8.1, is dimensionless. Analysis with units can be easily performed in a similar fashion.

Discussion

The filter-based approach facilitates near real-time surface heat flux estimation and only requires temperature data from a limited number of previous and future time steps to estimate the surface heat flux at a given time. For Examples 8.1 and 8.2, the m_f was found as 9 and 10, respectively, suggesting that the algorithm can respond with a delay as small as 9 and 10 time

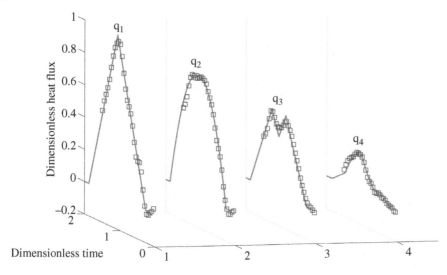

Figure 8.13 Estimated heat flux values for Example 8.2 with 0.5% error present in the temperature data (q_1, q_2, q_3, and q_4).

steps. This feature of the filter-based approach admits their use for monitoring and control purposes of the thermal processes.

8.4 Chapter Summary

The solution of a heat conduction problem in a two-dimensional domain with time- and space-varying heat fluxes on the surface can be constructed, for linear problems, using superposition of a series of time-varying pulses of finite width on the boundary. This model of the forward heat conduction problem facilitates representation of the IHCP solution to determine the time- and space-varying heat fluxes in digital filter form through Tikhonov regularization. A case study of a rectangular domain partially subjected to transient heat flux on one side and fully insulated on other surfaces was analyzed in this chapter. The solution to the direct problem was used to calculate the sensitivity coefficients, and the filter coefficients were calculated accordingly. The characteristics of digital filter for such problem were discussed in detail. The approach described in this chapter can be extended to IHCPs in different multidimensional geometries with multiple unknown heat fluxes.

Problems

8.1 Consider a rectangular domain (L × W) with an aspect ratio of 0.25 which is perfectly insulated from all sides other than the bottom side (Figure 8.14). On the bottom side, the domain is subject to two different transient heat fluxes including $\mathbf{q_1}$ (on $0 < y < W/2$) and $\mathbf{q_2}$ (on $W/2 < y < W$). The analysis is dimensionless and therefore all material properties (thermal conductivity, thermal diffusivity, and density) may be considered as 1. Calculate the filter coefficients associated with each unknown heat flux, and plot sample rows of the filter matrix. Discuss your observations about characteristics of the coefficients.

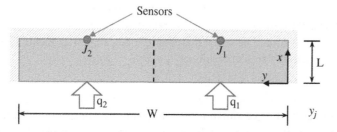

Figure 8.14 Schematic of an IHCP with two unknown heat fluxes and two temperature sensors (Problem 8.1).

8.2 Use the temperature data from the two sensors attached to the top surface of the domain, and calculate q_1 and q_2 using filter form of TR method. The numerical values of the temperature measurement are shown in Table 8.1.

Table 8.1 Temperature measurement data for Problem 8.2.

$\tilde{t}$	y/W = 1/4	y/W = 3/4	$\tilde{t}$	y/W = 1/4	y/W = 3/4	$\tilde{t}$	y/W = 1/4	y/W = 3/4
0	0	0	0.7	0.186176	0.058507	1.4	0.63533	0.292136
0.05	3.68E-05	6.10E-06	0.75	0.213665	0.069739	1.45	0.66075	0.306753
0.1	0.000658	3.77E-05	0.8	0.247348	0.085752	1.5	0.68617	0.321371
0.15	0.003447	0.000324	0.85	0.28103	0.101764	1.55	0.709112	0.335288
0.2	0.008482	0.001195	0.9	0.314713	0.117777	1.6	0.722952	0.346633
0.25	0.016782	0.002917	0.95	0.348759	0.134361	1.65	0.736792	0.357978
0.3	0.025082	0.004638	1	0.384137	0.153032	1.7	0.750632	0.369323
0.35	0.035448	0.007097	1.05	0.419515	0.171703	1.75	0.763076	0.380245
0.4	0.05336	0.012249	1.1	0.454893	0.190374	1.8	0.770394	0.389616
0.45	0.071273	0.017402	1.15	0.489215	0.20876	1.85	0.777711	0.398987
0.5	0.089185	0.022554	1.2	0.519657	0.226095	1.9	0.781478	0.407334
0.55	0.108791	0.028732	1.25	0.5501	0.24343	1.95	0.782954	0.415022
0.6	0.134586	0.038657	1.3	0.580542	0.260765	2	0.78443	0.42271
0.65	0.160381	0.048582	1.35	0.60991	0.277518			

8.3 Adjust the values of Tikhonov regularization parameter for time and space, and discuss how they impact the surface heat flux estimation. The exact numerical values for q_1 and q_2 are available in the companion website.

References

Dowding, K. J. and Beck, J. V. (1999) A sequential gradient method for the inverse heat conduction problem (IHCP), *Journal of Heat Transfer*. American Society of Mechanical Engineers Digital Collection, 121(2), pp. 300–306. https://doi.org/10.1115/1.2825980.

García, J. A. M., Cabeza, J. M. G. and Rodríguez, A. C. (2009) Two-dimensional non-linear inverse heat conduction problem based on the singular value decomposition, *International Journal of Thermal Sciences*, 48(6), pp. 1081–1093. https://doi.org/10.1016/J.IJTHERMALSCI.2008.09.002.

Imber, M. (2012) Temperature extrapolation mechanism for two-dimensional heat flow, *AIAA Journal* 12(8), pp. 1089–1093. https://doi.org/10.2514/3.49417.

Monde, M. *et al.* (2003) An analytical solution for two-dimensional inverse heat conduction problems using Laplace transform, *International Journal of Heat and Mass Transfer*. Pergamon, 46(12), pp. 2135–2148. https://doi.org/10.1016/S0017-9310(02)00510-0.

Najafi, H., Woodbury, K. A. and Besck, J. V. (2015) Real time solution for inverse heat conduction problems in a two-dimensional plate with multiple heat fluxes at the surface, *International Journal of Heat and Mass Transfer*, 91, pp. 1148–1156. https://doi.org/10.1016/j.ijheatmasstransfer.2015.08.020.

Osman, A. M. *et al.* (1997) Numerical solution of the general two-dimensional inverse heat conduction problem (IHCP), *Journal of Heat Transfer*. American Society of Mechanical Engineers Digital Collection, 119(1), pp. 38–45. https://doi.org/10.1115/1.2824098.

Wang, G., Zhu, L. and Chen, H. (2011) A decentralized fuzzy inference method for solving the two-dimensional steady inverse heat conduction problem of estimating boundary condition, *International Journal of Heat and Mass Transfer*, 54(13–14), pp. 2782–2788. https://doi.org/10.1016/J.IJHEATMASSTRANSFER.2011.01.032.

9

Heat Transfer Coefficient Estimation

9.1 Introduction

The determination of the heat transfer coefficient (HTC), h, is a challenging task because it is a derived quantity that cannot be measured directly. Transient temperature measurements can be used for estimating heat transfer coefficient, and inverse heat conduction problem (IHCP) solution techniques may be used for this purpose.

9.1.1 Recent Literature

Several studies have used the solution of IHCPs for evaluating the heat transfer coefficient. Chantasiriwan (1999) used function specification method to determine the surface heat flux and surface temperature using temperature measurements in a one-dimensional domain and estimated the time-dependent Biot number as well as heat transfer coefficient accordingly.

Sablani et al. (2005) used artificial neural network to determine the heat transfer coefficient at the interface of solid–fluid for two different cases including a cube with constant material properties and a semi-infinite plate with temperature-dependent material properties. They used Bi as the output and slope of the temperature–time curve as well as the dimensionless location as the input parameters and successfully trained and tested the network. Bi was then used to evaluate the heat transfer coefficient.

Fan et al. (2008) used thermographic temperature measurement data and solved an IHCP in two-dimensional and three-dimensional domain to evaluate the heat transfer coefficient. Zhang and Delichatsios (2009) studied determination of convective heat transfer coefficient in a three-dimensional IHCP for fire experiments.

Malinowski et al. (2014) studied evaluation of surface heat flux as well as heat transfer coefficient through the solution of IHCP in a three-dimensional domain for a metal plate that is cooled by water. Farahani et al. (2014) studied the local convective boiling heat transfer coefficient in a mini-channel through solution of IHCP using conjugate gradient method. The estimation of local heat transfer coefficient in coiled tubes using solution of IHCPs is presented by Bozzoli et al. (2014). They used the temperature measurements on the external surface of the coil as input data for the IHCP in the wall and solved it through Tikhonov regularization method. Wang et al. (2017) used weighted least squares Levenberg–Marquardt method and solved a nonlinear IHCP by minimizing a cost function, representing the difference between measured and estimated surface temperatures, to calculate the heat transfer coefficients in continuous casting process. Inglese and Olmi (2017) used the solution of IHCP and calculated internal heat transfer coefficient for a pipe, from temperature maps measured on the external side. Hadała et al. (2017) developed a solution technique for inverse evaluation of heat transfer coefficient on a water cooled plate. They achieved their target by minimizing the error between the estimated surface temperatures using the IHCP solution and the exact values of temperature by adjusting the heat transfer coefficients which were the functions of time and space. Mobtil et al. (2018) solved an IHCP to calculate the time-dependent heat transfer coefficient distribution over the fins of the second row of a staggered finned tube heat exchanger using surface temperatures obtained by thermography. Simultaneous estimation of heat transfer coefficient as well as reference temperature from impinging flame jets are studied by Kadam et al. (2018). They used surface temperature data obtained from infrared imaging and solved an IHCP in a three-dimensional domain through the approach proposed by Feng et al. (2011).

Razzaghi et al. (2019) solved an IHCP in a two-dimensional domain using temperature data through adjoint method to estimate the time- and space-varying convective heat transfer coefficient of cooling in the mixed convection regime. Mohebbi and Evans (2020) performed a study on simultaneous estimation of heat flux and heat transfer coefficient in

Inverse Heat Conduction: Ill-Posed Problems, Second Edition. Keith A. Woodbury, Hamidreza Najafi, Filippo de Monte, and James V. Beck.
© 2023 John Wiley & Sons, Inc. Published 2023 by John Wiley & Sons, Inc.

an eccentric hollow cylinder through the solution of IHCP using conjugate gradient method. Tourn et al. (2021) estimated the transient heat transfer coefficient for the surface of an infinitely long plate by solving a nonlinear IHCP through sequential quasi-Newton method (SQNM). Helmig and Kneer (2021) presented a novel approach for the estimation of heat transfer coefficients in rotating bearings through the solution of IHCP using temperature measurements from infrared thermography. Zhao-Hui et al. (2021) calculated the heat transfer coefficient in the cooling regions during the continuous casting process through the solution of IHCP. They used particle swarm optimization method to minimize the errors between the estimated and measured values of surface temperatures which led to accurate estimation of heat transfer coefficient. Abdelhamid et al. (2021) presented an investigation of estimating the heat transfer coefficient using the solution of IHCP. They explored two problems including one with spatial-dependent heat transfer coefficient for which they used Levenberg–Marquardt connected with the surrogate functional method and the modified conjugate gradient method and the second problem with space–time-dependent heat flux using modified conjugate gradient method.

9.1.2 Basic Approach

The estimation of the heat transfer coefficient, h, from transient temperature measurements has aspects of both the IHCP and parameter estimation.

An example of the treatment as an IHCP is that of a one-dimensional case with a known ambient temperature, $T_\infty(t)$, such as the transient determination of boiling heat transfer coefficients using an initially hot spherical copper solid suddenly immersed in water at its saturation temperature. From transient temperatures measured inside or at the surface of the copper body, the methods of the IHCP can be used to estimate the surface heat flux, q_M, and the surface temperature, $T_{0,M}$; the definition of the heat transfer coefficient, h, can be used to obtain the estimate of h given by:

$$\hat{h}_M = \frac{\hat{q}_M}{T_{\infty,M} - 0.5\left(\hat{T}_{0,M} + \hat{T}_{0,M-1}\right)} \tag{9.1}$$

In this expression, $\hat{T}_{0,M}$ is the estimated surface temperature at time t_M; $\hat{q}_M$ and $T_{\infty,M}$ are usually most accurately evaluated at $t_{M-1/2}$ associated with a piecewise constant assumption for the heat flux variation.

The estimated surface heat flux may be obtained using any of the methods that were discussed in Chapter 4 and illustrated in Chapter 7. As discussed in Chapter 5, in general, the vector of heat flux, $\mathbf{q}$, during the experiment can be found as:

$$\mathbf{q} = \mathbf{FY} \tag{9.2}$$

where $\mathbf{F}$ is the filter matrix which may be calculated for different IHCP solution techniques. The vector of surface temperatures, $\mathbf{T}_0$, can be evaluated as:

$$\mathbf{T}_0 = \mathbf{X}_0\mathbf{q} \tag{9.3}$$

where $\mathbf{X}_0$ in Eq. (9.3) is the sensitivity matrix evaluated at the surface (i.e. $\tilde{x} = 0$).

An example of a case that can be treated as a parameter estimation problem is a flat plate over which a fluid is flowing at a constant temperature, $T_{0,M}$ (Figure 9.1). If the plate is suddenly heated by some electric heaters inside the plate, the plate temperature begins to rise and the heat transfer coefficient is a strong function of position from the leading edge of the plate. In some cases, the time variation is small and the basic form of h is a function of x; that is, $h = h(x)$, is known, such as

$$h = \beta x^{-1/2} \tag{9.4}$$

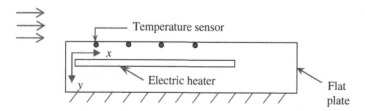

Figure 9.1 Electrically heated flat plate.

The determination of β utilizing various time- and space-dependent measurements of T in the solid and T_∞ in the fluid is a parameter estimation problem.

For cases similar to the two just given, basic solution techniques are known. These examples illustrate the large diversity of problems associated with determining the heat transfer coefficient. Accordingly, a discussion of some of these problems is given in the remainder of this section.

The heat transfer coefficient can be:

1) A constant (independent of x and t)
2) A function of t only; that is, $h(t)$
3) A function of x only; that is, $h(x)$
4) A function of x and t; that is, $h(x, t)$

In this list, x is a coordinate parallel to the heated surface; it can also be generalized to two surface coordinates such as $h = h(x, z)$, where both x and z are parallel to the heated surface. In these problems, the ambient temperature, T_∞, is assumed to be known but it can also be a function of x and t.

There are many concepts and techniques for IHCPs and parameter estimation problems that can be utilized in the solution for the heat transfer coefficient. One concept is that the sensitivity coefficients can be employed to gain insight into the estimation problems. Another concept is that the use of a sequential procedure can be advantageous in terms of computation speed and for online application (process monitoring). These concepts are briefly explored in this chapter.

Due to the large variety of cases in connection with the determination of the heat transfer coefficient, only a few can be treated. The case of $h = h(t)$ for lumped capacitance and one-dimensional applications is the main one covered in this chapter. The basic concepts and examples serve to illustrate procedures which can be modified for different conditions.

9.2 Sensitivity Coefficients

In this section, the sensitivity coefficients for the heat transfer coefficient, h, are investigated for a lumped body – that is, one in which the temperature is a function of time only. Sensitivity coefficients are given for h constant over the complete time domain and for finite time intervals.

The differential equation for a lumped body which is suddenly exposed to a fluid at a temperature T_∞ is

$$\rho c V \frac{dT}{dt} = hA(T_\infty - T) \tag{9.5}$$

where V is the volume and A is the heated surface area of the lumped body. For convenience in notation, the ratio of V to A is denoted L,

$$L = \frac{V}{A} \tag{9.6}$$

For an initial temperature of T_{in} and with both h and T_∞ independent of time, the solution of Eq. (9.5) is

$$T^+ \equiv \frac{T - T_{in}}{T_\infty - T_{in}} = 1 - \exp\left(\frac{-ht}{\rho cL}\right) \tag{9.7}$$

Note that T^+ can be plotted as a function of the single dimensionless time t^+, where

$$t^+ \equiv \frac{ht}{\rho cL} = \frac{hL}{k} \frac{kt}{\rho cL^2} = Bi_L Fo \tag{9.8}$$

Here Bi_L is the Biot number based on the characteristic length L, and Fo is the Fourier number, $Fo = \alpha t / L^2$. For a constant h, the h step sensitivity coefficient, Z_h, is given by

$$Z_h = \frac{\partial T}{\partial h} = (T_\infty - T_{in}) \frac{t}{\rho cL} \exp(-t^+) = (T_\infty - T_{in}) Fo \frac{L}{k} \exp(-Bi_L Fo) \tag{9.9}$$

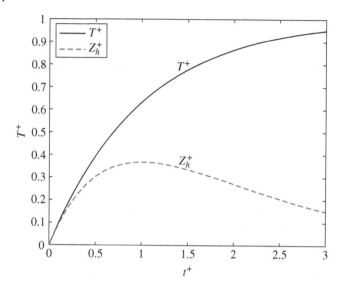

Figure 9.2 Temperatures and sensitivity coefficients for convective heat transfer from a lumped body.

A dimensionless h step sensitivity coefficient is

$$Z_h^+ \equiv \frac{h}{T_\infty - T_{in}} \frac{\partial T}{\partial h} = t^+ \exp(-t^+) = Bi_L Fo \exp(-Bi_L Fo) \tag{9.10}$$

which is also a function of t^+.

Figure 9.2 displays the dimensionless temperature response T^+ from Eq. (9.7). Also shown in Figure 9.2 is the dimensionless heat transfer sensitivity coefficient. For the small values of $t^+ < 0.5$, Z_h^+ increases and is nearly equal to T^+; in the range of $0.5 < t^+ < 1.5$, Z_h^+ attains a maximum; and for $t^+ > 1.5$, Z_h^+ decreases toward zero. There are several ramifications of these behaviors of Z_h^+. First, the optimal times for measuring temperature in order to estimate h are near $t^+ = 1$, and measurements at the early and late times contain less information regarding h. Second, the average magnitude of Z_h^+ is considerably less than T^+, and hence h is relatively sensitive to measurement errors in temperature.

The last expression in Eq. (9.8) helps interpret the time base in Figure 9.2. The dimensionless t^+ is based on the heat transfer coefficient, but $Fo = at/L^2$ is a conventional dimensionless time. From Eq. (9.8), $Fo = t^+/Bi_L$, so when Bi_L is small (h is small), the time required for the measurement is longer than for large Bi_L. Recall that a common assumption for validity of a lumped capacitance model is $Bi_L \leq 0.1$ or so.

Although not used directly to estimate time variable $h(t)$, some insights can be seen by considering the sensitivity coefficients for the time variable case. This case of a lumped body is sufficiently tractable that the sensitivity coefficients for time variable h and T_∞ can be obtained in algebraic form. The case where $h(t)$ and $T_\infty(t)$ are constant over the time segments is considered:

$$0 \leq t < t_1, \qquad t_1 \leq t < t_2, \qquad t_2 \leq t < t_3, \dots \tag{9.11}$$

This is illustrated in Figure 9.3. The heat transfer coefficient components are $h_1, h_2, \dots$ and the corresponding components for T_∞ are $T_{\infty,1}, T_{\infty,2}, \dots$. The equations for T for three time intervals are for $0 \leq t < t_1$

$$T = T_{\infty,1} + (T_0 - T_{\infty,1}) \exp\left(\frac{-h_1 t}{\rho c L}\right) \tag{9.12}$$

for $t_1 \leq t < t_2$,

$$T = T_{\infty,2} + (T_1 - T_{\infty,2}) \exp\left[\frac{-h_2(t - t_1)}{\rho c L}\right] \tag{9.13}$$

for $t_2 \leq t < t_3$,

$$T = T_{\infty,3} + (T_2 - T_{\infty,3}) \exp\left[\frac{-h_3(t - t_2)}{\rho c L}\right] \tag{9.14}$$

and the T_1 and T_2 values are those evaluated at t_1 and t_2,

$$T_1 = T_{\infty,1} + (T_0 - T_{\infty,1}) \exp\left(\frac{-h_1 t_1}{\rho c L}\right) \tag{9.15}$$

$$T_2 = T_{\infty,2} + (T_1 - T_{\infty,2}) \exp\left[\frac{-h_2(t_2 - t_1)}{\rho c L}\right] \tag{9.16}$$

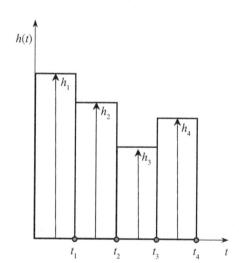

Figure 9.3 Heat transfer coefficient history approximated by constant segments.

The pulse sensitivity coefficients can be found by differentiation as previously illustrated. The pulse sensitivity coefficient for h_i is

$$X_{h_i} \equiv \frac{\partial T}{\partial h_i} \tag{9.17}$$

To reduce the number of dimensionless groups, after the differentiations are completed, the h_i components are made equal,

$$h_1 = h_2 = h_3 = \cdots = h_0 \tag{9.18}$$

and the same is done for $T_{\infty,i}$,

$$T_{\infty,1} = T_{\infty,2} = T_{\infty,3} = \cdots = T_\infty \tag{9.19}$$

Figure 9.4 displays the heat transfer coefficient sensitivity coefficients for the two cases of dimensionless time steps of 0.25 and 0.5.

The $X_{h_i}^+$ values are uncorrelated as can be seen by comparing the values at $t^+ = h_0 A t/\rho c V = 0.25$, 0.5, 0.75, and 1 of $X_{h_1}^+$ which are 0.19, 0.15, 0.12, and 0.09 and those for $X_{h_2}^+$ which are 0, 0.15, 0.12, and 0.09. These values are not proportional for all times and hence are uncorrelated. The first values of $X_{h_1}^+$ and $X_{h_2}^+$ are different, but the succeeding values are identical. This suggests that a sequential procedure (which emphasizes "recent" values) of estimating $h = h(t)$ would be more effective

Figure 9.4 Heat transfer sensitivity coefficients for h_i constant over segments and $h_1 = h_2 = \ldots = h_0$ and $T_{\infty,1} = T_{\infty,2} = \ldots T_\infty$. (a) $\Delta t^+ = 0.25$ and (b) $\Delta t^+ = 0.50$.

(a)

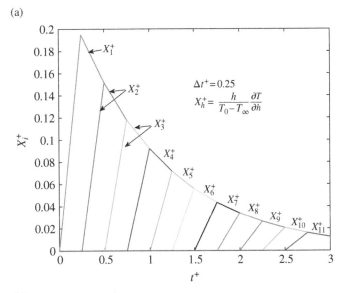

(b)

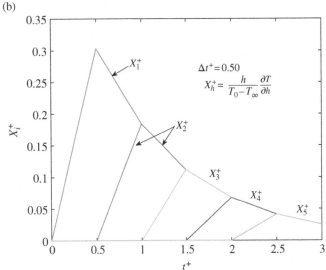

than a whole domain estimation procedure which uses all the data simultaneously. Another conclusion that can be drawn from Figure 9.4 is that the h_i components become more difficult to estimate as i increases. This is a consequence of the X_{hi} values becoming smaller in magnitude as i increases.

9.3 Lumped Body Analyses

In this section, the determination of the heat transfer coefficient for the case of a lumped body is discussed. A thermally lumped body is one in which the temperature is uniform in the body but varies with time. Transient calorimeters based on this approach have been used for measuring heat transfer coefficients and heat fluxes for a number of applications including determination of the boiling curve.

The case of a lumped body involves a simple model, Eq. (9.5), and thus the details of the estimation procedure can be readily seen. Three procedures are discussed in this section. In Section 9.3.1, exact matching of the model temperature is employed. Section 9.3.2 presents a sequential method based on Tikhonov regularization and filter concepts. Lastly, in Section 9.3.3, direct estimation of a constant heat transfer coefficient is developed using parameter estimation and future times estimation techniques.

9.3.1 Exact Matching of the Measured Temperatures

If a constant heat flux at the surface replaces the heat transfer coefficient, the energy balance of Eq. (9.5) becomes

$$\rho c V \frac{dT}{dt} = q_c A \tag{9.20}$$

and the solution to Eq. (9.20) is, with initial temperature $T(0) = T_{in}$, is

$$T(t) = \frac{q_c}{\rho c L} t + T_{in} \tag{9.21}$$

or in dimensionless form

$$\tilde{T}(\tilde{t}) = \frac{T(t) - T_{in}}{q_c L / k} = \tilde{t} \tag{9.22}$$

Eq. (9.21) can be used with superposition to find the solution at any time t_M due to a sequence of piecewise constant heat fluxes, analogous to Eq. (3.49) for the finite body case, as

$$\tilde{T}_M = \sum_{i=1}^{M} \tilde{q}_i \Delta \phi_{M-i} \tag{9.23}$$

Here ϕ is the response of the lumped body to a unit step increase in the heat flux, $\phi = \tilde{T} = \tilde{t}$, $\Delta \phi_i = \phi_{i+1} - \phi_i$, as before, and $\tilde{q} = q/q_c$. Because of the linear-in-time nature of ϕ, the difference $\Delta \phi_1 = \Delta \phi_2 = \Delta \phi_M = \Delta \tilde{t}$ for uniform time increments. The solution can be represented in a matrix form as

$$\begin{Bmatrix} T_1 - T_{in} \\ T_2 - T_{in} \\ T_3 - T_{in} \\ \vdots \\ T_M - T_{in} \end{Bmatrix} \frac{1}{q_{ref} L / k} = \begin{bmatrix} \Delta \tilde{t} & 0 & 0 & \cdots & 0 \\ \Delta \tilde{t} & \Delta \tilde{t} & 0 & \cdots & 0 \\ \Delta \tilde{t} & \Delta \tilde{t} & \Delta \tilde{t} & \ddots & \vdots \\ \vdots & \vdots & \ddots & \Delta \tilde{t} & 0 \\ \Delta \tilde{t} & \Delta \tilde{t} & \Delta \tilde{t} & \Delta \tilde{t} & \Delta \tilde{t} \end{bmatrix} \begin{Bmatrix} \tilde{q}_1 \\ \tilde{q}_2 \\ \tilde{q}_3 \\ \vdots \\ \tilde{q}_M \end{Bmatrix} \tag{9.24a}$$

or

$$\tilde{\mathbf{T}} = \tilde{\mathbf{X}}_{LC} \tilde{\mathbf{q}} \tag{9.24b}$$

where the subscript "LC" indicates lumped capacitance. Now any of the filter-based methods from Chapter 4 can be used to solve for the heat flux history using Eq. (9.1). Once the heat flux history is known, the corresponding heat transfer coefficients can be found using Eq. (9.1). In this case, the "estimated surface temperature" is the same as the measured temperature (lumped capacitance); therefore,

$$\hat{h}_M = \frac{\hat{q}_M}{T_{\infty,M} - 0.5(Y_M + Y_{M-1})} \tag{9.25}$$

Of course, the temperature driving the heat transfer, $T_{\infty,M}$, must be known, but it can vary with time.

Notice in Eqs. (9.22) and (9.24a) the thermal conductivity is used for non-dimensionalization. This physical parameter has no significance in the lumped capacitance analysis and hence solutions are insensitive to the choice of value for k – it is arbitrary.

Example 9.1 A solid copper billet that is 0.0462 m long and 0.0254 m in diameter is heated in a furnace and then removed. Two thermocouples are attached to the billet. For this test, $\rho = 8890$ kg/m^3 and $c_p = 385$ J/kg-K. The first eleven measured temperatures are given in Table 9.1. These data are from an actual experiment presented on p. 243 of Beck and Arnold (1977). Find the heat transfer coefficient for the ambient temperature of 27.5 °C using exact matching of data.

Solution

Eq. (9.24a) is used to compute the heat fluxes using exact matching. This solution can be performed in either dimensional or dimensionless variables. The characteristic length is $L = V/A = 0.00498$ m^3 (includes both ends of the cylinder).

The coefficient matrix in Eq. (9.24a), in dimensionless variables, is

$$\frac{k}{\rho c L^2}\begin{bmatrix} \Delta t & 0 & 0 & \cdots & 0 \\ \Delta t & \Delta t & 0 & \cdots & 0 \\ \Delta t & \Delta t & \Delta t & \ddots & \vdots \\ \vdots & \vdots & \ddots & \Delta t & 0 \\ \Delta t & \Delta t & \Delta t & \Delta t & \Delta t \end{bmatrix}_{10\times10} = \begin{bmatrix} 440.93 & 0 & 0 & \cdots & 0 \\ 440.93 & 440.93 & 0 & \cdots & 0 \\ 440.93 & 440.93 & 440.93 & \ddots & \vdots \\ \vdots & \vdots & \ddots & 440.93 & 0 \\ 440.93 & 440.93 & 440.93 & 440.93 & 440.93 \end{bmatrix}_{10\times10}$$

The condition number of this matrix is 13.2 which is not large and indicates that the system of equations is not ill-conditioned. The solution for the **q** vector is found directly by taking the inverse of the coefficient matrix and solving Eq. (9.24a). The resulting values are shown in fourth column of Table 9.1. These values correspond to the "half" time step consistent with the $q = const$ heat flux variation assumption.

The corresponding estimates for the heat transfer coefficients are computed using Eq. (9.25). These values are listed in the fifth column of Table 9.1.

Table 9.1 Data for the copper billet example and exact matching results.

M	t_M(s)	Y_M, C	$\hat{q}_M$,W/m^2	$\hat{h}_M$,W/m^2-K	$\hat{T}_M$,C	$\hat{q}_{Mf}$ W/m^2
0		137.55				
1	96	129.37	−1452.6	13.71	129.37	−1449.9
2	192	121.96	−1315.9	13.40	121.96	−1313.8
3	288	115.17	−1205.8	13.24	115.17	−1203.9
4	384	108.99	−1097.4	12.98	108.99	−1095.7
5	480	102.91	−1079.7	13.76	102.91	−1078.0
6	576	97.70	−925.2	12.71	97.70	−923.7
7	672	92.98	−838.2	12.36	92.98	−837.2
8	768	88.69	−761.8	12.03	88.69	−759.6
9	864	84.69	−710.3	12.00	84.69	−710.1
10	960	80.91	−671.2	12.14	80.91	−669.8

9.3.2 Filter Coefficient Solution

The condition number of the coefficient matrix $\mathbf{X}_{LC}$ for lumped capacitance depends on the time step and can be a moderate value, in which case the matrix can be readily inverted for exact matching. However, noisy data may require some regularization in the solution, and any of the methods of Chapter 4 may be used to solve for the heat fluxes before computing the heat transfer coefficient.

As an example, consider zeroth-order Tikhonov regularization for the estimation of the heat flux components in Example 9.1. Figure 9.5 shows the middle row of the dimensionless Tikhonov filter matrix computed using $q = const$ assumption for various regularization coefficients. These coefficients seem large in absolute magnitude but should be gauged in comparison to the entries of the $\mathbf{X}^T\mathbf{X}$ matrix, whose entries are on the order of 1E6. For small values of α_0, such as $\alpha_0 = 100$, the amount of regularization is insignificant and has negligible effect, and the filter coefficients are the same as if direct inversion of the $\mathbf{X}$ matrix is used in exact matching. Notice that these filter coefficients are only nonzero for the current and the immediately preceding time step. In fact, these values are of equal magnitudes and opposite signs. It is not very surprising that the heat flux at time t_M depends only on the difference between the current and previous data reading – a finite approximation to the first derivative (cf. Eq. (9.20)).

$$\hat{q}_M \approx C\frac{Y_M - Y_{M-1}}{\Delta t} = (2.264\mathrm{E}-3)(Y_M - Y_{M-1}) \tag{9.26}$$

The constant on the right side of Eq. (9.26) is the value of the peak of the $\alpha_0 = 100$ curve in Figure 9.5.

Example 9.2 Calculate the surface heat flux for Example 9.1 using Eq. (9.26).

Solution

The value of heat flux over the first time interval is

$$q_1 = 0.002264 \times \frac{(129.37 - 137.55)\mathrm{C}}{(1\,\mathrm{W/m}^2 \times 0.00498\,\mathrm{m}/390\,\mathrm{W/m-C})} = -1450.3\,\mathrm{W/m}^2 \tag{9.27}$$

$$q_2 = 0.002264 \times \frac{(121.96 - 129.37)\mathrm{C}}{(1\,\mathrm{W/m}^2 \times 0.00498\,\mathrm{m}/390\,\mathrm{W/m-C})} = -1313.8\,\mathrm{W/m}^2 \tag{9.28}$$

The remaining values are included in the last column labeled "$q_{M,f}$" in Table 9.1.

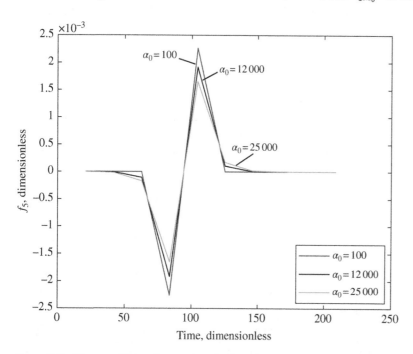

Figure 9.5 Filter coefficient for zeroth-order Tikhonov regularization for heat flux estimation in Example 9.1.

Discussion

The values from the simple filter computation $q_{M,f}$ compared to three significant figures (only three significant figures can be expected since there are only three in $L = 0.00498$ m) with the values obtained from exact matching. These results demonstrate that a simple online method for computing heat fluxes and/or heat transfer coefficients using a lumped capacitance probe can be easily implemented. Only data from the current, t_M, and immediately past time step, t_{M-1}, are needed to compute q_M and the corresponding value of the heat transfer coefficient.

Larger values of α_0 shown in Figure 9.5 affect the filter coefficients by increasing the dependence of the heat flux on an additional future and past time step. This is seen in Figure 9.5 as the values for f_{M+2} and f_{M-2} begin to deviate from zero. The effect of these larger α_0 values on the estimated heat transfer coefficient is seen in Figure 9.6. Relatively large values of the parameter cannot help smooth the peak in the curve at around 400 seconds. These large values affect the estimated values at the end of the data time interval (recall biasing effect of Tikhonov regularization).

9.3.3 Estimating Constant Heat Transfer Coefficient

Sometimes only a constant or average value of heat transfer coefficient is of interest. This can be estimated easily with a lumped capacitance model Eq. (9.5) and the exact solution for a constant value of h, Eq. (9.7). This is a parameter estimation problem to determine this single value of h and is a nonlinear estimation problem because the sensitivity coefficients depend on the unknown parameter (see Eqs. (9.9) and (9.10)). Due to this nonlinearity, an iterative solution is needed. One such procedure is the Gauss method (Beck and Arnold, 1977; Crassidis and Junkins, 2004).

In the Gauss linearization method, first an estimate of $h_c^{(\nu-1)}$ is assumed known for the $(\nu-1)$st iteration and then an improved value is sought. The sensitivity coefficient given explicitly in Eq. (9.9) is evaluated with h equal to $h_c^{(\nu-1)}$. The calculated temperature is approximated by the two-term Taylor series,

$$\mathbf{T}^{(\nu)} = \mathbf{T}^{(\nu-1)} + \mathbf{Z}^{(\nu-1)} \Delta h_c^{(\nu)} \tag{9.29}$$

Here $\mathbf{T}$ and $\mathbf{Z}$ are vectors of length n corresponding to the number of measurements, and $\Delta h_c^{(\nu)}$ is the scalar correction to be determined in the iterations. The sensitivity vector $\mathbf{Z}$ is evaluated at the measurement times using Eq. (9.9).

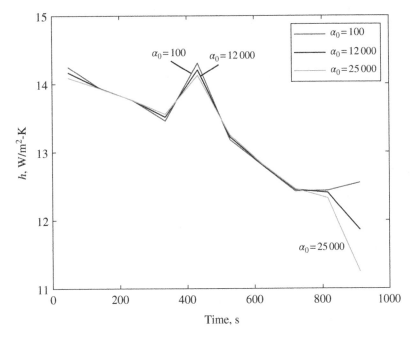

Figure 9.6 Effect of regularization parameter for zeroth-order Tikhonov on heat transfer estimates in Example 9.1.

Now the sum of squared errors between the model Eq. (9.7) and the measurements is minimized in the usual way by taking the derivative with respect to the unknown increment $\Delta h_c^{(\nu)}$ and setting the result equal to zero and solving for the unknown. The result is

$$\Delta h^{(\nu)} = \left(\mathbf{Z}^T\mathbf{Z}\right)^{-1}\mathbf{Z}^T\left(\mathbf{Y}-\mathbf{T}^{(\nu-1)}\right) \tag{9.30}$$

Notice that $\mathbf{Z}^T\mathbf{Z}$ in this application is a scalar result, and so the arithmetic in Eq. (9.30) is very simple since $(\mathbf{Z}^T\mathbf{Z})^{-1} = 1/(\mathbf{Z}^T\mathbf{Z})$. This equation is used in an iterative manner until the changes in $\Delta h_c^{(\nu)}$ are less than some small amount, such as

$$\left|\Delta h^{(\nu)}\right| < 10^{-4} \tag{9.31}$$

The Gauss procedure outlined above is simple to program and can be easily implemented in any computer code. However, many computational packages and environments, including MATLAB, have inbuilt algorithms to do the heavy lifting of minimizing a user-defined function. The sum of squares function can be written directly as a function of the heat transfer coefficient as

$$S(h) = (\mathbf{T}(h)-\mathbf{Y})^T(\mathbf{T}(h)-\mathbf{Y}) \tag{9.32}$$

where $\mathbf{T}(h)$ follows from Eq. (9.7). This scalar function can be minimized using any suitable algorithm, such as fminbnd() in MATLAB.

Eq. (9.32) and fminbnd() (or the Gaussian minimization scheme) can be applied over the whole time domain, which will result in an effective or average heat transfer coefficient, but may also be applied in sequential overlapping time windows using future information to introduce some smoothing into the estimation. Consider limiting the model and data vectors in Eq. (9.32) to the current time, M, and $r-1$ future times. The values of h and T_∞ will be held constant over these times and designated h_M and $T_{\infty,M}$. The r components of the model vector are computed using Eq. (9.7)

$$T_{M+i-1} = \left(T_{\infty,M}-Y_{M-1}\right)\left(1-\exp\left(-\frac{h_M(i\Delta t)}{(\rho c L)}\right)\right) + Y_{M-1} \qquad i = 1,2,\dots,r \tag{9.33}$$

and the corresponding measurement values are used for the $\mathbf{Y}$ vector. After minimizing the sum of squares and finding the optimal h_M, the time step is incremented by one and the procedure repeated. Notice that the "initial temperature" in Eq. (9.8), Y_{M-1}, is updated at each time step. For $M = 1$, the initial temperature measurement Y_{in} is used.

Example 9.3 Use the data from Example 9.1 to calculate the average heat transfer coefficient for the billet cooling using Gaussian iteration. Use an initial guess of $h^{(0)} = 1.0\,\text{W/m}^2\text{-}°\text{C}$ for the iterations.

Solution

For the first iteration, with $h^{(0)} = 1.0\,\text{W/m}^2\text{-}°\text{C}$, the sensitivity coefficients using Eq. (9.9) are

$$\mathbf{Z}^0 = \lfloor -0.616 \;\; -1.23 \;\; -1.83 \;\; -2.42 \;\; -3.01 \;\; -3.60 \;\; -4.17 \;\; -4.74 \;\; -5.30 \;\; -5.86 \rfloor^T$$

The units on $\mathbf{Z}$ are $\text{m}^2\text{-}\text{C}^2/\text{W}$.
The model-calculated temperatures from Eq. (9.7) are

$$\mathbf{T}^0 = \lfloor -0.618 \;\; -1.23 \;\; -1.84 \;\; -2.45 \;\; -3.06 \;\; -3.66 \;\; -4.25 \;\; -4.85 \;\; -5.44 \;\; -6.03 \rfloor^T$$

These are actually temperature differences $T - T_{in}$ in degrees Celsius. Corresponding values of $Y - Y_{in}$ are used in the calculations. The numerical value of $\mathbf{Z}^T(\mathbf{Y} - \mathbf{1}T_{in}) = 1273.1\,\text{m}^2\text{-}\text{C/W}$ and the value of $\mathbf{Z}^T\mathbf{Z} = 135.4\,\text{m}^4\text{-}\text{C}^4/\text{W}^2$, and these values in Eq. (9.30) result in a correction of $\Delta h^{(1)} = 9.4\,\text{W/m}^2\text{-}°\text{C}$. Table 9.2 lists the values for the corrections for subsequent iterations. These values have been rounded for presentation.

Discussion

The converged value of the heat transfer coefficient is $h = 13.12\,\text{W/m}^2\text{-}°\text{C}$, which is in the middle of the range of the values obtained from exact matching in Example 9.1. The Gaussian iteration converges quickly in five iterations.

Table 9.2 Convergence history for Example 9.3.

ν	$h^{(\nu-1)}$	$Z^T(Y - T^0)$	Z^TZ	$\Delta h^{(\nu)}$
1	1.0	1273.1	135.4	9.4
2	10.4	155.7	60.7	2.6
3	12.97	7.50	49.2	0.15
4	13.12	0.04	48.6	0.001
5	13.12	0	48.6	0

Example 9.4 Use the data from Example 9.1 to calculate the heat transfer coefficient over each time interval using $r = 2$ future times for each calculation.

Solution

For the first estimate, only the first two values of data are used ($M = 1$, $M + r - 1 = 2$). Either temperature values or temperature differences can be used in the solution. Here the temperature values will be used ($\mathbf{Y}_{M=1} = \begin{bmatrix} 129.37 & 121.96 \end{bmatrix}^T$). For each time step, two model values will be computed using Eq. (9.7). At the first time step, $Y_{M-1} = Y_{in} = 137.55$. For example, if $h_M = 10$ W/m²-°C, then the two model-computed values are $\mathbf{T}_{M=1,\, h=10} = \begin{bmatrix} 131.5 & 125.8 \end{bmatrix}^T$.

The problem is coded in MATLAB and solved using fminbnd() at each time step. The coding looks something like this:

```
time_r = [ dtime:dtime:r*dtime]';
T0_nu = T0;
for M = 1:nt-r+1
    y = Y(M:M+r-1)
    S = @(h) ((1-exp(-time_r*h/rho/cp/L))* ...
            (Tinf-T0_nu) +T0_nu - y)'*...
            ((1-exp(-time_r*h/rho/cp/L))* ...
            (Tinf-T0_nu) +T0_nu - y);
    h_M(M) = fminbnd( S, 0.01, 1000 );
    T0_nu = Y(it);
end
```

For each value of M from 1 to $n - r + 1$, a value of h_M is computed by minimizing the autonomous function S, which is appropriately redefined at each time step. Importantly, the "initial time" for the next sliding window of the calculations is updated at the bottom of the loop. The results of the calculations are listed in Table 9.3.

Table 9.3 Converged results for h_M for Example 9.4 using $r = 2$.

M	Time (s)	h_M, W/m²-°C
1	96	13.595
2	192	13.345
3	288	13.142
4	384	13.284
5	480	13.358
6	576	12.576
7	672	12.228
8	768	12.015
9	864	12.063

Discussion

These results compare favorably with those from the previous examples. Including future time information in the estimates introduces bias but can smooth the results. Larger values of r bias estimates toward the values of the future (in this case, will reduce magnitude of each h_M) and lessen the variation in the h vs. t curve.

9.4 Bodies with Internal Temperature Gradients

For bodies with internal temperature gradients, the suggested approach to computing the surface heat transfer coefficient is to first compute the surface heat flux history using the $q = const$ assumption, then compute the corresponding surface temperatures using this heat flux history, and finally compute the resulting heat transfer coefficient using Eq. (9.1). This approach is suggested because it does not require iteration due to nonlinearity of the sensitivity coefficients.

Eq. (9.1) requires the temperature history at the surface. Once the heat flux components have been found, the temperatures at the surface are computed using

$$\hat{\mathbf{T}}_0 = \mathbf{X}_0 \hat{\mathbf{q}} \tag{9.34}$$

$\mathbf{X}_0$ is the same as the sensitivity matrix $\mathbf{X}$ but is evaluated at the surface ($x = 0$).

An example is given to illustrate the approach. The example is for a one-dimensional, linear heat conduction problem, and the methods introduced in Chapter 4 are used for estimating surface heat flux.

Example 9.5 A 1.0-m-thick plate insulated on one side has thermal properties $k = 1$ W/m-°C and $\alpha = 1$ m^2/s. The plate is heated on the exposed face with a fluid at temperature $T_\infty = 800$ °C. Temperature measurements from the insulated surface are in Table 9.4. Estimate the heat transfer coefficient that causes these measurements.

Solution

Zeroth-order Tikhonov regularization is used to estimate the heat fluxes, although any of the methods from Chapter 4 can be utilized. These time steps are relatively large, and the $\Delta\phi$ entries in the $\mathbf{X}$ matrix (see Chapter 4) become nearly constant after a few time steps. The first few entries of the $\mathbf{X}$ matrix at the measurement location are

$$\mathbf{X} = \begin{bmatrix} 0.2372 & 0 & 0 & 0 & 0 & \cdots \\ 0.3962 & 0.2372 & 0 & 0 & 0 & \cdots \\ 0.3999 & 0.3962 & 0.2372 & 0 & 0 & \cdots \\ 0.4000 & 0.3999 & 0.3962 & 0.2372 & 0 & \ddots \\ 0.4000 & 0.4000 & 0.3999 & 0.3962 & 0.2372 & \ddots \\ \vdots & \vdots & \ddots & \ddots & \ddots & \ddots \end{bmatrix} \tag{9.35}$$

Using a value of $\alpha_0 = 4.94$E-2 for the Tikhonov regularization, the heat flux components in Table 9.5 are obtained (values are rounded for presentation).

Table 9.4 Measured data (°C) for Example 9.5.

t_i	0.4	0.8	1.2	1.6	2.0	2.4	2.8	3.2
Y_i	3.7	−0.4	2.7	19.2	55.9	100.7	169.0	245.4
t_i	3.6	4.0	4.4	4.8	5.2	5.6	6.0	6.4
Y_i	309.0	353.5	383.8	397.2	399.5	399.7	403.8	401

Table 9.5 Estimated heat flux components (W/m²) for Example 9.5.

t_i	0.4	0.8	1.2	1.6	2.0	2.4	2.8	3.2
$\hat{q}_i$	6.2	−6.4	20.2	62.3	100.9	135.0	183.7	181.7
t_i	3.6	4.0	4.4	4.8	5.2	5.6	6.0	6.4
$\hat{q}_i$	139.6	95.8	58.2	20.2	1.5	4.6	4.1	−5.9

The sensitivity coefficient for the location at the surface are needed. The first few entries of this matrix are

$$
\mathbf{X}_0 = \begin{bmatrix}
0.7294 & 0 & 0 & 0 & 0 & \cdots \\
0.4038 & 0.7294 & 0 & 0 & 0 & \cdots \\
0.4001 & 0.4038 & 0.7294 & 0 & 0 & \cdots \\
0.4000 & 0.4001 & 0.4038 & 0.7294 & 0 & \ddots \\
0.4000 & 0.4000 & 0.4001 & 0.4038 & 0.7294 & \ddots \\
\vdots & \vdots & \ddots & \ddots & \ddots & \ddots
\end{bmatrix}
\tag{9.36}
$$

Multiplying this matrix into the estimated heat fluxes gives the estimated surface temperatures in Table 9.6.

Finally, Eq. (9.1) is used to estimate the heat transfer coefficients over each time interval. The resulting values are shown in Table 9.7 and graphed in Figure 9.7 using the open circle symbols.

Discussion

The data for this example were generated using a triangular heat flux input. Exact values of the heat transfer coefficients were computed using Eq. (9.1) using the known heat fluxes and computed surface temperatures from the analytical solution. These exact values of h are plotted in Figure 9.7 as the dashed line.

The results obviously depend on the choice of the regularization parameter. For this solution, the nominal value was selected using knowledge of the actual surface heating, which, of course, is not possible in practice. For comparison, additional results are generated using one-half and two-times the nominal optimal value. These results are shown in Figure 9.7 as the diamond and triangle symbols, respectively. These results suggest that precise knowledge of the optimal regularization coefficient is not necessary, and a reasonable value of the correct order of magnitude will yield satisfactory results.

Table 9.6 Estimated surface temperatures (°C) for Example 9.5.

t_i	0.4	0.8	1.2	1.6	2.0	2.4	2.8	3.2
$\hat{T}_{0,i}$	4.5	−2.1	14.6	53.5	106.8	172.1	261.8	334.0
t_i	3.6	4.0	4.4	4.8	5.2	5.6	6.0	6.4
$\hat{T}_{0,i}$	376.0	399.7	410.4	405.8	400.1	402.9	404.4	398.7

Table 9.7 Estimated heat transfer coefficients (W/m²-°C) for Example 9.5.

t_i	0.4	0.8	1.2	1.6	2.0	2.4	2.8	3.2
$\hat{h}_i$	0.0078	−0.008	0.0254	0.0813	0.1402	0.2044	0.3151	0.3619
t_i	3.6	4.0	4.4	4.8	5.2	5.6	6.0	6.4
$\hat{h}_i$	0.3137	0.2325	0.1473	0.0514	0.0038	0.0115	0.0105	−0.0148

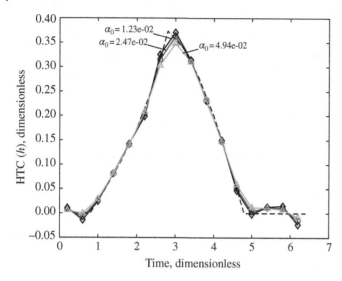

Figure 9.7 Estimate heat transfer coefficients for Example 9.5 for various regularization parameters. Dashed line indicates the exact values.

9.5 Chapter Summary

Time-varying heat transfer coefficients can be estimated by first estimating the heat flux and then using the measured or computed surface temperature and knowledge of the environment temperature to compute the desired heat transfer coefficient. Direct estimation of the heat transfer coefficient is a nonlinear estimation problem because the sensitivity coefficients needed to compute the unknown HTCs depend on these unknown HTCs.

Lumped capacitance probes can be used to measure heat transfer coefficients for quenching and other applications. Several analysis techniques can be applied for lumped capacitance bodies. (Lumped capacitance is a good assumption when the thermal conductivity of the material is very high compared to the external heat transfer coefficient. A rule of thumb for this condition is $Bi_L = hL/k < 0.1$, where L is the volume to surface ratio for the body.) Both whole domain and sequential computational schemes can be used to compute the heat fluxes for lumped bodies, and the measured data serves as the surface temperature for these conditions, so the heat transfer coefficient can be computed.

For lumped capacitance problems, direct computation of the HTCs is possible. This nonlinear estimation can be completed using an algorithm such as Gaussian linearization and iteration or using inbuilt minimization routines such as MATLAB's fminbnd(). These solutions can be performed either over the whole domain, wherein a single, average value of h is obtained, or sequentially over each time step using r future time steps of data at each step.

The filter concept can also be applied for lumped capacitance problems. The filter coefficients for estimating the heat flux for the lumped capacitance analysis have only two nonzero values corresponding to the current and most previous time step. Calculation of the heat flux from these filter coefficients is equivalent to using backward difference approximation for the first derivative in Eq. (9.20) and using exact matching of the data. This gives the possibility of an online computational scheme for computing h sequentially as each new time observation is acquired.

For one-dimensional bodies with internal temperature gradients, techniques from Chapter 4 can be used to estimate surface heat flux. This heat flux can then be used to compute the temperature at the surface of the body. The heat flux and average surface temperature over each time interval can then be used to compute the heat transfer coefficient.

Problems

9.1 Analyze the data from Problem 7.1 treating the plate as a lumped body.

   a) Use Eq. (9.24a) to compute the heat flux history. Does exact matching produce acceptable results? Add some Tikhonov regularization, if needed.

b) Use results from part (a) and the measured data to compute the heat transfer coefficient history if the environment temperature is constant 200 °C.

9.2 Process the data from Problem 7.1 sequentially to determine the heat transfer coefficient over each time step and considering $r = 2$ time steps of data. Consider the steel plate as a lumped body.

9.3 Analyze the data from Problem 7.1 considering the plate with temperature gradients to estimate the heat fluxes. Generalized Cross Validation (GCV) can be used to choose the amount of regularization. Use Eq. (9.1) to compute the heat transfer coefficients, assuming the environment temperature is constant at 200 °C.

References

Abdelhamid, T., Elsheikh, A.H., Omisore, O.M., Saeed, N.A., et al. (2021) Reconstruction of the heat transfer coefficients and heat fluxes in heat conduction problems, *Mathematics and Computers in Simulation*, 187, pp. 134–154. https://doi.org/10.1016/j.matcom.2021.02.011.

Beck, J. V. and Arnold, K. J. (1977) *Parameter Estimation in Engineering and Science*. New York: John Wiley and Sons.

Bozzoli, F., Cattani, L., Rainieri, S., Viloche Bazán,F.S. and Borges, L.S. (2014) Estimation of the local heat-transfer coefficient in the laminar flow regime in coiled tubes by the Tikhonov regularisation method, *International Journal of Heat and Mass Transfer*, 72, pp. 352–361. https://doi.org/10.1016/j.ijheatmasstransfer.2014.01.019.

Chantasiriwan, S. (1999) Inverse heat conduction problem of determining time-dependent heat transfer coefficient, *International Journal of Heat and Mass Transfer*, 42(23), pp. 4275–4285. https://doi.org/10.1016/S0017-9310(99)00094-0.

Crassidis, J. L. and Junkins, J. L. (2004) *Optimal Estimation of Dynamic Systems*. 2nd edn. Taylor & Francis. https://doi.org/10.1201/b11154.

Fan, C., Sun, F. and Yang, L. (2008) A numerical method on inverse determination of heat transfer coefficient based on thermographic temperature measurement', *Chinese Journal of Chemical Engineering*, 16(6), pp. 901–908. https://doi.org/10.1016/S1004-9541(09)60014-8.

Farahani, S. D., Kowsary, F. and Jamali, J. (2014) Direct estimation of local convective boiling heat transfer coefficient in mini-channel by using conjugated gradient method with adjoint equation, *International Communications in Heat and Mass Transfer*, 55, pp. 1–7. https://doi.org/10.1016/j.icheatmasstransfer.2014.03.004.

Feng, Z.C., Chen, J.K., Zhang, Y. and Griggs, J.L. (2011) Estimation of front surface temperature and heat flux of a locally heated plate from distributed sensor data on the back surface, *International Journal of Heat and Mass Transfer*, 54(15–16), pp. 3431–3439. https://doi.org/10.1016/J.IJHEATMASSTRANSFER.2011.03.043.

Hadała, B., Malinowski, Z. and Szajding, A. (2017) Solution strategy for the inverse determination of the specially varying heat transfer coefficient, *International Journal of Heat and Mass Transfer*, 104, pp. 993–1007. https://doi.org/10.1016/j.ijheatmasstransfer.2016.08.093.

Helmig, T. and Kneer, R. (2021) A novel transient infrared-thermography based experimental method for the inverse estimation of heat transfer coefficients in rotating bearings, *International Journal of Thermal Sciences*, 167(June 2020), p. 107000. https://doi.org/10.1016/j.ijthermalsci.2021.107000.

Inglese, G. and Olmi, R. (2017) Nondestructive evaluation of spatially varying internal heat transfer coefficients in a tube, *International Journal of Heat and Mass Transfer*, 108, pp. 90–96. https://doi.org/10.1016/j.ijheatmasstransfer.2016.12.001.

Kadam, A. R., Prabhu, S. V. and Hindasageri, V. (2018) Simultaneous estimation of heat transfer coefficient and reference temperature from impinging flame jets, *International Journal of Thermal Sciences*, 131(May), pp. 48–57. https://doi.org/10.1016/j.ijthermalsci.2018.05.017.

Malinowski, Z., Telejko, T., Hadała, B., Cebo-Rudnicka,A. and Szajding, A. (2014) Dedicated three dimensional numerical models for the inverse determination of the heat flux and heat transfer coefficient distributions over the metal plate surface cooled by water, *International Journal of Heat and Mass Transfer*, 75, pp. 347–361. https://doi.org/10.1016/j.ijheatmasstransfer.2014.03.078.

Mobtil, M., Bougeard, D. and Russeil, S. (2018) Experimental study of inverse identification of unsteady heat transfer coefficient in a fin and tube heat exchanger assembly, *International Journal of Heat and Mass Transfer*, 125, pp. 17–31. https://doi.org/10.1016/j.ijheatmasstransfer.2018.04.028.

Mohebbi, F. and Evans, B. (2020) 'Simultaneous estimation of heat flux and heat transfer coefficient in irregular geometries made of functionally graded materials', *International Journal of Thermofluids*, 1–2, pp. 1–13. https://doi.org/10.1016/j.ijft.2019.100009.

Razzaghi, H., Kowsary, F. and Ashjaee, M. (2019) Derivation and application of the adjoint method for estimation of both spatially and temporally varying convective heat transfer coefficient, *Applied Thermal Engineering*, 154, pp. 63–75. https://doi.org/10.1016/j.applthermaleng.2019.03.068.

Sablani, S.S., Kacimov, A., Perret, J., Mujumdar, A.S.and Campo, A. (2005) Non-iterative estimation of heat transfer coefficients using artificial neural network models, *International Journal of Heat and Mass Transfer*, 48(3–4), pp. 665–679. https://doi.org/10.1016/j.ijheatmasstransfer.2004.09.005.

Tourn, B. A., Álvarez Hostos, J. C. and Fachinotti, V. D. (2021) A modified sequential gradient-based method for the inverse estimation of transient heat transfer coefficients in non-linear one-dimensional heat conduction problems, *International Communications in Heat and Mass Transfer*, 127. https://doi.org/10.1016/j.icheatmasstransfer.2021.105488.

Wang, Y., Luo, X., Yu, Y. and Yin, Q. (2017) Evaluation of heat transfer coefficients in continuous casting under large disturbance by weighted least squares Levenberg–Marquardt method, *Applied Thermal Engineering*, 111, pp. 989–996. https://doi.org/10.1016/j.applthermaleng.2016.09.154.

Zhang, J. and Delichatsios, M. A. (2009) Determination of the convective heat transfer coefficient in three-dimensional inverse heat conduction problems, *Fire Safety Journal*, 44(5), pp. 681–690. https://doi.org/10.1016/j.firesaf.2009.01.004.

Zhao-Hui, R. *et al.* (2021) Determining the heat transfer coefficient during the continuous casting process using stochastic particle swarm optimization, *Case Studies in Thermal Engineering*, 28, p. 101439. https://doi.org/10.1016/j.csite.2021.101439.

10

Temperature Measurement

Random measurement errors and their impact on inverse heat conduction problem (IHCP) solutions were considered in previous chapters. However, in all physical measurement systems, a disturbance is created by the presence of the sensing element. This introduces a bias error in the measurement that can result in incorrect estimation of the surface disturbance in the IHCP.

In this chapter, the concept of temperature correction kernels is introduced. This idea can be used in at least two ways. One application is to directly use a response function that includes the dynamics of the sensing system in the convolution or superposition procedure to solve the IHCP. Another application is to directly correct the indicated temperature from the sensor to get a better estimate of the true temperature in the undisturbed region away from the sensing element.

Two specific types of thermocouple installations are considered in this chapter. One very common application results when a temperature sensor is embedded within a surrounding medium that has much lower thermal conductivity than the sensor and wires. Another application, which is often referred to as an intrinsic thermocouple, results when a thermocouple junction is formed by directly joining thermocouple wire to a metallic surface.

The plan for this chapter is as follows. First, a discussion of the effect of sensor wires on temperature measurement will be given. The subsequent section will introduce the idea of the temperature correction kernels and explain how they can be used either directly in the solution of the IHCP or to correct temperature readings. The final section discusses the unsteady surface element (USE) method which can be used to generate information needed for correction kernels in certain applications, such as the intrinsic thermocouple application.

10.1 Introduction

The very act of measurement necessarily alters the system which is being measured. This principle is true in high-energy particle physics as attested by the Heisenberg uncertainty principle. It is also especially true for temperature measurements using wired sensors, including thermocouples, thermistors, and other sensing devices.

A thermocouple or other temperature sensing device can only indicate its own temperature. Unless the sensor's wiring is negligibly small or has material properties closely matched to the thermal properties of the surrounding medium, the temperature field around the sensing element will be significantly altered by the presence of the sensor and wiring. In most cases, the sensor temperature cannot be the same as the material which it is supposed to measure.

10.1.1 Subsurface Temperature Measurement

Figure 10.1a depicts a common situation of a sensor and wire imbedded in a medium with lower thermal conductivity. The isotherms suggest the distortion of the temperature field and these isotherms are decreasing in magnitude from the top of the figure to the bottom of the figure. Distortion in the temperature field is most pronounced when a sensor is installed perpendicular to the heated surface, as in Figure 10.1a, and can be reduced, but not eliminated, by installing the sensor parallel to the isotherms (Beck 1961b; Woolley 2008; Pope et al. 2022). The presence of the sensor and wire results in a depression in the temperature at the sensing element causing a biased temperature reading.

Inverse Heat Conduction: Ill-Posed Problems, Second Edition. Keith A. Woodbury, Hamidreza Najafi, Filippo de Monte, and James V. Beck.
© 2023 John Wiley & Sons, Inc. Published 2023 by John Wiley & Sons, Inc.

(a)

(b)

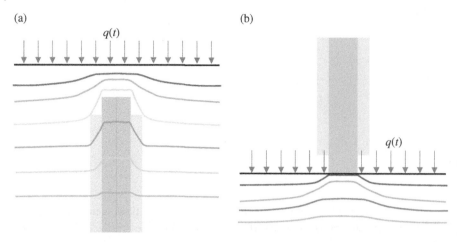

Figure 10.1 Isotherms around an idealized temperature sensor: (a) sensor imbedded in low conductivity domain and (b) intrinsic thermocouple on metal surface.

Errors in temperature measurement associated with subsurface thermocouple installations have been investigated since the early 1960s. Simple cylindrical wires or voids are typically used to model the presence of the sensor. Beck and others (Beck and Hurwicz 1960; Beck 1961a, 1962) performed numerical calculations to quantify the error in the measurement using cylindrical models considering intimate thermal contact of the sensor with the surrounding medium. Attia and Kops (1986, 1988, 1993; Attia et al. 2002) used cylindrical models to examine the effect of air-filled voids for installation of sensors on heat flow in the material. Woolley (2008; Woolley and Woodbury 2011) also considered cylindrical models but concentrated on detailed three-dimensional finite element models to characterize distortion in temperature fields surrounding thermocouple installations. Most recently Pope et al. (2022) used cylindrical models allowing for contact resistance between the sensor and surrounding medium to develop a correction method for temperature measurement.

10.1.2 Surface Temperature Measurement

Figure 10.1b suggests an idealized intrinsic thermocouple which is created when a wire is connected directly via welding or brazing to a metal surface. The presence of the wire provides a path for heat flow from the surface, resulting in a temperature depression around the thermocouple. The subsurface isotherms in the sketch suggest this effect.

The response of intrinsic thermocouples has been investigated through numerical calculations since the 1960s (Henning and Parker 1967; Gat et al. 1975). Significant interest in these sensors in high-temperature heating investigations, such as fires and quenching, has prompted continued research (Litkouhi and Beck 1985; Nakos 2003; Tszeng and Saraf 2003; Xu and Gadala 2007; Feng et al. 2020).

10.2 Correction Kernel Concept

The correction kernel idea follows from superposition or convolution concepts for linear problems. Figure 10.2 illustrates a simplified model of a subsurface sensor imbedded in a domain. The temperature indicated by the sensor is designated T_p and the figure suggests this value associated with the tip of the sensor. The subsurface sensor is used here for illustration but these concepts also apply to the intrinsic thermocouple depicted in Figure 10.1b. The corresponding undisturbed temperature is designated $T_{p\infty}$ and is the temperature of the surrounding domain at the same depth $z = E$ but "far away" from the sensor, as suggested in Figure 10.2. For an intrinsic thermocouple, $T_{p\infty}$ is the temperature of the surface "far away" from the thermocouple junction.

10.2.1 Direct Calculation of Surface Heat Flux

For the domain and sensor at a uniform initial temperature T_i, the temperature response of the sensor $T_p(t)$ to an arbitrary surface heat flux input $q(t)$ can be found, for linear problems, by superposition using the convolution integral

$$T_p(t) - T_i = \int_{\tau=0}^{\tau=t} q(\tau) \frac{\partial \phi_p(t-\tau)}{\partial t} d\tau \tag{10.1}$$

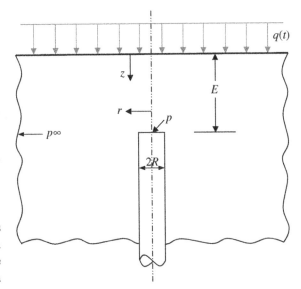

Here, $\phi_p(t)$ is the response of the sensing thermocouple to a unit step increase in the surface heat flux. Typically, $\phi_p(t)$ is determined from an appropriate numerical model simulated with a constant heat flux boundary condition, but could be determined experimentally (see Section 10.2.5).

Eq. (10.1) can be represented in discrete form, similar to Eq. (3.54), as

$$\mathbf{T}_p - T_i\{\mathbf{1}\} = \mathbf{X}_{\phi_p}\mathbf{q} \tag{10.2}$$

If measurements are available for the temperatures $\mathbf{T}_p$, then this equation can be used directly to solve for the unknown excitation $\mathbf{q}$. Of course, this relation is also inherently ill-posed, hence some form of regularization is necessary. Any of the IHCP solution methods presented in Chapter 4 can be applied to invert Eq. (10.2).

Figure 10.2 Schematic of idealized subsurface thermocouple installation and nomenclature of Beck (1968) and Woolley and Woodbury (2008).

10.2.2 Temperature Correction Kernels

In some applications, determination of the surface heat flux may not be the objective of an experiment, but an improved estimate for the true temperature of the domain surrounding the sensor is desired. A convolution integral of the correction kernel, $H(\tau)$, and the indicated sensor temperature, T_p, can be derived by considering the dependence of the heat flux on the undisturbed temperature (Eq. (3.54)) in addition to that of the sensor in Eq. (10.1). By eliminating the unknown heat flux from these two expressions, the convolution integral in Eq. (10.3) results (see (Beck 1968) or (Woolley et al. 2008) for derivation).

$$T_{p\infty}(t) - T_p(t) = \int_{\tau=0}^{\tau=t} H(\tau) \frac{\partial T_p(t-\tau)}{\partial t} d\tau \tag{10.3}$$

Eq. (10.3) has two applications. One application is to determine the correction kernel, $H(\tau)$. Once the correction kernel is known for a particular sensor installation, it can be used directly to correct experimental temperature measurements.

10.2.2.1 Determining the Correction Kernel

To determine the correction kernel, "data" are needed for the sensed temperature and the corresponding far-field temperature, $T_{p\infty}$. These data generally are obtained through numerical simulation of the sensor installation and surrounding medium. As noted by Beck (Beck 1968), the $H(\tau)$ kernel is independent of the imposed heating on the surface, as it only depends on the difference between the far field and sensed temperatures as a function of time. Therefore, any external heating excitation may be used, including the simple case of a step (constant) heat flux input. Only the simultaneous histories of $T_p(t)$ and $T_{p\infty}(t)$ are needed.

When a constant unit heat flux is used to generate these data, the response function $T_p(t)$ in Eq. (10.3) is the same as $\phi_p(t)$ in Eq. (10.1). The convolution integral in Eq. (10.3) can then be written in discrete form using the notation in Eq. (10.2):

$$\mathbf{T}_{p\infty} - \mathbf{T}_p = \mathbf{X}_{\phi_p}\mathbf{H} \tag{10.4}$$

Here, $\mathbf{H}$ is a vector of the correction kernel function components at the discrete times corresponding to the data on the left side, and $\mathbf{X}_{\phi_p}$ is the sensitivity matrix based on the sensed temperature from the simulation, and $T_p=\phi_p$ for the unit step heat flux. Eq. (10.4) is inherently ill-posed, and any of the regularization methods from Chapter 4 may be used to invert the equation to determine $\mathbf{H}$.

In addition to ill-posedness of Eq. (10.4), the "data" on the left side of the equation are typically computed from simulation with finite precision. Further, the data for the solution are the difference between two computer simulation outputs, which amplifies errors in the individual values for $\mathbf{T}_{p\infty}$ and $\mathbf{T}_p$, especially when time steps in the data are small. Woolley (2008) found that pre-smoothing of these data using mollification and generalized cross-validation (Murio 1993; Murio et al. 1998; Zhan et al. 2001) was needed, in addition to a relatively high degree of regularization in his solution of Eq. (10.4) using the function specification method.

10.2.2.2 Correcting Temperature Measurements

The second application of Eq. (10.3) is to determine $T_{p\infty}(t)$ from measurements of $T_p(t)$ after the correction kernel $\mathbf{H}=\{H(t_i)\}$ has been determined. This can be accomplished directly by integration of the equation using the known correction kernel. This problem is not ill-posed and can be computed directly in discrete terms.

$$T_{p\infty}(t_M) = T_p(t_M) + \sum_{i=1}^{M} H_i \Delta T_{p,M-i} \tag{10.5a}$$

where

$$\Delta T_{p,k} = T_{p,k+1} - T_{p,k} \tag{10.5b}$$

Eq. (10.5a) can be expressed in matrix form as

$$\mathbf{T}_{p\infty} = \mathbf{T}_p + \left[\Delta \mathbf{T}_p\right]\{\mathbf{H}\} \tag{10.5c}$$

Here, $\Delta \mathbf{T}_p$ is a full lower triangular matrix with identical structure and entries as the sensitivity matrices $\mathbf{X}$ used in solution of the IHCP. The entries in the matrix are differences of the sensed temperatures at subsequent times:

$$\left[\Delta \mathbf{T}_p\right] = \begin{bmatrix} \Delta T_{p,0} & 0 & 0 & 0 & 0 \\ \Delta T_{p,1} & \Delta T_{p,0} & 0 & \cdots & 0 \\ \Delta T_{p,2} & \Delta T_{p,1} & \Delta T_{p,0} & \ddots & \vdots \\ \vdots & \ddots & \ddots & \ddots & 0 \\ \Delta T_{p,n-1} & \cdots & \Delta T_{p,2} & \Delta T_{p,1} & \Delta T_{p,0} \end{bmatrix} \tag{10.5d}$$

10.2.3 2-D Axisymmetric Model

A simple axisymmetric model created in ANSYS Workbench for solution with ANSYS Mechanical Transient Thermal analysis is used to demonstrate these correction kernel concepts. Representative temperature contours from the ANSYS model are illustrated in Figure 10.3. Examples in this section make use of this model.

Example 10.1 The axisymmetric model is used to generate T_p and $T_{p\infty}$ data for the parameters shown in Table 10.1. Data, to five significant figures, are obtained as shown in Table 10.2.

Use these data and first-order Tikhonov regularization to compute the correction kernel vector $\mathbf{H}$. Use a Tikhonov regularization coefficient $\alpha_{Tik} = 10$.

Solution

The sensitivity matrix, $\mathbf{X}$, is constructed from the differences between the entries of the T_p vector, prepended with a "0" (just like the $\Delta\phi$ values used to compute the sensitivity matrix for heat flux estimation). The first few values of this vector of differences are:

$$\begin{aligned} \Delta\phi_p &= \lfloor 7.7962-0 \quad 20.404-7.7962 \quad 33.028-20.404 \quad 45.188-33.028 \quad \cdots \rfloor^T \\ &= \lfloor 7.7962 \quad 12.608 \quad 12.624 \quad 12.160 \quad \cdots \rfloor^T \end{aligned} \tag{10.6}$$

Figure 10.3 ANSYS mesh and typical simulation results.

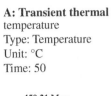

A: Transient thermal
temperature
Type: Temperature
Unit: °C
Time: 50

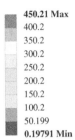

450.21 Max
400.2
350.2
300.2
250.2
200.2
150.2
100.2
50.199
0.19791 Min

Table 10.1 Simulation parameters for Example 10.1.

Sensor depth, E, mm	0.75	Sensor radius, R, mm	0.375
Wire length, L, mm	20	Domain radius, R_D, mm	8
k_w, W/m-K	25	k, W/m-K	0.625
C_w, J/m³-K	2.1E6	C, J/m³-K	4.2E6
q_0, W/m²	1E5		

Table 10.2 Sensor and undisturbed temperature data for heating with $q = const = 10^5$ W/m².

t,s	T_p,C	T_p,C	t,s	T_p,C	T_p,C
1	7.796	21.66	11	118.95	220.16
2	20.404	50.63	12	128.31	234.05
3	33.028	76.399	13	137.46	247.41
4	45.188	99.423	14	146.40	260.28
5	56.856	120.35	15	155.16	272.72
6	68.074	139.64	16	163.74	284.76
7	78.892	157.62	17	172.16	296.43
8	89.354	174.51	18	180.44	307.78
9	99.498	190.48	19	188.56	318.81
10	109.35	205.66	20	196.56	329.57

These values are used to create the 20×20 matrix $\mathbf{X}_{\phi_p}$. The first few rows of this matrix are

$$
\mathbf{X}_{\phi_p} = \begin{bmatrix}
7.7962 & 0 & 0 & 0 & 0 & 0 & \cdots \\
12.608 & 7.7962 & 0 & 0 & 0 & 0 & \cdots \\
12.624 & 12.608 & 7.7962 & 0 & 0 & 0 & \cdots \\
12.160 & 12.624 & 12.608 & 7.7962 & 0 & 0 & \cdots \\
\vdots & \vdots & \ddots & \ddots & \ddots & \ddots
\end{bmatrix}_{20 \times 20}
\tag{10.7}
$$

The $\mathbf{X}_{\phi_p}$ matrix is used with first-order Tikhonov regularization to generate the filter matrix:

$$
\mathbf{F}_{\phi_p} = \left(\mathbf{X}_{\phi_p}^T \mathbf{X}_{\phi_p} + \alpha_{\mathrm{Tik}} \mathbf{H}_1^T \mathbf{H}_1 \right)^{-1} \mathbf{X}_{\phi_p}^T
\tag{10.8}
$$

The first few entries of this matrix are:s

$$
\mathbf{F}_{\phi_p} = \begin{bmatrix}
0.0584 & 0.0368 & -0.0027 & -0.0033 & 0.0002 & 0.0003 & \cdots \\
-0.0576 & 0.0180 & 0.0369 & 0.0007 & -0.0033 & -0.0001 & \cdots \\
-0.0041 & -0.0562 & 0.0207 & 0.0370 & 0.0005 & -0.0033 & \cdots \\
0.0056 & -0.0003 & -0.0563 & 0.0204 & 0.0370 & 0.0005 & \cdots \\
\vdots & \vdots & \vdots & \vdots & \vdots & \vdots & \vdots
\end{bmatrix}_{20 \times 20}
\tag{10.9}
$$

The $\mathbf{F}_{\phi_p}$ matrix multiplies into the "data" of $\mathbf{T}_{p\infty} - \mathbf{T}_p$ in the inversion of Eq. (10.4) to find the kernel vector $\mathbf{H}$:

$$
\mathbf{H} = \left(\mathbf{X}_{\phi_p}^T \mathbf{X}_{\phi_p} + \alpha_{\mathrm{Tik}} \mathbf{H}_1^T \mathbf{H}_1 \right)^{-1} \mathbf{X}_{\phi_p}^T \{ \mathbf{T}_{p\infty} - \mathbf{T}_p \} = \mathbf{F}_{\phi_p} \{ \mathbf{T}_{p\infty} - \mathbf{T}_p \}
\tag{10.10}
$$

The result of these calculations is the following sequence:

H_{1-7}	1.6613	1.1935	0.9624	0.8790	0.8156	0.7598	0.7151
H_{8-15}	0.6780	0.6468	0.6195	0.5945	0.5726	0.5532	0.5362
H_{16-20}	0.5199	0.5053	0.4902	0.4775	0.4663	0.4574	

Discussion

Unlike many examples in this text, these calculations are carried out using dimensional data. Units on the entries of $\mathbf{X}_{\phi_p}$ are Kelvin, and those on $\mathbf{F}_{\phi_p}$ are K^{-1}. The correction kernel $\mathbf{H}$ is dimensionless.

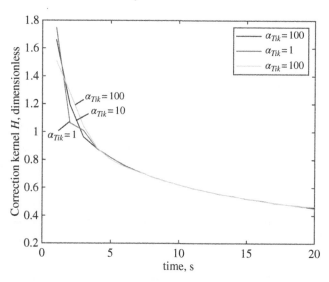

Figure 10.4 Correction kernel **H** for various regularization coefficients.

Selection of the proper amount of regularization to use in the inversion is difficult. In this example, the value of $\alpha_{Tik} = 10$ was specified. This is a relatively large value in comparison to those used in solution of the IHCP in Chapter 7. In practice, the regularization should be large enough to produce a smooth $H(t)$ curve. Experience suggests values from a rather broad range of the regularization parameter produce similar results.

Figure 10.4 displays results for **H** obtained using a range of regularization parameters. All of the values are much larger than used for IHCP solution. The results for the $\alpha_{Tik} = 1$ display irregularities in the result, indicating the regularization coefficient is too small. Results for $\alpha_{Tik} = 100$ are very smooth, but values are seen diminished compared to the nominal curve.

Another tool for selection of the regularization coefficient is consideration of a known heat flux input into the simulation model. See Example 10.2.

Example 10.2 Woolley (2008; Woolley et al. 2008) devised a simple heat flux model to simulate heating at the surface of a mold during solidification of casting. This heating function is

$$q(t) = \begin{cases} q_0 t & 0 \leq t \leq 1.0 \\ q_0(1 - e^{-0.013t}) & t > 1.0 \end{cases}, \quad t \text{ in seconds} \tag{10.11}$$

This heating function with $q_0 = 10^5 \, \text{W/m}^2$ is used in the ANSYS model for a 20-second time interval and the same parameters as Example 10.1. The data obtained are shown in Table 10.3.

Use these data and knowledge of the exact heat flux to determine an appropriate value for the regularization coefficient, α_{Tik}.

Solution

For this application, Eq. (10.2) is utilized. Data for "T_p" and "$T_{p\infty}$" from a step change in surface are needed to formulate the $\mathbf{X}_{\phi_p}$ matrix. These are the same data as in Example 10.1, and the resulting $\mathbf{X}_{\phi_p}$ matrix is also the same as before. The $\mathbf{F}_{\phi_p}$ is calculated in the same way as in Example 10.1 (using Eq. (10.8)) and now only depends on the value of the regularization coefficient α_{Tik}.

Figure 10.5 displays the exact heat flux along with three reconstructions of this heat flux. For each curve, the estimated heat flux is computed from inversion of Eq. (10.2) using Tikhonov regularization:

$$\hat{\mathbf{q}} = \left[\mathbf{F}_{\phi_p}(\alpha_{Tik}) \right] \{ \mathbf{T}_p - \mathbf{T}_0 \} \tag{10.12}$$

Table 10.3 Sensor and undisturbed temperature data for heating with Eq. (10.11).

t,s	T_p,C	T_p,C	t,s	T_p,C	T_p,C
1	3.2190	9.3354	11	106.36	197.00
2	14.492	37.377	12	114.33	208.43
3	26.868	63.784	13	121.98	219.15
4	38.740	86.828	14	129.32	229.24
5	49.971	107.24	15	136.37	238.74
6	60.594	125.58	16	143.16	247.72
7	70.662	142.26	17	149.70	256.20
8	80.230	157.55	18	155.99	264.23
9	89.344	171.68	19	162.07	271.85
10	98.045	184.78	20	167.92	279.08

Figure 10.5 Heat flux and reconstructions for varying values of α_{Tik} in Example 10.2.

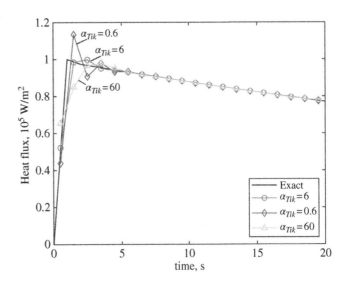

In Eq. (10.12), $\mathbf{T}_p$ refers to data from the known heating function simulation in Table 10.3, and $\mathbf{F}_{\phi_p}(\alpha_{Tik})$ indicates the dependence of the matrix on the parameter α_{Tik}. These results are obtained using piecewise constant assumption about the heat flux (see calculation of the $\mathbf{X}_{\phi_p}$ matrix in Eqs. (10.6) and (10.7)), hence the estimated heat flux components q_i are shown at times $t_i - \Delta t/2$.

Discussion

The "optimal" results in Figure 10.5 are for $\alpha_{Tik} = 6$ and were determined holistically by varying this parameter until the estimated heat fluxes best matched the exact curve. The other curves for $\alpha_{Tik} = 0.6$ and $\alpha_{Tik} = 60$ demonstrate effects of under-regularization and over-regularization on the estimates.

Since the filter matrix $\mathbf{F}_{\phi_p}$ is common to both the heat flux estimation and correction kernel computation, the idea of using a companion simulation with a known heat flux, demonstrated in this example, can be used to guide the selection of amount of regularization needed for determining the correction kernel $\mathbf{H}$.

The final application of the correction kernel concept is to directly correct measured temperatures using a previously determined $\mathbf{H}$ for particular sensor installation. The following example illustrates this application.

Example 10.3 Use the correction kernel found in Example 10.1 to correct the sensor readings T_p in Example 10.2 and compare the corrected values to the undisturbed $T_{p\infty}$ values.

Solution

Eqs. (10.5c) and (10.5d) are used to compute the corrected values. The matrix of $\Delta \mathbf{T}_p$ in Eq. (10.5d) is computed using the column of T_p values in Example 10.2 in a manner identical to the calculation of $\mathbf{X}_{\phi_p}$ in Example 10.1. First, the ΔT_p vector is computed using differences of the T_p vector, just as the $\Delta \phi_p$ values were computed in Example 10.1. The first few values are

$$
\begin{aligned}
\Delta T_p &= \lfloor 3.219 - 0 \quad 14.492 - 3.219 \quad 26.868 - 14.492 \quad 38.740 - 26.868 \quad \cdots \rfloor^T \\
&= \lfloor 3.219 \quad 11.273 \quad 12.376 \quad 11.231 \quad \cdots \rfloor^T
\end{aligned}
\tag{10.13}
$$

These values are used to build the $\Delta \mathbf{T}_p$ matrix. The first part of this matrix is

$$
\Delta \mathbf{T}_p =
\begin{bmatrix}
3.219 & 0 & 0 & 0 & a & 0 & \cdots \\
11.273 & 3.219 & 0 & 0 & 0 & 0 & \cdots \\
12.376 & 11.273 & 3.219 & 0 & 0 & 0 & \cdots \\
11.231 & 12.376 & 11.273 & 3.219 & 0 & 0 & \cdots \\
\vdots & \vdots & \ddots & \ddots & \ddots & \ddots
\end{bmatrix}_{20 \times 20}
\tag{10.14}
$$

Finally, Eq. (10.5c) can be used to compute the correction by matrix multiplication and vector addition using the T_p data from Example 10.2:

$$
\mathbf{T}_{p\infty} = \mathbf{T}_p + \left[\Delta \mathbf{T}_p\right]\{\mathbf{H}\}
\tag{10.15}
$$

The results of these calculations are the following temperatures:

$T_{p\infty,1-7}$	8.57	37.06	63.98	86.91	107.24	125.59	142.27
$T_{p\infty,8-15}$	157.57	171.69	184.80	197.01	208.44	219.17	229.25
$T_{p\infty,16-20}$	238.74	247.73	256.23	264.23	271.86	279.07	

The root mean squared error between these values and the $T_{p\infty}$ values in Example 10.2 is 0.1921 K. The comparison is shown graphically in Figure 10.6.

Discussion

The temperature data from Example 10.2 are from the same numerical model but with a different heat flux excitation than Example 10.1, which provides a fair test of the ability of the results from the step heat flux to generate the correction kernel to apply to a different "experiment." However, no measurement errors are considered in this calculation. Random errors with zero mean and Gaussian distribution having standard deviation $\sigma_{noise} = 1.0$ K were added to the T_p data from Example 10.2 and the calculations repeated. For this case, the RMS error in the corrected temperature is 2.4092 K, and the graphical comparison is shown in Figure 10.7.

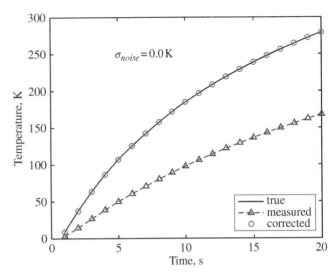

Figure 10.6 Measured and corrected sensor readings in Example 10.3 using exact data.

The preceding three examples illustrate how a numerical model of a sensor installation can improve estimates for surface heat flux and/or indicated sensor temperature. Of course, the improvement can be no better than the representation of the physical system by the numerical model. For example, the previous simple axisymmetric model does not account for contact resistance between the imbedded sensor and the surrounding medium. If this resistance is significant, then the model will be biased and the results will be correspondingly distorted. However, any reasonable model to incorporate the sensor dynamics will improve results over those obtained by completely ignoring such effects.

10.2.4 High Fidelity Models and Thermocouple Measurement

Axisymmetric models are relatively simple to create and exercise to generate results for heat flux determination or temperature correction. However, more realistic detailed models of sensor installation can provide better information for improving these calculations.

Woolley studied thermocouple installations for investigating heat transfer to sand molds during casting (Woolley 2008). Two significant aspects of his work will be reviewed here: utilization of detailed (high fidelity) three-dimensional computer models to generate correction kernels and determination of the location of the "sensed" temperature in a conventional thermocouple bead weld.

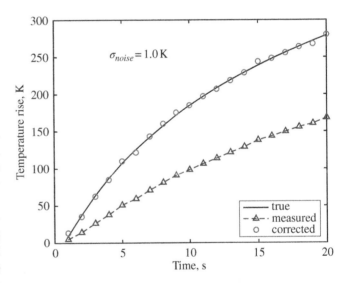

Figure 10.7 Measured and corrected sensor readings in Example 10.3 using noisy data with $\sigma_{noise} = 1.0$ K.

10.2.4.1 Three-Dimensional Models

The three-dimensional geometry investigated by Woolley is shown in Figure 10.8 and includes separate wires for each leg of a thermocouple and an ellipsoidal weld bead. The ellipsoid geometry for the bead was chosen rather than a sphere based on measurements of commercially available 24 gage thermocouples.

Simulation using a constant heat input $q(t)$ for both the three-dimensional geometry of Figure 10.8 and an "equivalent" axisymmetric model as in Figure 10.2 produced the temperature histories displayed in Figure 10.9a. The "equivalent" axisymmetric model incorporated a single wire with the same cross-sectional area as the two thermocouple wires ($R = \sqrt{2}r_w$) with thermal properties equal to the average of those of the two thermocouple wires.

Figure 10.9b displays the difference between the undisturbed temperature and the "sensed" temperature of the models. The errors predicted by the axisymmetric model are significant (60–80 K), but the three-dimensional model suggests the actual errors are much larger (80–115 K).

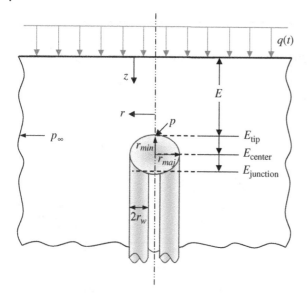

Figure 10.8 Three-dimensional geometry used by Woolley et al. (2008).

(a)

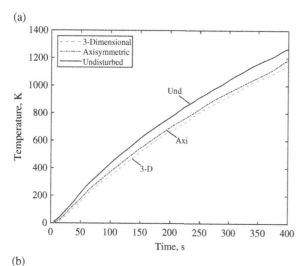

(b)

Figure 10.9 Comparison of axisymmetric and three-dimensional sensor models: (a) calculated temperatures and (b) resulting apparent errors.

Three-dimensional models also allow investigation of different sensor installations, such as parallel to the surface rather than perpendicular. Using three-dimensional geometries, Woolley was able to demonstrate the significant reduction in sensor error for sensor wire installation parallel to isotherms. In Woolley's simulations using the exponential heat flux input (see Example 10.2), sensor errors for parallel installations were on the order of 30 K, while those for the perpendicular arrangement were about 80 K.

10.2.4.2 Location of Sensed Temperature in Thermocouples

An important but generally neglected question regarding thermocouple thermometry is "where is the sensed temperature in/on the thermocouple bead?" The answer may be unimportant for very small thermocouples, but in many applications (e.g. metal casting and fire research) the size of thermocouple required to survive the experimental procedure is painfully finite. The location of the measurement must be known, whether, or not, simulations are performed to account for sensor errors.

Woolley performed detailed analysis of commercially available 24-gage chromel–alumel thermocouples to determine the composition of the thermocouple bead and performed experimentation to determine *where* on the thermocouple temperature is measured (Woolley et al. 2009). A heated thermocouple was allowed to cool in still air while an infrared camera was used to record the temperature of the surface of the thermocouple. Simultaneously, a data acquisition (DAQ) system was used to record the indicated temperature of the thermocouple, and the measured temperature history was compared to the infrared imaging. Three primary sites on the thermocouple bead, indicated in Figure 10.8 as the tip, the center, and the junction, were investigated. Comparison of the measured thermocouple signal from the DAQ to time histories recovered from the infrared observations at these sites concluded that the junction site correlated most closely with the indicated temperature. This conclusion is important for determining the correct "sensor location" for IHCP or kernel correction analysis.

10.2.5 Experimental Determination of Sensitivity Function

In Section 10.2, the response of a temperature sensor to a step heat flux, ϕ_p, was determined through numerical modeling and simulation. However, with suitable heating sources, a step change in surface heat flux can be imposed on a domain with a thermal sensor, and this response function ϕ_p may be determined experimentally. With knowledge of this response, the same physical environment with sensor may be exposed to an unknown transient heat flux, and the surface heat flux can be determined from inversion of Eq. (10.2). However, it is not possible to further determine the correction kernel, **H**, unless a reliable, independent means of determining the corresponding temperature of the "undisturbed" medium is available.

Frankel and coworkers promoted the idea of calibrating IHCP sensors experimentally (Chen et al. 2016; Frankel et al. 2017; Frankel and Keyhani 2018). Although their preferred method of IHCP analysis is quite different from techniques presented in this text, their work is based on the notion of constructing an IHCP "sensor" and calibrating it by exposing the sensing surface to a constant surface heat flux and capturing the sensor response. This response is subsequently used to solve the convolution integral (Eq. (10.1)) for the heat flux, $q(t)$, when the sensor is later exposed to an unknown heating source. This practice removes the need for exact knowledge of the location of a subsurface sensor or the precise thermal properties of the medium. All the necessary information is contained in the experimentally measured step heat flux response function, ϕ_p.

In a comprehensive report, Myrick et al. (2017) used a high-powered laser to experimentally determine the step heat flux response. The base material of their IHCP "sensor" was 304 stainless steel, which was instrumented with two 30 gage surface thermocouples and two subsurface sensors of the same size at different depths. The surface of the stainless steel was coated with a high absorptivity black paint.

Pope et al. (2021, 2022) used an experimental apparatus combined with detailed thermocouple modeling to determine a measurement correction technique. Although their focus is on establishing the correction methodology, their experimental apparatus is used to impose a step heat flux. In their work, a Mass Loss Calorimeter is used to impose a radiant flux of either 5 or 60 kW/m^2 onto a surface. Axisymmetric models of Inconel sheathed thermocouples installed perpendicular to the heated surface were used to quantify the error in measurement. Surprisingly, measurements from experiments showed that modeling the thermocouple as a solid cylinder of Inconel produced "sensed" temperatures very close to the experimental values, whereas models using cylinders with properties based on weighted averages of composition are more approximate.

During calibration experiments, the heating power of the laser source is known, but the net flux is needed for the calibration, not the incident source, so care must be taken to estimate emitted radiation and convection. The following relation can be used to compute the net heat flux when radiative and convective properties of the surface are known:

$$q_{net} = \alpha q_{incident} - \varepsilon\sigma\left(T_{sur}^4 - T_\infty^4\right) - h_c\left(T_{sur} - T_\infty\right) \tag{10.16}$$

Here, α is the absorptivity of the surface for the wavelength of incident radiation, ε is the emissivity of the surface for long-wavelength reradiation, and h_c is the convection coefficient for the exposed surface. To get a true (net) step heat flux input, the magnitude of the incident radiation must be modulated during the heating in response to the increasing surface temperature, which is very challenging.

If a constant net heat flux can be imposed, the measured sensor output from this experiment can be used to construct the $\mathbf{X}_{\phi_p}$ for use in Eq. (10.2) to estimate a surface heat flux from a subsequent experiment with an arbitrary surface heating.

10.3 Unsteady Surface Element Method

The USE method was introduced by Keltner and coworkers (Keltner and Bickle 1976; Keltner and Beck 1981, 1983) and later improved by Litkouhi and Beck(1985). Cole et al. (2011) devotes a chapter to this topic. Keltner's motivation is the intrinsic thermocouple problem and the need to characterize errors in measurements made with thermal sensors attached directly to a metal surface.

A brief outline of the USE method follows in order to introduce the results for the intrinsic thermocouple, which are presented in the next subsection. Interested readers can consult Cole et al. (2011, Chapter 12) for additional details on the USE method.

The USE method will determine the interface conditions (either temperature or heat flux) for two bodies in contact. The contact may be thermally perfect or imperfect, and the bodies may have different geometries. The fundamental idea of the method is based on superposition principles (which can be interpreted as Duhamel's summation or Green's function solutions) so the method is restricted to linear problems. By constructing a superposition expression for the response of each body to a disturbance over the contacting area and enforcing continuity of temperature and heat flux across the interface, the unknown interfacial condition can be determined.

In its simplest application, the USE method considers a single temperature or heat flux at the interface, and for these applications, analytical solutions can be found for some geometries. More generally, contacting regions between bodies are subdivided into smaller regions, and a system of equations requiring numerical solution results. See (Litkouhi and Beck 1985; Cole et al. 2011) for more detail on these applications.

The USE method can be formulated either in terms of the interfacial heat flux or the interfacial temperature. Only the flux-based method will be discussed here. For either approach, once the flux (or temperature) is determined from the USE solution, then the corresponding temperature (or flux) can be determined, if desired.

Consider two bodies in contact and that the bodies are insulated on all surfaces except the common contact area. For each body, the *influence function*, $\phi_i(\mathbf{r}, t)$, is needed (here, $\mathbf{r}$ denotes generalized spatial coordinate: either rectangular, cylindrical, or spherical). The influence function is the temperature response within the body "i" due to a unit step heat flux into the body at the interface. For body "i" initially at temperature T_{0i}, the response in the body to an arbitrary heat flux at the contacting region $q(t)$ may be found using the superposition integral:

$$T_i(\mathbf{r}, t) = T_{0i} + \int_{\tau=0}^{\tau=t} q(\tau)\frac{\partial \phi_i(\mathbf{r}, t-\tau)}{\partial t}d\tau = T_{0i} + \frac{\partial}{\partial t}\int_{\tau=0}^{\tau=t} q(\tau)\phi_i(\mathbf{r}, t-\tau)d\tau \tag{10.17}$$

By writing an expression in the form of Eq. (10.17) for each body, and enforcing continuity of temperature between the bodies, one expression with the unknown interfacial heat flux will result. In some cases, this resulting expression can be solved analytically, often through the use of Laplace transforms. In other cases, numerical solution will be required.

The following subsection reviews the development of the USE method for the intrinsic thermocouple problem. The solution of that problem will result in the time-varying temperature of the sensor due to a step change in the surrounding substrate temperature. These temperature histories will then be used to compute the correction kernel $\mathbf{H}$ which can subsequently be used to correct the surface temperature readings due to an arbitrary change in the surface temperature.

10.3.1 Intrinsic Thermocouple

The geometry for the intrinsic thermocouple is seen in Figure 10.10. The subscripts "w" and "b" have been adopted to refer to the wire and body, respectively. The thermal properties are thermal conductivity, k, and volumetric heat capacity, $C = \rho c_p$. All surfaces except the contact area are assumed insulated. The wire is assumed to be a semi-infinite cylinder with a one-dimensional temperature distribution; its temperature at $z = 0$ is of interest. The body is a semi-infinite solid, and the area-averaged temperature over the contact area is of interest.

An equation for the interface heat flux is found beginning by writing an expression using Eq. (10.17) for the temperature at the interface for the cylinder, $T_w(0,t)$, and another for the average temperature of the disk at the interface on the body, $T_b(0,t)$. For intimate (perfect) contact, these temperatures will be equal. After taking the Laplace transform of each of the resulting expressions and setting the two relations equal, an equation for the Laplace transform of the interfacial heat flux will result:

$$\bar{q} = \mathcal{L}\left[\frac{q_b(t)}{k_b(T_{b0} - T_{w0})/r_w}\right] = \frac{1}{s^2\left(\bar{\phi}_b k_b/r_w + \bar{\phi}_w k_b/r_w\right)} \tag{10.18}$$

With suitable expressions for the influence functions, $\phi_b(t)$ and $\phi_w(t)$, this equation may be inverted. Once $q(t)$ is known, Eq. (10.17) may be used to find the interface temperature.

Functions for ϕ_b for large times are analytically challenging (Beck 1981). To work around this, a piecemeal approach to the intrinsic thermocouple analysis, using both temperature-based and heat flux-based USE expressions, and utilizing appropriate relations for ϕ_b for short and long times, gives excellent results (Keltner and Beck 1983):

$$T_w^+(0, t) = \begin{cases} \dfrac{1}{1+\beta}\left[1+\exp\left(\left(\dfrac{2\beta}{\pi(1+\beta)}\right)^2 t^+\right)\text{erf}\left(\dfrac{2\beta}{\pi(1+\beta)}\sqrt{t^+}\right)\right] & t < 0.1 \\[4mm] 1 - \dfrac{\beta}{8/\pi^2+\beta}\text{erfcx}\left(\dfrac{4\sqrt{t^+}}{8/\pi+\beta\pi}\right) & t \geq 0.1 \end{cases}$$

$$\tag{10.19}$$

In Eq. (10.19), erfcx(x)=exp(x^2)erfc(x) is the scaled complementary error function in MATLAB, and in this equation the variables appearing are:

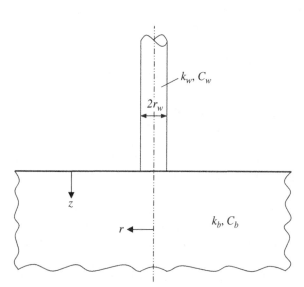

Figure 10.10 Intrinsic thermocouple geometry.

$$t^+ = \frac{k_b}{C_b}\frac{t}{r_w^2}; \quad T_w^+ = \frac{T_w}{T_{b0}}; \quad \beta = \sqrt{\frac{k_w C_w}{k_b C_b}} \tag{10.20}$$

Eq. (10.19) gives the temperature of the contact between a substrate and an attached cylindrical sensor relative to the constant temperature of the surrounding surface. The response is the same as a wire at zero temperature brought into contact with an infinite medium at temperature T_{b0} at time zero. This information can be used to generate the correction kernel for the intrinsic thermocouple, as shown in the following example.

Example 10.4 Consider an intrinsic thermocouple having $\beta = 1.33$. For a step change in substrate temperature, determine the correction kernel **H**. Use a time step of $\Delta t^+ = 0.01$ and consider $0 \le t^+ \le 10$.

Solution

Eq. (10.19) is used to compute the temperature of the wire surface relative to the constant substrate temperature $T_b = 1$ K. The first several values (at every other time step) are shown in Table 10.4, and the values are plotted in Figure 10.11.

These data are used in Eq. (10.2) to determine the correction kernel **H**. Recall that only corresponding temperature values for the sensed and undisturbed locations are needed for application of Eq. (10.2), and it is not necessary that these result from a step change in heat flux. Hence, the (T_w, T_{b0}) pairs will be used. The ΔT_w values are used to compute the $\mathbf{X}_{\phi_p}$ matrix. The first few entries (for every time step) are:

$$
\mathbf{X}_{\phi_p} = \begin{bmatrix}
0.4468 & 0 & 0 & 0 & 0 & 0 & \cdots \\
0.0073 & 0.4468 & 0 & 0 & 0 & 0 & \cdots \\
0.0056 & 0.0073 & 0.4468 & 0 & 0 & 0 & \cdots \\
0.0048 & 0.0056 & 0.0073 & 0.4468 & 0 & 0 & \cdots \\
\vdots & \vdots & \ddots & \ddots & \ddots & \ddots &
\end{bmatrix}
\tag{10.21}
$$

The difference $T_{b0} - T_w$ is the data for solving for **H**. The first few values are

$$
\{\mathbf{T}_{b0} - \mathbf{T}_w\} = \lfloor 0.5532 \quad 0.5459 \quad 0.5403 \quad 0.5355 \quad 0.5313 \quad \cdots \rfloor^T
\tag{10.22}
$$

Table 10.4 Temperatures for Example 10.4.

t	0.01	0.03	0.05	0.07	0.09	0.11	0.13	0.15	0.17	0.19
T_w	0.4468	0.4597	0.4687	0.4760	0.4864	0.4960	0.5045	0.5122	0.5192	0.5256
T_{b0}	1.0000	1.0000	1.0000	1.0000	1.0000	1.0000	1.0000	1.0000	1.0000	1.0000

Figure 10.11 Temperatures for Example 10.4.

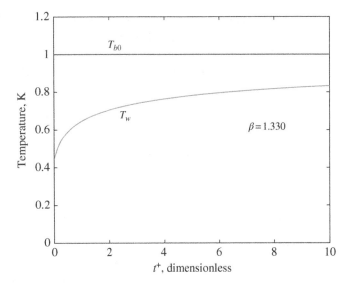

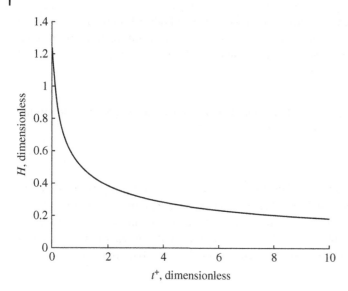

Figure 10.12 **H** vector for Example 10.4.

With the matrix in Eq. (10.21) and the data in Eq. (10.22), Eq. (10.2) is not ill-conditioned. This is because the matrix $\mathbf{X}_{\phi_p}$ is not ill-conditioned (its condition number is only 1.73!) owing primarily to the relatively large diagonal terms resulting from the rapid change in the wire temperature at time zero. The solution for **H** can be found without regularization, and so the vector **H** can be found directly as $\mathbf{H} = \left[\mathbf{X}_{\phi_p}\right]^{-1}\{\mathbf{T}_{b0} - \mathbf{T}_w\}$. The first few values of the vector **H** are

$$\mathbf{H} = \lfloor 1.2381 \quad 1.2015 \quad 1.1739 \quad 1.1510 \quad 1.1310 \quad \cdots \rfloor^T \tag{10.23}$$

A plot of the **H** vector is shown in Figure 10.12.

Example 10.5 Consider the same intrinsic thermocouple as in Example 10.4 with $\beta = 1.33$. Suppose that "measurements" are made using the thermocouple, and that these measurements are

$$Y_i = 20\exp\left(-0.15t^+\right) + \varepsilon_i \tag{10.24}$$

where ε_i is a random number from a Gaussian distribution with zero mean and standard deviation $\sigma = 0.5$ K. For this example, the same random number sequence from examples in Chapter 7 will be used.

Solution

The measurements Y are computed using the specified formula and the random errors based on the sequence in Chapter 7. The first 30 values are shown in Table 10.5.

With the **H** vector already determined, the corrected temperatures can be computed directly using Eq. (10.5c). The $\Delta\mathbf{T}_p$ matrix, based on the "sensed" temperatures, Y, is needed. The first few entries of this 100×100 matrix are

$$[\Delta\mathbf{T}_p] = \begin{bmatrix} 20.903 & 0 & 0 & 0 & 0 & \cdots \\ -1.062 & 20.903 & 0 & 0 & 0 & \cdots \\ -0.067 & -1.062 & 20.903 & 0 & 0 & \cdots \\ -0.236 & -0.067 & -1.062 & 20.903 & 0 & \cdots \\ \vdots & \vdots & \ddots & \ddots & \ddots & \ddots \end{bmatrix} \tag{10.25}$$

Direct application of Eq. (10.5c) results in the desired corrected temperatures, $\mathbf{T}_{p\infty}$. The first 30 values of the corrected temperatures are listed in Table 10.6, and a plot showing the measured and corrected values is shown in Figure 10.13.

Discussion

Although the correction process is not ill-posed, the corrected temperatures in Figure 10.13 appear to contain more noise than in the original data. This is because the correction is computed from the noisy data resulting in a "noisy" correction, which is then added to the original noised data.

Table 10.5 Temperature measurements for Example 10.5.

Y_{1-10}	20.903	19.842	19.775	19.539	20.341	19.172	19.887	20.236	20.467	19.891
Y_{11-20}	19.783	19.421	19.487	19.506	20.493	19.777	19.506	19.419	18.671	19.552
Y_{21-30}	20.375	18.827	19.135	20.028	19.700	19.300	19.462	19.270	19.803	20.444

Table 10.6 Corrected Temperatures for Example 10.5.

$T_{p\infty 1-10}$	46.785	43.642	42.954	41.979	43.392	40.402	41.710	42.009	42.014	40.256
$T_{p\infty 11-20}$	39.609	38.419	38.234	37.954	39.856	37.923	37.044	36.603	34.698	36.477
$T_{p\infty 21-30}$	38.095	34.383	34.900	36.721	35.785	34.719	34.926	34.327	35.377	36.652

Figure 10.13 Simulated measurements and corrected temperatures for Example 10.5.

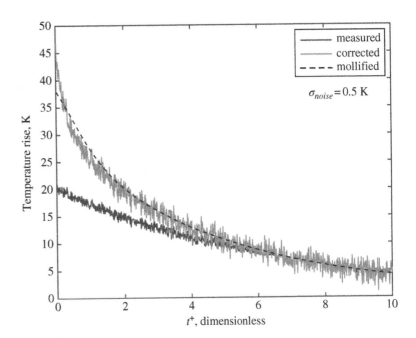

The issue of noisy data can be addressed by smoothing data, either before correction, or after correction, or both. Smoothing can be accomplished by moving average or other means, such as mollification (Murio 1993) using generalized cross-validation (Murio et al. 1998). The dark line in Figure 10.13 shows the result of the mollification of the corrected temperatures only using generalized cross-validation. As seen in the figure, very smooth results for the corrected temperature are obtained but at the expense of some bias in the values introduced by the smoothing. The bias is most evident at early times, as the initial temperature is reduced by about 10 K, and a smaller difference between the smooth curve and the computed results of about 2 K is seen up until about $t^+ = 2$.

10.4 Chapter Summary

Any temperature measurement using a contacting sensor will produce a biased reading relative to the undisturbed region away from the sensor. Both embedded (subsurface) and surface temperature sensors with connecting signal wires are susceptible. While this effect can be reduced through use of small sensors and wires, or careful matching of sensor/wire thermal properties to those of the parent domain, the effect cannot be eliminated.

Numerical or analytical modeling of sensor installations can provide information to determine the temperature correction kernel, $\mathbf{H}$, to correct the biased readings. Data required to compute the correction kernel consists of the "sensed" temperature, T_p, and the corresponding undisturbed temperature, $T_{p\infty}$. The nature of the heat flux used to generate these values in the modeling is unimportant (arbitrary), and so a constant surface heating condition may be used, if desired.

If a constant heat flux is used to generate the T_p data, this same data can be used to compute the filter matrix, **F**, which can then be used to directly invert temperature data from an experiment to determine the unknown heating action, **q**.

Depending on the quality of data for T_p and the size of the time step in the data, regularization may be needed to compute the correction kernel **H**. Regularization will always be needed to find the unknown surface heating using the filter matrix, **F**. A good means of determining the appropriate amount of regularization to apply is to have two data cases available from the numerical or analytical model: one case of $(T_p, T_{p\infty})$ pairs to determine **F** and **H**, and a second set of data resulting from a known, nonconstant heat flux. This latter data set can be used to select the amount of regularization needed to apply in developing **H** and **F**.

Surface thermocouples can be modeled using numerical methods, but a class of analytical solutions has been developed based on the USE method for application to the intrinsic thermocouple. These analytical solutions can be used to generate "data" $((T_p, T_{p\infty})$ data pairs) to determine the correction kernel, **H**, for the intrinsic thermocouple. (These data cannot be used to determine **F** to find the heat flux vector **q**; however, since they do not result from a step change in surface heating.) Once determined, the correction kernel can be used to correct data from subsequent experiments.

Problems

10.1 A simple resistance–capacitance model is used to characterize bias error in a thermocouple installation and the following data are obtained.

t	0.00	0.06	0.12	0.18	0.24	0.30	0.36	0.42
$T_{p\infty}$	25	29.76	34.06	37.96	41.48	44.67	47.56	50.17
T_p	25	26.20	28.89	32.20	35.68	39.10	42.34	45.34
t	0.48	0.54	0.60	0.66	0.72	0.78	0.84	0.90
$T_{p\infty}$	52.53	54.67	56.61	58.36	59.94	61.37	62.67	63.84
T_p	48.10	50.64	52.92	55.02	56.91	58.62	60.18	61.58
t	0.96	1.02	1.08	1.14	1.20	1.26	1.32	1.38
$T_{p\infty}$	64.91	65.87	66.74	67.52	68.23	68.88	69.46	69.99
T_p	62.85	64.01	65.04	65.98	66.84	67.61	68.31	68.94
t	1.44	1.50	1.56	1.62	1.68	1.74	1.80	
$T_{p\infty}$	70.46	70.90	71.29	71.64	71.96	72.25	72.51	
T_p	69.51	70.03	70.50	70.92	71.30	71.65	71.97	

Use these data to determine the correction kernel vector, **H**, for this installation.

10.2 A simulation of a constant heat flux into the surface of an insulated plate produced the following data.

t	0.00	0.06	0.12	0.18	0.24	0.30	0.36	0.42
$T_{p\infty}$	0.000	0.023	0.079	0.138	0.198	0.258	0.318	0.378
T_p	0.000	0.004	0.026	0.064	0.113	0.166	0.223	0.281
t	0.48	0.54	0.60	0.66	0.72	0.78	0.84	0.90
$T_{p\infty}$	0.438	0.498	0.558	0.618	0.678	0.738	0.798	0.858
T_p	0.339	0.399	0.458	0.518	0.578	0.638	0.698	0.758
t	0.96	1.02	1.08	1.14	1.20	1.26	1.32	1.38
$T_{p\infty}$	0.918	0.978	1.038	1.098	1.158	1.218	1.278	1.338
T_p	0.818	0.878	0.938	0.997	1.057	1.117	1.177	1.237
t	1.44	1.50	1.56	1.62	1.68	1.74	1.80	
$T_{p\infty}$	1.398	1.458	1.518	1.578	1.638	1.698	1.758	
T_p	1.297	1.357	1.417	1.477	1.537	1.597	1.657	

During a subsequent experiment, the following measurements are obtained

t	0.00	0.06	0.12	0.18	0.24	0.30	0.36	0.42
T_p	−0.001	0.003	−0.001	−0.001	−0.001	−0.001	−0.002	0.000
t	0.48	0.54	0.60	0.66	0.72	0.78	0.84	0.90
T_p	−0.001	0.001	0.000	0.000	0.004	0.012	0.029	0.058
t	0.96	1.02	1.08	1.14	1.20	1.26	1.32	1.38
T_p	0.095	0.141	0.187	0.223	0.251	0.272	0.286	0.293
t	1.44	1.50	1.56	1.62	1.68	1.74	1.80	
T_p	0.294	0.298	0.299	0.301	0.300	0.299	0.299	

Use the simulation results and the measured data to estimate the surface heat flux.

10.3 Use the simulation results in Problem 7.2 to correct the measured data from Problem 7.2.

References

Attia, M. H. and Kops, L. (1986) Distortion in thermal field around inserted thermocouples in experimental interfacial studies, *Journal of Engineering for Industry*, 108(4), pp. 241–246. https://doi.org/10.1115/1.3187073.

Attia, M. H. and Kops, L. (1988) Distortion in thermal field around inserted thermocouples in experimental interfacial studies – Part II: effect of the heat flow through the thermocouple, *Journal of Engineering for Industry*, 110(1), pp. 7–14. https://doi.org/10.1115/1.3187847.

Attia, M. H. and Kops, L. (1993) Distortion in the thermal field around inserted thermocouples in experimental interfacial studies – Part 3: experimental and numerical verification, *Journal of Engineering for Industry*, 115(4), pp. 444–449. https://doi.org/10.1115/1.2901788.

Attia, M. H., Cameron, A. and Kops, L. (2002) Distortion in thermal field around inserted thermocouples in experimental interfacial studies, Part 4: end effect, *Journal of Manufacturing Science and Engineering*, 124(1), pp. 135–145. https://doi.org/10.1115/1.1419199.

Beck, J. V. (1961a) *Calculation of Transient Thermocouple Temperature Measurements in Heat-Conducting Solids, Part II, The Calculation of Transient Heat Fluxes Using the Inverse Convolution*. Wilmington, MA: AVCO Corp. Research and Advanced Development Division.

Beck, J. V. (1961b) *Correction of Transient Thermocouple Temperature Measurements in Heat-Conducting Solids, Part II, The Calculation of Transient Heat Fluxes Using the Inverse Convolution*. Wilmington, MA: AVCO Corp. Research and Advanced Development Division.

Beck, J. V (1962) Thermocouple temperature disturbances in low conductivity materials, *Journal of Heat Transfer*, 84(2), pp. 124–131. https://doi.org/10.1115/1.3684310.

Beck, J. V (1968) Determination of undisturbed temperatures from thermocouple measurements using correction kernels, *Nuclear Engineering and Design*, 7(1), pp. 9–12. https://doi.org/10.1016/0029-5493(68)90122-2.

Beck, J. V (1981) Large time solutions for temperatures in a semi-infinite body with a disk heat source, *International Journal of Heat and Mass Transfer*, 24(1), pp. 155–164. https://doi.org/10.1016/0017-9310(81)90104-6.

Beck, J. V and Hurwicz, H. (1960) Effect of thermocouple cavity on heat sink temperature, *Journal of Heat Transfer*, 82(1), pp. 27–36. https://doi.org/10.1115/1.3679876.

Chen, H., Frankel, J. I. and Keyhani, M. (2016) Two-dimensional formulation for inverse heat conduction problems by the calibration integral equation method (CIEM), *Applied Mathematical Modelling*, 40(13–14), pp. 6588–6603. https://doi.org/10.1016/j.apm.2016.02.003.

Cole, K.D., Beck, J. V, Haji-Sheikh, A. and Litkouhi,B. (2011) *Heat Conduction using Green's Functions*, 2nd edn. Boca Raton, FL: CRC Press.

Feng, Z. C., Chen, J. K. and Zhang, Y. (2020) Intrinsic thermal couples for measurement in high temperature and high heat flux environment, *Proceedings of the 15th IEEE Conference on Industrial Electronics and Applications, ICIEA 2020*, Kristiansand, Norway (9–13 November 2020), pp. 475–477. https://doi.org/10.1109/ICIEA48937.2020.9248226.

Frankel, J. I. and Keyhani, M. (2018) Response function formulation for inverse heat conduction: concept, *Journal of Engineering Mathematics*, 110(1), pp. 75–95. https://doi.org/10.1007/s10665-017-9932-8.

Frankel, J. I., Chen, H. C. and Keyhani, M. (2017) New step response formulation for inverse heat conduction, *Journal of Thermophysics and Heat Transfer*, 31(4), pp. 988–995. https://doi.org/10.2514/1.T5067.

Gat, U., Kammer, D. S. and Hahn, O. J. (1975) The effect of temperature dependent properties on transients measurement with intrinsic thermocouple, *International Journal of Heat and Mass Transfer*, 18(12), pp. 1337–1342. https://doi.org/10.1016/0017-9310(75)90246-X.

Henning, C. D. and Parker, R. (1967) Transient response of an intrinsic thermocouple, *Journal of Heat Transfer*, 89(2), pp. 146–152. https://doi.org/10.1115/1.3614337.

Keltner, N. R. and Beck, J. V. (1981) Unsteady surface element method, *Journal of Heat Transfer*, 103(4), pp. 759–764. https://doi.org/10.1115/1.3244538.

Keltner, N. R. and Beck, J. V. (1983) Surface temperature measurement errors', *Journal of Heat Transfer*, 105(2), pp. 312–318. https://doi.org/10.1115/1.3245580.

Keltner, N. R. and Bickle, L. W. (1976) 'Intrinsic Thermocouple Measurement Errors', in *American Society of Mechanical Engineers 76 -HT-65. American Society of Mechanical Engineers*, pp. 1–8.

Litkouhi, B. and Beck, J. V. (1985) Intrinsic thermocouple analysis using multinode unsteady surface element method, *AIAA Journal*, 23(10), pp. 1609–1614. https://doi.org/10.2514/3.9131.

Murio, D. A. (1993) *The Mollification Method and the Numerical Solution of Ill-Posed Problems*. Wiley.

Murio, D. A., Mejía, C. E. and Zhan, S. (1998) Discrete mollification and automatic numerical differentiation, *Computers & Mathematics with Applications*, 35(5), pp. 1–16. https://doi.org/10.1016/S0898-1221(98)00001-7.

Myrick, J. A., Keyhani, M. and Frankel, J. I. (2017) Determination of surface heat flux and temperature using in-depth temperature data – Experimental verification, *International Journal of Heat and Mass Transfer*, 111, pp. 982–998. https://doi.org/10.1016/j.ijheatmasstransfer.2017.04.029.

Nakos, J. T. (2003) Understanding the systematic error of a mineral-insulated, metal sheathed (MIMS) thermocouple attached to a heated flat surface, *ASTM Special Technical Publication*, 2003(1427), pp. 32–50.

Pope, I., Hidalgo, J. P. and Torero, J. L. (2021) A correction method for thermal disturbances induced by thermocouples in a low-conductivity charring material, *Fire Safety Journal*, 120, p. 103077. https://doi.org/10.1016/j.firesaf.2020.103077.

Pope, I., Hidalgo, J.P., Hadden, R.M. and Torero, J.L. (2022) 'A simplified correction method for thermocouple disturbance errors in solids', *International Journal of Thermal Sciences*, 172, 107324. https://doi.org/10.1016/j.ijthermalsci.2021.107324.

Tszeng, T. C. and Saraf, V. (2003) 'A study of fin effects in the measurement of temperature using surface-mounted thermocouples', *Journal of Heat Transfer*, 125(5), pp. 926–935. https://doi.org/10.1115/1.1597622.

Woolley, J. W. (2008) *Accounting for Transient Temperature Measurement Error with a High Fidelity Thermocouple Model and Application to Metal/Mold Interfacial Heat Flux Estimation*. Tuscaloosa, Alabama, USA: The University of Alabama.

Woolley, J. W. and Woodbury, K. A. (2008) Accounting for sensor errors in estimation of surface heat flux by an inverse method, in *ASME SUMMER HEAT TRANSFER CONFERENCE*, Jacksonville, Florida, USA (10–14 August 2008). American Society of Mechanical Engineers, pp. 69–76. https://doi.org/10.1080/01457632.2011.525468.

Woolley, J. W. and Woodbury, K. A. (2011) Thermocouple data in the inverse heat conduction problem, *Heat Transfer Engineering*, 32(9), pp. 811–825. https://doi.org/10.108001457632.2011.525468.

Woolley, J. W., Wilson, H. B. and Woodbury, K. A. (2008) Incorporation of measurement models in the IHCP: validation of methods for computing correction kernels, *Journal of Physics: Conference Series*, 135(1), p. 012103. https://doi.org/10.1088/1742-6596/135/1/012103.

Woolley, J. W. *et al.* (2009) Obtaining the sensed temperatures from a detailed model of a welded thermocouple', *ASME International Mechanical Engineering Congress and Exposition, Proceedings*, 10(PART A), pp. 649–655. https://doi.org/10.1115/IMECE2008-68031.

Xu, F. and Gadala, M. S. (2007) On the application of intrinsic thermocouples in temperature measurements in water jet cooling', *Numerical Heat Transfer, Part A*, 51, pp. 343–362. https://doi.org/10.1080/10407780600829456.

Zhan, S., Coles, C. and Murio, D. A. (2001) Automatic numerical solution of generalized 2-D IHCP by discrete mollification, *Computers & Mathematics with Applications*, 41(1–2), pp. 15–38. https://doi.org/10.1016/S0898-1221(01)85003-3.

Appendix A

Numbering System

The number system has several components to identify the elements of a conventional heat conduction problem including dimensionality, coordinate system, boundary conditions, initial temperature distribution, volumetric heat source/sink term, fin-type, and moving body terms. For the sake of brevity, only the rectangular coordinate system is here considered. For cylindrical and spherical frames of reference as well as an exhaustive treatment on this subject, the reader can refer to Chapter 2 of the book by Cole et al. (2011).

A.1 Dimensionality, Coordinate System, and Types of Boundary Conditions

For the rectangular coordinate system, here of interest, the symbol X is used to denote the *x*-coordinate; Y is used to indicate the *y*-coordinate; and Z is utilized to denote the *z*-coordinate. As an example, for a two-dimensional (2D) problem involving *x*- and *y*-coordinates, X and Y are used. Four different boundary conditions are given and are numbered 1, 2, 3, and 0. The boundary condition of the first kind refers to a prescribed temperature history at a boundary; it is also called the Dirichlet condition. The boundary condition of the second kind is for a prescribed heat flux (a special case is zero heat flux, the thermal insulation condition) and is also denoted the Neumann condition. The third kind of boundary condition is for surface heat convection/long-wavelength thermal radiation and is called the Robin condition. The use of a number to identify boundary conditions of the first, second, and third kinds has been in common usage for many years. See, for example, Luikov (1968).

As far as the zeroth-kind boundary condition is concerned, it is required where a physical boundary does not exist, but the temperature is bounded (that is, less than infinite). The type zero boundary (also called of Beck type) occurs in a body in which one dimension goes to infinity and the temperature must be bounded there; this applies to bodies described by rectangular, cylindrical, and spherical coordinates. As an example, in rectangular coordinates, the one-dimensional semi-infinite body ($x \geq 0$) has a type zero boundary condition at $x \to \infty$. The zeroth kind of boundary condition may also occur at the center of a body described by a radial coordinate. That is, at the center of a solid cylinder or solid sphere there is a coordinate boundary ($r = 0$) but no physical boundary, and the temperature is bounded there. The zeroth boundary condition is important in the number system because it allows diverse geometries to be indexed together so that the relationship among them may be made clear.

A summary of the types of boundary conditions, along with the name, description, and equation defining each one, is given in Table A.1, where k is the thermal conductivity of the body, while h_i is the heat transfer coefficient on the *i*-th boundary. Also, f_i is the space- and/or time-dependent boundary condition function that accounts for the possible space and/or time variation of the same condition. The f_i function can also be space- and time-independent (uniform and constant boundary condition) or even zero. In the latter case, the related boundary condition is termed as homogeneous (see Section A.2). When the solid is subject to a boundary condition of Dirichlet type on the *i*-th boundary, the f_i function has units of K (or °C). However, when the boundary condition is of Neumann or Robin type, f_i has units of W m^{-2}.

As an example, for one-dimensional bodies in the rectangular coordinate system, notation X11 represents a slab with boundary conditions of the first kind on both sides, X30 represents a semi-infinite body that is convectively heated or cooled at $x = 0$, and X00 represents the infinite body. There are 16 possible cases (XIJ, with I, J = 0,1,2,3), as shown by Table A.2, but only 10 of them are distinct cases. For example, X13 can be found from X31 (or vice versa) by a simple change of coordinate (i.e., $x \to L - x$, where L is the slab thickness). Similarly, X02 can be found from X20 (or vice versa) by changing $x \to -x$.

Inverse Heat Conduction: Ill-Posed Problems, Second Edition. Keith A. Woodbury, Hamidreza Najafi, Filippo de Monte, and James V. Beck.
© 2023 John Wiley & Sons, Inc. Published 2023 by John Wiley & Sons, Inc.

Table A.1 Types of boundary conditions included in the heat conduction numbering system. n_i is the direction of the outward vector on the i-th boundary.

Number	Name	Description	Equation
0	Beck	No physical boundary	$T = \text{finite}$
1	Dirichlet	Prescribed temperature	$T = f_i$
2	Neumann	Prescribed heat flux	$k(\partial T/\partial n_i) = f_i$
3	Robin	Convection/long-wavelength thermal radiation	$k(\partial T/\partial n_i) + h_i T = f_i$

Table A.2 Cases (XIJ) for one-directional rectangular bodies.

I\J	0	1	2	3
0	X00	X01	X02	X03
1	X10	X11	X12	X13
2	X20	X21	X22	X23
3	X30	X31	X32	X33

The number system also describes two- and three-dimensional geometries. For example, X22Y10 denotes a rectangular domain, finite along x with the boundary conditions both of the second kind, and semi-infinite in the y-direction with a prescribed temperature at $y = 0$. In addition, it can also describe a two-layer configuration, with "C1" and "C3" denoting a perfect and imperfect thermal contact between two adjacent layers, respectively. As an example, notation X2C12 represents a two-layer slab with boundary conditions of the second kind on both sides whose layers are in perfect contact.

A.2 Boundary Condition Information

Beyond the type of boundary condition, the number system must also address the temporal variations at the boundary if any. In addition, for multidimensional problems, the possibility of spatial variations on the boundaries must be included too. This means that the function f_i appearing in the equation defining the boundary conditions given in Table A.1 must somehow be described. For one-dimensional cases, f_i can be only a function of time, that is $f_i = f_i(t)$. In detail, it includes zero (denoted by B0, where B signifies boundary information, while 0 indicates homogeneous, i.e. zero constant value), constant with time (B1), linear with time (B2), and power variation other than 1 (B3). See Table A.3 where B followed by a dash line (B-) indicates an arbitrary time variation. Some examples of power-law time variations in surface conditions (B3) are given by Beck et al. (2008), Woodbury and Beck (2013) and Cole et al. (2014).

The notations listed in the second column of Table A.3 for 1D cases modify for multidimensional problems where space variations of f_i on the boundaries are also possible. Therefore, B0, B1, B2, B3, and B- become B0 (this does not modify), Bt1, Bt2, Bt3, and Bt-, respectively, as shown by the last column of the same table. In such cases, it is assumed that the boundary function f_i be expressible only as a product of single space- and time-variable functions. For example, for a 2D rectangular

Table A.3 Types of time boundary condition function on the i-th boundary.

f_i function	Notation for 1D problems	Notation for 2D and 3D problems
$f_i(t) = 0$	B0	B0
$f_i(t) = c$	B1	Bt1
$f_i(t) = ct$	B2	Bt2
$f_i(t) = ct^p, \quad p > 1$	B3	Bt3
Arbitrary $f_i(t)$	B-	Bt-

Table A.4 Types of space boundary condition function for a 2D problem involving x- and y-coordinates on the i-th boundary.

$u_i(x)$ function	Notation at a y_i boundary	Notation at a x_i boundary, $u_i(x) \to v_i(y)$
$u_i(x) = c$	Bx1	By1
$u_i(x) = cx$	Bx2	By2
$u_i(x) = cx^p, \quad p > 1$	Bx3	By3
Arbitrary $u_i(x)$	Bx-	By-

problem involving x- and y-coordinates and at a $y = y_i$ boundary, f_i might be a function of t alone, a function of x alone, or a function of both x and t in the form $f_i(x, t) = u_i(x)\varphi_i(t)$. In this case, f_i in Table A.3 has to be replaced with φ_i. As regards the $u_i(x)$ function, it includes uniform with space (Bx1), linear with space (Bx2), and so on. See Table A.4 for other notations where Bx3 denotes a power variation with x other than 1 along the y_i boundary. Also, Bx- indicates an arbitrary space variation with x. Note that Bx0 is not needed here. Table A.4 also gives the notation at a $x = x_i$ boundary where $f_i(y, t) = v_i(y)\varphi_i(t)$, as shown by the last column.

As an example, for 1D rectangular solids, notation X12B20 represents a slab subject to a boundary condition of the first kind ("1" of X12) on the left side, where the assigned temperature varies linearly with time ("2" of B20), and thermally insulated on the other side (in fact, "2" of X12 and "0" of B20 indicate zero heat flux there). For 2D solids, notation X12B20Y21B(x2t2)0 denotes a finite rectangular domain. The X12B20 part of the alphanumeric string was described right before. The second part Y21B(x2t2)0 indicates that the rectangle is subject to a heat flux at a boundary along y ("2" of Y21 and "Y" of Y21) with a linear variation with x and varying linearly with time ("x2" and "t2" in B(x2t2)0, respectively). Also, it is subject to a zero temperature on the opposite side ("1" of Y21 and "0" of B(x2t2)0).

If the boundary function f_i is not expressible only as a product of single space- and time-variable functions, but is in the form of $f_i(x, t) = \sum_{j=1}^{N} u_{i,j}(x)\varphi_{i,j}(t)$, where $u_{i,j}(x)$ and/or $\varphi_{i,j}(t)$ can also be equal to 0 or 1, the problem can be split up into N simpler problems by using the principle of superposition (Özişik 1993; Woodbury and Beck 2013). See also Section 3.2 of the current book. Thus, the notation described above can still be utilized for either simpler problem having the j-th boundary function $f_{i,j}(x, t) = u_{i,j}(x)\varphi_{i,j}(t)$.

A.2.1 Finite-in-time Boundary Condition

The notation given in Table A.3 is valid when the $f_i(t)$ boundary function is applied for an unlimited period of time. However, it can occur that the heating (or cooling) at a boundary has a finite-in-time period (Woodbury and Beck 2013; D'Alessandro and de Monte 2019; D'Alessandro et al. 2019). In such a case, the lowercase letter "p" (portion of time) can indicate this situation.

For one-dimensional cases, when f_i is only a function of time, B1p denotes constant with time for a while (vs B1), B2p linear with time for a finite period (vs B2), and so on. (Note that B0p is not needed here.) For example, notation X21B1p0 represents a slab subject to a boundary condition of the second kind on the left side, where the prescribed heat flux is constant but applied for a finite duration ("1p" of B1p0), and with zero temperature on the other side ("0" of B1p0 indicates zero value there).

For multidimensional problems where space variations of f_i on the boundaries are also possible, Bt1, Bt2, Bt3, and Bt- of Table A.3 become Bt1p, Bt2p, Bt3p, and Bt-p, respectively. As an example, notation X21B(y1t1p)0Y22B(x1t2p)0 denotes a heat-conducting rectangle. The X21B(y1t1p)0 part of the alphanumeric string is equal to X21B1p0 analyzed above. The second part Y22B(x1t2p)0 indicates that the 2D rectangular body is subject to a heat flux at a boundary along y (the first "2" of Y22) that is uniform with x and varying linearly with time for a finite-in-time period ("x1" and "t2p" in B(x1t2p)0, respectively). Also, it is thermally insulated on the opposite side (the second "2" of Y22 and "0" of B(x1t2p)0).

A.2.2 Partial-in-space Boundary Condition

The notation given in Table A.4 is right when the $u_i(x)$ boundary function is applied to the entire boundary surface $y = y_i$ (second column). Similarly, when $v_i(y)$ is applied to the entire boundary $x = x_i$ (third column). However, it can happen that the body is subject to a partial heating or cooling at a boundary (Woodbury et al. 2017; McMasters et al. 2018, 2019, 2020). In other words, a portion of the boundary is "active" (nonhomogeneous boundary condition); while the remaining part is "inactive" (homogeneous boundary condition) and of the same kind. In such a case, the lowercase letter "p" (portion of boundary) can still be chosen to indicate this situation.

Therefore, Bx1, Bx2, Bx3, and Bx- of Table A.4 become Bx1p, Bx2p, Bx3p, and Bx-p, respectively. Similarly, By1, By2, By3, and By- modify as By1p, By2p, By3p, and By-p. For example, Bx2p denotes a linear-in-space partial heating/cooling at a $y = y_i$ boundary through a boundary condition of the first, second, or third kind. The remaining part of the boundary surface is homogeneous ($u_i(x) = 0$) and of the same kind. Thus, the heat conduction problem described by X21B1p0Y20B(x1pt2p) denotes a 2D rectangular solid semi-infinite along y ("0" of Y20). The X21B(y1t1p)0 part of the alphanumeric string is equal to X21B1p0 described in Section A.2.1. As regards the second part Y20B(x1pt2p), it indicates that the 2D rectangular body is subject to a uniform heat flux applied to a portion of the only boundary along y ("x1p" in B(x1pt2p)); also, it varies linearly with time for a finite duration ("t2p" in B(x1pt2p)). Moreover, the remaining part of the boundary along y is subject to a homogeneous boundary condition of the second kind, that is, is thermally insulated.

A.3 Initial Temperature Distribution

When there is an initial condition (only for transient problems), its spatial variation must also be addressed for one-dimensional and multidimensional cases as part of a complete and comprehensive number system for heat conduction.

For one-dimensional transient cases with x being the space coordinate, the initial condition is expressed by $T(x, t = 0) = F(x)$, where $F(x)$ includes zero (denoted by T0, where T signifies initial temperature, while 0 indicates homogeneous, i.e. zero value), uniform with space (T1), linear with x (T2), and power space-variation other than 1 (T3). See Table A.5 where T followed by a dash line (T-) indicates an arbitrary temperature distribution at time $t = 0$. An interesting case of nonuniform initial temperature was analyzed by McMasters et al. (1999).

The notations listed in the second column of Table A.5 modify for multidimensional problems where space variations of the initial temperature along y- and z-coordinates are also possible. In such cases, it is assumed that the initial condition function F be expressible only as a product of single space-variable functions. For example, for a 2D rectangular problem, it might be a function of x alone, a function of y alone, or a function of both x and y in the form $F(x, y) = F_1(x)F_2(y)$. In this case, F in Table A.5 has to be replaced with F_1. Therefore, T0, T1, T2, T3, and T- become T0 (this does not modify), Tx1, Tx2, Tx3, and Tx-, respectively, as shown by the last column of the same table. Similarly, it results in Ty1, Ty2, Ty3, and Ty- when the variation of the initial condition occurs along y and regards F_2. As an example, notation T(x1y2z2) indicates an initial temperature distribution uniform along x and changing linearly in the y- and z-directions.

If the initial condition for a 2D transient problem, $T(x, y, t = 0) = F(x, y)$, is not expressible only as a product of single space-variable functions, but is in the form of $F(x, y) = \sum_{j=1}^{N} F_{1,j}(x)F_{2,j}(y)$, where $F_{1,j}(x)$ and/or $F_{2,j}(y)$ can also be equal to 0 or 1, the problem can be split up into N simpler problems by using the principle of superposition. Thus, the notation described above can still be utilized for either simpler problem having the j-th initial condition function $F_j(x, y) = F_{1,j}(x)F_{2,j}(y)$.

Table A.5 Types of space-variable initial condition.

$F(x)$ function	Notation for 1D problems	Notation for 2D and 3D problems
$F(x) = 0$	T0	T0
$F(x) = c$	T1	Tx1
$F(x) = cx$	T2	Tx2
$F(x) = cx^p, \quad p > 1$	T3	Tx3
Arbitrary $F(x)$	T-	Tx-

A.3.1 Partial-in-space Initial Condition

The notation of Table A.5 is valid when the $F(x)$ initial condition function is applied to the entire body (second column). Also, for multidimensional cases, when $F_1(x)$ is applied to the entire dimension of the solid along x (third column). However, it can happen that the initial temperature of the body is piecewise. In other words, a portion of the body has initially temperature other than 0 (nonhomogeneous initial condition); while the remaining part is initially at zero temperature (homogeneous initial condition). In such a case, the lowercase letter "p" (portion of body) can still be chosen to indicate this situation.

Therefore, T1, T2, T3, and T- of Table A.5 become T1p, T2p, T3p, and T-p, respectively. Similarly, Tx1, Tx2, Tx3, and Tx- modify as Tx1p, Tx2p, Tx3p, and Tx-p. For example, T2p denotes a linear-in-space initial temperature only for a portion of a one-dimensional body, being the remaining portion at zero value.

References

Beck, J. V., Wright, N. T. and Haji-Sheikh, A. (2008) Transient power variation in surface conditions in heat conduction for plates, *International Journal of Heat and Mass Transfer*, 51(9–10), p. 2553. doi:https://doi.org/10.1016/j.ijheatmasstransfer.2007.07.043.

Cole, K. D., Beck, J. V., Haji-Sheikh, A. and Litkouhi, B. (2011) *Heat Conduction using Green's Functions*, 2nd edn. Boca Raton, FL: CRC Press.

Cole, K. D., Beck, J. V., Woodbury, K. A. and de Monte, F. (2014) Intrinsic verification and a heat conduction database, *International Journal of Thermal Sciences*, 78, pp. 36–47. http://dx.doi.org/10.1016/j.ijthermalsci.2013.11.002.

D'Alessandro, G. and de Monte, F. (2019) Optimal experiment design for thermal property estimation using a boundary condition of the fourth kind with a time-limited heating period, *International Journal of Heat and Mass Transfer*, 134, pp. 1268–1282. https://doi.org/10.1016/j.ijheatmasstransfer.2019.02.035.

D'Alessandro, G., de Monte, F. and Amos, D. E. (2019) Effect of heat source and imperfect contact on simultaneous estimation of thermal properties of high-conductivity materials, *Mathematical Problems in Engineering*, 2019, p. 15. https://doi.org/10.1155/2019/5945413.

Luikov, A. V. (1968) *Analytical Heat Diffusion Theory*. New York: Academic Press.

McMasters, R. L., Beck, J. V., Dinwiddie, R. and Wang, S. (1999) Accounting for penetration of laser heating in flash thermal diffusivity experiments, *ASME Journal of Heat Transfer*, 121(1), pp. 15–21. https://doi.org/10.1115/1.2825929.

McMasters, R. L., de Monte, F. and Beck, J. V. (2019) Generalized solution for two-dimensional transient heat conduction problems with partial heating near a corner, *ASME Journal of Heat Transfer*, 141(7), pp. 071301-1-071301-8. doi: https://doi.org/10.1115/1.4043568.

McMasters, R.L., de Monte, F., Beck, J. V. and Amos, D. E. (2018) Transient two-dimensional heat conduction problem with partial heating near corners, *ASME Journal of Heat Transfer*, 140(2), pp. 021301–1–021301–10. https://doi.org/10.1115/1.4037542.

McMasters, R. L., de Monte, F. and Beck, J. V. (2020) Generalized Solution for Rectangular Three-Dimensional Transient Heat Conduction Problems with Partial Heating, *AIAA Journal of Thermophysics and Heat Transfer*, 34(3), pp. 516–521.

Özişik, M.N. (1993) *Heat Conduction*. 2nd edn. New York: Wiley.

Woodbury, K. A. and Beck, J. V. (2013) Heat conduction in a planar slab with power-law and polynomial heat flux input, *International Journal of Thermal Sciences*, 71, pp. 237–248. https://doi.org/10.1016/j.ijthermalsci.2013.04.009.

Woodbury, K. A., Najafi, H. and Beck, J. V. (2017) Exact analytical solution for 2-D transient heat conduction in a rectangle with partial heating on one edge, *International Journal of Thermal Sciences*, 112, pp. 252–262. doi:https://doi.org/10.1016/j.ijthermalsci.2016.10.014.

Appendix B

Exact Solution X22B(y1pt1)0Y22B00T0

The solution to this problem is detailed in the following sections. There are two primary forms for the solution: the short-time form (Section B.1) and the long-time form (Section B.2). However, for the short-time form, there are several applicable representations, and the best one to use depends on the location in the domain relative to the heating source/sink in addition to the time variable. Similarly, when dealing with the large-time form; whose quasi-steady part has two different expressions, and the best one to use from a computational viewpoint depends on the point of interest.

B.1 Exact Analytical Solution: Short-Time Form

The *short-time* form comes from the application of Laplace transform (LT) approach to the governing equations (Özişik 1993). Also, from the application of Green's function method, when using "*short-cotime*" Green's functions (Cole et al. 2011). It is valid at any time, but it is computationally convenient at short times, that is, requires less terms than the large-time form (Section 2.4.3). However, when dealing with two-dimensional transient problems, two different expressions of the short-time form of the temperature solution are also available, depending on the location of interest. By using "*short-cotime*" Green's functions, they are

$$\tilde{T}_e(\tilde{x}, \tilde{y}, \tilde{t}) = \tilde{T}_{X22Y20}(\tilde{x}, \tilde{y}, \tilde{t}) + \sum_{m=1}^{\infty} \sum_{n=1}^{\infty} \tilde{T}_{x,mn}^{(S)}(\tilde{x}, \tilde{y}, \tilde{t}) \tag{B.1a}$$

$$\tilde{T}_e(\tilde{x}, \tilde{y}, \tilde{t}) = \tilde{T}_{X20Y22}(\tilde{x}, \tilde{y}, \tilde{t}) + \sum_{m=1}^{\infty} \sum_{n=1}^{\infty} \tilde{T}_{y,mn}^{(S)}(\tilde{x}, \tilde{y}, \tilde{t}) \tag{B.1b}$$

whose terms are discussed afterward.

Both are valid at any point within the rectangular domain, but Eq. (B.1a) is computationally convenient (that is, requires less terms than Eq. (B.1b)) when the point $(\tilde{x}, \tilde{y})$, with $\tilde{x} \in [0, 1]$ and $\tilde{y} \in [0, \tilde{W}]$, satisfies the condition: $\tilde{y} \leq \tilde{x} + [2(\tilde{W} - 1) - \tilde{W}_0]$. This indicates that the location is more affected by the inactive boundary condition at $\tilde{x} = 1$ than by the one at $\tilde{y} = \tilde{W}$. On the contrary, Eq. (B.1b) is computationally convenient (that is, requires less terms than Eq. (B.1a)) for $\tilde{y} \geq \tilde{x} + [2(\tilde{W} - 1) - \tilde{W}_0]$, that is, the disturbance due to the inactive boundary condition at $\tilde{y} = \tilde{W}$ on the temperature at $(\tilde{x}, \tilde{y})$ is stronger than the one due to the boundary at $\tilde{x} = 1$. (The above constraint may be derived using the concept of 1D deviation time due to an inactive boundary condition defined in Section 2.3.5.2). If the partial heating/cooling at the $\tilde{x} = 0$ boundary of the domain occurs at time $\tilde{t} = \tilde{t}_0 > 0$, the temperature solution can still be derived by using the short-time forms of the solution listed before, where $\tilde{t}$ has however to be replaced with $\tilde{t} = \tilde{t}_0$. Therefore, $\tilde{T}_e(\tilde{x}, \tilde{y}, \tilde{t}) \rightarrow \tilde{T}_e(\tilde{x}, \tilde{y}, \tilde{t} - \tilde{t}_0)$, with $\tilde{t}_0 = \alpha t_0/L^2$. This is of great concern when using the principle of superposition related to the piecewise-constant approximation of the time-dependent part of the boundary function as proposed in Section 3.6.2 of Chapter 3.

B.1.1 Short-Time Form with Semi-infinite Solution Along y

Equation (B.1a) consists of two parts. The first part denoted by $\tilde{T}_{X22Y20}(\tilde{x}, \tilde{y}, \tilde{t})$ is the temperature solution of the associate two-dimensional problem finite along x (X22) but semi-infinite along y (Y20) and denoted by X22B(y1pt1)0Y20B0T0. Therefore, this part is not influenced by the boundary condition of zero heat flux at $y = W$. The second part accounts for the

Inverse Heat Conduction: Ill-Posed Problems, Second Edition. Keith A. Woodbury, Hamidreza Najafi, Filippo de Monte, and James V. Beck.
© 2023 John Wiley & Sons, Inc. Published 2023 by John Wiley & Sons, Inc.

thermal deviation effects on temperature due to the $y = W$ inactive boundary that affects the semi-infinite behavior along y when the time increases. It is in a form of a double-series consisting of an infinite number of terms denoted by $\tilde{T}_{x,mn}^{(S)}(\tilde{x}, \tilde{y}, \tilde{t})$ that are available only in an integral form. As this form can be solved only numerically, the second part of Eq. (B.1a) (which can play a role for large times) will here not be utilized. In fact, it will be replaced with the companion large-time form of the solution (Section B.2), as explained in Section 2.4.3. For this reason, the integral expression of $\tilde{T}_{x,mn}^{(S)}(\tilde{x}, \tilde{y}, \tilde{t})$ is not given in this book.

In turn, the former part may be split into two parts as

$$\tilde{T}_{X22Y20}(\tilde{x}, \tilde{y}, \tilde{t}) = \tilde{T}_{X20Y20}(\tilde{x}, \tilde{y}, \tilde{t}) + \sum_{m=1}^{\infty} \tilde{T}_{x,m}^{(S)}(\tilde{x}, \tilde{y}, \tilde{t}) \tag{B.2}$$

where $\tilde{T}_{X20Y20}(\tilde{x}, \tilde{y}, \tilde{t})$ is the temperature solution of the associate two-dimensional problem semi-infinite in both the x- and y- directions and denoted by X20B(y1pt1)Y20B0T0. Therefore, it is not influenced by the boundary condition of zero heat flux at both $x = L$ and $y = W$. The following single m-summation of Eq. (B.2) accounts for the *thermal deviation effects* on temperature due to the $x = L$ inactive boundary that affects the semi-infinite behavior along x when the time increases. It consists of an infinite number of terms denoted by $\tilde{T}_{x,m}^{(S)}(\tilde{x}, \tilde{y}, \tilde{t})$ that are in an integral form. As this form cannot be solved analytically, its expression is not given in this book, and the related m-summation (which can play a role for large times) will here not be utilized. It will in fact be replaced with the companion large-time form of the solution (Section B.2), as explained in Section 2.4.3. The use of Eq. (B.2) will be limited to only the former term $\tilde{T}_{X20Y20}(\tilde{x}, \tilde{y}, \tilde{t})$ at the boundary surface $\tilde{x} = 0$ for which an exact analytical solution is available (Section B.1.3).

B.1.2 Short-Time Form with Semi-infinite Solution Along x

The first part of Eq. (B.1b) denoted by $\tilde{T}_{X20Y22}(\tilde{x}, \tilde{y}, \tilde{t})$ is the solution of the associate 2D problem semi-infinite along x (X20) but finite along y (Y22) and denoted by X20B(y1pt1)Y22B00T0. Therefore, this part is not influenced by the boundary condition of zero heat flux at $x = L$. The second part accounts for the *thermal deviation effects* on temperature due to the $x = L$ inactive boundary that disturbs the semi-infinite behavior along x when the time increases. It is in a form of a double-series consisting of an infinite number of terms denoted by $\tilde{T}_{y,mn}^{(S)}(\tilde{x}, \tilde{y}, \tilde{t})$ that are available only in an integral form. As this form cannot be solved analytically, the second part of Eq. (B.1b) (which becomes more relevant for large times) will here not be used. In fact, it will be replaced with the companion large-time form of the solution (Section B.2), as explained in Section 2.4.3. For this reason, the integral expression of $\tilde{T}_{y,mn}^{(S)}(\tilde{x}, \tilde{y}, \tilde{t})$ is not given in this book.

In turn, the former part may be split into two parts as

$$\tilde{T}_{X20Y22}(\tilde{x}, \tilde{y}, \tilde{t}) = \tilde{T}_{X20Y20}(\tilde{x}, \tilde{y}, \tilde{t}) + \sum_{n=1}^{\infty} \tilde{T}_{y,n}^{(S)}(\tilde{x}, \tilde{y}, \tilde{t}) \tag{B.3}$$

where $\tilde{T}_{X20Y20}(\tilde{x}, \tilde{y}, \tilde{t})$ is the same temperature appearing in Eq. (B.2). The following single n-summation accounts for the *thermal deviation effects* on temperature due to the $y = W$ inactive boundary that disturbs the semi-infinite behavior along y when the time increases. It involves an infinite number of terms denoted by $\tilde{T}_{y,n}^{(S)}(\tilde{x}, \tilde{y}, \tilde{t})$ that are in an integral form. As this form can be solved only numerically, its expression is not given in this book, and the related n-summation (which can play a role for large times) will here not be utilized. It will in fact be replaced with the companion large-time form of the solution (Section B.2), as explained in Section 2.4.3. The use of Eq. (B.3) will be limited to only the former term $\tilde{T}_{X20Y20}(\tilde{x}, \tilde{y}, \tilde{t})$ at $\tilde{x} = 0$, as shown in the following section.

B.1.3 Two-Dimensional Semi-infinite Solution

The $\tilde{T}_{X20Y20}(\tilde{x}, \tilde{y}, \tilde{t})$ solution appearing in Eqs. (B.2) and (B.3) may be derived using the procedure proposed by Cole et al. (2011; see section 6.8, p. 218) based on short-cotime Green's functions (the ones coming from LT method). By using the dimensionless variables defined in Eq. (2.49), it is found in an integral form that

$$\tilde{T}_{\text{X20Y20}}(\tilde{x}, \tilde{y}, \tilde{t}) = \frac{1 + \text{sign}(\tilde{W}_0 - \tilde{y})}{2} 2\sqrt{\tilde{t}} \, \text{ierfc}\left(\frac{\tilde{x}}{2\sqrt{\tilde{t}}}\right)$$

$$- \frac{1}{2\sqrt{\pi}} \left[\text{sign}(\tilde{W}_0 - \tilde{y}) \int\limits_{\tilde{u}=0}^{\tilde{t}} \frac{e^{-\frac{\tilde{x}^2}{4\tilde{u}}}}{\sqrt{\tilde{u}}} \text{erfc}\left(\frac{|\tilde{W}_0 - \tilde{y}|}{2\sqrt{\tilde{u}}}\right) d\tilde{u} + \int\limits_{\tilde{u}=0}^{\tilde{t}} \frac{e^{-\frac{\tilde{x}^2}{4\tilde{u}}}}{\sqrt{\tilde{u}}} \text{erfc}\left(\frac{\tilde{W}_0 + \tilde{y}}{2\sqrt{\tilde{u}}}\right) d\tilde{u} \right] \qquad (\text{B.4})$$

$$= \frac{1 + \text{sign}(\tilde{W}_0 - \tilde{y})}{2} \tilde{T}_{\text{X20B1T0}}(\tilde{x}, \tilde{t}) - \tilde{T}_{2D}(\tilde{x}, \tilde{y}, \tilde{t})$$

where, for $z > 0$, $\text{sign}(z) = 1$ and $\text{sign}(-z) = -1$; while $\text{sign}(0) = 0$, as introduced by Oldham et al. (2009).

The first term is related to the temperature solution of the well-known 1D semi-infinite body subject to a constant surface heat flux, that is, $\tilde{T}_{\text{X20B1T0}}(\tilde{x}, \tilde{t})$ (Section 2.3.8). The second term, $\tilde{T}_{2D}(\tilde{x}, \tilde{y}, \tilde{t})$, accounts for the *two-dimensional* deviation effects on temperature due to the step change in heat flux along y at $y = W_0$ that disturbs the first 1D term, that is, the one-dimensional behavior of the thermal field along x when the time increases. It is in an integral form that can be computed using series expressions (Cole et al. 2011; see section 6.8.3, p. 219) with the only exception of the heated or cooled boundary surface $\tilde{x} = 0$ that is frequently the surface of greatest interest. At this boundary, in fact, Eq. (B.4) reduces to an algebraic expression as

$$\tilde{T}_{\text{X20Y20}}(0, \tilde{y}, \tilde{t}) = \frac{1 + \text{sign}(\tilde{W}_0 - \tilde{y})}{2} 2\sqrt{\frac{\tilde{t}}{\pi}} - \sqrt{\frac{\tilde{t}}{\pi}} \left\{ \text{sign}(\tilde{W}_0 - \tilde{y}) \text{erfc}\left(\frac{|\tilde{W}_0 - \tilde{y}|}{2\sqrt{\tilde{t}}}\right) \right.$$

$$\left. - \frac{(\tilde{W}_0 - \tilde{y})}{2\sqrt{\pi\tilde{t}}} E_1\left[\frac{(\tilde{W}_0 - \tilde{y})^2}{4\tilde{t}}\right] + \text{erfc}\left(\frac{\tilde{W}_0 + \tilde{y}}{2\sqrt{\tilde{t}}}\right) - \frac{(\tilde{W}_0 + \tilde{y})}{2\sqrt{\pi\tilde{t}}} E_1\left[\frac{(\tilde{W}_0 + \tilde{y})^2}{4\tilde{t}}\right] \right\} \qquad (\text{B.5a})$$

where the function $E_1(.)$ is the exponential integral (Cole et al. 2011; see App. I, table I.1) available in computer libraries (as an example, MATLAB). Note that Eq. (B.5a) was derived by applying an integration by parts to Eq. (B.4) for $\tilde{x} = 0$ and using the integral # 3.2, App. I, Table I.1, given by Cole et al. (2011).

Two locations are of greater concern at the heated or cooled surface: (i) the corner point $(\tilde{x}, \tilde{y}) = (0, 0)$, where the temperature variation is maximum at any time; and (ii) the edge point $(\tilde{x}, \tilde{y}) = (0, \tilde{W}_0)$ of the same heating or cooling. At these locations, Eq. (B.5a) simplifies, respectively, to

$$\tilde{T}_{\text{X20Y20}}(0, 0, \tilde{t}) = 2\sqrt{\frac{\tilde{t}}{\pi}} \text{erf}\left(\frac{\tilde{W}_0}{2\sqrt{\tilde{t}}}\right) + \frac{\tilde{W}_0}{\pi} E_1\left(\frac{\tilde{W}_0^2}{4\tilde{t}}\right) \qquad (\text{B.5b})$$

$$\tilde{T}_{\text{X20Y20}}(0, \tilde{W}_0, \tilde{t}) = \sqrt{\frac{\tilde{t}}{\pi}} \text{erf}\left(\frac{\tilde{W}_0}{\sqrt{\tilde{t}}}\right) + \frac{\tilde{W}_0}{\pi} E_1\left(\frac{\tilde{W}_0^2}{\tilde{t}}\right) \qquad (\text{B.5c})$$

B.2 Exact Analytical Solution. Large-Time Form

The *large-time* form comes from the application of separation-of-variables (SOV) method to the governing equations (Özişik 1993). Also, from the application of Green's function method when using "*large-cotime*" Green's functions (Cole et al. 2011). It is valid at any time, but it is computationally convenient at large times, that is, requires less terms than the companion short-time form. The large-time form of the solution may be seen as summation of four different problems (Najafi et al. 2015) whose physical significance is given in the work by Woodbury et al. (2017) to which the reader can refer. Alternatively, it may also be seen as summation of two parts, the quasi-steady part, and the complementary transient part, as shown by McMasters et al. (2019) for a generalized 2D problem, where a very low convection coefficient at both the "active" and "inactive" boundaries leads to the solution to the current problem. In a dimensionless form,

$$\tilde{T}_e(\tilde{x}, \tilde{y}, \tilde{t}) = \tilde{T}_{qs}(\tilde{x}, \tilde{y}, \tilde{t}) + \tilde{T}_{ct}(\tilde{x}, \tilde{y}, \tilde{t}) \qquad (\text{B.6})$$

where, by using large-cotime Green's functions, the quasi-steady part is given by the sum of three algebraic terms and a double summation (Najafi et al. 2015; Woodbury et al. 2017) as

$$
\begin{aligned}
\tilde{T}_{qs}(\tilde{x}, \tilde{y}, \tilde{t}) &= \frac{\tilde{W}_0}{\tilde{W}}\tilde{t} + \frac{\tilde{W}_0}{\tilde{W}}\tilde{T}_x^{(L)}(\tilde{x}) + \tilde{T}_y^{(L)}(\tilde{y}) + \sum_{m=1}^{\infty}\sum_{n=1}^{\infty}\tilde{T}_{xy,mn}^{(L)}(\tilde{x}, \tilde{y}) \\
&= \frac{\tilde{W}_0}{\tilde{W}}\tilde{t} + \frac{\tilde{W}_0}{\tilde{W}}\left(\frac{\tilde{x}^2}{2} - \tilde{x} + \frac{1}{3}\right) + \frac{1}{12\tilde{W}} \\
&\quad \cdot \left\{ \left[(\tilde{y} + \tilde{W}_0)^3 - 3\tilde{W}(\tilde{y} + \tilde{W}_0)^2 + 2\tilde{W}^2(\tilde{y} + \tilde{W}_0) \right] \right. \\
&\quad \left. - \left[(\tilde{y} - \tilde{W}_0)^3 - \text{sign}(\tilde{y} - \tilde{W}_0)3\tilde{W}(\tilde{y} - \tilde{W}_0)^2 + 2\tilde{W}^2(\tilde{y} - \tilde{W}_0) \right] \right\} \\
&\quad + 4\sum_{m=1}^{\infty}\sum_{n=1}^{\infty}\sin\left(\eta_n\frac{\tilde{W}_0}{\tilde{W}}\right)\frac{F_{x,m}(\tilde{x})F_{y,n}(\tilde{y})}{\eta_n\left[\beta_m^2 + (\eta_n/\tilde{W})^2\right]}
\end{aligned}
\tag{B.7}
$$

The $\eta_n = n\pi$ eigenvalues, with $n = 1, 2, \ldots$, are the eigenvalues in the homogeneous direction (y); while $F_{y,n}(\tilde{y}) = \cos\left(\eta_n\tilde{y}/\tilde{W}\right)$ is the corresponding n-th eigenfunction along y. They are typical of the Y22 case. Similarly, $\beta_m = m\pi$ and $F_{x,m}(\tilde{x}) = \cos(\beta_m\tilde{x})$, with $m = 1, 2, \ldots$, are the m-th eigenvalue and the corresponding m-th eigenfunction along x, respectively, that is the nonhomogeneous direction. They are typical of the X22 slab.

When $\tilde{W}_0 = \tilde{W}$, the heating or cooling is uniform, and the problem becomes 1D. In fact, in Eq. (B.7), the quantity between braces vanishes as well as the double summation due to $\sin\left(\eta_n\tilde{W}_0/\tilde{W}\right) = 0$. Thus, Eq. (B.7) reduces to

$$
\tilde{T}_{qs}(\tilde{x}, \tilde{t}) = \tilde{t} + \tilde{T}_x^{(L)}(\tilde{x}) = \tilde{t} + \left(\frac{\tilde{x}^2}{2} - \tilde{x} + \frac{1}{3}\right)
\tag{B.8}
$$

that is the quasi-steady part of the X22B10T0 case treated in Section 2.3.8. This is an application of symbolic intrinsic verification (Cole et al. 2014; D'Alessandro and de Monte 2018). Note that Eq. (B.8) divided by a factor of 2 is also valid when using Eq. (B.7) for $\tilde{y} = \tilde{W}_0 = \tilde{W}/2$. In such a case, in fact, $\sin\left(\eta_n\tilde{W}_0/\tilde{W}\right)F_{y,n}(\tilde{y}) = \left[\sin(n\pi)\right]/2 = 0$, and, hence, the double summation vanishes in Eq. (B.7) as well as $\tilde{T}_y^{(L)}(\tilde{y})$.

Now, the double summation in Eq. (B.7) can be reduced to a single summation in two different ways. For this reason, two different expressions of the quasi-steady part of the large-time form of the temperature solution are available: (i) the "*standard*" expression using eigenvalues in the homogeneous direction (y); and the "*nonstandard*" expression using eigenvalues in the nonhomogeneous direction (x). Both are valid at any point within the rectangular domain, but the computational efficiency (related to the number of terms) depends on the location of interest, as shown in Sections 2.4.3.4 and 2.4.3.5.

As regards the complementary transient part of Eq. (B.6), it does have only one expression that is analyzed in Section B.2.3. Also, as already said in Section B.1, if the partial heating/cooling at the $\tilde{x} = 0$ boundary of the domain occurs at time $\tilde{t} = \tilde{t}_0 > 0$, the temperature solution can still be derived by using the large-time forms of the solution listed before, where $\tilde{t}$ has however to be replaced with $\tilde{t}_0$. Therefore, $\tilde{T}_e(\tilde{x}, \tilde{y}, \tilde{t}) \to \tilde{T}_e(\tilde{x}, \tilde{y}, \tilde{t} - \tilde{t}_0)$, with $\tilde{t}_0 = \alpha t_0/L^2$. This is relevant when using the principle of superposition related to the piecewise-constant approximation of the time-dependent part of the surface heat flux as proposed in Section 3.6.2 of Chapter 3.

B.2.1 Quasi-Steady Part with Eigenvalues in the Homogeneous Direction

Following the procedure proposed by McMasters et al. (2018) for a companion problem, and still by McMasters et al. (2019) for a generalized problem, the double summation in Eq. (B.7) can be rearranged to

$$
\sum_{m=1}^{\infty}\sum_{n=1}^{\infty}\tilde{T}_{xy,mn}^{(L)}(\tilde{x}, \tilde{y}) = 2\sum_{n=1}^{\infty}\sin\left(\eta_n\frac{\tilde{W}_0}{\tilde{W}}\right)\frac{\cos\left(\eta_n\tilde{y}/\tilde{W}\right)}{\eta_n}2\sum_{m=1}^{\infty}\frac{\cos(\beta_m\tilde{x})}{\left[\beta_m^2 + (\eta_n/\tilde{W})^2\right]}
\tag{B.9}
$$

where the following algebraic identity can be used (Beck and Cole 2007; see Eq. (29b) and app. B; McMasters et al. 2019; see Eq. (23))

$$2 \sum_{m=1}^{\infty} \frac{\cos(\beta_m \tilde{x})}{\left[\beta_m^2 + \left(\eta_n / \tilde{W} \right)^2 \right]} = \frac{1}{(\eta_n / \tilde{W})} \frac{e^{-(\eta_n/\tilde{W})\tilde{x}} + e^{-(\eta_n/\tilde{W})(2-\tilde{x})}}{1 - e^{-2(\eta_n/\tilde{W})}} - \frac{1}{(\eta_n/\tilde{W})^2} \tag{B.10}$$

By substituting Eq. (B.10) in Eq. (B.9), the double summation reduces to two single summations with eigenvalues in the homogeneous direction plus an additional algebraic term as

$$\sum_{m=1}^{\infty} \sum_{n=1}^{\infty} \tilde{T}_{xy,mn}^{(L)}(\tilde{x}, \tilde{y}) = 2\tilde{W} \sum_{n=1}^{\infty} \sin\left(\eta_n \frac{\tilde{W}_0}{\tilde{W}} \right) \frac{\cos(\eta_n \tilde{y}/\tilde{W})}{\eta_n^2} \frac{e^{-(\eta_n/\tilde{W})\tilde{x}} + e^{-(\eta_n/\tilde{W})(2-\tilde{x})}}{1 - e^{-2(\eta_n/\tilde{W})}}$$
$$- \tilde{T}_y^{(L)}(\tilde{y}) \tag{B.11}$$

where $\tilde{T}_y^{(L)}(\tilde{y})$ is defined in Eq. (B.7). Also, the following identity was used (Cole et al. 2011; see App. F, #4, p. 503; Woodbury et al. 2017; see Eq. (13)):

$$2\tilde{W}^2 \sum_{n=1}^{\infty} \sin\left(\eta_n \frac{\tilde{W}_0}{\tilde{W}} \right) \frac{\cos(\eta_n \tilde{y}/\tilde{W})}{\eta_n^3} = \tilde{T}_y^{(L)}(\tilde{y}) \tag{B.12}$$

to obtain Eq. (B.11).

The single n-summation on the RHS of Eq. (B.11) has two parts. The former converges exponentially (fast) for all values of $\tilde{x}$ and $\tilde{y}$ except at $\tilde{x} = 0$, which is the heated or cooled surface, where converges algebraically (slowly; that is, requires many terms) due to $1/\eta_n^2$. Also, it converges slowly near $\tilde{x} = 0$. The latter converges exponentially for any location. The convergence of the above two single n-summations is discussed in Section 2.4.3.4.

Substituting Eq. (B.11) in Eq. (B.7) yields the "*standard*" expression of the quasi-steady part of the large-time solution, denoted by $\tilde{T}_{qs,y}(\tilde{x}, \tilde{y}, \tilde{t})$, where the subscript "$y$" indicates the homogeneous direction. The standard expression is efficient for any location provided $\tilde{x}$ is not equal to zero or near zero, as shown in Section 2.4.3.4. Note that applying the SOV method (Özişik 1993) to the governing equations leads the same expression.

B.2.2 Quasi-Steady Part with Eigenvalues in the Nonhomogeneous Direction

Following the procedure proposed by Woodbury et al. (2017), the double summation in Eq. (B.7) can be rearranged to

$$\sum_{m=1}^{\infty} \sum_{n=1}^{\infty} \tilde{T}_{xy,mn}^{(L)}(\tilde{x}, \tilde{y}) = 2 \sum_{m=1}^{\infty} \cos(\beta_m \tilde{x}) 2\tilde{W}^2 \sum_{n=1}^{\infty} \sin\left(\eta_n \frac{\tilde{W}_0}{\tilde{W}} \right) \frac{\cos(\eta_n \tilde{y}/\tilde{W})}{\eta_n \left[(\beta_m \tilde{W})^2 + \eta_n^2 \right]} \tag{B.13}$$

where the following algebraic identity can be used (Beck 2011; see Appendix C; Woodbury et al. 2017; see Eq. (19))

$$2\tilde{W}^2 \sum_{n=1}^{\infty} \sin\left(\eta_n \frac{\tilde{W}_0}{\tilde{W}} \right) \frac{\cos(\eta_n \tilde{y}/\tilde{W})}{\eta_n \left[(\beta_m \tilde{W})^2 + \eta_n^2 \right]} = \frac{1 - \text{sign}(\tilde{y} - \tilde{W}_0) - 2\tilde{W}_0/\tilde{W}}{2\beta_m^2}$$
$$+ \text{sign}(\tilde{y} - \tilde{W}_0) \frac{e^{-\beta_m|\tilde{y} - \tilde{W}_0|} - e^{-\beta_m(2\tilde{W} - |\tilde{y} - \tilde{W}_0|)}}{2\beta_m^2 \left(1 - e^{-2\beta_m \tilde{W}} \right)}$$
$$- \frac{e^{-\beta_m(\tilde{y} + \tilde{W}_0)} - e^{-\beta_m[2\tilde{W} - (\tilde{y} + \tilde{W}_0)]}}{2\beta_m^2 \left(1 - e^{-2\beta_m \tilde{W}} \right)} \tag{B.14}$$

By substituting Eq. (B.14) in Eq. (B.13), the double summation reduces to four single summations with eigenvalues in the nonhomogeneous direction plus an additional term as

$$
\sum_{m=1}^{\infty} \sum_{n=1}^{\infty} \tilde{T}_{xy,mn}^{(L)}(\tilde{x}, \tilde{y}) = \frac{1 - 2\tilde{W}_0/\tilde{W} - \mathrm{sign}\left(\tilde{y} - \tilde{W}_0\right)}{2} \tilde{T}_x^{(L)}(\tilde{x})
$$

$$
+ \mathrm{sign}\left(\tilde{y} - \tilde{W}_0\right) \sum_{m=1}^{\infty} \cos\left(\beta_m \tilde{x}\right) \frac{e^{-\beta_m |\tilde{y} - \tilde{W}_0|} - e^{-\beta_m \left(2\tilde{W} - |\tilde{y} - \tilde{W}_0|\right)}}{\beta_m^2 \left(1 - e^{-2\beta_m \tilde{W}}\right)}
\tag{B.15}
$$

$$
- \sum_{m=1}^{\infty} \cos\left(\beta_m \tilde{x}\right) \frac{e^{-\beta_m \left(\tilde{y} + \tilde{W}_0\right)} - e^{-\beta_m \left[2\tilde{W} - \left(\tilde{y} + \tilde{W}_0\right)\right]}}{\beta_m^2 \left(1 - e^{-2\beta_m \tilde{W}}\right)}
$$

where $\tilde{T}_x^{(L)}(\tilde{x})$ is defined in Eq. (B.7). Also, the following identity was used (Cole et al. 2011; see Eq. (6.95), p. 207):

$$
2 \sum_{m=1}^{\infty} \frac{\cos\left(\beta_m \tilde{x}\right)}{\beta_m^2} = \tilde{T}_x^{(L)}(\tilde{x})
\tag{B.16}
$$

to obtain Eq. (B.15).

The first m-summation on the RHS of Eq. (B.15) has two parts. The former converges exponentially for all values of $\tilde{x}$ and $\tilde{y}$ but converges slowly for $\tilde{y}$ near (but not equal to) $\tilde{W}_0$. In fact, for $\tilde{y} = \tilde{W}_0$ this summation vanishes as $\mathrm{sign}(\tilde{y} - \tilde{W}_0) = 0$. The latter converges exponentially for all values of $\tilde{x}$ and $\tilde{y}$. Similarly, the second m-summation on the RHS of Eq. (B.15) has two parts, and both of them converge exponentially for all values of $\tilde{x}$ and $\tilde{y}$. The convergence of the above four m-summations is discussed in Section 2.4.3.5.

Substituting Eq. (B.15) in Eq. (B.7) yields the "*nonstandard*" expression of the quasi-steady part of the large-time solution, denoted by $\tilde{T}_{qs,x}(\tilde{x}, \tilde{y}, \tilde{t})$, where the subscript "$x$" indicates the nonhomogeneous direction. The nonstandard expression is efficient for any location provided $\tilde{y}$ is not near $\tilde{W}_0$, as shown in Section 2.4.3.5. Therefore, the nonstandard expression is complementary to the standard one given in the previous section but much more important than this because the heated or cooled surface, $\tilde{x} = 0$, is of primary interest. However, the weakness of the standard expression at the active surface is not commonly recognized in the specialized heat conduction literature.

B.2.3 Complementary Transient Part

This part is given by (Najafi et al. 2015; Woodbury et al. 2017)

$$
\tilde{T}_{ct}(\tilde{x}, \tilde{y}, \tilde{t}) = -2 \frac{\tilde{W}_0}{\tilde{W}} \sum_{m=1}^{\infty} \frac{F_{x,m}(\tilde{x})}{\beta_m^2} e^{-\beta_m^2 \tilde{t}} - 2 \sum_{n=1}^{\infty} \sin\left(\eta_n \frac{\tilde{W}_0}{\tilde{W}}\right) \frac{F_{y,n}(\tilde{y})}{\eta_n \left(\eta_n/\tilde{W}\right)^2} e^{-\left(\frac{\eta_n}{\tilde{W}}\right)^2 \tilde{t}}
$$

$$
- 4 \sum_{m=1}^{\infty} \sum_{n=1}^{\infty} \sin\left(\eta_n \frac{\tilde{W}_0}{\tilde{W}}\right) \frac{F_{x,m}(\tilde{x}) F_{y,n}(\tilde{y})}{\eta_n \left[\beta_m^2 + \left(\eta_n/\tilde{W}\right)^2\right]} e^{-\left[\beta_m^2 + \left(\frac{\eta_n}{\tilde{W}}\right)^2\right]\tilde{t}}
\tag{B.17}
$$

where the two single summations and the double summation converge exponentially.

As they tend to have the eigenvalues to the second power in the exponentials, in contrast to quasi-steady cases where the eigenvalues in the exponent are to the first power, as shown in the previous two sections, the complementary transient solution converges faster than the quasi-steady one, that is, it requires less terms.

When $\tilde{W}_0 = \tilde{W}$, the heating or cooling is uniform, and the problem becomes 1D. In fact, in Eq. (B.17), the n-summation vanishes as well as the double summation due to $\sin\left(\eta_n \tilde{W}_0/\tilde{W}\right) = 0$. Thus, Eq. (B.17) reduces to

$$
\tilde{T}_{ct}(\tilde{x}, \tilde{t}) = -2 \sum_{m=1}^{\infty} \frac{\cos\left(\beta_m \tilde{x}\right)}{\beta_m^2} e^{-\beta_m^2 \tilde{t}}
\tag{B.18}
$$

that is the complementary transient part of the X22B10T0 case treated in Section 2.3.8. This is an application of symbolic intrinsic verification (Cole et al. 2014; D'Alessandro and de Monte 2018). Note that Eq. (B.18) divided by a factor of 2 is also valid when using Eq. (B.17) for $\tilde{y} = \tilde{W}_0 = \tilde{W}/2$. In such a case, in fact, $\sin\left(\eta_n \tilde{W}_0/\tilde{W}\right) F_{y,n}(\tilde{y}) = [\sin(n\pi)]/2 = 0$, and, hence, the double summation vanishes as well as the single n-summation in Eq. (B.17).

References

Beck, J. V. (2011) Transient three-dimensional heat conduction problems with partial heating, *International Journal of Heat and Mass Transfer*, 54(11–12), pp. 2479–2489. https://doi.org/10.1016/j.ijheatmasstransfer.2011.02.014.

Beck, J. V. and Cole, K. D. (2007) Improving convergence of summations in heat conduction, *International Journal of Heat and Mass Transfer*, 50(1–2), pp. 257–268. https://doi.org/10.1016/j.ijheatmasstransfer.2006.06.032.

Cole, K. D., Beck, J. V., Haji-Sheikh, A. and Litkouhi, B. (2011) *Heat Conduction using Green's Functions*, 2nd edn. Boca Raton, FL: CRC Press.

Cole, K. D., Beck, J. V., Woodbury, K. A. and de Monte, F. (2014) Intrinsic verification and a heat conduction database, *International Journal of Thermal Sciences*, 78, pp. 36–47. http://dx.doi.org/10.1016/j.ijthermalsci.2013.11.002.

D'Alessandro, G. and de Monte, F. (2018) Intrinsic verification of an exact analytical solution in transient heat conduction, *Computational Thermal Sciences*, 10(3), pp. 251–272. http://dx.doi.org/10.1615/ComputThermalScien.2017021201.

McMasters, R. L., de Monte, F. and Beck, J. V. (2019) Generalized solution for two-dimensional transient heat conduction problems with partial heating near a corner, *ASME Journal of Heat Transfer*, 141(7), pp. 071301-1–071301-8. https://doi.org/10.1115/1.4043568.

McMasters, R. L., de Monte, F., Beck, J. V. and Amos, D.E. (2018) Transient two-dimensional heat conduction problem with partial heating near corners, *ASME Journal of Heat Transfer*, 140(2), pp. 021301-1–021301-10. https://doi.org/10.1115/1.4037542.

Najafi, H., Woodbury, K. A. and Beck, J. V. (2015) Real time solution for inverse heat conduction problems in a two-dimensional plate with multiple heat fluxes at the surface, *International Journal of Heat and Mass Transfer*, 91, pp. 1148–1156. http://dx.doi.org/10.1016/j.ijheatmasstransfer.2015.08.020.

Oldham, K., Myland, J. and Spanier, J. (2009) *An Atlas of Functions*. 2nd edn. New York: Springer.

Özişik, M.N. (1993) *Heat Conduction*. 2nd edn. New York: Wiley.

Woodbury, K. A., Najafi, H. and Beck, J. V. (2017) Exact analytical solution for 2-D transient heat conduction in a rectangle with partial heating on one edge, *International Journal of Thermal Sciences*, 112, pp. 252–262. https://doi.org/10.1016/j.ijthermalsci.2016.10.014.

Appendix C

Green's Functions Solution Equation

C.1 Introduction

In this appendix, it is proven that the superposition-based numerical approximation of the temperature solution of a body subject to an arbitrary boundary condition of the first and second kinds, as derived in Chapter 3 using piecewise changes of the boundary function, can also be interpreted using Green's functions (GFs).

C.2 One-Dimensional Problem with Time-Dependent Surface Temperature

Consider the X12B-0T1 problem of Section 3.3. The solution to this linear problem may be derived in an integral form of convolution-type by using the well-established Green's function solution equation (GFSE) for one-dimensional transient rectangular problems (Cole et al. 2011; see Eq. (3.16)). Its application to the X12B-0T1 case gives

$$
T(x,t) = T_{in} \int_{x'=0}^{L} G_{X12}(x,x',t-\tau)\big|_{\tau=0} dx' + \alpha \int_{\tau=0}^{t} T_0(\tau)\frac{\partial}{\partial x'} G_{X12}(x,x',t-\tau)\bigg|_{x'=0} d\tau \tag{C.1}
$$

where the first term on the right-hand side (RHS) of the equation is the contribution of the uniform initial condition to the temperature $T(x,t)$, while the second term represents the contribution due to the nonhomogeneous boundary condition of the first kind at $x=0$. The BC at $x=L$ does not give any contribution as it is homogeneous.

As regards $G_{X12}(x,x',t-\tau)$ having units of m^{-1}, it is the GF of the X12 slab where the point of interest is at (x,t) and the local instantaneous source is at (x',τ) according to the GF significance (Cole et al. 2011). Also, $t-\tau=u$ is the convolution time or, simply, "cotime." For the X12 GF, two different expressions are available in the specialized heat conduction literature (Cole et al. 2011; see Appendix X, pp. 588-590):

- The *short-cotime* form. It is valid at any cotime, but it is computationally convenient at short cotimes, that is, when computing the temperature $T(x,t)$ given by Eq. (C.1) at early times. It comes from Laplace transform and is defined as

$$
G_{X12}(x,x',t-\tau) = \frac{1}{\sqrt{4\pi\alpha(t-\tau)}} \sum_{m=-\infty}^{\infty} (-1)^m \left[e^{-\frac{(2mL+x-x')^2}{4\alpha(t-\tau)}} - e^{-\frac{(2mL+x+x')^2}{4\alpha(t-\tau)}} \right] \tag{C.2a}
$$

- The *large-cotime* form. It is valid at any cotime, but it is computationally convenient at large cotimes, that is, when determining $T(x,t)$ at large times. It comes from the application of separation-of-variables method and may be taken as

$$
G_{X12}(x,x',t-\tau) = \frac{2}{L} \sum_{m=1}^{\infty} e^{-\beta_m^2 \alpha(t-\tau)/L^2} \sin\left(\beta_m \frac{x}{L}\right) \sin\left(\beta_m \frac{x'}{L}\right) \tag{C.2b}
$$

where $\beta_m = (m-1/2)\pi$ is the m-th eigenvalue of the X12 case.

Inverse Heat Conduction: Ill-Posed Problems, Second Edition. Keith A. Woodbury, Hamidreza Najafi, Filippo de Monte, and James V. Beck.
© 2023 John Wiley & Sons, Inc. Published 2023 by John Wiley & Sons, Inc.

By some algebra, Eq. (C.1) may be rearranged as

$$T(x, t_M) = T_M(x) = T_{in} + \alpha \int_{\tau=0}^{t_M} \theta_0(\tau) \left| \frac{\partial}{\partial x'} G_{X12}(x, x', t_M - \tau) \right|_{x'=0} d\tau \tag{C.3}$$

where $\theta_0(\tau) = T_0(\tau) - T_{in}$ and $t = t_M$ is the time of interest.

C.2.1 Piecewise-Constant Approximation

This approximation gives the piecewise profile of Figure 3.1, fully described in Section 3.3.1. The temperature solution $T_i(x, t_M)$ at time t_M within the flat plate due only to the i-th surface temperature rise component $\theta_{0,i}$ applied at $x = 0$ for the finite period of time between t_{i-1} and t_i, with $i \leq M$, as shown in Figure 3.2, when the plate is initially at zero temperature, may be calculated by applying Eq. (C.3) that is valid for any $\theta_0(t)$ function. It is found that

$$T_i(x, t_M) = T_{i,M}(x) = \alpha \int_{\tau=0}^{t_M} \theta_{0,i}(\tau) \frac{\partial}{\partial x'} G_{X12}(x, x', t_M - \tau) \Big|_{x'=0} d\tau \tag{C.4}$$

where $i = 1, 2, ..., M$.

By applying the principle of superposition (based on subtraction) to $\theta_{0,i}(t)$, it results in

$$\theta_{0,i}(t) = \underbrace{H(t - t_{i-1}) \times \theta_{0,i}}_{\theta_{0,i}^{(a)}(t)} - \underbrace{H(t - t_i) \times \theta_{0,i}}_{\theta_{0,i}^{(b)}(t)} = \theta_{0,i}^{(a)}(t) - \theta_{0,i}^{(b)}(t) \tag{C.5}$$

where $H(.)$ is the Heaviside unit step function in time, while $\theta_{0,i}^{(a)}(t)$ and $\theta_{0,i}^{(b)}(t)$ are shown in Figure C.1a and Figure C.1b, respectively.

Therefore, Eq. (C.4) becomes

$$T_{i,M}(x) = \theta_{0,i} \left[\alpha \int_{\tau=t_{i-1}}^{t_M} \frac{\partial}{\partial x'} G_{X12}(x, x', t_M - \tau) \Big|_{x'=0} d\tau \right. \\ \left. - \alpha \int_{\tau=t_i}^{t_M} \frac{\partial}{\partial x'} G_{X12}(x, x', t_M - \tau) \Big|_{x'=0} d\tau \right] \tag{C.6}$$

where the first quantity between square brackets (dimensionless) is the temperature rise at time t_M due to a "*unit*" (1 °C) step increase of the surface temperature rise applied at $t = t_{i-1}$, with $i \leq M$; while the second quantity between the same brackets (dimensionless) is the temperature rise at the same time t_M, still due to a "*unit*" step increase of the surface temperature, but applied at $t = t_i$, with $i \leq M$.

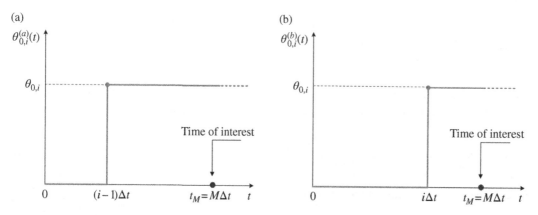

(a)

(b)

Figure C.1 Principle of superposition applied to the constant surface temperature-rise component $\theta_{0,i}(t)$ over the time interval $t = (i-1)\Delta t \rightarrow i\Delta t$ defined by Eq. (C.5) and shown in Fig. 3.2: (a) $\theta_{0,i}^{(a)}(t)$ with $t \in [(i-1)\Delta t, \infty)$; and (b) $\theta_{0,i}^{(b)}(t)$ with $t \in [i\Delta t, \infty)$.

Therefore, the first quantity between square brackets is the dimensionless temperature solution $\tilde{T}_{\text{X12B10T0}}$ of the X12B10T0 problem treated in Section 2.3.6 calculated at $t = t_M$ when the sudden variation of the surface temperature rise occurs at $t = t_{i-1}$, that is, $\tilde{T}_{\text{X12B10T0}}(\tilde{x}, \tilde{t}_M - \tilde{t}_{i-1})$, where $\tilde{t}_M - \tilde{t}_{i-1} = \tilde{t}_{M-i+1} = (M-i+1)\Delta\tilde{t}$. Similarly, the second quantity between square brackets in Eq. (C.6) is the dimensionless temperature solution of the same case computed at $t = t_M$ when the sudden variation of the surface temperature rise occurs at $t = t_i$, that is, $\tilde{T}_{\text{X12B10T0}}(\tilde{x}, \tilde{t}_M - \tilde{t}_i)$, where $\tilde{t}_M - \tilde{t}_i = \tilde{t}_{M-i} = (M-i)\Delta\tilde{t}$. For the sake of brevity, the above two dimensionless temperatures are denoted in Section 3.3.1 by $\varphi(\tilde{x}, \tilde{t}_{M-i+1}) = \varphi_{x,M-i+1}$ and $\varphi(\tilde{x}, \tilde{t}_{M-i}) = \varphi_{x,M-i}$, respectively.

Then, Eq. (C.6) becomes

$$T_{i,M}(x) = \theta_{0,i}\left(\varphi_{x,M-i+1} - \varphi_{x,M-i}\right) = \theta_{0,i}\Delta\varphi_{x,M-i} \tag{C.7}$$

C.2.1.1 Numerical Approximation of the GF Equation

To compute the temperature $T(x, t)$ at time $t = t_M$ due to $\theta_0(t)$ applied up to $t = t_M$, the principle of superposition valid for linear problems can now be used. Therefore, the temperature solution $T(x, t_M)$ may be calculated summing up all the contributions $T_{i,M}(x)$ due to the corresponding surface temperature rise components $\theta_{0,i}(t) = \theta_{0,i}$, each applied over its own time range $[(i-1)\Delta t, i\Delta t]$, with $i \leq M$. Bearing also in mind that the plate is initially at the T_{in} uniform temperature that is in general other than zero, it results exactly in Eq. (3.12) derived without using GFs.

For this reason, Eq. (3.12) may be considered as the numerical form of the "*temperature-based*" GF solution Eq. (C.3) when dealing with a "*piecewise-constant*" approximate-in-time profile of the applied surface temperature rise.

C.2.2 Piecewise-Linear Approximation

This approximation gives the piecewise profile of Figure 3.3, fully described in Section 3.3.2. In such a case, the i-th surface temperature rise component, $\theta_{0,i}$, depends linearly on the time as given by Eq. (3.25) and companion expressions Eqs. (3.26a)-(3.26c).

The temperature solution $T_i(x, t)$ at time $t = t_M$, that is, $T_i(x, t_M) = T_{i,M}(x)$, due only to the surface temperature rise component $\theta_{0,i}(t)$ (with $i \leq M$) given by Eq. (3.25) and shown in Figure 3.4, when the plate is initially at zero temperature, may be calculated by applying Eq. (C.3) that is valid for any $\theta_0(t)$ function. It results in Eq. (C.4), where however $\theta_{0,i}(\tau)$ is defined by Eq. (3.25) with $t \to \tau$. Also, $i = 1, 2, \ldots, M$.

By applying the principle of superposition (based on subtraction) to $\theta_{0,i}(t)$, it results in

$$\theta_{0,i}(t) = \theta_{0,i}^{(a)}(t) - \theta_{0,i}^{(b)}(t), \qquad t \in [(i-1)\Delta t, i\Delta t] \tag{C.8}$$

where $\theta_{0,i}^{(a)}(t)$ and $\theta_{0,i}^{(b)}(t)$ are shown in Figure C.2a and Figure C.2b, respectively, and are defined as

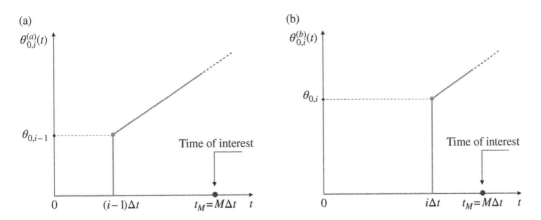

Figure C.2 Principle of superposition applied to the linear-in-time surface temperature rise component $\theta_{0,i}(t)$ defined by Eq. (3.25) and shown in Figure 3.4: (a) $\theta_{0,i}^{(a)}(t)$ with $t \in [(i-1)\Delta t, \infty)$ defined by Eq. (C.9a); and (b) $\theta_{0,i}^{(b)}(t)$ with $t \in [i\Delta t, \infty)$ defined by Eq. (C.9b).

$$\theta_{0,i}^{(a)}(t) = H(t-t_{i-1}) \times \left(\theta_{0,i-1} + \Delta\theta_{0,i}\frac{t-t_{i-1}}{\Delta t}\right) \tag{C.9a}$$

$$\theta_{0,i}^{(b)}(t) = H(t-t_i) \times \left(\theta_{0,i} + \Delta\theta_{0,i}\frac{t-t_i}{\Delta t}\right) \tag{C.9b}$$

Therefore, Eq. (C.4) becomes

$$T_{i,M}(x) = \alpha \underbrace{\int\limits_{\tau=t_{i-1}}^{t_M} \theta_{0,i}^{(a)}(\tau)\frac{\partial}{\partial x'}G_{X12}(x,x',t_M-\tau)\bigg|_{x'=0} d\tau}_{I_{i-1,M}}$$

$$-\alpha \underbrace{\int\limits_{\tau=t_i}^{t_M} \theta_{0,i}^{(b)}(\tau)\frac{\partial}{\partial x'}G_{X12}(x,x',t_M-\tau)\bigg|_{x'=0} d\tau}_{I_{i,M}} \tag{C.10}$$

where, bearing in mind Eqs. (C.9a) and (C.9b), either integral may be split into two parts as follows

$$I_{i-1,M} = \theta_{0,i-1}\left[\alpha \int\limits_{\tau=t_{i-1}}^{t_M} \frac{\partial}{\partial x'}G_{X12}(x,x',t_M-\tau)\bigg|_{x'=0} d\tau\right]$$

$$+ \Delta\theta_{0,i}\left[\alpha \int\limits_{\tau=t_{i-1}}^{t_M} \frac{(\tau-t_{i-1})}{\Delta t}\frac{\partial}{\partial x'}G_{X12}(x,x',t_M-\tau)\bigg|_{x'=0} d\tau\right] \tag{C.11}$$

$$I_{i,M} = \theta_{0,i}\left[\alpha \int\limits_{\tau=t_i}^{t_M} \frac{\partial}{\partial x'}G_{X12}(x,x',t_M-\tau)\bigg|_{x'=0} d\tau\right]$$

$$+ \Delta\theta_{0,i}\left[\alpha \int\limits_{\tau=t_i}^{t_M} \frac{(\tau-t_i)}{\Delta t}\frac{\partial}{\partial x'}G_{X12}(x,x',t_M-\tau)\bigg|_{x'=0} d\tau\right] \tag{C.12}$$

C.2.2.1 Superposition of Solutions at the (*i*-1)-th Time Step

The first quantity between square brackets in Eq. (C.11) (dimensionless) is the temperature rise at time t_M due to a "*unit*" (1 °C) step increase of the surface temperature rise applied at $t = t_{i-1}$, with $i \leq M$, as shown in Figure C.3a setting $\theta_{0,i-1} = 1\,°C$. Therefore, it is the dimensionless temperature solution $\tilde{T}_{X12B10T0}$ of the X12B10T0 problem calculated at $t = t_M$ when the sudden variation of the surface temperature occurs at $t = t_{i-1}$, that is, $\tilde{T}_{X12B10T0}(\tilde{x},\tilde{t}_M-\tilde{t}_{i-1})$, where $\tilde{t}_M-\tilde{t}_{i-1} = \tilde{t}_{M-i+1} = (M-i+1)\Delta\tilde{t}$ with $\Delta\tilde{t} = \alpha\Delta t/L^2$. It is denoted by $\varphi_{x,M-i+1}$ according to Section 3.3.1.

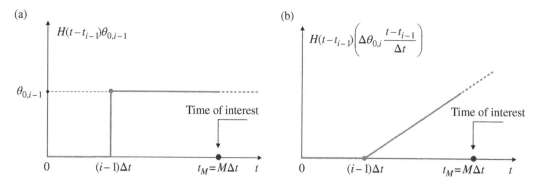

Figure C.3 Basic functions deriving from superposition applied to the linear-in-time function $\theta_{0,i}^{(a)}(t)$ of Figure C.2a, with $t \geq (i-1)\Delta t$: (a) constant; and (b) linear.

As regards the second quantity between square brackets in Eq. (C.11) (dimensionless), it is the temperature rise at time t_M due to a "*unit*" (1 °C) linear-in-time increase of the surface temperature rise over the time Δt (s) applied at $t = t_{i-1}$, with $i \leq M$, as shown in Figure C.3b setting $\Delta\theta_{0,i} = 1$ °C (i.e., the heating rate is of 1 °C/Δt s). Therefore, it is the dimensionless temperature solution $\tilde{T}_{\text{X12B20T0}}$ calculated at $t = t_M$ of the X12B20T0 problem (see Section 2.3.7) when the linear-in-time variation of the surface temperature occurs at $t = t_{i-1}$ with $t_{ref} = \Delta t$, that is, $\tilde{T}_{\text{X12B20T0}}(\tilde{x}, \tilde{t}_M - \tilde{t}_{i-1}, \tilde{t}_{ref} = \Delta\tilde{t})$, where $\tilde{t}_M - \tilde{t}_{i-1} = \tilde{t}_{M-i+1} = (M - i + 1)\Delta\tilde{t}$. It is denoted by $\mu(\tilde{x}, \tilde{t}_{M-i+1}) = \mu_{x,M-i+1}$ according to Section 3.3.2.

Thus, Eq. (C.11) becomes

$$
\begin{aligned}
I_{i-1,M} &= \theta_{0,i-1}\varphi_{x,M-i+1} + \Delta\theta_{0,i}\mu_{x,M-i+1} \\
&= \theta_{0,i-1}\left(\varphi_{x,M-i+1} - \mu_{x,M-i+1}\right) + \theta_{0,i}\mu_{x,M-i+1}
\end{aligned}
\tag{C.13}
$$

where $\Delta\theta_{0,i} = \theta_{0,i} - \theta_{0,i-1}$ was used to get the last expression.

Note that the superposition of the two surface temperature rises shown in Figure C.3 gives $\theta_{0,i}^{(a)}(t)$ depicted in Figure C.2a.

C.2.2.2 Superposition of Solutions at the *i*-th Time Step

The first quantity between square brackets in Eq. (C.12) (dimensionless) is the temperature rise at time t_M due to a "*unit*" (1 °C) step increase of the surface temperature rise applied at $t = t_i$, with $i \leq M$, as depicted in Figure C.4a setting $\theta_{0,i} = 1$ °C. Therefore, it is the dimensionless temperature solution $\tilde{T}_{\text{X12B10T0}}$ of the X12B10T0 case calculated at $t = t_M$ when the sudden variation of the surface temperature occurs at $t = t_i$, that is, $\tilde{T}_{\text{X12B10T0}}(\tilde{x}, \tilde{t}_M - \tilde{t}_i)$, where $\tilde{t}_M - \tilde{t}_i = \tilde{t}_{M-i} = (M - i)\Delta\tilde{t}$. It is denoted by $\varphi(\tilde{x}, \tilde{t}_{M-i}) = \varphi_{x,M-i}$.

As regards the second quantity between square brackets appearing in Eq. (C.12) (dimensionless), it is the temperature rise at time t_M due to a "*unit*" (1 °C) linear-in-time increase of the surface temperature rise over the time Δt (s) applied at $t = t_i$, with $i \leq M$, as shown in Figure C.4b setting $\Delta\theta_{0,i} = 1$ °C (i.e., the heating rate is of 1 °C/Δt s). In other words, it is the dimensionless temperature solution $\tilde{T}_{\text{X12B20T0}}$ calculated at $t = t_M$ of the X12B20T0 problem when the linear-in-time variation of the surface temperature occurs at $t = t_i$ with $t_{ref} = \Delta t$, that is, $\tilde{T}_{\text{X12B20T0}}(\tilde{x}, \tilde{t}_M - \tilde{t}_i, \tilde{t}_{ref} = \Delta\tilde{t})$, where $\tilde{t}_M - \tilde{t}_i = \tilde{t}_{M-i} = (M - i)\Delta\tilde{t}$. It is denoted by $\mu(\tilde{x}, \tilde{t}_{M-i}) = \mu_{x,M-i}$.

Thus, Eq. (C.12) becomes

$$
I_{i,M} = \theta_{0,i}\varphi_{x,M-i} + \Delta\theta_{0,i}\mu_{x,M-i} = \theta_{0,i}\left(\varphi_{x,M-i} + \mu_{x,M-i}\right) - \theta_{0,i-1}\mu_{x,M-i}
\tag{C.14}
$$

where $\Delta\theta_{0,i} = \theta_{0,i} - \theta_{0,i-1}$ was used to get the last expression of Eq. (C.14).

Note that the superposition of the two surface temperature rises shown in Figure C.4 gives the $\theta_{0,i}^{(b)}(t)$ boundary function depicted in Figure C.2b.

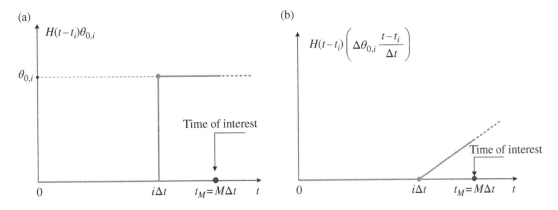

Figure C.4 Basic functions deriving from superposition applied to the linear-in-time function $\theta_{0,i}^{(b)}(t)$ of Figure C.2b, with $t \geq i\Delta t$: (a) constant; and (b) linear.

C.2.2.3 Numerical Approximation of the GF Equation

Substituting Eqs. (C.13) and (C.14) in Eq. (C.10) yields

$$T_{i,M}(x) = \theta_{0,i-1}\left(\varphi_{x,M-i+1} - \Delta\mu_{x,M-i}\right) + \theta_{0,i}\left(\Delta\mu_{x,M-i} - \varphi_{x,M-i}\right) \tag{C.15}$$

where

$$\Delta\mu_{x,M-i} = \mu_{x,M-i-1} - \mu_{x,M-i} \tag{C.16}$$

To compute now the temperature $T(x, t)$ at time $t = t_M$ due to $\theta_0(t)$ applied up to $t = t_M$, the principle of superposition can be utilized. Thus, the temperature solution $T(x, t_M) = T_M(x)$ may be calculated summing up all the contributions $T_{i,M}(x)$ given by Eq. (C.15) and due to the corresponding surface temperature rise components $\theta_{0,i}(t)$, defined by Eq. (3.25) and related Eqs. (3.26a)-(3.26c), each applied over its own time range $[(i-1)\Delta t, i\Delta t]$, with $i \leq M$. Bearing also in mind that the plate is initially at the T_{in} uniform temperature, it results exactly in Eq. (3.30) derived without using GFs and, hence, in Eq. (3.31).

For this reason, Eq. (3.31) may be considered as the numerical form of the "*temperature-based*" GF solution Eq. (C.3) when dealing with a "*piecewise-linear*" approximate-in-time profile of the applied surface temperature rise.

C.3 One-Dimensional Problem with Time-Dependent Surface Heat Flux

Consider the X22B-0T1 problem of Section 3.4. The solution to this linear problem may be derived by using the GFSE for 1D transient rectangular problems (Cole et al. 2011; see Eq. (3.16), pp. 66–67). Its application to the current X22B-0T1 problem gives

$$T(x,t) = T_{in}\int_{x'=0}^{L} G_{X22}(x,x',t-\tau)|_{\tau=0}\,dx' + \frac{\alpha}{k}\int_{\tau=0}^{t} q_0(\tau)G_{X22}(x,x'=0,t-\tau)\,d\tau \tag{C.17}$$

where the first term on the RHS of the equation is the contribution of the uniform initial condition to the temperature $T(x, t)$, while the second term is the contribution due to the nonhomogeneous boundary condition of the second kind at $x = 0$.

As regards $G_{X22}(x, x', t - \tau)$, it is the GF (units of m^{-1}) of the X22 slab for which two different expressions are available in the heat conduction literature (Cole et al. 2011; see appendix X, pp. 596–599):

- The *short-cotime* form. It is valid at any cotime $u = t - \tau$, but it is computationally convenient at short cotimes. It comes from LT approach and is defined as

$$G_{X22}(x,x',t-\tau) = \frac{1}{\sqrt{4\pi\alpha(t-\tau)}}\sum_{m=-\infty}^{\infty}\left[e^{-\frac{(2mL+x-x')^2}{4\alpha(t-\tau)}} + e^{-\frac{(2mL+x+x')^2}{4\alpha(t-\tau)}}\right] \tag{C.18a}$$

- The *large-cotime* form. It is valid at any cotime, but it is computationally convenient at large cotimes. It comes from SOV method and may be taken as

$$G_{X22}(x,x',t-\tau) = \frac{1}{L} + \frac{2}{L}\sum_{m=1}^{\infty} e^{-\beta_m^2\alpha(t-\tau)/L^2}\cos\left(\beta_m\frac{x}{L}\right)\cos\left(\beta_m\frac{x'}{L}\right) \tag{C.18b}$$

where $\beta_m = m\pi$ is the m-th eigenvalue of the X22 case.

As the former integral on the RHS of Eq. (C.17) is equal to 1 (Cole et al. 2011; see Eq. (X22.8), p. 599), the temperature solution Eq. (C.17) simplifies to

$$T(x,t_M) = T_{in} + \frac{\alpha}{k}\int_{\tau=0}^{t_M} q_0(\tau)G_{X22}(x,x'=0,t_M-\tau)\,d\tau \tag{C.19}$$

where $t = t_M$ is the time of interest.

C.3.1 Piecewise-Constant Approximation

This approximation gives the piecewise profile of Figure 3.6, that is a sequence of constant segments where $q_{0,i}$ (i-th surface heat flux component) is an approximation for $q_0(t)$ between $t_{i-1} = (i-1)\Delta t$ and $t_i = i\Delta t$ (with $i = 1, 2, ..., M$, being $t_M = M\Delta t$ the time of evaluation). The i-th surface heat flux component, $q_{0,i}$, is time-independent and may be identified with the time coordinate $t_{i-1/2} = (i-1/2)\Delta t$, that is, $q_0(t = t_{i-1/2}) = q_{0,i}$.

The temperature solution $T_i(x, t_M)$ at time t_M within the flat plate due only to the surface heat flux component $q_{0,i}$ of Figure 3.6, when the plate is initially at zero temperature, may be calculated by applying Eq. (C.19) that is valid for any $q_0(t)$ function. It is found that

$$T_i(x, t_M) = T_{i,M}(x) = \frac{\alpha}{k} \int_{\tau=0}^{t_M} q_{0,i}(\tau) G_{X22}(x, x' = 0, t_M - \tau) d\tau \tag{C.20}$$

where $i = 1, 2, ..., M$. Also, the subscript "M" used for $T_{i,M}(x)$ denotes that this temperature is calculated at time $t_M = M\Delta t \geq i\Delta t$.

By using the principle of superposition described in Section C.2.1 for the companion problem of surface temperature assigned (it is here sufficient to replace $\theta_{0,i}$ with $q_{0,i}$), Eq. (C.20) may be taken as

$$T_i(x, t_M) = T_{i,M}(x) = q_{0,i} \left[\frac{\alpha}{k} \int_{\tau=(i-1)\Delta t}^{M\Delta t} G_{X22}(x, x' = 0, t_M - \tau) d\tau \right. $$
$$\left. - \frac{\alpha}{k} \int_{\tau=i\Delta t}^{M\Delta t} G_{X22}(x, x' = 0, t_M - \tau) d\tau \right] \tag{C.21}$$

where the first quantity between square brackets (units of °C/(W m^{-2})) represents the temperature rise at time t_M due to a "*unit*" (1 W m^{-2}) step increase of the surface heat flux at $t = (i-1)\Delta t$, with $i \leq M$; while the second quantity between the same brackets (units of °C/(W m^{-2})) is the temperature rise at the same time t_M still due to a "*unit*" step increase of the surface heat flux but applied at $t = i\Delta t$, with $i \leq M$.

Therefore, the first quantity between square brackets in Eq. (C.21) is the temperature solution $\tilde{T}_{X22B10T0}$ of the X22B10T0 problem (apart from the factor L/k), treated in Section 2.3.8, calculated at $t = t_M$ when the sudden variation of the surface heat flux occurs at $t = t_{i-1}$, that is, $\tilde{T}_{X22B10T0}(\tilde{x}, \tilde{t}_M - \tilde{t}_{i-1}) \times (L/k)$, where $\tilde{t}_M - \tilde{t}_{i-1} = \tilde{t}_{M-i+1} = (M-i+1)\Delta\tilde{t}$. Similarly, the second quantity between square brackets in Eq. (C.21) is the temperature solution of the same case computed at $t = t_M$ when the sudden variation of the surface heat flux occurs at $t = t_i$, that is, $\tilde{T}_{X22B10T0}(\tilde{x}, \tilde{t}_M - \tilde{t}_i) \times (L/k)$, where $\tilde{t}_M - \tilde{t}_i = \tilde{t}_{M-i} = (M-i)\Delta\tilde{t}$. The above two temperatures per unit of surface heat flux (units of °C/(W m^{-2})) are denoted by $\phi(\tilde{x}, \tilde{t}_{M-i+1}) = \phi_{x,M-i+1}$ and $\phi(\tilde{x}, \tilde{t}_{M-i}) = \phi_{x,M-i}$, respectively, according to Section 3.4.1.

Then, Eq. (C.21) becomes

$$T_{i,M}(x) = q_{0,i}(\phi_{x,M-i+1} - \phi_{x,M-i}) = q_{0,i}\Delta\phi_{x,M-i} \tag{C.22}$$

C.3.1.1 Numerical Approximation of the GF Equation

To compute the temperature $T(x, t_M) = T_M(x)$ due to $q_0(t)$ applied up to $t_M = M\Delta t$, the principle of superposition valid for linear problems can now be used. Therefore, the temperature solution $T_M(x)$ may be calculated summing up all the contributions $T_{i,M}(x)$ given by Eq. (C.22) due to the corresponding surface heat flux components $q_{0,i}$, each applied over its own time range $[(i-1)\Delta t, i\Delta t]$, with $i \leq M$. Bearing also in mind that the plate is initially at the T_{in} uniform temperature, it results exactly in Eq. (3.47) derived without using GFs.

For this reason, Eq. (3.47) may be considered as the numerical form of the "*heat flux-based*" GF solution Eq. (C.19) when dealing with a "*piecewise-constant*" approximate-in-time profile of the surface heat flux.

C.3.2 Piecewise-Linear Approximation

This approximation gives the piecewise profile of Figure 3.7, where $q_{0,i}$ (*i*-th surface heat flux component) is a linear approximation for $q_0(t)$ between $t_{i-1} = (i-1)\Delta t$ and $t_i = i\Delta t$ (with $i = 1, 2, ..., M$, being $t_M = M\Delta t$ the time of interest).

Therefore, contrary to the piecewise-constant approximation of Section C.3.1, now $q_{0,i}$ depends linearly on the time as given by Eq. (3.57).

The temperature solution $T_i(x, t)$ at time $t = t_M$, that is, $T_i(x, t_M) = T_{i,M}(x)$, due only to the surface heat flux component $q_{0,i}(t)$ (with $i \leq M$) given by Eq. (3.57) and shown in Figure 3.7, when the plate is initially at zero temperature, may be derived by applying Eq. (C.19) that is valid for any $q_0(t)$ function. It results in Eq. (C.20), where however $q_{0,i}(\tau)$ is defined by Eq. (3.57) with $t \to \tau$. Also, $i = 1, 2, ..., M$.

By applying the principle of superposition (based on subtraction) described in Section C.2.2 for the companion problem of surface temperature prescribed (it is here sufficient to replace $\theta_{0,i}(t)$ with $q_{0,i}(t)$), Eq. (C.20) becomes

$$
\begin{aligned}
T_{i,M}(x) = \underbrace{\frac{\alpha}{k} \int\limits_{\tau = t_{i-1}}^{t_M} q_{0,i}^{(a)}(\tau) G_{X22}(x, x' = 0, t_M - \tau) d\tau}_{J_{i-1,M}} \\[2mm]
- \underbrace{\frac{\alpha}{k} \int\limits_{\tau = t_i}^{t_M} q_{0,i}^{(b)}(\tau) G_{X22}(x, x' = 0, t_M - \tau) d\tau}_{J_{i,M}}
\end{aligned}
\tag{C.23}
$$

where $q_{0,i}^{(a)}(t)$ and $q_{0,i}^{(b)}(t)$ may be taken, respectively, as

$$
q_{0,i}^{(a)}(t) = H(t - t_{i-1}) \times \left(q_{0,i-1} + \Delta q_{0,i} \frac{t - t_{i-1}}{\Delta t} \right),
\tag{C.24a}
$$

$$
q_{0,i}^{(b)}(t) = H(t - t_i) \times \left(q_{0,i} + \Delta q_{0,i} \frac{t - t_i}{\Delta t} \right),
\tag{C.24b}
$$

Thus, either integral of Eq. (C.23) may be split into two parts as follows

$$
\begin{aligned}
J_{i-1,M} = q_{0,i-1} \left[\frac{\alpha}{k} \int\limits_{\tau = t_{i-1}}^{t_M} G_{X22}(x, 0, t_M - \tau) d\tau \right] \\[2mm]
+ \Delta q_{0,i} \left[\frac{\alpha}{k} \int\limits_{\tau = t_{i-1}}^{t_M} \left(\frac{\tau - t_{i-1}}{\Delta t} \right) G_{X22}(x, 0, t_M - \tau) d\tau \right]
\end{aligned}
\tag{C.25}
$$

$$
\begin{aligned}
J_{i,M} = q_{0,i} \left[\frac{\alpha}{k} \int\limits_{\tau = t_i}^{t_M} G_{X22}(x, 0, t_M - \tau) d\tau \right] \\[2mm]
+ \Delta q_{0,i} \left[\frac{\alpha}{k} \int\limits_{\tau = t_i}^{t_M} \left(\frac{\tau - t_i}{\Delta t} \right) G_{X22}(x, 0, t_M - \tau) d\tau \right]
\end{aligned}
\tag{C.26}
$$

C.3.2.1 Superposition of Solutions at the (*i*-1)-th Time Step

The first quantity between square brackets in Eq. (C.25) is the temperature rise at time t_M due to a "*unit*" (1 W m^{-2}) step increase of the surface heat flux applied at $t = t_{i-1}$, with $i \leq M$. Therefore, this temperature has units of °C/(W m^{-2}) and (apart from the factor L/k) is the temperature solution $\tilde{T}_{X22B10T0}$ of the X22B10T0 problem calculated at $t = t_M$ when the sudden variation of the surface heat flux occurs at $t = t_{i-1}$, that is, $\tilde{T}_{X22B10T0}(\tilde{x}, \tilde{t}_M - \tilde{t}_{i-1}) \times (L/k)$, where $\tilde{t}_M - \tilde{t}_{i-1} = \tilde{t}_{M-i+1} = (M - i + 1)\Delta\tilde{t}$. It is denoted by $\phi(\tilde{x}, \tilde{t}_{M-i+1}) = \phi_{x,M-i+1}$.

As regards the second quantity between square brackets in Eq. (C.25) (units of °C/(W m^{-2})), it is the temperature rise at time t_M due to a "*unit*" (1 W m^{-2}) linear-in-time increase of the surface heat flux over the time Δt (s) applied at $t = t_{i-1}$, with

$i \leq M$ (i.e., due to 1 W m^{-2}/Δt s). Therefore, apart from the factor L/k, it is the temperature solution $\tilde{T}_{\text{X22B20T0}}$ calculated at $t = t_M$ of the X22B20T0 problem (see Section 2.3.9) when the linear-in-time variation of the surface heat flux occurs at $\tilde{t} = \tilde{t}_{i-1}$ with $\tilde{t}_{ref} = \Delta\tilde{t}$, that is, $\tilde{T}_{\text{X22B20T0}}(\tilde{x}, \tilde{t}_M - \tilde{t}_{i-1}, \tilde{t}_{ref} = \Delta\tilde{t}) \times (L/k)$, where $\tilde{t}_M - \tilde{t}_{i-1} = \tilde{t}_{M-i+1} = (M - i + 1)\Delta\tilde{t}$. It is here denoted by $\gamma(\tilde{x}, \tilde{t}_{M-i+1}) = \gamma_{x,M-i+1}$, according to Section 3.4.2.

Thus, Eq. (C.25) becomes

$$
\begin{aligned}
J_{i-1,M} &= q_{0,i-1}\phi_{x,M-i+1} + \Delta q_{0,i}\gamma_{x,M-i+1} \\
&= q_{0,i-1}(\phi_{x,M-i+1} - \gamma_{x,M-i+1}) + q_{0,i}\gamma_{x,M-i+1}
\end{aligned}
\tag{C.27}
$$

where $\Delta q_{0,i} = q_{0,i} - q_{0,i-1}$ was used to get the last expression.

C.3.2.2 Superposition of Solutions at the *i*-th Time Step

By using the same procedure of the previous section, Eq. (C.26) becomes

$$
J_{i,M} = q_{0,i}\phi_{x,M-i} + \Delta q_{0,i}\gamma_{x,M-i} = q_{0,i}(\phi_{x,M-i} - \gamma_{x,M-i}) + q_{0,i-1}\gamma_{x,M-i}
\tag{C.28}
$$

where

$$
\phi_{x,M-i} = \phi(\tilde{x}, \tilde{t}_{M-i}) = \frac{L}{k}\tilde{T}_{\text{X22B10T0}}(\tilde{x}, \tilde{t}_M - \tilde{t}_i)
\tag{C.29a}
$$

$$
\gamma_{x,M-i} = \gamma(\tilde{x}, \tilde{t}_{M-i}) = \frac{L}{k}\tilde{T}_{\text{X22B20T0}}(\tilde{x}, \tilde{t}_M - \tilde{t}_i, \tilde{t}_{ref} = \Delta\tilde{t})
\tag{C.29b}
$$

C.3.2.3 Numerical Approximation of the GF Equation

Substituting Eqs. (C.27) and (C.28) in Eq. (C.23) yields

$$
T_{i,M}(x) = q_{0,i-1}(\phi_{x,M-i+1} - \Delta\gamma_{x,M-i}) + q_{0,i}(\Delta\gamma_{x,M-i} - \phi_{x,M-i})
\tag{C.30}
$$

By using the principle of superposition and bearing in mind that the plate is initially at T_{in} uniform temperature, the solution $T(x, t_M) = T_M(x)$ due to $q_0(t)$ applied up to $t_M = M\Delta t$ may be calculated summing up all the contributions $T_{i,M}(x)$ given by Eq. (C.30). It results exactly in Eq. (3.61) obtained without using GFs and, hence, in Eq. (3.62).

For this reason, Eq. (3.62) may be considered as the numerical form of the "*heat flux-based*" GF solution Eq. (C.19) when dealing with a "*piecewise-linear*" approximate-in-time profile of the applied surface heat flux.

C.4 Two-Dimensional Problem with Space- and Time-Dependent Surface Heat Flux

Consider the X22B(y-t-)0Y22B00T1 problem of Section 3.6. Its solution may be derived by using the GFSE for 2D transient rectangular problems (Cole et al. 2011; see Section 3.3). Its application to the current problem gives

$$
T(x,y,t) = T_{in} + \frac{\alpha}{k}\int_{\tau=0}^{t}G_{\text{X22}}(x, x'=0, t-\tau)\int_{y'=0}^{W}q(y',\tau)G_{\text{Y22}}(y, y', t-\tau)dy'd\tau
\tag{C.31}
$$

where $G_{\text{X22}}(x, x', t-\tau)$ and $G_{\text{Y22}}(y, y', t-\tau)$ are transient GFs given in Section C.3.

Applying the piecewise-uniform heat flux approximation of Figure 3.9 to Eq. (C.31) at time $t = t_M$ gives

$$
T(x,y,t_M) = T_M(x,y) = T_{in} + \sum_{j=1}^{N}\underbrace{\left[\frac{\alpha}{k}\int_{\tau=0}^{t}q_{0,j}(\tau)G_{\text{X22}}(x,0,t-\tau)\int_{y'=(j-1)\Delta y}^{j\Delta y}G_{\text{Y22}}(y,y',t-\tau)dy'd\tau\right]}_{= T_j(x,y,t)}
\tag{C.32}
$$

The quantity between square brackets in Eq. (C.32) is the temperature rise of the plate, initially at zero temperature, at time t due to the j-th uniform heat flux component $q_{0,j}(t)$ applied to the region $y \in [(j-1)\Delta y, j\Delta y]$, with $j = 1, 2, ..., N$ and $y_N = W$. It is denoted by $T_j(x, y, t)$.

Applying the superposition in space of Figure 3.10 to $T_j(x, y, t)$ yields

$$
T_{j,M}(x,y) = \underbrace{\frac{\alpha}{k} \int_{\tau=0}^{t_M} q_{0,j}(\tau) G_{X22}(x,0,t_M-\tau) \int_{y'=0}^{j\Delta y} G_{Y22}(y,y',t_M-\tau) dy' d\tau}_{\vartheta_j(x,y,t) = \vartheta_{j,M}(x,y)}
$$

$$
- \underbrace{\frac{\alpha}{k} \int_{\tau=0}^{t_M} q_{0,j}(\tau) G_{X22}(x,0,t_M-\tau) \int_{y'=0}^{(j-1)\Delta y} G_{Y22}(y,y',t-\tau) dy' d\tau}_{\vartheta_{j-1}(x,y,t) = \vartheta_{j-1,M}(x,y)}
$$

$$
= \vartheta_j(x,y) - \vartheta_{j-1,M}(x,y) = \Delta\vartheta_{j,M}(x,y) \tag{C.33}
$$

where the former, $\vartheta_{j,M}$ is the temperature due to a uniform and time-dependent heat flux having $q_{0,j}(t)$ as a magnitude and applied to the region $y \in [0, j\Delta y]$; while the latter, $\vartheta_{j-1,M}$, is the temperature due to a uniform and time variable heat flux still having $q_{0,j}(t)$ as a magnitude but applied to the region $y \in [0, (j-1)\Delta y]$.

Then, applying the piecewise-constant heat flux approximation of Figure 3.6 to both terms of Eq. (C.33) at time t_M gives

$$
\vartheta_{j,M}(x,y) = \sum_{i=1}^{M} \underbrace{\left[q_{0,i,j} \frac{\alpha}{k} \int_{\tau=t_{i-1}}^{t_i} G_{X22}(x,0,t_M-\tau) \int_{y'=0}^{j\Delta y} G_{Y22}(y,y',t_M-\tau) dy' d\tau \right]}_{\vartheta_{i,j,M}(x,y)} \tag{C.34a}
$$

$$
\vartheta_{j-1,M}(x,y) = \sum_{i=1}^{M} \underbrace{\left[q_{0,i,j} \frac{\alpha}{k} \int_{\tau=t_{i-1}}^{t_i} G_{X22}(x,0,t_M-\tau) \int_{y'=0}^{(j-1)\Delta y} G_{Y22}(y,y',t_M-\tau) dy' d\tau \right]}_{\vartheta_{i,j-1,M}(x,y)} \tag{C.34b}
$$

The quantity between square brackets in Eq. (C.34a) is the temperature solution at time t_M within the flat plate due only to the uniform and constant surface heat flux component $q_{0,i,j}$ applied over the region $y \in [0, j\Delta y]$ between t_{i-1} and t_i, when the plate is initially at zero temperature, with $i = 1, 2, ..., M$. It is denoted by $\vartheta_{i,j,M}(x, y)$.

Similarly, the quantity between square brackets in Eq. (C.34b) is the temperature solution at time t_M within the flat plate due only to the uniform and constant surface heat flux component $q_{0,i,j}$ applied to the region $y \in [0, (j-1)\Delta y]$ between t_{i-1} and t_i, when the plate is initially at zero temperature. It is indicated by the $\vartheta_{i,j-1,M}(x, y)$ symbol.

Lastly, applying the superposition in time to the temperatures $\vartheta_{i,j,M}(x, y)$ and $\vartheta_{i,j-1,M}(x, y)$ that appear in Eqs. (C.34a) and (C.34b) yields, respectively,

$$
\vartheta_{i,j,M}(x,y) = q_{0,i,j} \frac{L}{k} \frac{\alpha}{L} \underbrace{\int_{\tau=t_{i-1}}^{t_M} G_{X22}(x,0,t_M-\tau) \int_{y'=0}^{j\Delta y} G_{Y22}(y,y',t_M-\tau) dy' d\tau}_{(L/k) \times \tilde{T}_{X22B(y1pt1)0Y22B00T0}(\tilde{x},\tilde{y},\tilde{t}_M-\tilde{t}_{i-1}, \tilde{W}_0 = j\Delta\tilde{y}) = \lambda_{j,xy,M-i+1}}
$$

$$
- q_{0,i,j} \frac{L}{k} \frac{\alpha}{L} \underbrace{\int_{\tau=t_i}^{t_M} G_{X22}(x,0,t_M-\tau) \int_{y'=0}^{j\Delta y} G_{Y22}(y,y',t_M-\tau) dy' d\tau}_{(L/k) \times \tilde{T}_{X22B(y1pt1)0Y22B00T0}(\tilde{x},\tilde{y},\tilde{t}_M-\tilde{t}_i, \tilde{W}_0 = j\Delta\tilde{y}) = \lambda_{j,xy,M-i}} \tag{C.35a}
$$

$$
= q_{0,i,j}\left(\lambda_{j,xy,M-i+1} - \lambda_{j,xy,M-i}\right) = q_{0,i,j}\Delta_{M-i}\left(\lambda_{j,xy,M-i}\right)
$$

$$\vartheta_{i,j-1,M}(x,y) = q_{0,i,j}\frac{L}{k}\frac{\alpha}{L}\underbrace{\int\limits_{\tau=t_{i-1}}^{t_M} G_{X22}(x,0,t_M-\tau)\int\limits_{y'=0}^{(j-1)\Delta y} G_{Y22}(y,y',t_M-\tau)dy'd\tau}_{(L/k)\times\tilde{T}_{X22B(y1pt1)0Y22B00T0}\left[\tilde{x},\tilde{y},\tilde{t}_M-\tilde{t}_{i-1},\tilde{W}_0=(j-1)\Delta\tilde{y}\right]=\lambda_{j-1,xy,M-i+1}}$$

$$-q_{0,i,j}\frac{L}{k}\frac{\alpha}{L}\underbrace{\int\limits_{\tau=t_i}^{t_M} G_{X22}(x,0,t_M-\tau)\int\limits_{y'=0}^{(j-1)\Delta y} G_{Y22}(y,y',t_M-\tau)dy'd\tau}_{(L/k)\times\tilde{T}_{X22B(y1pt1)0Y22B00T0}\left[\tilde{x},\tilde{y},\tilde{t}_M-\tilde{t}_{i},\tilde{W}_0=(j-1)\Delta\tilde{y}\right]=\lambda_{j-1,xy,M-i}}$$

$$= q_{0,i,j}\left(\lambda_{j-1,xy,M-i+1}-\lambda_{j-1,xy,M-i}\right) = q_{0,i,j}\Delta_{M-i}\left(\lambda_{j-1,xy,M-i}\right)$$

(C.35b)

where $\lambda_{j,xy,M-i+1}$ (°C-m²/W) is the temperature rise per unit of surface heat flux of the X22B(y1pt1)0Y22B00T0 heat conduction case (apart from the factor L/k) at a generic location $(\tilde{x},\tilde{y})$ and time $\tilde{t}_M - \tilde{t}_{i-1} = (M-i+1)\Delta\tilde{t} = \tilde{t}_{M-i+1}$ when $\tilde{W}_0 = j\Delta\tilde{y}$, that is, the heated surface region is between the corner point $(0,0)$ and the j-th grid point $\left(0,\tilde{y}_j = j\Delta\tilde{y}\right)$ and the sudden variation of the surface heat flux occurs at time $t = 0$. Similarly, $\lambda_{j,xy,M-i}$ and so on.

Substituting Eqs. (C.33)–(C.35b) in Eq. (C.32) gives Eq. (3.94) that was derived without using GFs.

For this reason, Eq. (3.94) may be interpreted as the approximate form in both space and time of the "*heat flux-based*" 2D transient GF solution Eq. (C.31) at time $t = t_M$ when dealing with a "*piecewise-uniform*" and "*piecewise-constant*" profile of the surface heat flux.

Reference

Cole, K. D., Beck, J. V, Haji-Sheikh, A. and Litkouhi, B. (2011) *Heat Conduction using Green's Functions*, 2nd edn. Boca Raton, FL: CRC Press.

Index

Note: page numbers referring to *figures are in italics* and those referring to **tables are in bold**.